4th International Conference on Food and Agriculture (ICOFA 2021)

AA007124

IOP Conference Series: Earth and Environmental Science
Volume 980

Jember, Indonesia
6 – 7 November 2021

ISBN: 978-1-7138-4961-2
ISSN: 1755-1307

Printed from e-media with permission by:

Curran Associates, Inc.
57 Morehouse Lane
Red Hook, NY 12571

proceedings.com

Some format issues inherent in the e-media version may also appear in this print version.

Institute of Physics
Dirac House, Temple Back
Bristol BS1 6BE UK

Phone: 44 1 17 929 7481
Fax: 44 1 17 920 0979

techtracking@iop.org

Additional copies of this publication are available from:

Curran Associates, Inc.
57 Morehouse Lane
Red Hook, NY 12571 USA
Phone: 845-758-0400
Fax: 845-758-2633
Email: curran@proceedings.com
Web: www.proceedings.com

TABLE OF CONTENTS

IOP Conf. Series: Earth and Environmental Science **980** (2022) 011001 doi:10.1088/1755-1315/980/1/011001

PREFACE

Dear colleagues,

We are glad to announce the release of the 4th International Conference of Food and Agriculture (ICOFA-2021). The ICOFA is an annual conference organized primarily by Politeknik Negeri Jember. In this year the theme of our conference is "Scientific and Technological Efforts in Sustainable Food and Agriculture"

Surely, we plan to conduct this conference physically just like the previous ICOFA two years ago in November 2019. However, due to the COVID-19 pandemic, the situation has not subsided, the 4th ICOFA 2021 was carried out virtually as same as ICOFA 2020 by the Zoom meeting platform. We took this option because this conference was already designated and funded. Keynote and invited speakers were also scheduled for this event. Many delegations and authors requested for this conference to be performed, even virtually, since they need it to cover their publication and sharing knowledge requirements.

The key directions of the conference covered the discussion of advanced achievements and touch upon the main aspects of agriculture engineering and biotechnology, organic agriculture, animal science related to food, food science and technology, food safety, food security and sovereignty, IT for agriculture and renewable and novel energy sources.

The conference was run as planned on 6th November 2021 for the plenary session to 7th November 2021 for the parallel session and supported by a virtual event organizer started from 9.00 am to 5.00 PM. Conference models are a Plenary session with keynotes (40 minutes for each keynote speech) and participants' presentations (20 minutes for each presentation). A total of 144 participants (120 presenters and 24 non-presenters) from Universities, Research Institutes, and also Government Departments joined the conference in over five countries from Japan, Thailand, India, South Korea, and Algeria.

Our honorable speakers were Dr. Nishu Gupta SRM from the Institute of Science and Technology, India, Prof. Kyung-Min Kim, Ph.D. from Kyungpook National University, Korea, Prof. Dr. Tosawat Seetawan from Sakon Nakhon Rajabhat University, Thailand and Associate Professor Bibin Bintang Andriana, Ph.D. from Kwansei Gakuin University, Japan.

More than 120 full papers were submitted to the 4th ICOFA 2021. After a pre-review on originality and language, the peer review process was arranged by Editorial Committee and 73 manuscripts were selected for publication in IOP Conference Series: Earth and Environmental Science (EES).

We also wish to thank the keynote speaker, invite speakers, presenters, and participants, and all authors for contributions to the 4th ICOFA 2021 and colleague members of the organizing committee, please accept my deep appreciation for your hard working in ensuring the success of the conference.

Editor-in-Chief, ICOFA 2021
Dr. Ir. Rosa Tri Hertamawati, M.Si, IPM

A list of Steering Committees and Organizing Committees are available in this pdf.

IOP Conf. Series: Earth and Environmental Science **980** (2022) 011001 doi:10.1088/1755-1315/980/1/011001

Steering Committee ICOFA 2021

Dr. Nishu Gupta	SRM Institute of Science and Technology, INDIA
Prof Kyung-Min Kim, Ph.D	Kyungpook National University, KOREA
Prof Dr Tosawat Seetawan	Sakon Nakhon Rajabhat University, THAILAND
Assistant Professor Bibin Bintang Andriana, Ph.D	Kwansei Gakuin University, JAPAN
Dr. Ir. Agus Susanto, MSc	Universitas Jendral Soedirman, INDONESIA
Dr. Ir. Parawita Dewanti, MP	Universitas Jember, INDONESIA
Dr. Ir. Dadik Pantaya, M.Si. IPU	Politeknik Negeri Jember, INDONESIA
Dr. Ir. Nurul Syamsiah, M.P.	Politeknik Negeri Jember, INDONESIA
Dr. Elly Kurniawati, S.TP., M.P.	Politeknik Negeri Jember, INDONESIA
Dr. Ir. Ujang Suryadi, M.P.	Politeknik Negeri Jember, INDONESIA
Dr. Ir. Ridwan Iskandar, M.T.	Politeknik Negeri Jember, INDONESIA
Dr. Bayu Rudianto, M.Si.	Politeknik Negeri Jember, INDONESIA
Dyah Nuning Erawati, S.P., M.P.	Politeknik Negeri Jember, INDONESIA
Refa Firgiyanto, S.P., M.Si.	Politeknik Negeri Jember, INDONESIA
Rizza Wijaya, S.TP., M.Sc.	Politeknik Negeri Jember, INDONESIA
Aulia Brilliantina, S.TP., M.P.	Politeknik Negeri Jember, INDONESIA
Trismayanti Dwi P., S.Kom., M.Cs.	Politeknik Negeri Jember, INDONESIA
Hermawan Arif Putranto, S.T., M.T.	Politeknik Negeri Jember, INDONESIA
Syamsiar Kautsar, S.T., M.T.	Politeknik Negeri Jember, INDONESIA

IOP Conf. Series: Earth and Environmental Science **980** (2022) 011001 doi:10.1088/1755-1315/980/1/011001

Organizing Committee ICOFA 2021

Saiful Anwar, S.TP, MP	Chairman of International Advisory Board
Surateno, S.Kom, M.Kom	Advisor
Ir. Abi Bakri, M.Si	Advisor
Wahyu Kurnia Dewanto, S.Kom, MT	Advisor
Dr. Ir. Budi Hariono, M.Si	Advisor
Dr. Ir. Nantil Bambang Eko S., MT	Chairman of Organizing Committee
Suluh Nusantoro, S.Pi, M.Sc	Vice Chairman of Organizing Committee
Prawidya Destarianto, S.Kom, MT	Secretary
Dr. Ir. Rosa Tri Hertamawati, M.Si	Editor in Chief
Retno Sari Mahanani, SP. MM	Financial Treasury Coordinator
Andik Susanto	Financial Treasury Support
Mery Hadiyah Wati, S.Kom	Financial Treasury Support
Khafidurrohman Agustianto, S.Pd. M.Eng	IT Publication System Coordinator
I Gede Wiryawan, S.Kom, M.Kom	IT Publication System Support
David Juli Ariyadi, S.Kom	IT Publication System Support
Ahmad Nuril Firdaus, SE	Publication and Book Program
Ahmad Nugroho Ardhianto, A.Md	Publication and Book Program
Afif Sugi Hendrianto, A.Md	Publication and Book Program
Dra. Yogyarsi Budiwiyanti	Supporting Staff
Suryadi	Supporting Staff

IOP Conf. Series: Earth and Environmental Science **980** (2022) 011001 doi:10.1088/1755-1315/980/1/011001

Conference Photograph

Plenary Session

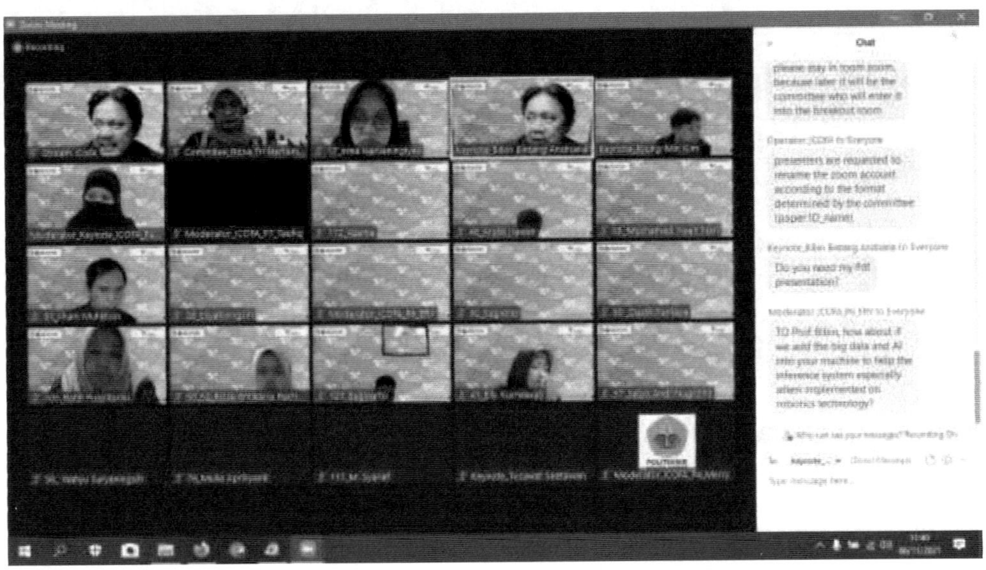

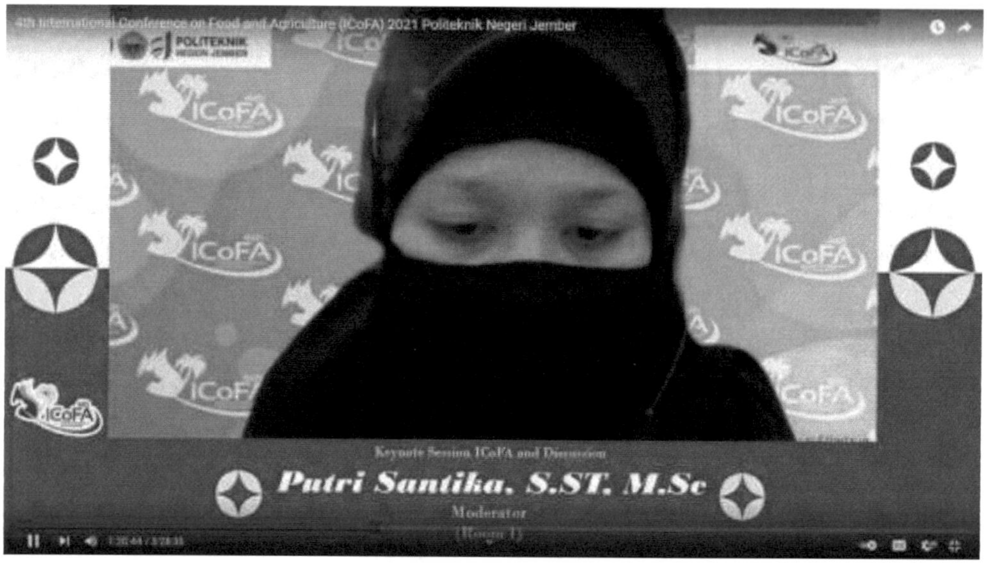
4th International Conference on Food and Agriculture (ICoFA) 2021 Politeknik Negeri Jember
POLITEKNIK NEGERI JEMBER
ICoFA
Keynote Session ICoFA and Discussion
Putri Santika, S.ST, M.Sc
Moderator

POLITEKNIK NEGERI JEMBER
ICoFA

IOP Conf. Series: Earth and Environmental Science **980** (2022) 011001 doi:10.1088/1755-1315/980/1/011001

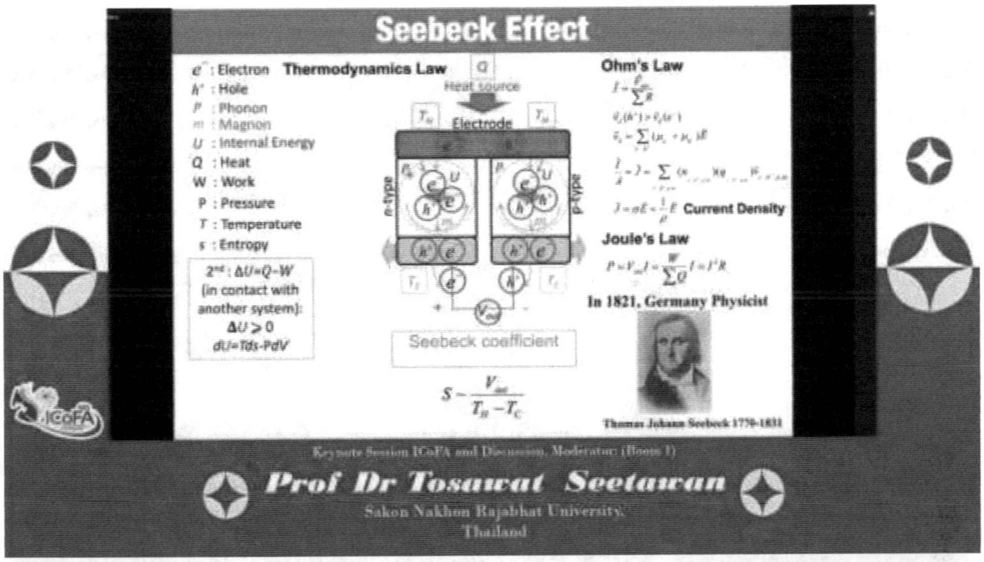

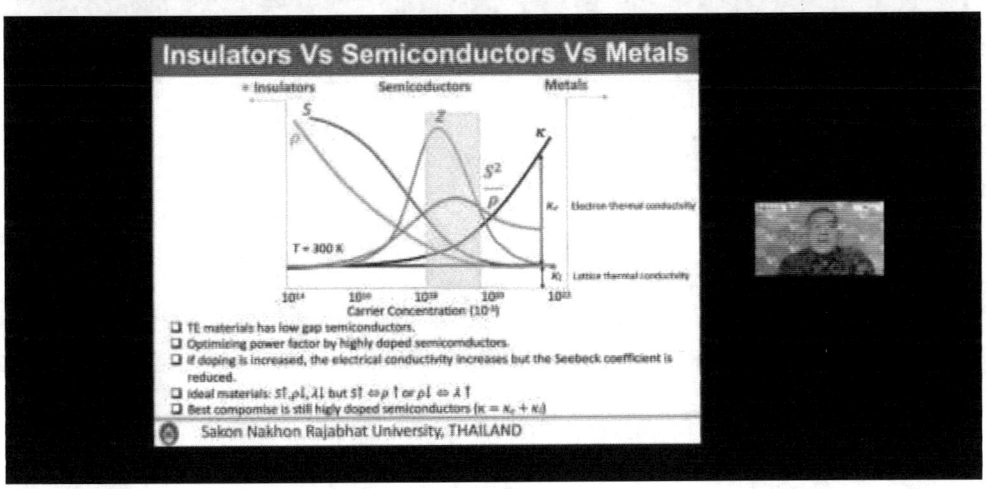

4th International Conference on Food and Agriculture (ICoFA) 2021 Politeknik Negeri Jember
Keynote Session ICoFA and Discussion, Moderator: (Room 1)
Assistant Professor Bibin Bintang Andriana
Departement of Biomedical Chemistry, Graduate School of Science and Technology,
Kwansei Gakuin University, Japan

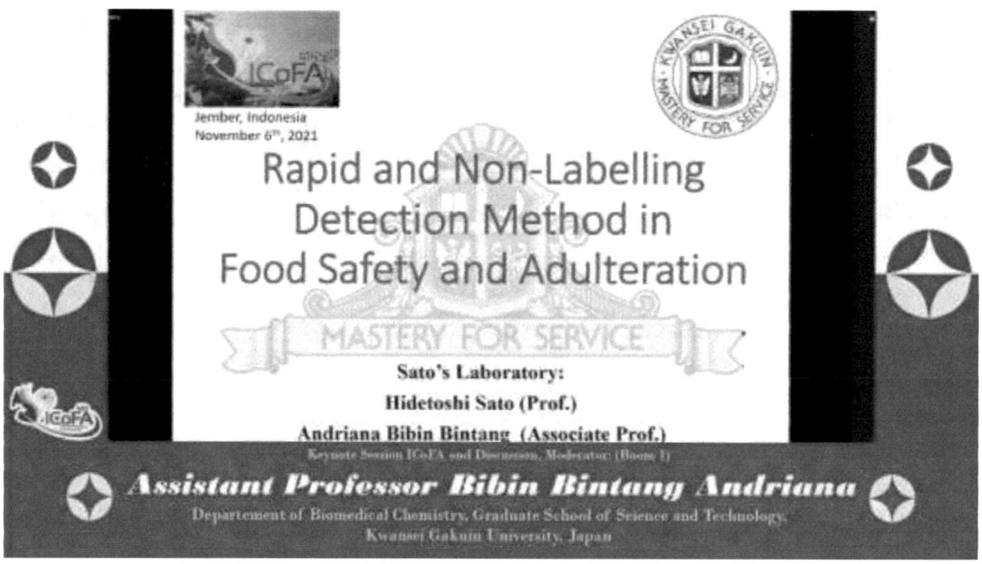
Jember, Indonesia
November 6th, 2021
KWANSEI GAKUIN
MASTERY FOR SERVICE
Rapid and Non-Labelling
Detection Method in
Food Safety and Adulteration
MASTERY FOR SERVICE
Sato's Laboratory:
Hidetoshi Sato (Prof.)
Andriana Bibin Bintang (Associate Prof.)
Keynote Session ICoFA and Discussion, Moderator: (Room 1)
Assistant Professor Bibin Bintang Andriana
Departement of Biomedical Chemistry, Graduate School of Science and Technology,
Kwansei Gakuin University, Japan

IOP Conf. Series: Earth and Environmental Science **980** (2022) 011001 doi:10.1088/1755-1315/980/1/011001

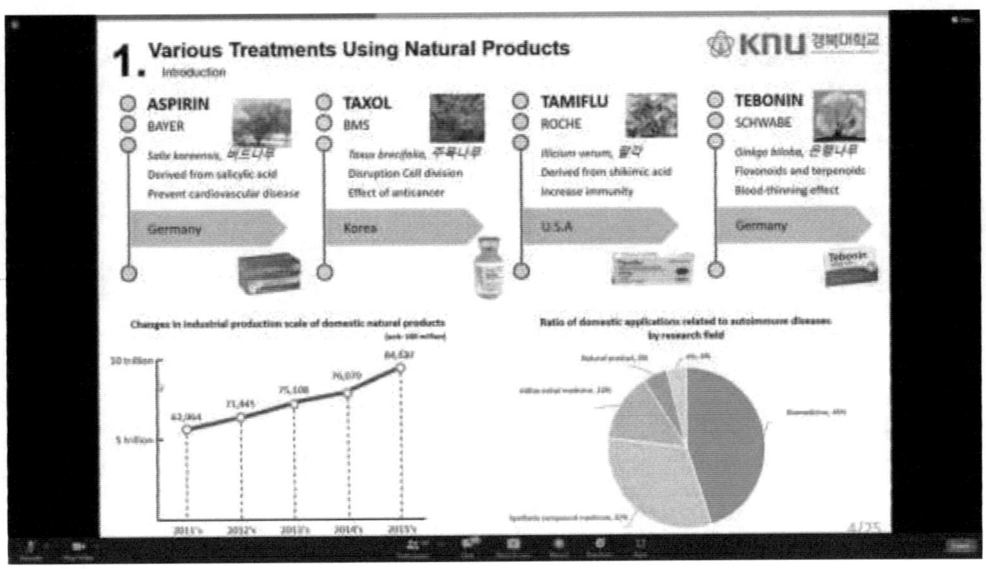

Parallel Session

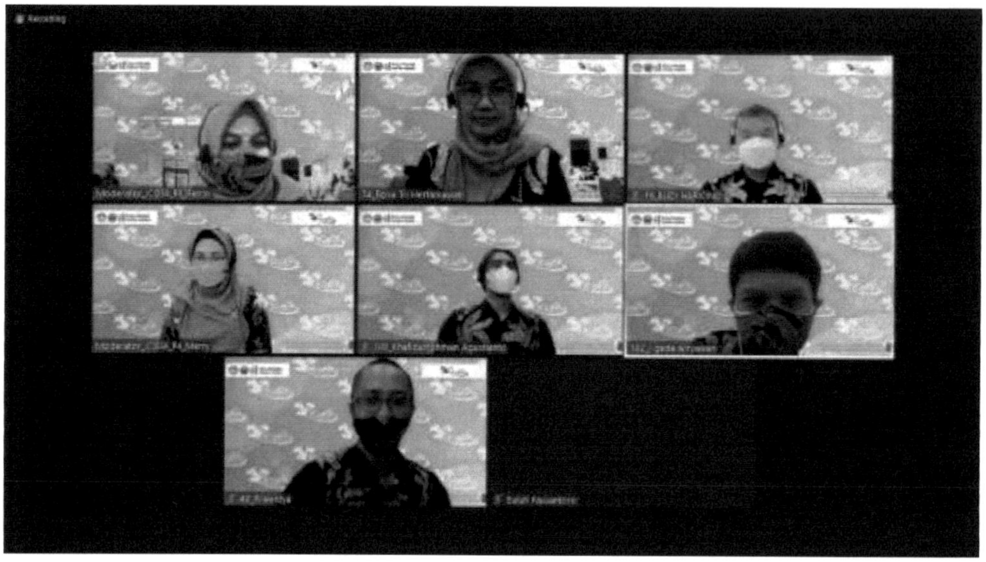

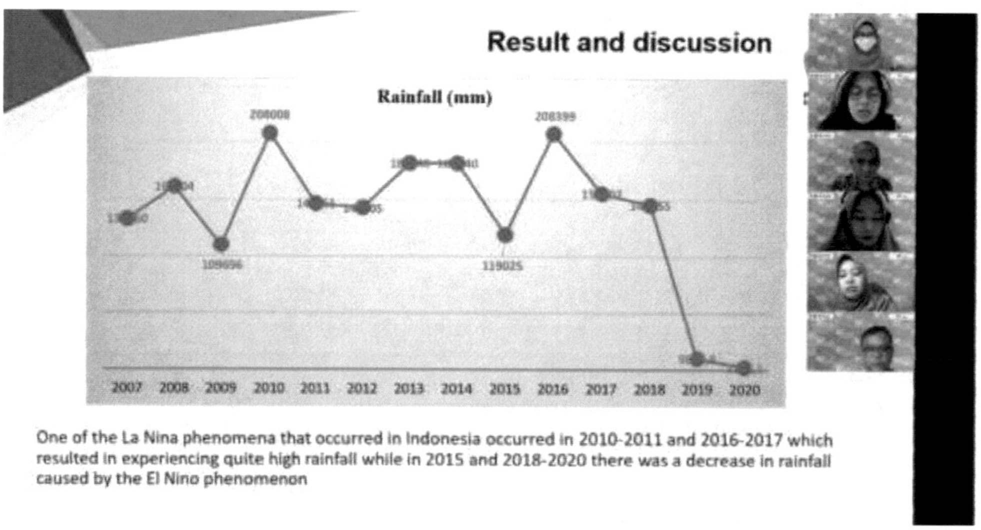

IOP Conf. Series: Earth and Environmental Science **980** (2022) 011001 doi:10.1088/1755-1315/980/1/011001

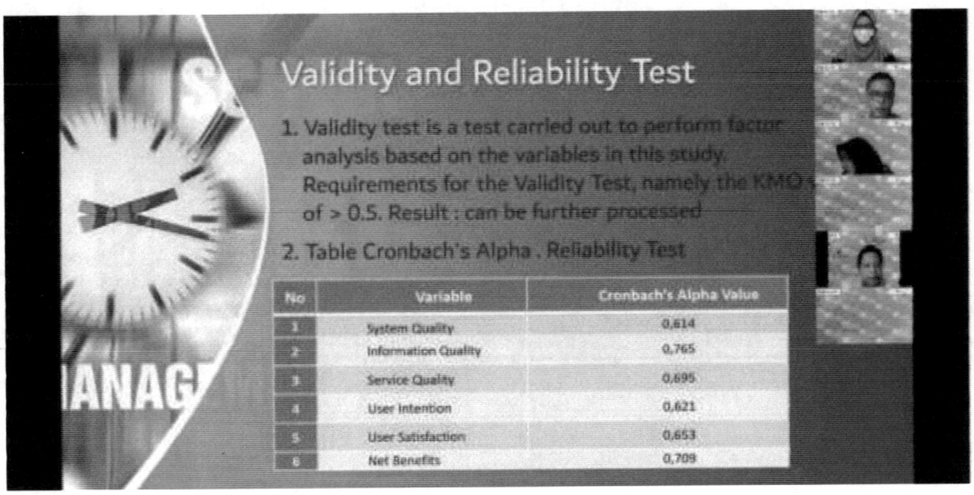

IOP Conf. Series: Earth and Environmental Science **980** (2022) 011001 doi:10.1088/1755-1315/980/1/011001

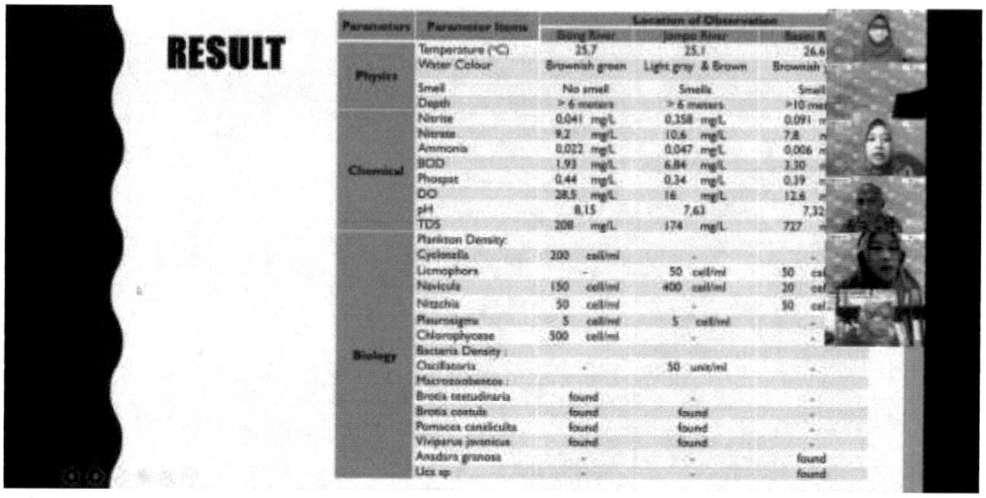

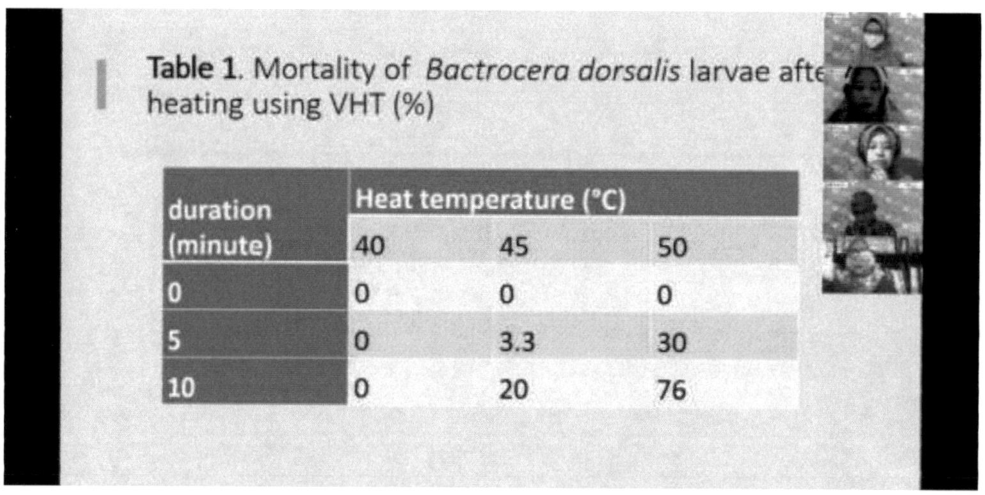

Peer Review Statement

All papers published in this volume have been reviewed through processes administered by the Editors. Reviews were conducted by expert referees to the professional and scientific standards expected of a proceedings journal published by IOP Publishing.

1. Type of peer review: Single Anonymous
2. Conference submission management system: Morressier
3. Number of submissions received: 122
4. Number of submissions sent for review: 92
5. Number of submissions accepted: 68
6. Acceptance Rate (Submissions Accepted / Submissions Received × 100): 55.7 %
7. Average number of reviews per paper: 3.00
8. Total number of reviewers involved: 11
9. Contact person for queries:

Name : Dr. Ir. Rosa Tri Hertamawati, M.Si
Affiliation : Politeknik Negeri Jember
Email : rosa_trihertamawati@polije.ac.id

IOP Conf. Series: Earth and Environmental Science **980** (2022) 012001 doi:10.1088/1755-1315/980/1/012001

Optimization of the sterilization method for leaf explant Robusta BP 308 coffee in vitro

S L Asmono[1*], R Wardana[1] and Rahmawati[1]

[1]Department of Agricultural Production, Politeknik Negeri Jember, Jalan Mastrip PO BOX 164, Jember, 68101, Indonesia

* sepdian@polije.ac.id

Abstract. The research aimed to know the optimal sterilizing agent for the success of the coffee callus inoculation from young leaf explant. This research was conducted at the Plant Tissue Culture Laboratory for 2 months. The explants are used from young leaves of Robusta clones BP 308. This experiment used a nonfactorial completely randomized design with three sterilization methods and was repeated 10 times. The treatments tested were as follows Method 1: Fungicide 2 gr/L, Bactericide 2 gr/L, Erythromycin 4 gr/L (20 min); 70% alcohol (3 sec); NaOCl 0.525% (10 min); NaOCl 1.05% (10 min). Method 2: 70% alcohol (3 sec), 0.1% HgCL2 (5 min). Method 3: Fungicide 2 gr/L, Bactericide 2 gr/L, Erythromycin 4 gr/L (20 min); 70% alcohol (3 sec); NaOCl 0.525% (10 min); NaOCl 1.05% (10 min); 0.1% HgCL2 (5 min). The results showed that the sterilization method used suppressed the rate of bacterial and fungal contamination, but there was no significant difference between the methods. A very significant effect was seen in the percentage of browning explants. In sterilizing agent treatment 1, there was no browning explant, while the other methods caused browning to reach 62.5%. Browning also causes differences in callus regeneration ability. The explants which were sterilized in method 1 showed a high percentage of callus (85%) and were also appear to callus faster at 11 days after inoculation.

1. Introduction

Robusta coffee clone BP 308 is a coffee genotype that has been confirmed as a superior planting material to have a high level of resistance to *P. Coffeae* nematode attack [1]. In its propagation, the type of robusta coffee is a cross-pollinated plant so it requires other types of robusta to produce seeds. Of course, it will affect the genetic uniformity of the seeds produced. One of the propagation methods that can produce genetically uniform seeds is through in vitro culture techniques. This technique is a method of plant propagation by isolating body parts or organs (cells or tissues) from plants and growing under sterile or aseptic conditions so that they can grow into whole plants. The advantage of this technique is that it produces plants that have the same or identical characteristics as the parent, producing large numbers of seeds in a relatively short time [2], [3].

In vitro culture on coffee, plants are usually used to produce Somatic Embryogenesis (SE) seeds [4], [5]. Cells from 1 piece of leaf explant measuring 1cm^2, were able to develop into hundreds of somatic embryos, each of which would later become seedlings with identical properties to the parent. However, the problem in the early stages of inoculation is the contamination of the explants. Contamination can inhibit the growth process so that the explants cannot grow or die. The problem arises because the inoculated explants are not sterile.

The solution to suppress contamination is to use an appropriate sterilization method. Sterilization of explants aims to prevent or kill microorganisms that may be carried or attached to the leaves when explants are taken. There are several sterilization methods used by previous researchers, namely by immersion using a solution of the Dithane fungicide and Agrimycine bactericide, dipping in 50% alcohol and soaking in 5.25% NaOCl solution [6]. Then in another study there were washing explants with liquid detergent, soaked in a solution of 2 g/L Dithane fungicides, immersed in 70% alcohol, and soaked in NaOCl solution [7]. In its application, the sterilization method from the research conducted has not optimally obtained sterile explants, especially if there are different types of explants and the origin of the explants. The novelty in this study is that the researchers will use Robusta clones BP 308. Therefore, the authors want to conduct research to test and optimize several methods of sterilizing coffee leaf explants, to obtain a suitable sterilization method for the success of Arabica and Robusta coffee explant cultures in vitro, especially the origin of coffee leaf explants from the Jember-Indonesia.

2. Methods

2.1 Preparation of plant material

The research was carried out for 2 month at the Culture Laboratory, Politeknik Negeri Jember, Indonesia. The explants used were young leaves from coffee seedlings aged 4-5 months.

2.2 Surface sterilization

Explants in the form of young leaves were cut and sterilized using 3 treatments of sterilization methods, as shown in **Table 1** below:

Table 1. Sterilization methods of coffee leaf explants

Treatment	Sterilizing agent	Concentration	Soaking time
Sterilization Method 1	Liquid detergent	2 ml/L	10 minutes
	Rinse sterile distilled water 3 times		1 minute
	Dithane Fungicide	2 gr/L	20 minutes
	Agrimycine Bactericide	2 gr/L	20 minutes
	Erythromycin	4 gr/L	20 minutes
	Rinse sterile distilled water 3 times		1 minute
	Alcohol	70%	3 seconds
	Rinse sterile distilled water 3 times		1 minute
	NaOCl (5.25%)	10%	10 minutes
	NaOCl (5.25%) 20%	20%	10 minutes
	Rinse sterile distilled water 3 times		1 minute
Sterilization Method 2	Liquid detergent	2 ml/L	10 minutes
	Rinse sterile distilled water 3 times		1 minute
	Alcohol	70%	3 seconds
	Rinse sterile distilled water 3 times		1 minute
	$HgCL_2$	0,1%	5 minutes
	Rinse sterile distilled water 3 times		1 minute
Sterilization Method 3	Liquid detergent	2 ml/L	10 minutes
	Rinse sterile distilled water 3 times		1 minute
	Dithane Fungicide	2 gr/L	20 minutes
	Agrimycine Bactericide	2 gr/L	20 minutes
	Erythromycin	4 gr/L	20 minutes
	Rinse sterile distilled water 3 times		1 minute
	Alcohol	70%	3 seconds
	Rinse sterile distilled water 3 times		1 minute
	NaOCl (5.25%)	10%	10 minutes

	NaOCl (5.25%) 20%	20%	10 minutes
	Rinse sterile distilled water 3 times		1 minute
	$HgCL_2$	0,1%	5 minutes
	Rinse sterile distilled water 3 times		1 minute

2.3Culture medium
Basal medium for callus induction using 1/2 MS medium [8] with 1ppm 2,4-D dan 2ppm Kinetin. MS medium was adjusted at pH 5.8, then autoclaved at 121 psi pressure for 30 minutes.

2.4Explant inoculation
Leaf explants were taken from the parent plant, then sterilized according to the tested method (**Table 1**). In Laminar Air Flow, leaf explants were cut into squares of 1 cm^2, then implanted in culture media and incubated in the dark. Each bottle contains 1 explant. Observations are made every day to find out if there is contamination.

2.5Parameters and data analysis
This study used a non-factorial Completely Randomized Design (CRD) with 3 treatments and was repeated 10 times. Data were analyzed by Analysis of Variance (ANOVA) and DMRT test ($p<0.05$). Observations were made every day until the 30th day after inoculation (DAI). Parameters observed included: fungal contamination, bacterial contamination, browning, and callus formation.

3. Results and discussion
3.1 Fungal contamination
The data presented in this parameter include the time of the appearance of the fungus and the percentage of fungal contamination. Observations of this fungal contamination were carried out every day until the age of 30 days after inoculation (DAI). Based on ANOVA data analysis, these fungal contamination parameters showed results that were not significantly different, as shown in **Table 2.**

Table 2. Data on fungal contamination of explants

Treatment	Fungi emergence (DAI)	Percentage of fungal contamination
Sterilization Method 1	5	7.50%
Sterilization Method 2	10	5.00%
Sterilization Method 3	5	7.50%

From these data, the three sterilant methods tested showed the appearance of the fungus starting on day 5 and day 10 after inoculation. This duration is also consistent with the results of studies on the incubation period of fungi which showed that 94% of fungi appeared on day 7 and 98% on day 14 [9]. In addition, the percentage value of fungal contamination is still relatively small or safe, namely at 5%-7.5%, or from a total of 40 bottles in each treatment, only 2 or 3 bottles were contaminated with fungi.

The fungal contamination that occurred was characterized by the presence of hyphae covering the surface of the explants and media [10] (**Figure 1.**). The presence of contamination is a common problem that poses a challenge to the success of in vitro culture. Based on observational data on fungal contamination in this study the results were not significantly different, it can be said that the sterilant materials used in all methods of sterilization treatment were optimal and proven to be able to suppress the growth of fungi. Especially Fungicide at a concentration of 2g/L for 20 minutes. The use of 70% alcohol and 0.1% HgCL2 for 2 minutes was also able to suppress the growth and development of fungi. This can be seen in sterilization method 2, the fungal contamination showed 5% with a slower emergence at 10 DAI. One study stated that HgCL2 (Mercury chloride) is highly toxic to microbes, and can cause certain concentrations [11].

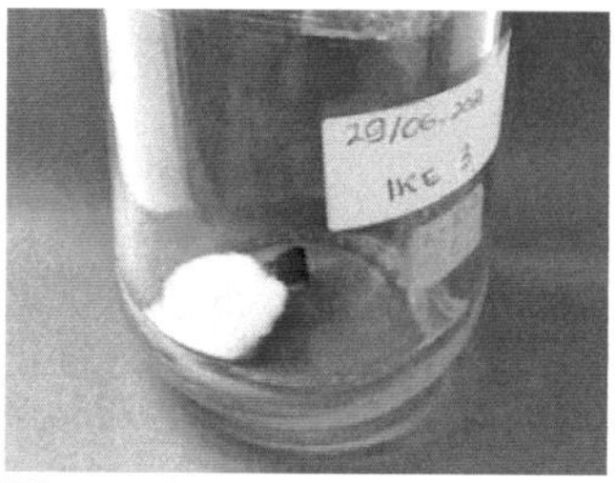

Figure 1. Explants contaminated with fungi

Figure 2. Explants contaminated with bacteria

3.2 Bacterial contamination

Observations of contamination were also seen from the data of explants contaminated with bacteria. That was characterized by the presence of white mucus around the explants as shown in **Figure 2**. The data on the percentage of bacteria showed results that were not significantly different according to the ANOVA test, as shown in **Table 3.**

Table 3. Data on bacterial contamination of explants

Treatment	Bacteria emergence (DAI)	Percentage of Bacteria contamination
Sterilization Method 1	0	0.00%
Sterilization Method 2	10	17.50%
Sterilization Method 3	5	7.50%

In **Table 3**, the emergence of bacteria started from day 5 to day 10 after inoculation. However, in sterilization method 1, the explants did not experience bacterial contamination. While in the second sterilization method the percentage of bacteria reached 17.5% and the third method of bacterial contamination showed a percentage of 7.5%. In sterilization method 2, only use liquid detergent 2 ml/L, 70% alcohol, 0.1% HgCL2. It was proven that the combination did not contain any additional bactericidal ingredients so that the percentage of bacterial contamination was the highest. The possibility of this material is only able to kill bacteria on the surface, not systemically kill endogenous bacteria contained in explant tissue.

The results of bacterial contamination data showed method 1 did not grow bacteria. In the first sterilization method using Liquid detergent 2 ml/L, Dithane Fungicide 2 g/L, Agrimycine bactericide 2 g/L, Erythromycin 4 g/L, alcohol 70%, NaOCl (5.25%) 10%, and 20%. The combination of these sterilants was able to eliminate bacteria on the surface of the explants. This can be seen from the absence of bacterial contamination that occurs. Bacterial growth on culture media and explants can occur rapidly if the sterilization method is not optimal, both in the use of materials and concentrations or due to an unhygienic explant inoculation process [12], [13].

3.3 Browning

Based on observations on browning explants until 30 days after inoculation, the results of data analysis showed very significant results, as shown in **Table 4** below.

Table 4. Data on the percentage of browning explants

Treatment	Percentage of Browning	
Sterilization Method 1	0.00%	a
Sterilization Method 2	20.00%	b
Sterilization Method 3	62.50%	c

[a] The number followed by the same letter is not significantly different at ($p<0.05$) level of Duncan's test.

In method 1, it is assumed that the concentration is optimal and safe for the cells in the leaves so that it does not cause death in the leaf cells **(Figure 3)**. The use of Sodium hypochlorite (NaOCl) is a sterile material that has low activity so that the explant cells do not become stressed and injured. However, if these materials are used at high concentrations, they can cause cell injury (necrosis) and eventually die and then visually turn brown or browning [14].

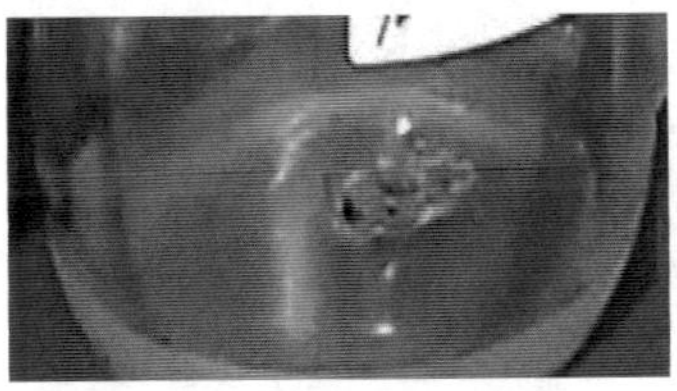

Figure 3. Healthy explants from sterilization method 1

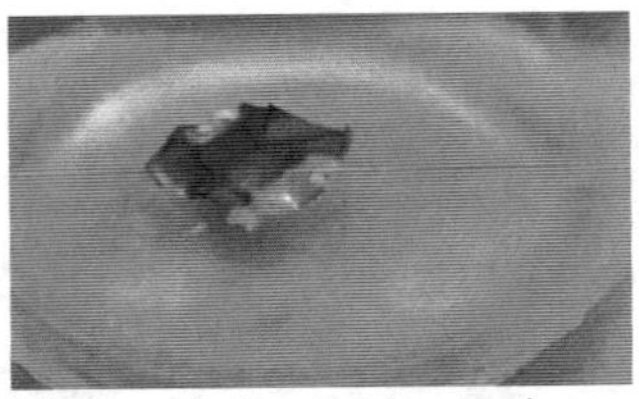

Figure 4. Partial browning explants in the sterilization method 2

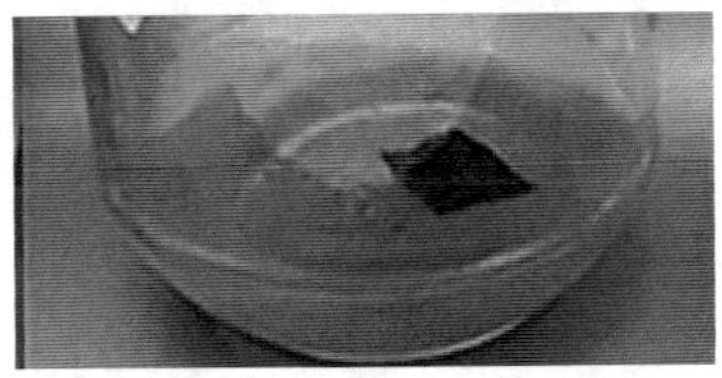

Figure 5. Overall browning explants in the sterilization method 3

In general, browning in plants occurs due to stress conditions and the oxidation process of phenolic compounds that come out due to explant injury [15], [16]. In addition, the type and concentration of sterilant ingredients are also able to stimulate the appearance of browning. As in the treatment tested, sterilization method 3 stimulated the highest explant browning, namely 62.5% **(Figure 5)**, and sterilization method 2 stimulated 20% browning **(Figure 4)**. The use of HgCL2 or Mercury Chloride was suspected to be the cause of browning in treatments 2 and 3 because both treatments used HgCL2 in the series of sterile materials. Mercury Chloride is a sterilant that is highly toxic not only to microbes but also to explants and humans. Explants that were sterilized in treatments 2 and 3 were seen to have browned in some of the explants. However, some had all of the explants browned. Browning is one sign that the explant cells are dead [17].

Mercury Chloride is highly corrosive and at high concentrations can inhibit enzyme activity and cause protein precipitation in microbial cells and plant cells, causing damage to plant tissues. The use of HgCl2 at high concentrations has also been shown to cause browning and death of sugarcane leaf explants [18].

3.4 Callus formation

Observations on callus formation were carried out every day until the 30th day after inoculation. The results of ANOVA data analysis showed that the percentage of explants that succeeded in forming callus and the score of callus formation on the circumference of the explants showed very significant results, as shown in the data listed in **Table 5.**

IOP Conf. Series: Earth and Environmental Science **980** (2022) 012001 doi:10.1088/1755-1315/980/1/012001

Table 5. Callus formation data

Treatment	Callus emergence (DAI)	Percentage of callus formation		Callus formation score on the circumference of the explant	
Sterilization Method 1	11.67	85.00%	a	3 (>50%-75%)	a
Sterilization Method 2	15.30	45.00%	b	1 (0-25%)	b
Sterilization Method 3	16.00	35.00%	b	1 (0-25%)	b

[a] The number followed by the same letter is not significantly different at ($p<0.05$) level of Duncan's test.

Observational data showed that the tested sterile materials also affected callus formation and growth. This can be seen in the treatment of sterilization method 1 which showed the appearance of callus for the first time at 11.67 DAI, and an average of 85% of the explant cells was able to regenerate to form a callus and the growth was relatively even, with a score of 3 or about 75% around the explants able to form a callus.

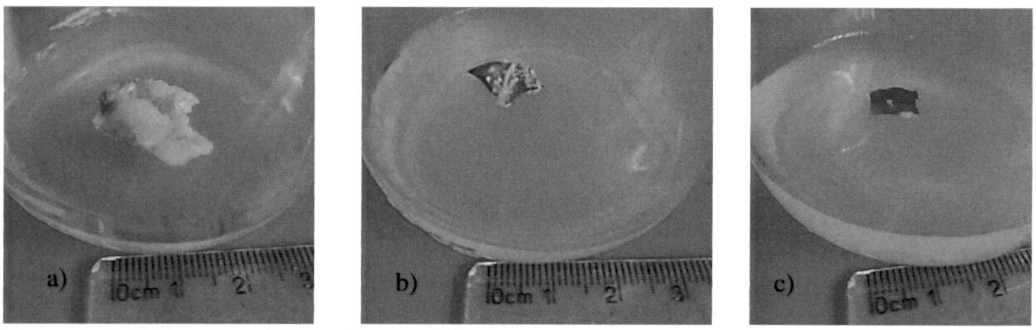

Figure 6. a) Callus growth score 3 on sterilization method 1; b) Callus growth score 1 on the 2nd sterilization method; c) Callus growth score 1 in the 3rd sterilization method

Different results were shown in the treatment of sterilization methods 2 and 3 which were relatively slower to produce callus, namely at 15-16 DAI with a percentage between 35%-45% of explants capable of forming a callus. However, the callus formed was uneven, only in small spots on the explants that did not experience browning.

The difference in the response of callus formation to the treatment of sterile materials was thought to be due to stress and cell necrosis caused by each treatment, especially from methods 2 and 3. The recovery process of explant cells affected cell formation. This can be seen in the treatment of method 1 which seems to be able to form callus faster.

In addition, browning on some of the explant surfaces also affected callus formation. Based on the results of observations callus appeared on the explants that did not experience browning. It is known from the data of this study that the treatment of sterilization methods 2 and 3 on average experienced browning of explants so that only certain parts of the explants grew callus. Based on the score, it only reached 25% of the total circumference of the explant piece. The browning effect on explants resulted in inhibited absorption of growth regulators and nutrients [19], so that the cells were unable to regenerate to form callus.

4. Conclusions

The conclusion of this study shows that sterilization method 1 is the optimal method to suppress fungal and bacterial contamination. On average, fungi and bacteria appeared on day 5 after inoculation. In addition, the treatment of several sterilization methods greatly affected the browning of the explants. In sterilization method 1 there was no browning of explants, while other methods caused browning to reach 62.5%. Browning also causes differences in callus regeneration ability. Explants sterilized by method 1 showed a high percentage of callus (85%) and callus appeared more quickly at 11 days after inoculation.

5. Acknowledgements

The authors acknowledged the financial support by PNBP Politeknik Negeri Jember 2021.

6. References

[1] Wiryadiputra S Santoso A . and Mawardi S, 1994 Ketahanan beberapa jenis dalam marga Coffea terhadap Nematoda Pratylenchus coffeae in *Prosiding Simposium Pemuliaan III.*

[2] Pierik R, 1987 International Symposium on Propagation of Ornamental Plants 226 in *In Vitro Culture Of Higher Plants As A Tool In The Propagation Of Horticultural Crops* p. 25–40.

[3] George E F Hall M A and De Klerk G J, 2007 *Plant Propagation by Tissue Culture: Volume 1. The Background* Springer Netherlands.

[4] Riyadi I, 2017 Pengaruh 2, 4-D terhadap induksi embrio somatik kopi arabika *Bul. Plasma Nutfah* **10**, 2 p. 82–89.

[5] Campos N A Panis B and Carpentier S C, 2017 Somatic Embryogenesis in Coffee: The Evolution of Biotechnology and the Integration of Omics Technologies Offer Great Opportunities *Front. Plant Sci.* **8** p. 1460.

[6] Ibrahim I Nasr M Mohammed B and El-Zefzafi M, 2008 Plant growth regulators affecting in vitro cultivation of Stevia rebaudiana *Sugar Tech* **10** p. 254–259.

[7] Hapsoro D Setiawan D Hamiranti R and Yusnita Y, 2019 Pengaruh 2-iP, BA, 2, 4-D, dan TDZ pada embriogenesis somatik in vitro kopi robusta unggul Lampung *J. Agrotek Trop.* **7**, 3 p. 527–537.

[8] Murashige T and Skoog F, 1962 A revised medium for rapid growth and bio assays with tobacco tissue cultures *Physiol. Plant.* **15**, 3 p. 473–497.

[9] Morris A J Byrne T C Madden J F and Reller L B, 1996 Duration of incubation of fungal cultures *J. Clin. Microbiol.* **34**, 6 p. 1583–1585.

[10] Oratmangun K M Pandiangan D and Kandou F E, 2017 Deskripsi Jenis-Jenis Kontaminan Dari Kultur Kalus Catharanthus roseus (L.) G. Donnaman *J. MIPA* **6**, 1 p. 47–52.

[11] Kataky A and Handique P J, 2010 Standardization of sterilization techniques prior to in vitro propagation of Andrographis paniculata (Burm. F) nees *Asian J. Sci. Technol.* **6** p. 119–122.

[12] Cassells A C, 1991, Problems in tissue culture: culture contamination, in *Micropropagation*, (Springer), p. 31–44.

[13] Leifert C Ritchie J Y and Waites W M, 1991 Contaminants of plant-tissue and cell cultures *World J. Microbiol. Biotechnol.* **7**, 4 p. 452–469.

[14] Mahmoud S N and Al-Ani N K, 2016 Effect of Different Sterilization Methods on Contamination and Viability of Nodal Segments of Cestrum nocturnumL *Int. J. Res. Stud. Biosci.* **4**, 1 p. 4–9.

[15] Ahmed W Feyissa T and Bitima T D, 2013 Somatic embryogenesis of a coffee (Coffea arabica L.) hybrid using leaf explants *J. Hortic. Sci. Biotechnol.* **88** p. 469–475.

[16] Beckman C H, 2000 Phenolic-storing cells: keys to programmed cell death and periderm formation in wilt disease resistance and in general defence responses in plants? *Physiol. Mol. Plant Pathol.* **57**, 3 p. 101–110.

[17] Krishna H *et al.*, 2008 Mango explant browning: Effect of ontogenic age, mycorrhization and pre-treatments *Sci. Hortic. (Amsterdam).* **118**, 2 p. 132–138.

[18] Tiwari A K Tripathi S Lal M and Mishra S, 2012 Screening of some chemical disinfectants for media sterilization during in vitro micropropagation of sugarcane *Sugar Tech* **14**, 4 p. 364–369.

[19] Macias F A Galindo J C G and Molinillo J M G, 2003 *Allelopathy: Chemistry and mode of action of allelochemicals* CRC Press.

Optimization of Production Botanily Seeds (TSS) Shallot (*Alliun cepa* var. *ascalonicum*) Biru Lancor Variety through improvement of hand pollination in the lowland

E Siswadi [1*], R R D Pertami [1], S A Nugroho [1]

[1] Department of Agricultiral Production, Politeknik Negeri Jember, Jl. Mastrip PO BOX 164 Jember, East Java, Indonesia

* email: edi_siswadi@polije.ac.id

Abstract. The productivity of shallots is influenced by bulb quality, disease, and low vigor characteristics, leading to many ideas for developing true shallot seed (TSS). TSS planting material advantage is the use of less seed, it can be said maintenance would be easier, minimization of seed-borne diseases, and has a fairly long shelf life compared to seed tubers. It is hoped that improved pollination through modification of Hand Artificial Pollination can replace or increase the success of flowers into fruit and be followed by the formation of true seeds or True Shallot Seed (TSS) of the biru lancor variety of red onion in the lowland. The research aims to analyze the treatment of Hand Artificial Pollination in replacing insect pollination in the production of true seed (TSS) of the lancor blue variety in the Lowlands. The research method was to prepare the bulbs of the biru lancor variety shallots given vernalization treatment by storing in the refrigerator (showcase) at temperatures of 5°C, 10°C, and room temperature for 4 weeks. Furthermore, the tuber seeds after undergoing vernalization are planted. Research results show vernalization with a temperature of 10°C has no significant effect compared to tubers without vernalization. Vernalization with 5°C temperature can increase flowering and fruit formation (number of umbels per clump, number of flowers per umbel, number of fruit per umbel), viability, and pollen count, production of TSS seeds (number of seeds per umbel, seed weight per umbel, seed weight per clump).

1. Introduction

Shallot cultivation in Indonesia is carried out in an agro-climatic and efficient manner, as well as growth optimization, is mostly cultivated in the lowlands when compared to the highlands [1]. One of the varieties of shallot (Allium cepa var. ascalonicum) is biru lancor. Based on the Decree of the Minister of Agriculture Number: 2830/KPTS/SR.120/7/2009 origin of the lancor blue onion from the Cabean Hamlet, Pabean Village, Dringu District, Probolinggo Regency, East Java Province. Probolinggo as the center of shallots in East Java has an average yield of 8.9 tons/ha, while the potential yield can reach more than 15 tons/ha. Prolonged use of tubers as seeds at the farm level can lead to asexual mutations [2]. The government needs to organize the distribution of production centers, distribution of crop yields between regions, and monitor and evaluate shallot price policies. These policies aim to ensure the adequacy and smooth distribution of shallots [3].

The low productivity of shallots is influenced by the low quality of tubers available at the farm level, the presence of diseases and low vigor characteristics, which have led to many ideas for developing True Shallot Seed (TSS). The advantages of TSS planting material are the use of less seeds, it can be said that

IOP Conf. Series: Earth and Environmental Science **980** (2022) 012002 doi:10.1088/1755-1315/980/1/012002

it will be easier to maintain, minimize congenital diseases from seeds, and have a fairly long shelf life compared to seed tubers [4].

The vernalization method of shallot bulbs at a temperature of 4oC for 2 months can increase flowering, umbel formation and seeds in several varieties tested, although all the seeds produced have not been able to germinate. Based on the results of the research that has been done, this research will try to develop a method for flowering by giving the effect of vernalization temperatures of 5oC and 10oC for 4 weeks and combined with growth regulators (ZPT) using benzyl amino purines (BAP) and BAP which has been proven can give a positive response to induce flowering in the Bima and Trisula varieties of shallots, given that endogenous hormone conditions often still need to be improved again to encourage flowering through the application of PGR, to be able to further encourage the flowering process on shallots, especially the Biru lancor variety which has been proven adapted well in the lowland area as one of the shallot centers in East Java. Based on this, this research was conducted to study the effect of vernalization and the application of ZPT benzyl amino purine and BAP on the flowering of the Biru lancor variety of shallots.

2. Methods

Research on the vernalization of tuber seedlings was conducted at the Jember State Polytechnic Plant Laboratory. The results obtained will then be planted in the Lowlands, Leces District, Probolinggo Regency at an altitude of 89 meters above sea level from June 2021 to October 2021. Preparation The bulbs of the biru lancor variety of shallot seeds were subjected to vernalization treatment by storing in a refrigerator (showcase) at temperatures of 5°C, 10°C, and room temperature for 4 weeks. Furthermore, the tuber seeds after undergoing vernalization are planted. Planting was carried out in 2 months in polybags with a diameter of 22.5 x 40 cm.

The planting media used were topsoil, compost, and manure in a ratio (1:1:1) and placed under a transparent plastic house to avoid direct exposure to rainwater, while drip irrigation was used for irrigation, considering the need for water is very vital. Fertilization using NPK fertilizer (16:16:16) at a dose of 600 kg/ha and manure at a dose of 10 tons/ha and dolomite at a dose of 1.0 tons/ha. Meanwhile, to help pollinate, assisted by insect pollinators *Apis mellifera* and hand pollination. Insect pest control is carried out using insecticides with the active ingredient abamectin following the recommendations.

Parameters Observations were carried out by observing the growth of shallot plants and the number of flowers, namely: a) Average Plant Height in Vernalization Treatment, b) Number of BAP Leaves, c) Number of BAP tillers, d) Flowering Time of Vernalization and BAP Interaction, e) Flowering Percentage of Plants, f) Number of Flowers per Umbel, g) Flowering Time on Vernalization Interaction and BAP Seed Weight Per Umbel, h) Seed Weight Per Plant on Vernalization Treatment and BAP Interaction.

3. Result and Discussion

The results showed that the vernalization and benzyl amino purine treatments affected plant height, number of leaves, number of tillers, and number of flowers of the biru lancor variety of shallots in the lowlands. Vernalization technology allows for the acceleration of the sprouting process which is much faster than without vernalization so that through this technology it is hoped that it will be able to provide sustainable onion seeds [5]. Benzyl amino purine or (BAP) triggers flowering, as well as cell regeneration. Benzyl amino purine or (BAP) is a cytokinin group of plant growth hormones that naturally promotes cell division, cell division, tissue differentiation, dormancy, flowering phase, and fertilization [6].

The table above shows that the vernalization treatment did not have a significant effect on the height of shallot plants. There was no difference in plant height because plant growth was determined by genetic factors rather than vernalization treatment. This is by the opinion of experts that vernalization, in general, is more directed to generative growth than growth [7]. Benzyl amino purine treatment effects 28 days after planting. Benzyl Amino Purines are included in the cytokinin group which plays a role in shoot growth. Benzyl Amino Purine is a type of cytokinin that is resistant to degradation [8]. The relationship between vernalization and BAP has a significant effect at 42 days after planting.

Table 1. Average Plant Height on Vernalization Treatment

Vernalization Treatment	Plant Height					
	7 DAP	14 DAP	21 DAP	28 DAP	35 DAP	42 DAP
V_o	9,81a	20,16a	26,53a	32,15a	33,16a	31,16a
V_1	7,63b	17,14b	23,25a	29,93b	33,12a	32,23a

Note: The numbers followed by the same letter in the same column show no significant difference in the 5% BNT test. V0 = Non-Vernalization, V1 = Vernalization.

Table 2. The average number of BAP leaves

Vernalization Treatment	Number of Leaves					
	7 DAP	14 DAP	21 DAP	28 DAP	35 DAP	42 DAP
B_o	11a	20b	26b	32b	35b	49c
B_1	13a	24a	31a	37ab	43a	69ab
B_2	12a	20b	30a	39a	33b	75a
B_3	11a	20b	25b	32b	38ab	61b
BNT	5%	5%	5%	5%	5%	5%

Note: The numbers followed by the same letter in the same column show no significant difference in the 5% BNT test. B0 = 0 ppm; B1 = 50 ppm; B2 = 100 ppm; and B3 = 150 ppm= 0 ppm; B1 = 50 ppm; B2 = 100 ppm; dan B3 = 150 ppm

Table 3. The average number of BAP tillers

Vernalization Treatment	Number of Tubers					
	7 DAP	14 DAP	21 DAP	28 DAP	35 DAP	42 DAP
B_o	0a	0a	0a	24b	25b	28a
B_1	0a	0a	0a	26a	27a	31a
B_2	0a	0a	0a	25ab	27a	31a
B_3	0a	0a	0a	20c	20c	23b
BNT	5%	5%	5%	5%	5%	5%

Note: The numbers followed by the same letter in the same column show no significant difference in the 5% BNT test. B0 = 0 ppm; B1 = 50 ppm; B2 = 100 ppm; and B3 = 150 ppm= 0 ppm; B1 = 50 ppm; B2 = 100 ppm; dan B3 = 150 ppm

Vernalization and BAP treatments had no significant effect on the number of leaves parameters, while vernalization and BAP treatments affected 42 days after planting. The vernalization treatment on day 42 can have a very significant effect because the vernalization treatment can stimulate plant growth, especially plant elongation growth through an acceleration of dormancy so that plants can grow faster normally. This growth acceleration was due to an increase in cell division activity and endogenous gibberellins and auxin content. The number of leaves will greatly determine the surface area to receive sunlight for the photosynthesis process. In this process the carbohydrates produced will be distributed throughout the plant organs for growth [9][10].

The vernalization treatment with BAP showed no significant effect on the number of shallot tillers. A significant effect occurred on the interaction of vernalization treatment with BAP at 42 days after planting. The number of shallot tillers correlated with the size of the bulbs caused by cell enlargement which was more dominant than cell division. The increase in the wet weight of the tubers was influenced by the amount of water absorption and the accumulation of photosynthetic products in the leaves to be translocated for tuber formation. So, the difference in water content will affect the wet weight of the tubers produced [11].

Table 4. Average Time of Interest in Vernalization and BAP Interaction

Vernalization Treatment x BAP	**Flower time**
	63 DAP
V_0B_0	0c
V_0B_1	0c
V_0B_2	0c
V_0B_3	0c
V_1B_0	29a
V_1B_1	25b
V_1B_2	22b
V_1B_3	24b
DMRT	5%

Note: The numbers followed by the same letter in the same column show that they are not significantly different in the multiple-distance test (DMRT) 5%. V0B0 = Non Vernalization + 0 ppm BAP; V0B1 = Non Vernalization + 50 ppm BAP; V0B2 = Non Vernalization + 100 ppm BAP; V0B3 = Non Vernalization + 150 ppm BAP; V1B0 = Vernalization + 0 ppm BAP; V1B1 = Vernalization + 50 ppm BAP; V1B2 = Vernalization + 100 ppm BAP; V1B3 = Vernalization + 150 ppm BAP

Based on the results of the F test (Table 4), it can be seen that the interaction between the vernalization factor and the BAP factor has a very significant different effect on the parameter of the percentage of flowering plants. While the single factor vernalization gave no significant effect on the parameters of the percentage of flowering plants, as well as the BAP gave an insignificant different effect on the parameters of the percentage of flowering plants.

The flowering process of plants occurs through four stages, namely induction, flower initiation, flower differentiation, maturation of flower parts, and anthesis, initiation of flowering is this stage where morphological changes occur to form generative buds and transition from vegetative shoots to generative buds that can be detected from changes in shape. as well as bud size, as well as subsequent processes that begin to form generative organs. Flowering response and yield of shallot seeds increased with a combination of vernalization treatment (10°C) for 4 weeks on seed tubers. BAP was found to stimulate flowering and increase production [12]. The increase in production is due to the presence of cytokinin in the plant which stimulates an increase in the size of the reproductive meristem so that it can produce more flowers [13].

Amount of Interest Per Umbel Based on the results of the F test (Table 4), the interaction between the vernalization factor and the BAP factor has a very significant effect on the parameter of the amount of interest per umbel. While the vernalization factor gave an insignificant different effect on the parameter of the amount of interest per umbel, as well as the BAP gave an insignificant different effect on the parameter of the amount of interest per umbel.

Table 5. Average Parameter of Flowering Plant Percentage

Treatment	Percentage of flowering plants
	63 DAP
V_0B_0	0c
V_0B_1	0c
V_0B_2	0c
V_0B_3	0c
V_1B_0	58ab
V_1B_1	71a
V_1B_2	67ab
V_1B_3	50b
DMRT	5%

Note: The numbers followed by the same letter in the same column show that they are not significantly different in the multiple-distance test (DMRT) 5%. V0B0 = Non Vernalization + 0 ppm BAP; V0B1 = Non Vernalization + 50 ppm BAP; V0B2 = Non Vernalization + 100 ppm BAP; V0B3 = Non Vernalization + 150 ppm BAP; V1B0 = Vernalization + 0 ppm BAP; V1B1 = Vernalization + 50 ppm BAP; V1B2 = Vernalization + 100 ppm BAP; V1B3 = Vernalization + 150 ppm BAP

Non-vernalization treatment resulted in a longer flower emergence time and BAP did not stimulate more tillers to produce umbel shoots. BAP is more involved in increasing the size of differentiated meristems producing the number of flowers per umbel rather than stimulating stumps of tillers. BAP application has no effect on flower emergence time because 0 and 50 ppm flowers appear first because low concentrations of BAP accelerate flower growth, the flowering period is not affected by BAP application, possibly because vernalized plant conditions can reduce the effect of BAP to the acceleration of flowering plants.

Cytokinins have an important role in meristem morphogenesis, thus allowing the development of larger flower meristems. However, the application of BAP at various concentrations did not affect the increase in the number of umbels per clump. This means that BAP does not stimulate more tillers to produce umbel shoots. It is possible that BAP plays a more important role in increasing the size of the differentiated meristems resulting in the number of flowers per umbel rather than stimulating swarming tillers. This happens because PGR is known to promote plant growth and development at low concentrations but at high concentrations, it can inhibit plant growth and development. This study showed that 50 ppm BAP was the optimum concentration to promote shallot flowering [14].

Number of Capsules Per Umbel Based on the results of the F test (Table 4) shows that the interaction between the vernalization factor and the BAP factor has a very significant different effect on the parameter of the number of capsules per umbel. While the single factor of vernalization gave an insignificant different effect on the parameter of the number of capsules per umbel, as well as the single BAP gave an insignificant different effect on the parameter of the number of capsules per umbel.

Table 6. The average parameter of the amount of interest per umbel

Vernalization Treatment x BAP	Percentage of flowering plants
	63 DAP
V_0B_0	0c
V_0B_1	1c
V_0B_2	2c
V_0B_3	3c
V_1B_0	107ab
V_1B_1	96ab
V_1B_2	90b
V_1B_3	112a
DMRT	5%

Note: The numbers followed by the same letter in the same column show that they are not significantly different in the multiple-distance test (DMRT) 5%. V0B0 = Non Vernalization + 0 ppm BAP; V0B1 = Non Vernalization + 50 ppm BAP; V0B2 = Non Vernalization + 100 ppm BAP; V0B3 = Non Vernalization + 150 ppm BAP; V1B0 = Vernalization + 0 ppm BAP; V1B1 = Vernalization + 50 ppm BAP; V1B2 = Vernalization + 100 ppm BAP; V1B3 = Vernalization + 150 ppm BAP

Table 7. Average Flowering Time on Vernalization and BAP interaction

Vernalization Treatment x BAP	flower time
	63 DAP
V_0B_0	0c
V_0B_1	0c
V_0B_2	0c
V_0B_3	0c
V_1B_0	31a
V_1B_1	23b
V_1B_2	22b
V_1B_3	29a
DMRT	5%

Note: The numbers followed by the same letter in the same column show that they are not significantly different in the multiple-distance test (DMRT) 5%. V0B0 = Non Vernalization + 0 ppm BAP; V0B1 = Non Vernalization + 50 ppm BAP; V0B2 = Non Vernalization + 100 ppm BAP; V0B3 = Non Vernalization + 150 ppm BAP; V1B0 = Vernalization + 0 ppm BAP; V1B1 = Vernalization + 50 ppm BAP; V1B2 = Vernalization + 100 ppm BAP; V1B3 = Vernalization + 150 ppm BAP

IOP Conf. Series: Earth and Environmental Science **980** (2022) 012002 doi:10.1088/1755-1315/980/1/012002

Table 8. Average Seed Weight Per Umbel and Seed Weight Per Plant in V x B Interaction Treatment

Vernalization Treatment x BAP	Botanical Seed Production (TSS)	
	Seed Weight Per Umbel (g)	Seed Weight Per Plant (g)
V_0B_0	0 b	0 b
V_0B_1	0,17 ab	0,17 b
V_0B_2	0,18 ab	0,18 b
V_0B_3	0 b	0 b
V_1B_0	0,39 ab	0,39 ab
V_1B_1	0,64 a	0,73 a
V_1B_2	0,42 ab	0,42 ab
V_1B_3	0,43 ab	0,52 ab

Note: The numbers followed by the same letter in the same column show that they are not significantly different in the multiple-distance test (DMRT) 5%. V0B0 = Non Vernalization + 0 ppm BAP; V0B1 = Non Vernalization + 50 ppm BAP; V0B2 = Non Vernalization + 100 ppm BAP; V0B3 = Non Vernalization + 150 ppm BAP; V1B0 = Vernalization + 0 ppm BAP; V1B1 = Vernalization + 50 ppm BAP; V1B2 = Vernalization + 100 ppm BAP; V1B3 = Vernalization + 150 ppm BAP

The results of the 5% DMRT test analysis showed that there was an interaction between vernalization treatment and BAP which significantly affected the parameters of seed weight per umbel and seed weight per plant. The interaction treatment between vernalization and BAP (V1B1), increased seed weight per umbel and seed weight per plant which was heavier than the control treatment. The vernalization treatment with BAP could produce an average seed weight per umbel of 0.64 grams and an average seed weight of 0.73 grams per plant, but the weight of 100 grains was low, there was no significant interaction between the vernalization treatment and the BAP treatment. single treatment also did not affect the weight parameter of 100 grains.

The comparison of the weight of shallot seeds was more clearly seen in the V1B1 treatment which gave the effect of increasing the weight of shallot seeds on the parameters of seed weight per umbel and seed weight per plant by producing a difference of 0.09 grams. The interaction treatment between vernalization and BAP at a concentration of 50 ppm could induce maximum shallot flowering, especially in the Biru Lancor variety because it produced the highest seed weight compared to the control and other treatments. The best concentration for yields giving the highest value for seed weight per clump was 200 ppm BAP. So that in the application of BAP the concentration is lower in the seed weight of the Biru Lancor variety of shallots.

4. Conclusion

The interaction between vernalization and Benzyl amino purine (BAP) gave a significantly different effect on plant height parameters and flower parameters in shallots. single factor vernalization and BAP gave no significant effect on all parameters.

5. Acknowledgment

The author would like to thank Kemendikbud Ristek for the PNBP research fund in 2021 through the Research and Community Service Center of Jember State Polytechnic.

IOP Conf. Series: Earth and Environmental Science **980** (2022) 012002 doi:10.1088/1755-1315/980/1/012002

References

[1] Suherman R, Basuki R. 1990. Strategi Pengembangan Luas Areal Usahatani Bawang Merah (*Allium cepa* var. *ascalonicum*) di Jawa Barat: Tinjauan dari Segi Biaya Usahatani. Bul Penel Hort Ed Khusus.

[2] Singh V, Kumar R. 2017. Study of Phytochemical Analysis and Antioxidant Activity of *Allium sativum* of Bundelkhand Region. *Int. J. Life. Sci. Scienti. Res.* 3(6), 1451-1458

[3] Sahara, Utari MH, Azijah Z. 2019. Volatilitas Harga Bawang Merah di Indonesia. *Buletin Ilmiah Litbang Perdagangan*. 13(2), 309–36.

[4] Sartono P. 2010. Pengujian Beberapa Klon Bawang Merah Dataran Tinggi (*Clones Testing Of Some Highlands Shallots*). *Bul Penel Hort*. 26(4):86–92.

[5] Edi S, Cholyubi Y, Muhammad Z.S, Refa F. 2019. Pengembangan Sentra Agribisnis Bawang Putih Di Kecamatan Sukapura Kabupaten Probolinggo. *J-Dinamika*. 4(2), 153-158.

[6] Amanullah MM, Sekar S, Vincent S. 2010. Plant growth substances in crop production: a review. *Asian J Plant Sci*. 6(3), 1-8.

[7] Sumarni, N. dan E. Sumiati. 2001. Pengaruh vernalisasi, giberelin dan auksin terhadap pembungaan dan hasil biji bawang merah. *Jurnal Hortikultura*.11(1), 1-8.

[8] Wattimena. 1988. Zat Pengatur Tumbuh Tanaman. Institut Pertanian Bogor. Bogor.

[9] Jain R., A.K. Shrivastava, S. Solomon, R.L. Yadav. 2007. Low temperature stress – induced biochemical changes affect stubble bud sprouting in sugarcane (*Saccharum* spp. *hybrid). Plant Growth Regul*. 53:17-23.

[10] Dinarti, D., B.S. Purwoko, A. Purwito, A.D. Susila. 2011. Perbanyakan tunas mikro pada beberapa umur simpan umbi dan pembentukan umbi mikro bawang merah pada dua suhu ruang kultur. *J. Agron. Indonesia*. 39: 97-102.

[11] Setiyowati, S. H. dan R. B. Hastuti. 2010. Pengaruh perbedaan konsentrasi pupuk organik cair tehadap produksi bawang merah (*Allium ascalonicum* L.) laboratorium biologi dan struktur fungsi tumbuhan fmipa undip. *BIOMA*. 12: 44-48.

[12] Sumarni N, Rosliani R, Suwandi. 2013. Optimasi Jarak Tanam dan Dosis Pupuk NPK untuk Produksi Bawang Merah dari Benih Umbi Mini di Dataran Tinggi. *J Hort*. 22(2), 147-154.

[13] Pangestuti R, Sulistyaningsih E. 2011. Potensi Penggunaan *True Seed Shallot* (TSS) Sebagai Sumber Benih Bawang Merah Di Indonesia. *Prosiding Semiloka Nasional*. 258-266. Semarang 14 Juli 2011.

[14] Werner, T, Motyka, V, Strnad, M, Schmulling, T. 2001, 'Regulation of plant by cytokinin', *Plant Biol*. 98(18), 87-92.

IOP Conf. Series: Earth and Environmental Science **980** (2022) 012003 doi:10.1088/1755-1315/980/1/012003

Design of Flash Dryer Cum-UV for Improving the Quality of Drying Cassava Chip

D Hermanuadi[1], A Brilliantina[2], EKN Sari[3]*
[1]Department of Food Engineering Technology, Politeknik Negeri Jember, Indonesia
[2]Department of Food Industry, Politeknik Negeri Jember, Indonesia
[3]Department of Agricultural Engineering, Politeknik Negeri Jember, Indonesia

Corresponding author : elok_kurnia@polije.ac.id

Abstract. Drying is a scientific operation, which includes the transient transfer of heat and mass at some rate of process. The purpose of this study was to design and create a flash system dryer, which is equipped with UV. The selection of flash system dryers is based on conditions where some materials to be dried are sensitive to hot air contact due to prolonged drying. As a heat source, this dryer uses a heater to heat the air. In order to spread the heat evenly in the dryer room, the blower is installed on the heater. The intake of air blower is 60 m/s, while the air temperature on the air heater is conditioned at 50 and 60^0C. In order to collect data of air temperature, a thermocouple is placed on heat channel pipe and dryer room. The performance test of the dryer by using cassava chips, showed that the lowest cassava chip water content was 7,2% at 50°C treatment temperature, and the highest water content by 18.5% at treatment temperature of 60°C.

1. Introduction

[1] stated that Indonesia is the second largest producer of cassava after Thailand. It's just that Indonesian cassava is more widely consumed in the country. In the future Indonesia has the opportunity to develop the production of yams, including processed products and derivatives, so that it becomes one of the local foods that can be used as a yam-based agricultural industry. According to [2] cassava is one of the rice substitute foodstuffs that is quite important in supporting the food security of a region. However, there are still many obstacles faced in changing the consumption patterns of the community that have been formed so far. In order to sustain the food security of a region, it is necessary to socialize the diversification of food made from sweet potatoes or cassava as an alternative food.

Fermented processed cassava, is one of the efforts to extend the shelf life of cassava. One of the efforts fermented yam, namely drying. This is because, microbes can survive long on products with high material water content. Cassava in the form of slices of drying, can then be processed further into flour. Cassava flour, is one of the best flours as a substitute for wheat flour [3].

Drying is a method used to remove most of the water content of the material by providing heat. Reduction of material water content, usually done in materials with wet solid conditions, semi-solid, and liquid materials that will be converted into solid products by the process of evaporation of material liquids. The usual drying method is drying by venting. However, this method takes quite a long time, thus extending the overall bending process time [4].

One drying method that can overcome this problem is flash dryer. Flash dryer is a dryer by utilizing high-speed hot air for the process of drying materials. This drying method is relatively simple in operation, requires little place, is suitable for drying heat-sensitive foodstuffs, and the control system can generally respond very quickly to changes in the operating conditions of the drying process, so in this study designed the design of flash dyer models with the addition of UV. The reason for the selection of flash type dryers is that the time needed to dry the material is relatively short. So that this dryer is suitable for materials that are sensitive to heat. Based on this, it is expected that this research can improve the quality and quantity of cassava in accordance with the wishes of the industry.

2. Literature review

Drying is a way to scatter or remove most of the water from the material using heat energy. The production of water from the material is carried out until the moisture content balances with certain environments where fungi, enzymes, microorganisms, and destructive insects become inactive [5]. Drying according to [6] has the understanding of the application of heating through regular conditions, so that it can remove most of the water in a material by evaporating. Removal of water in a material by drying has a different unit of operation with dehydration. Dehydration will decrease the activity of water contained in the material by removing or removing more water, so that the shelf life of food becomes longer or longer.

When wet objects are thermally drained, there are two processes that take place simultaneously, namely:

1. Transfer of energy from the environment to evaporate water found on the surface of solid objects. This transfer of energy from the environment can take place by conduction, convection, radiation, or a combination of the three. This process is influenced by temperature, humidity, the rate and direction of airflow, the physical shape of the solids, the surface area of contact with air and pressure. This process is an important process during the early stages of drying when unbound water is removed. Evaporation that occurs on the surface of a solid is controlled by the diffusion of steam from the surface of the solid into the environment through a thin film layer of air.
2. Transfer of the mass of water contained in the object to the surface. When there is evaporation on the surface of the solid, there is a difference in temperature so that water flows from the inside of the solid object to the surface of the solid object. The structure of the solid object will determine the mechanism of internal flow of water.

Flash Dryer

Flash dryer is a dryer installation used to dry flour that has a certain moisture content. The purpose of using this tool is to dry the product and reduce the moisture content of the product that was originally high to low. This machine dries the flour by continuing to drain hot air. The drying process that occurs in flash dryers takes place quickly. The time needed by the material that is dried from starting to enter the dryer vessel until it comes out becomes the result of fast drying products, therefore flash dryer is also called a fast dryer. Flash dryer type dryers have been widely applied to the chemical industry, food factories, agricultural products, ceramics industry, cement factory, paper mill, pharmaceutical industry and flour factory [7].

According to [8] to achieve an efficient drying process, the dryer airflow rate used must be greater than the minimum speed treated to move the item, so determining the speed of air to be used to dry a material becomes important to note. The speed of air exhaled by the blower or fan must be greater than the fast fall of the particles to be drained (wet material). The speed of air flow during the drying process using pneumatic (flash dryer) should not be too low or too high. At speeds that are too low, the material particles cannot be lifted by airflow, so the drying process cannot run perfectly, while the flow of air speed that is too high will cause the heat contact between dry air and the material will become too short, as a result of which the drying process is not effective, because the evaporated water is only small and the final water content of the product is usually still high.

3. Material and methods

The design, manufacture and testing of dryers was carried out at the Metal Laboratory of the Department of Agricultural Technology, Polytechnic State of Jember in July to August 2021.The initial stage in the design and manufacture of dryers is to determine the size / dimensions of the dryer needed. Flash Dryer is planned to be equipped with

1. Materials Room
   The material room in a flash type dryer has a design like a cabinet dryer type dryer. The material space consists of shelves made in multi-storey. The base where the material is placed, made perforated so that the heat that enters the material space will intersect directly with the material to be dried.
2. Ultraviolet
   The addition of UV serves to improve the quality of drying materials. UV serves to reduce microbial contamination, which may be investigated during the dispelling process before drying so that it attaches to the surface of the material to be dried. UV laying is on the entire dryer room and on each shelf where the material will be dried. So that the process of UV administration can run optimally.
3. Blower and Heater
   Blower placed on the side of the dryer room serves to supply air. In a flash dryer system, the heat exhaled is in the form of air that has been heated. Therefore, in this dryer is equipped with an air heater (air heater). The air heater serves to heat the air exhaled by the blower, so that the air that enters the dryer room is hot air.
4. Heating Room Temperature Controller
   Thermocouples are placed on the heating room, to read the temperature conditions of the heating room. This thermocouple is connected with a temperature reader display placed on top of the dryer room, so that the temperature of the heating room can be read. Thermocontrol is also placed on the pipe that connects the heater and the heating chamber. So, if the temperature has reached the desired amount, then the water heater will die automatically. And, if the input temperature is less than the desired amount, then the water heater will turn on again.

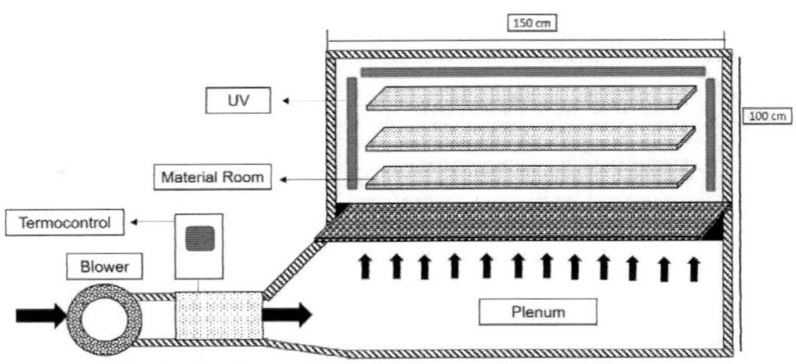

Fig 1. Design of Flash Dryer-Cum UV

The speification of Flash Dryer :

1. Capacity	: 4 shelves
2. Number of Trays	: 8 Trays
3. Tray Size	: 50 x 50 x 3 cm
4. Inner and outer body	: Stainless steel plate
5. Tray	: Stainless steel plate
6. Machine Frame	: Iron
7. Heating	: Electricity
8. Temperature Regulator	: Automatic
9. Hot Air Distribution	: Blower
10. Overall Dimensions of Dryer	:180 x 60 x 100 cm
11. Material weight	: 5 kg of wet materials

Fig 2. Flash Dryer-Cum UV Design Result

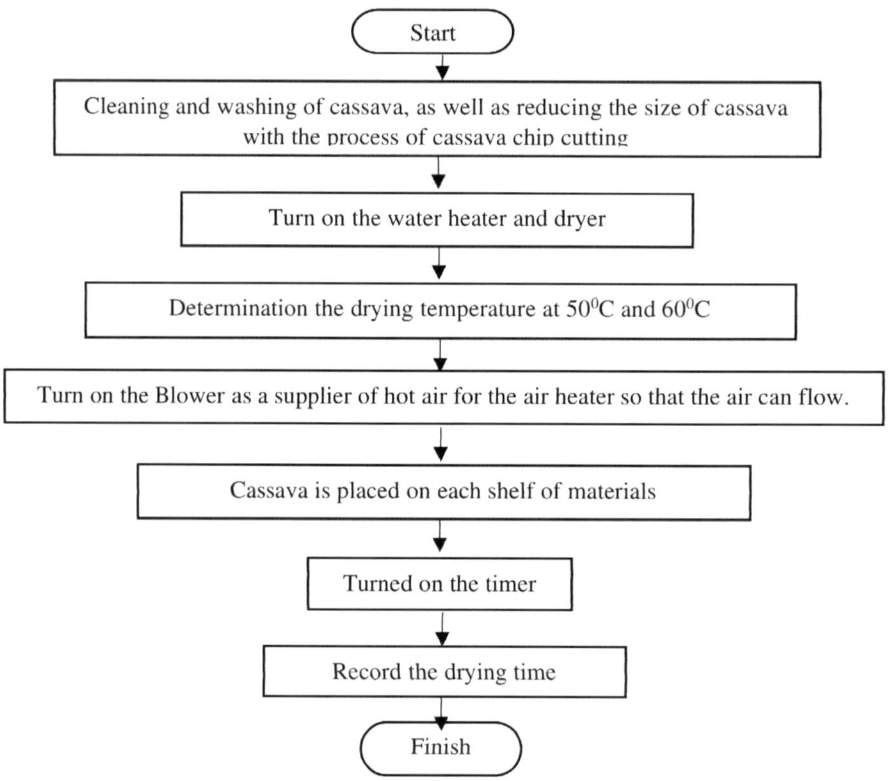

Fig 3. Diagram of Flash Dryer-Cum UV Test

Calculating the drying rate

The dring rate can be calculated by using the formula :

a. The total dry mass of the cassava chip, can be calculated by the formula:

$$Mk = \frac{M_0}{1 + MC_0(1)}$$

b. The mass total of cassava chip, can be calculated by the formula :
$DR = (1 + MC)Mk$
c. Drying rate
$$DR = \frac{M_0 - M_s}{time\ x\ surface\ area\ of\ cassava\ chip}$$

Drying efficiency

According to [9] drying efficiency is the result of a comparison between the heat theoretically needed with the actual use of heat in drying. The amount of heat energy required for drying can be calculated using the formula:

$$Q = Q_1 + Q_2 + Q_3$$

a. Q_1 is the amount of heat used to heat a material, derived from the equation :
$Q_1 = m_k.c_p.(T_p - T_{\sim})$
b. Q_2 is a sensible of hot water, which is heat used to raise the temperature of water in the material, which is obtained from the equation:
$Q_2 = m_a.c_a.(T_p - T_{\sim})$
c. Q_3 is latent heat of water evaporation
$Q_3 = mw.hfg$

To determine the amount of heat (heat) given by hot air in the drying material, the formula :

$$q = \rho.V.C_p(T_1 - T_2)$$

To determine the drying efficiency and efficiency of dryers, obtained by the use of the formula:

$$\eta = \frac{Q}{q} x100\%$$

Where:
m : mass of dried material (Kg)
Cp : heat of the type of drying material (kJ/Kg.0C)
ΔT : increase in material temperature (0C)
V : Volume of air
Cp : dryer air mass (kg/m3)
η : drying efficiency (100%)

4. Result and Discussion

4.1 *Drying Rate and Drying Efficiency*

The drying rate in flash type dryers is greatly affected by the dryer's airflow. The dryer airflow rate used should be greater than the water transfer of the surface of the material during the drying process. The speed of airflow exhaled by the blower is used to spread the heat generated by the water heater into the dryer room. This distribution of hot air is used to evaporate the water content of cassava chips. The speed of airflow exhaled by the blower, measured using an anemometer. The measurements showed the speed of air flow in the dryer room by 60 m / s.

Based on the results of the measurement, the speed of air flow exhaled by the blower is quite high. Thus, the contact of the material with the flow of hot air is short. Therefore, flash dryer can be applied to materials that are sensitive to heat. Based on the results of the calculation of the drying rate for cassava chips for 4 hours of drying, which is 0.00885 gr / s at 50^{0}C and 0.00985 gr/s at 60^{0}C. Drying efficiency is the amount of energy needed to evaporate water from the material divided by the energy produced during the drying process. Drying efficiency with the use of temperature 50^{0}C, by 65.76% while at 60^{0}C at 64.39%

4.2 *Effect of Dryer Air to Cassava Chip Mass Decline*

Flash dryer, utilizing hot air exhaled by the blower. Hot air is the result of heating by the water heater. Blowers placed side by side with the water heater will distribute and spread hot air throughout the dryer room. This hot air that intersects directly with the material, to evaporate the water content of the material. Hot air will cause evaporation, so the weight of cassava chips is also reduced. The drying process is determined for 4 hours, so the decrease in the mass of the material is measured every 30-minute span.

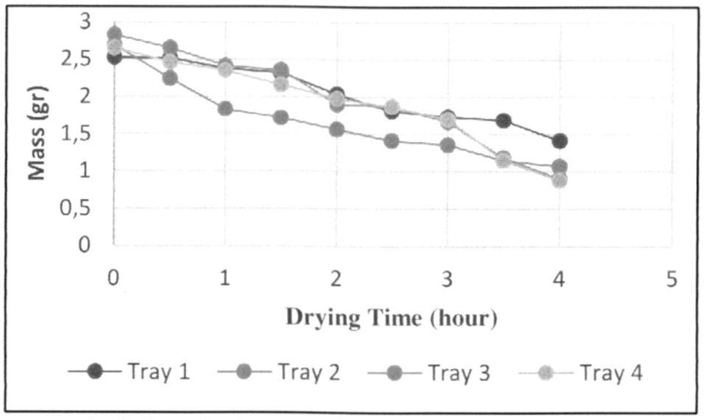

Fig 4. Decrease in Mass (gr) material at a Drying Temperature of 50^{0}C

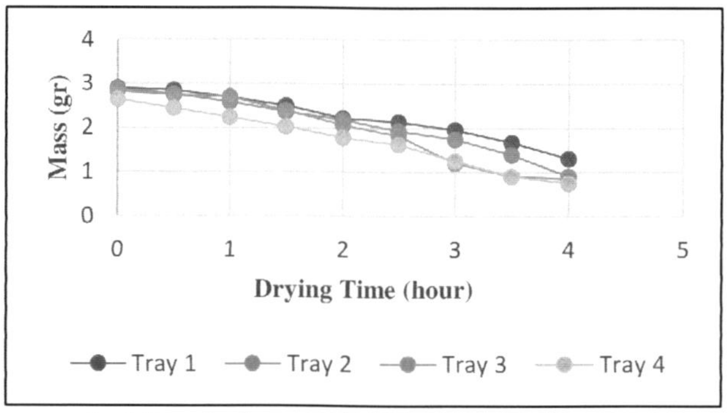

Fig 5. Decrease in Mass (gr) material at a Drying Temperature of 60^{0}C

4.3 *Effect of Dryer Air to Cassava Chip Water Content Decline*

Drying temperature greatly affects the decrease in cassava chip water content. In the performance test of this flash type dryer used 2 magnitudes of temperature, namely 50^{0}C and 60^{0}C. This temperature difference is intended, to see the effect of temperature on the water content of the material, measurement of the water content of the material, done by the oven method. Determination of cassava chip moisture content is done 30 minutes with a drying length of 4 hours

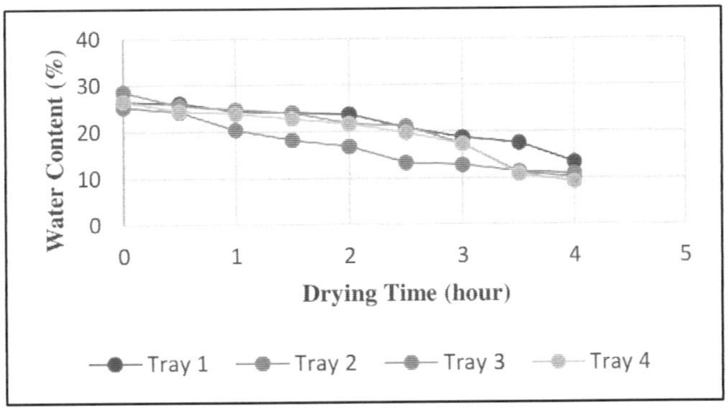

Fig 6. Decrease in Material Water Content at Drying Temperature of 50⁰C

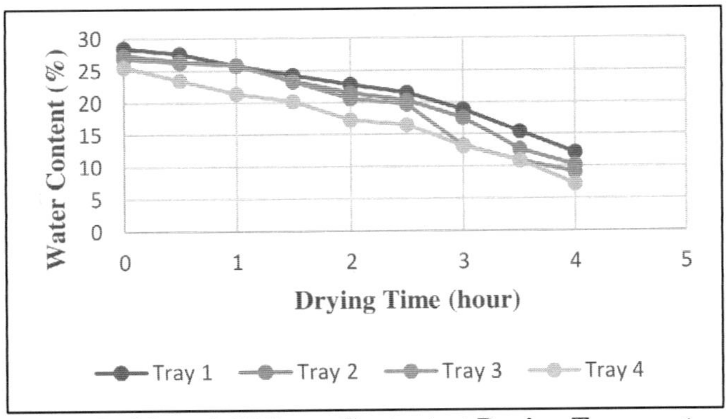

Fig 7. Decrease in Material Water Content at Drying Temperature of 60⁰C

In Figure 6 showed a graph of the drop in cassava chip moisture content at a drying temperature of 50⁰C. While in Figure 7 showed a graph of the decrease in the moisture content of the material at a drying temperature of 60⁰C. Based on both graphs, it can be observed that the length of drying has an effect on the decrease in the water content of the material. The selection of temperatures of 50⁰C and 60⁰C, based on conditions where, there are several types of materials that are sensitive to heat. So, it is determined that the use of drying temperature is not too high. Generally, drying materials with the use of temperatures of 50⁰C and 60⁰C, it takes a drying length of 8 - 10 hours to reach the appropriate cassava chip water content as a raw material for cassava flour.

Based on Indonesia's national standard on the quality requirements of cassava flour, the maximum cassava flour water content is 12%. Therefore, the water content of cassava chips before refining, the moisture content should not be more than 12%. Cassava chip water content values that conform to standard moisture content values, achieved at a drying temperature of 50⁰C for 210 minutes and require a drying length of 180 minutes at a drying temperature of 60⁰C.

Based on [10] research, The drying by using pneumatic dryer could reduce the water content of cassava grater up to 8-12% although the time of the drying process in the drying duct is quite short. The decrease in water content of cassava grater took place over time.

4.4 Degree of colour

The degree of white is the ability of a material to reflect light that hits the surface of the material [11]. The whiter the colour of flour, the higher the level of consumer acceptance of the flour. The purpose of measuring white degrees is to find out whether UV irradiation during the drying process affects the colour quality of cassava chips. To find out, the effect of UV administration on the colour quality of cassava chips, measurements of cassava chips without UV irradiation during drying and cassava chips with irradiation during the drying process.

The measurements showed a brightness level of 81.29 for drying cassava chips without UV irradiation. Meanwhile, for cassava chips with UV irradiation during the drying process showed a brightness level of 79.4. The measurement result is close to the value of 100, which is assumed to be the best value. This value is below the standard for yam flour, where the value of vaginal discharge is at least 85. According to [12], a shorter drying time is able to minimize the formation of brown colour on cassava chips due to the heating process. UV irradiation is intended as a substitute for irradiation from sunlight. Based on the results of measurements, UV irradiation during the drying process has no real effect on the browning of cassava chips.

5. Conclusion

Flash dryers utilize the flow of hot air, to reduce the water content of the material. So that the decrease in the water content of the material, depending on the speed of hot air flow. The flow of hot air is quite high, bringing the drying time to a short time. Thus, flash dryers are suitable for heat-sensitive materials.

Acknowledment

Author wishing to acknowledge assistance or encouragement from Polytecnik State of Jember, because this research was carried out on the research grant of PNBP Polytechnic State of Jember Budget Year in 2021

References

[1] Hermanto, M. A. (2015). Dinamika Konsumsi Pangan. In E. A. Editor : Bambang Irawan, Panel Petani Nasional : Rekonstruksi Agenda Peningkatan Kesejahteraan Petani (pp. 101 - 123). IAARD Press.

[2] The Center for Data and Information System of the Ministry of Agriculture of the Republic of Indonesia. 2016. Outlook for onion commodities. Jakarta: Pusdatin Ministry of Agriculture.

[3] Hermanuadi, D. Brilliantina, A, and Sari, E.K.N, Sari. 2018. Determination of Agro-Industry Area Based on Cassava Commodity in Bondowoso Regency. *IOP Conf. Series: Earth and Environmental Science 207*: 012010

[4] Nurin'in, Devi Nihayati. 2013. **Prospek Pengembangan Agroindustri Tape Singkong di Kabupaten Jember**. Skripsi. Universitas Jember: Jember

[5] Hermanuadi, D. Brilliantina, A, and Sari, E.N. 2020. Decision Support System For Selecting Strategy Of Agroindustry Development Based On "Tape" In Bondowoso Regency. **IOP Conf. Series: Earth and Environmental Science 411** (2020) 012016

[6] Robinson. 1997. **Manajemen Strategic**. Jakarta Barat. Binarupa Aksara

[7] Buckley, J. J.. 1985. "Fuzzy Hierarchical Analysis". **Fuzzy sets and systems** 17:233- 247

[8] Falatehan, A. Faroby. 2016. **Analytical Hierarchy Process (AHP) Teknik Pengambilan Keputusan untuk Pembangunan Daerah**. Yogyakarta: Indomedia Pustaka

[9] Taufiq. 2004. Pengeringan Lapisan Tipis Kentang. (Solanum tuberosum. L) Varietas Granola. Skripsi pada Universitas Hasanuddin.

[10] Wijaya, R. Hariono, B. 2019. The Mathematical Analysis of the Drying of Cassava Grater by Using Pneumantic (flash) Dryer with Heat Recirculation Method. IOP Conf. Series : Journal of Physics 1569(2020) 042061

[11] Iswari, K., Astuti, H. F., dan Srimaryati. 2016. Pengaruh Lama Fermentasi Terhadap Mutu Tepung Cassava Termodifikasi. BPTP Sumatera Barat.

[12] Hidayat, B., Kalsum, N., dan Surfiana. 2009. Karakterisasi Ubi Kayu Modifikasi yang diproses dengan Menggunakan Metode Pragelatinsasi Parsial. Jurnal Teknologi Industri dan Hasil Pertanian,14 (2): 148-159.

Utilization of foliar fertilizer as an alternative medium for enlargement of Vanda orchid plantlets before acclimatization

K Kasutjianingati[1]*, T R Koesparwanti[1], E Eliyatiningsih[1]

[1]Department of Agricultural Production, Politeknik Negeri Jember, Jalan Mastrip PO BOX 164 Jember, Indonesia

*kasutjianingati@gmail.com

Abstract: This study aims to obtain an appropriate and inexpensive alternative media combination for the growth of Vanda hybrid orchids prior to acclimatization. Experiment using Factorial Randomized Block Design. The first factor is the type of media (Vacin -Went and foliar fertilizer) and the second factor is the addition of organic matter (coconut water composition, sweet corn composition, tomato composition, BAP, NAA and without organic matter or PGR). The results showed that the interaction of treatment on plant height was only seen at 18 weeks after treatment, while the number of leaves could be seen earlier at 16 weeks after treatment. The independent effect of organic matter and PGR on the number of leaves has been seen since the age of the plant 10 weeks after treatment, while the plant height was only seen 16 weeks after treatment. Independent treatment of growing media had no effect on height and number of leaves up to 18 weeks after planting. The use of foliar fertilizer media with the type of organic material composition being tested can be suggested as an alternative growth medium for plantlet growth, not significantly different from vacin-went with BAP media.

1. Introduction

The orchid (*Orchidaceae*) business is a solution that should be considered because it is a plant that has high demand. Market demand is never empty, one of the favorite types of orchids in Indonesia is Vanda which has 5000 species with a variety of flower shapes and colors [1]. Based on data from the Central Statistics Agency for ornamental plants, orchid production in Indonesia has fluctuated, orchid production was 20,777,071 stalks (2013), 19,739,627 stalks (2014), 21,514,789 stalks (2015), in 2016 as many as 19,978,078 stalks, in 2017 as many as 20,045,577 stalks, in 2018 24,717,840 stalks, and in 2019 as many as 18,608,657 stalks.

The solution to increase the production of Vanda orchids is still more appropriate through in-vitro propagation, growing explants from one part of the plant under aseptic and accentic conditions. The main need for in vitro propagation is the presence of a growth medium equipped with complete nutrition, growth regulators and controlled lighting. This means that the determinants of the success of tissue culture techniques are the suitability of the explant characters with the nutrient composition of the culture media, aseptic/acsenic environmental conditions and the presence of growth regulators. The use of growth regulators (PGR) in tissue culture is very important, namely controlling the morphogenesis of the formation and development of shoots, roots, and callus. The formation of shoots generally uses cytokinins, while the formation of roots or callus uses auxin. [2]. PGR consists of natural and synthetic PGR. Natural ZPT can be used by Ambon banana, sweet corn, tomatoes, coconut water, and potatoes, replacing the role of synthetic ZPT Benzyl Aamino Purine (BAP) and Naphthalene Acetic Acid (NAA).

VW media (Vacin-Went) formulated by E. Vacin and F. Went in 1949 consisted of macro and micro nutrients in the form of inorganic salts suitable for the growth and development of orchids in vitro. The high cost of chemicals spurs the idea of using alternative materials to develop, the use of foliar fertilizers on the market as a nutrient source, for example Gandasil D. Consideration by taking into account that in the vegetative growth phase a lot of fertilizers with nitrogen content are needed as a protein constituent for cell division. utilize organic materials in the manufacture of media compositions. Alternative substitute for synthetic growth regulators (auxins/cytokinins) using tomato water (*Solanum lycopersicum*) contains the hormone auxin which functions to accelerate shoot growth in the process of cell division. According to [3] the content of auxin in tomato extract can stimulate organogenesis, somatic embryogenesis and shoot growth in micropropagation in various plant species. The content of auxin in tomatoes in some plants needs to be balanced with the provision of synthetic cytokinins (BAP). Cytokinins function to stimulate shoot growth, cell metabolism, cell division, reduce apical dominance, and encourage lateral shoot initiation. Based on the results of research [4] on the number of shoots given BAP 3 ppm + 100 ml of tomato extract in Vanda orchid culture media showed a positive response. Likewise, coconut water and sweet corn in several studies were able to affect cell elongation, tissue differentiation and initiation of root formation, while potatoes and bananas, apart from containing carbohydrates, also contributed some suitable vitamin content to support plantlet growth and development, so these alternatives were tried in this research.

2. Methods

The experiment was conducted at the Laboratory of Culture in Vitro at the State Polytechnic of Jember Laboratory in Jember (80 m above sea level). Planting material is plantlets from the propagation of Sanderiana vanda orchid (hybrid) 1 month old in media 0. Vacin Went media materials, foliar fertilizer (Gandasil D), BAP, NAA, sugar and culture glass The environmental design used was Random Group Design, consisting of two factors. The first factor is the type of planting media (M1=Vacin-Went and M2=Gandasil D). The second factor is six organic matter kompotition (Z1. Potato 150gr/L+ Banana 150gr/L+Coconut Water 150ml/L, Z2= Potato 150gr/L+ Banana 150gr/L+ Sweet Corn 100ml/L, Z3= Potato 150gr/L+ Banana 150gr /L+Tomato 100ml/L, Z4 = BAP 1 ppm, Z5= NAA 0,1 ppm, Z6 = media without organic matter or PGR). The experiment was repeated 3 times. Each trial unit consists of 3 bottle, each botlle consists 5 eksplan. Variables observed included survival (%), plant height (cm), number of leaves (strands), number of shoots. Data were analyzed using the F test at 5% with the BNJ test at 5%. The research was started by preparing all the ingredients, plant in media 0 for a month before entering the treatment media.

3. Result and Discussion

3.1. Plant height

The increase in plant height describes the pattern of growth rate during the culture period. The growth and development of vanda orchid seedlings depends on the ability of these plants to respond to the treatment of planting media. The interaction of treatment types of media (compotition of nutrient and hormone) in this experiment was observed at week 18, while the role of a single factor hormone had actually been seen since the observation at week 16. VW media with BAP remained a good composition of media which had a positive effect on growth. tall. Plant height growth using organic matter composition of sweet corn on VW media had the same effect as the treatment without PGR and synthetic PGR (BAP and NAA) (see Table 1). Treatment of coconut water, sweet corn and BAP, gave no significant difference in height growth on Gandasil media. Tomato extract gave the lowest height growth not significantly different on VW and Gandasil media (Table 1). In Table 2, the single effect without ZPT had no significant effect on growth height with synthetic PGR (NAA and BAP). The best plant height variable in the 1 ppm BAP treatment was 1.47 cm, while the lowest plant height variable was in the tomato composition treatment (Potato 150 gr/L + Banana 150 gr/L + Tomato 100ml/L. At 18 WAP observation, VW media showed that media is suitable for orchid growth because it has been formulated for orchid explants, while Gandasil D media is an alternative medium used for orchid culture.According to research from [5] the use of VW media and several growth regulators showed different results. on the

IOP Conf. Series: Earth and Environmental Science **980** (2022) 012004 doi:10.1088/1755-1315/980/1/012004

formation of organogenesis of Dendrobium orchid explants.Plantlet responses tended to show fresh green leaves and wider, growth of shoot organs, leaves and roots that grew more, while the roots grew longer

Table 1. Plant height in the treatment of media and growth regulators

Treatment	Plant height (cm) (week after planting)								
	2	4	6	8	10	12	14	16	18
M1Z1	0.51	0.54	0.55	0.57	0.59	0.66	0.71	0.74	0.76 a
M1Z2	0.48	0.51	0.54	0.56	0.58	0.60	0.81	1.21	1.35 bcde
M1Z3	0.67	0.67	0.68	0.69	0.73	0.73	0.73	0.74	0.84 ab
M1Z4	0.74	0.82	0.89	0.98	1.02	1.06	1.30	1.70	1.78 e
M1Z5	0.37	0.56	0.67	0.82	1.10	1.14	1.21	1.37	1.50 cde
M1Z6	0.55	0.61	0.70	0.86	1.07	1.33	1.40	1.51	1.67 de
M2Z1	0.51	0.57	0.63	0.66	0.75	0.79	0.93	1.07	1.14 abcd
M2Z2	0.51	0.58	0.61	0.64	0.69	0.74	0.94	1.11	1.15 abcd
M2Z3	0.48	0.50	0.52	0.54	0.59	0.65	0.79	0.98	1.01 abc
M2Z4	0.50	0.55	0.58	0.62	0.74	0.85	1.03	1.10	1.15 abcd
M2Z5	0.55	0.64	0.69	0.71	0.78	0.84	1.04	1.22	1.39 cde
M2Z6	0.49	0.54	0.57	0.60	0.64	0.68	0.88	1.04	1.07 abc

Note: Numbers followed by different letters in one column show a significant difference based on the DMRT test at the -5% level. M1 (Media VW), M2 (Media Gandasil D 2gr/L), Z1 (Potato 150gr/L+ Banana 150gr/L+Coconut Water 150ml/L), Z2 (Potato 150gr/L+Banana 150gr/L+ Sweet Corn 100ml/L), Z3 (Potato 150gr/L+ Banana 150gr/L+Tomato 100ml/L), Z4 (BAP 1 ppm), Z5 (NAA 0.1 ppm), Z6 (Control)

Table 2. Plant height in the single treatment of growth regulators and media treatment

Treatment	Plant height (cm) (week after planting)								
	2	4	6	8	10	12	14	16	18
Z1	0.51	0.56	0.59	0.62	0.67	0.73	0.82	0.95 a	0.95 a
Z2	0.50	0.54	0.58	0.60	0.63	0.67	0.87	1.25 ab	1.25 ab
Z3	0.58	0.59	0.60	0.61	0.66	0.69	0.76	0.92 a	0.92 a
Z4	0.62	0.69	0.74	0.80	0.88	0.95	1.17	1.47 b	1.47 b
Z5	0.46	0.60	0.68	0.76	0.94	0.99	1.13	1.44 b	1.44 b
Z6	0.52	0.57	0.64	0.73	0.85	1.01	1.14	1.37 b	1.37 b
M1	0.55	0.62	0.67	0.75	0.85	0.92	1.03	1.21	1.32
M2	0.51	0.56	0.60	0.63	0.70	0.76	0.93	1.09	1.15

Note: Numbers followed by different letters in one column show a significant difference based on the DMRT test at the -5% level. M1 (Media VW), M2 (Media Gandasil D 2gr/L), Z1 (Potato 150gr/L+ Banana 150gr/L+Coconut Water 150ml/L), Z2 (Potato 150gr/L+Banana 150gr/L+ Sweet Corn 100ml/L), Z3 (Potato 150gr/L+ Banana 150gr/L+Tomato 100ml/L), Z4 (BAP 1 ppm), Z5 (NAA 0.1 ppm), Z6 (Control)

2 r

Leaves are vegetative organs, their growth is influenced by the nitrogen content in the media. In addition, leaves are the site of photosynthesis which is important in plant growth, namely the process of forming carbohydrates from CO2 and H2O with the help of sunlight. The large number of leaves indicates better growth of explants [6]. The increase in the number of leaves is influenced by genotype factors and the environment itself. Plant growth and development is supported by various external and internal plant factors that work together in a harmonious balance [7].

Table 3 Number of leaves in the interaction of media and growth regulators

Treatment	Number of leaves (week after planting)								
	2	4	6	8	10	12	14	16	18
M1Z1	2.37	2.37	2.37	2.43	2.56	2.70	2.74	2.85 a	3.07 ab
M1Z2	2.15	2.30	2.44	2.74	3.20	3.55	3.74	6.11 bcd	7.67 cde
M1Z3	2.72	2.72	2.72	2.81	2.85	2.85	2.85	2.85 a	2.85 a
M1Z4	2.65	2.85	3.24	3.65	5.02	5.69	6.61	9.28 d	10.61 e
M1Z5	1.94	2.33	2.67	2.91	3.54	3.74	4.02	4.72 ab	6.33 cd
M1Z6	3.72	3.81	4.07	4.85	5.94	6.36	6.67	7.42 bcd	8.50 de
M2Z1	2.07	2.28	2.80	3.24	3.70	3.96	4.98	6.74 bcd	7.43 cde
M2Z2	2.07	2.20	2.50	2.69	3.43	3.63	4.72	5.33 abc	6.02 bcd
M2Z3	1.89	2.02	2.13	2.31	3.02	3.22	3.41	4.57 ab	4.60 abc
M2Z4	2.52	2.78	3.41	4.00	4.02	4.48	5.11	6.56 bcd	7.07 cd
M2Z5	2.31	2.50	3.00	3.37	4.11	4.46	4.89	8.00 cd	8.20 de
M2Z6	2.26	2.37	2.76	2.93	3.81	4.30	5.43	7.48 bcd	7.72 cde

Note: Numbers followed by different letters in one column show a significant difference based on the DMRT test at the -5% level. M1 (Media VW), M2 (Media Gandasil D 2gr/L), Z1 (Potato 150gr/L+ Banana 150gr/L+Coconut Water 150ml/L), Z2 (Potato 150gr/L+Banana 150gr/L+ Sweet Corn 100ml/L), Z3 (Potato 150gr/L+ Banana 150gr/L+Tomato 100ml/L), Z4 (BAP 1 ppm), Z5 (NAA 0.1 ppm), Z6 (Control)

Based on Table 3 the interaction of media and growth regulators showed significantly different results at 16 WAP and 18 WAP. The best number of leaves was treated with VW media and 1 ppm BAP as many as 10.61 leaves, while the lowest number of leaves was found in VW media treatment and tomato composition was 2.85 leaves. The increase in plant height was followed by an increase in the number of leaves. According to research from [8] giving BAP in low concentrations (1 ppm) can accelerate the growth of Vanda orchid shoots compared to high concentrations (See Table 3). Exogenous PGR is used to provide a balance for endogenous hormones so that they can affect the dedifferentiation of meristematic cells again, influence the response of shoot morphogenesis and root formation, and determine the physiological response as a driver of cell division and elongation during multiplication, and during ([9]

Based on Table 4. the number of leaves in the treatment of growth regulators showed results that were not significantly different at 2 WAP - 8 WAP, significantly different at 10 WAP - 18 WAP. The best treatment for the variable number of leaves was Z4 (BAP 1 ppm) which was 8.84 strands, while the lowest treatment was the coconut water composition treatment of 5.25 strands. BAP is a growth regulator of the cytokinin group, where one of the purposes of using cytokinins is to induce the formation of the main organs. Observation of the number of leaves variable with the media factor showed results that were not significantly different from 2 WAP - 18 WAP. In the vegetative phase, it is necessary to provide fertilizer with nitrogen (N) content, because the element N is the main ingredient for compiling proteins needed for cell division. Young orchid plants fed fertilizer with N content showed better and faster growth. Elemental N is the main ingredient of amino acids, proteins, nucleic acids, various enzymes and as a greening agent [10]. One of the fertilizers containing N is Gandasil D. Leaf fertilizer Gandasil D contains nutrients N (20 %), P (15 %), K (15 %) and additional micro elements Mg, Mn, B, Cu, Co, and Zn

Table 4. Number of leaves on single treatment of growth regulators and media treatment

Treatment	Number of leave (sheet) (week after planting)								
	2	4	6	8	10	12	14	16	18
Z1	2.22	2.32	2.58	2.83	3.13 a	3.33 a	3.86 a	4.80 ab	5.25 ab
Z2	2.11	2.25	2.47	2.71	3.31 ab	3.59 a	4.23 ab	5.72 abc	6.84 bc
Z3	2.31	2.37	2.43	2.56	2.94 a	3.04 a	3.13 a	3.71 a	3.73 a
Z4	2.58	2.81	3.32	3.82	4.52 bc	5.08 b	5.86 bc	7.92 c	8.84 c
Z5	2.13	2.42	2.83	3.14	3.82 abc	4.10 ab	4.45 abc	6.36 bc	7.27 bc
Z6	2.99	3.09	3.42	3.89	4.88 c	5.33 b	6.05 c	7.45 c	8.11 c
M1	2.59	2.73	2.92	3.23	3.85	4.15	4.44	5.54	6.51
M2	2.19	2.36	2.77	3.09	3.68	4.01	4.76	6.45	6.84

Note: Numbers followed by different letters in one column show a significant difference based on the DMRT test at the -5% level. M1 (Media VW), M2 (Media Gandasil D 2gr/L), Z1 (Potato 150gr/L+ Banana 150gr/L+Coconut Water 150ml/L), Z2 (Potato 150gr/L+Banana 150gr/L+ Sweet Corn 100ml/L), Z3 (Potato 150gr/L+ Banana 150gr/L+Tomato 100ml/L), Z4 (BAP 1 ppm), Z5 (NAA 0.1 ppm), Z6 (Control)

3.3. Percentage of life

Based on Figure 1. the live percentage variable at 18 WAP showed the highest percentage of survival in the M1Z3 treatment (VW Media and Potatoes 150 gr/L + Bananas 150 gr/L + Tomatoes 100 ml/L) of 89%. According to research from [11] giving 100 ml/L tomato extract added to the media can meet the needs of explants to form shoots in a fast time and a high percentage of life. Tomato extract contains the hormone auxin which plays a role in the formation of primordial shoot cells which causes cell elongation. Auxin hormone produced from tomato extract causes primary cell wall material to be produced and transferred to both cell walls, then the cell structure is stretched so that it will form more cell walls. Tomato extract contains the IAA auxin hormone so that it can accelerate plant growth and increase growth potential.

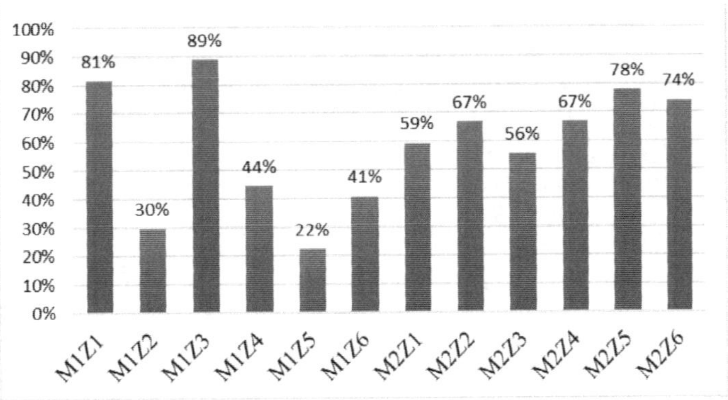

Picture 1. The percentage of explant life due to the influence of media and growth regulators
Note: M1 (Media VW), M2 (Media Gandasil D 2gr/L), Z1 (Potato 150gr/L+ Banana 150gr/L+Coconut Water 150ml/L), Z2 (Potato 150gr/L+Banana 150gr/L+ Sweet Corn 100ml/L), Z3 (Potato 150gr/L+ Banana 150gr/L+Tomato 100ml/L), Z4 (BAP 1 ppm), Z5 (NAA 0.1 ppm), Z6 (Control)

4. Conclusion

The real interaction between the media and the best growth regulators on plant height variables was in the VW and BAP 1 ppm media treatment with an average plant height of 1.78 cm. The variable number of leaves was in the treatment of VW and 1 ppm BAP media as many as 10.61 strands, single VW and Gandasil D media were not significantly different

Acknowledgment

The author would like to thank Ristekdikti for the PNBP research fund in 2021 through the Research and Community Service Center of Jember State Polytechnic.

References

[1] Yanti L P N, I K Suamba, IAU Dewi.2004. Perkembangan Usaha Agribisnis Bunga Anggrek Vanda Potong pada Kembang Batur Anggrek Collection di Desa Sanur Kaja Denpasar *Kamus Biologi* (Jakarta Indonesia: Balai Pustaka). E-Jurnal Agribisnis dan Agrowisata ISSN: 2301-6523 Vol. 7, No. 1, Januari 2018 https://ojs.unud.ac.id/index.php/JAA 101 Aidha N N 2013 *J Kimia Kemasan* **31** 58 – 64.

[2] Lestari EG. 2011. Peranan zat pengatur tumbuh dalam perbanyakan tanaman melalui kultur jaringan. J AgroBiogen 7(1):63-68

[3] Dwiyani, R., A. Purwantoro., A. Indrianto, E. Semiarti. 2009. Peningkatan kecepatan pertumbuhan embrio anggrek Vanda tricolor Lindl. pada medium diperkaya dengan ekstrak tomat. Prosiding Seminar Biologi Nasional XX, UIN Malang 24-25 Juli 2009.

[4] Setiawati T, A L Astuti, M Nurzaman, N R Kurniasih , N Alyaa, N K Sulviyani . 2017. Analisis Pertumbuhan dan Kandungan Total Flavonoid Kultur Kalus Krisan (*Chrysanthemum Morifolium*) dengan Pemberian Asam 2,4-Diklorofenoksiasetat (2,4-D) daan Air Kelapa. *J.l Bioma* **6** 29-41

[5] Rahmah S, T Rahayu, AH Winarti. 2017. Kajian Penambahan Bahan Organik Pada Media Tanam VW Pada Organogenesis Anggrek Dendrobium Secara In Vitro Study Adding Organic Substances on VW Media for Organogenesis the Dendrobium Orchid in Vitro *J. Simki-Techsain* **1**

[6] George EF, Sherrington. 1984. Plant Propagation by Tissue Culture. Exegetic Ltd. England.

[7] Endang S A 2017 *Optimasi Media Untuk Perkecambahan Biji Dan Pertumbuhan Seedling In Vitro Serta Pengaruh Media Dan Benziladenin Terhadap Keberhasilan Aklimatisasi Planlet Phalaenopsis Hibrida* (Bandar Lampung Indonesia: Universitas Lampung)

[8] Sutriana S, H B Jumin, M Mardaleni. 2014. Interaction of BAP and NAA on Explant Growth of Vanda Orchid In-vitro Jurnal Dinamika Pertanian. Vol. 29 No. 1 (2014): Edisi April 2014

[9] Kasutjianingati K, and I Rudi. 2013. *Media Alternative Perbanyakan In-Vitro Anggrek Bulan (Phalaenopsis Amabilis).* Jurnal Agroteknos, 3 (3). pp. 184-189. ISSN 2087-7706

[10] Sumiati A, A Astutik, 2019. Upaya Meningkatkan Produksi Tanaman Tomat Dengan Aplikasi Gandasil . BUANA SAINS; Vol 18, No 2 (2018)

[11] Serliana, Mukarlina, dan R. Linda. 2017. Pertumbuhan anggrek hitam (Coelogyne pandurata Lindl.) secara in vitro dengan penambahan ekstrak tomat dan benzylaminopurine (BAP). Probiont. 6 (3) : 310 – 315

Modeling the effect of climate on na-oogst tobacco production in Jember

I Harlianingtyas[1]*, M Azizah[1], S I Kusuma[1], Muksin[2], S Humaida[1]

[1]Department of Agriculture Production, Politeknik Negeri Jember, Indonesia
[2]Department of Agribusiness, Politeknik Negeri Jember, Indonesia

* *Email: irma@polije.ac.id*

Abstract. Tobacco (Nicotiana tabacum L.) makes a significant contribution to the source of farmers' income and the provision of employment opportunities. Tobacco plants are very susceptible to being influenced by the climatology and geography of the region. Na-Oogst tobacco is specifically cultivated in East Java in the Besuki Ex-Karesiden area which includes Jember and Bondowoso regencies, its mainly used for making cigars outside (dekblad), cigar dressings (omblad), and filling cigars (filler). Tobacco plants are sensitive to environmental factors including climatic factors, such as rainfall, humidity, and temperature. The climatic conditions of rainfall, both the amount and the distribution are very diverse so that it affects the growth, production, and quality of tobacco. Based on this, this study aims to identify characteristics, analyze and determine the modeling of the influence of climate factors on tobacco productions in Jember Regency. This is very important to do to determine recommendations and policies to increase tobacco productivity. The mathematical model of the analysis on Na-Oogst tobacco production in Jember Regency is Y = -362629 + 0.1285 rainfall + 2.93 rainy days + 16630 temperature – 1377 humidity with the coefficient of determination is 76.33%.

1. Introduction

Tobacco (*Nicotiana tabacum* L.) is one of the plantation crop commodities that has an important role in Indonesia. Based on statistical data, tobacco exports tend to increase, while tobacco imports have decreased. This has a positive impact on the government and tobacco farmers to increase productivity and quality so that they can compete with foreign producers [1]. Tobacco plants are very susceptible to being influenced by certain areas, such as requiring regional conditions that are climatologically and geographically suitable for the type of tobacco [2].

In 2018, the largest tobacco producers in Indonesia according to the size of their cultivation area were the Provinces of East Java, West Nusa Tenggara, Central Java, West Java with a total contribution of 94.76% to Indonesia's total tobacco production, while the largest contribution was East Java Province [3]. Tobacco cultivation and its industry are commodities that are very influential in the economy, not only for East Java but also for global economic activities [4].

Tobacco has several varieties including Voor-Ogst tobacco which is used for making white cigarettes as well as clove cigarettes and Na-Oogst tobacco, which is the type of tobacco used for the basic ingredients of making cigars and cigarillos, in addition to smoking or chewing tobacco. Besuki Na-Oogst Tobacco is mainly used for making cigars outside (dekblad), cigar dressings (omblad), and filling cigars (filler) [5]. Voor-Oogst tobacco is tobacco with a planting period at the end of the rainy season and a picking period in the dry season. Meanwhile, Na-Oogst tobacco is tobacco with a planting period at the end of the dry season and a picking period at the beginning of the rainy season [6]. Na-Oogst tobacco is specifically cultivated in East Java in the Besuki Ex-Karesiden area which includes Jember and Bondowoso regencies, therefore this tobacco is more often known as Besuki Na-Oogst tobacco [7].

Production of tobacco plants with the Besuki Na-Oogst tobacco type in Jember Regency has decreased due to uncertain weather, high rainfall, volcanic ash from Mount Raung, and the lack of production facilities [8]. One of the main problems in the production Na-Oogst tobacco is uncontrolled weather and the production environment [9]. Tobacco plants are plants that are sensitive to environmental factors including climatic factors, such as rainfall, humidity, and temperature. The optimal temperature for tobacco plant growth is 27°C or ranges from 22-33°C [10]. While the optimal humidity for plant growth ranges from 70%-80% [6]. The climatic conditions of rainfall, both in number and distribution, are very diverse so that it affects the growth, production, and quality of tobacco [11]. Meanwhile, the optimal temperature for the growth of Besuki Na-Oogst tobacco is between 24 – 27°C with an average relative humidity of 75 – 80%. Meanwhile, moderate rainfall of about 300 mm during the growing season has the best effect on plant growth so that it can produce maximum production [11].

Based on the previous description, this study aims to identify characteristics, analyze and determine the modeling of the influence of climate factors on Na-Oogst tobacco production in Jember Regency. This is very important to do to determine recommendations and policies for increasing tobacco production in Jember Regency, both produced by farmers and by companies.

2. Method

The method of analysis in this study uses descriptive analysis and multiple linear regression analysis [12].

1. Data Source

The source of data used in this study is secondary data taken from the database of the Central Bureau of Statistics of Jember Regency including data of uncontrolled weather [9] that is rainfall, rainy days, temperature, humidity, and tobacco production from 2007 to 2020 [13].

2. Operational Definition and Research Variables

The variables used in this study are divided into two, namely the independent variable and the dependent variable. The independent variables consist of rainfall, rainy days, humidity, and air temperature. While the dependent variable is tobacco production (quintal).

3. Construct Model

Multiple linear regression is a regression equation that describes the relationship between more than one independent variable (X) and one dependent variable (Y) [11]. The hypothesis that must be tested in multiple linear regression analysis is

$H_0 : \beta_1 = \beta_2 = \ldots = \beta_{p-1} = 0$

H_1 : Not all β_k (k=1,2,…,p−1) equal to zero

If x1, x2, . . ., xp is predictor variables as p that have a relationship with a response variable Y, then in general a linear regression model with one response variable is

$$Y = \beta_0 + \beta_1 x_1 + \beta_2 x_2 + \ldots + \beta_p x_p + \varepsilon$$

With an error condition has the following characteristics:

- ✓ $E(\varepsilon_j) = 0$;
- ✓ $Var(\varepsilon_j) = \sigma^2$ (constant);
- ✓ $Cov(\varepsilon_j, \varepsilon_k) = 0$; and
- ✓ $\varepsilon \sim n(0, \sigma^2)$, for $j \neq k$

If presented in matrix notation, the formula above becomes

$$\begin{bmatrix} Y_1 \\ Y_2 \\ \vdots \\ Y_n \end{bmatrix} = \begin{bmatrix} 1 & x_{11} & x_{12} & \cdots & x_{1p} \\ 1 & x_{21} & x_{22} & \cdots & x_{2p} \\ \vdots & \vdots & \vdots & \ddots & \vdots \\ 1 & x_{n1} & x_{n2} & \cdots & x_{np} \end{bmatrix} \begin{bmatrix} \beta_0 \\ \beta_1 \\ \vdots \\ \beta_p \end{bmatrix} + \begin{bmatrix} \varepsilon_1 \\ \varepsilon_2 \\ \vdots \\ \varepsilon_p \end{bmatrix}$$

or $\boldsymbol{Y} = \boldsymbol{\beta} \boldsymbol{X} + \varepsilon$

With the error condition being

- ✓ $E(\varepsilon) = \mathbf{0}$;
- ✓ $Cov(\varepsilon) = E(\varepsilon\varepsilon') = \sigma^2\mathbf{I}$ [12].

The assumptions that must be met in multiple regression analysis are:

- No multicollinearity (correlation between independent variables)
- There is no heteroscedasticity (constant error variance)
- Normality (normally distributed error)
- There is no autocorrelation (the error is random)

3. Result and discussion

3.1 Production of Na-Oogst Tobacco

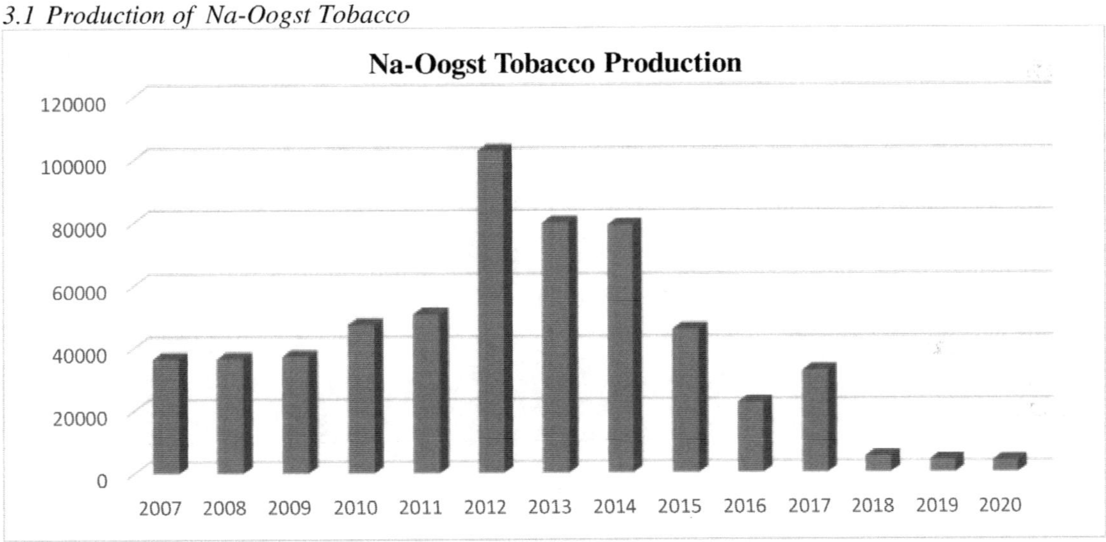

Figure 1. Production of Na-Oogst Tobacco

The results of the analysis of Na-Oogst tobacco production in Jember Regency for the last 14 years in 2007-2020 show that tobacco production during the 2007-2020 period fluctuated by having an average production Na-Oogst tobacco of 41,648.19 quintals. It can be seen in Figure 1 that the production of Na-Oogst tobacco above the highest average was in 2012 with a production of 102,770,0 and the lowest production was in 2020 with a production of 3,637.48 [5]. The development of Na-Oogst tobacco production in Jember Regency has decreased over a while, this is caused by several factors such as farmers not fully using the recommended technology in cultivation, lack of production facilities and information, limited knowledge, and limited land ownership. Besuki Na-Oogst tobacco production in Jember Regency experienced a drastic decline in the period, one of which was in 2015, this is caused by the erratic weather, the volcanic ash of Mount Raung and the lack of provision of production facilities, and the prices of some production inputs that have not been reached by farmers so that not many farmers grow the Besuki Na-Oogst commodity [8].

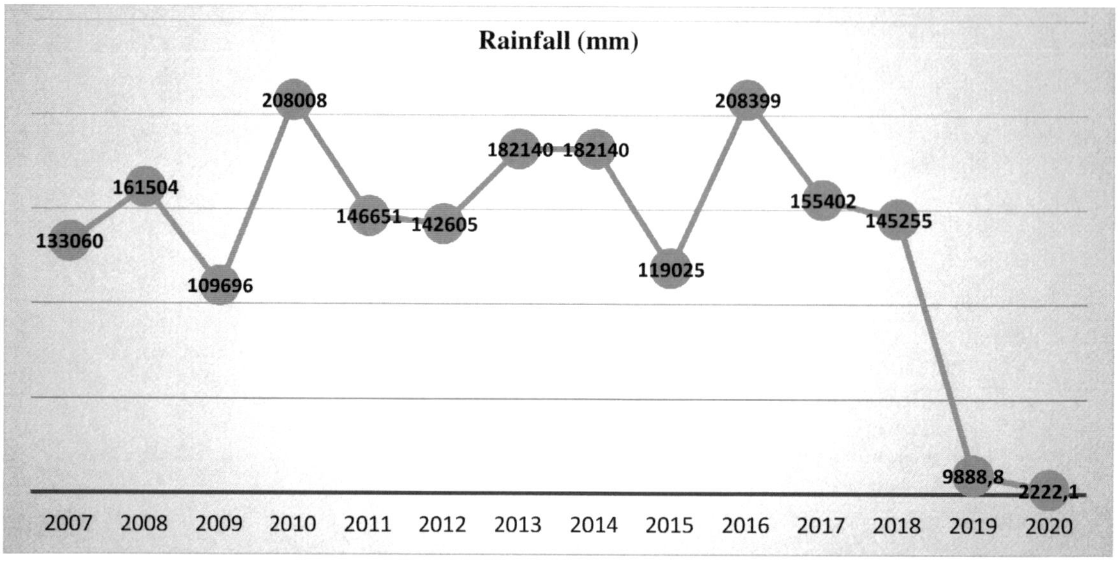

Figure 2. Rainfall

Rainfall for 14 years fluctuated with an average rainfall of 136,143 mm/year. It can be seen in Figure 2 that the highest above-average rainfall occurred in 2016 with a total rainfall of 208,339 mm/year and the lowest amount of rainfall occurred in 2020 with a total rainfall of 2222.1 mm/year. This is in line with [16] that rainfall can also be influenced by El Nino and La Nina phenomena. El Nino phenomena are climate anomalies followed by a decrease in rainfall and an increase in air temperature, while the La Nina phenomenon stimulates an increase in rainfall above normal rainfall [17]. One of the La Nina phenomena that occurred in Indonesia occurred in 2010-2011 and 2016-2017 which resulted in experiencing quite high rainfall while in 2015 and 2018-2020 there was a decrease in rainfall caused by the El Nino phenomenon [18].

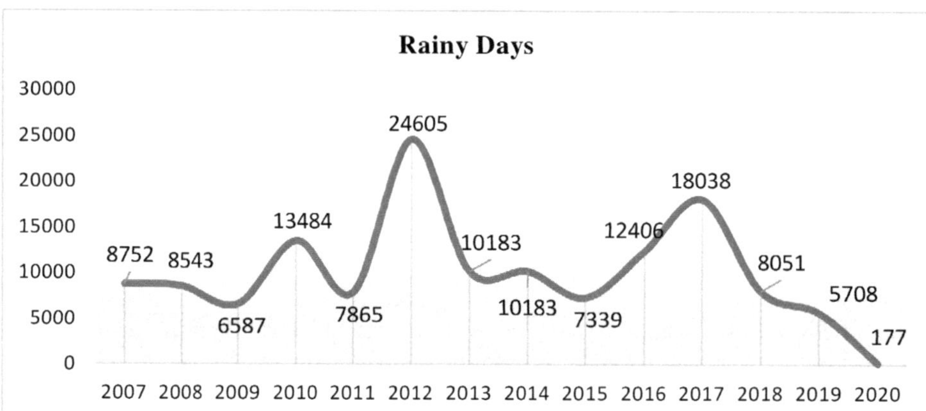

Figure 3. Rainy Days

Meanwhile, the results of the analysis of rainy days for the last 14 years in 2007-2020 show that the number of rainy days during the period 2007-2020 has fluctuated with an average rainy day of 10137.21 rainy days. It can be seen in Figure 2 that the highest number of rainy days above the average was in 2012 with 24,605 rainy days and the lowest number of rainy days was in 2020 with 177 rainy days.

Rainy days in Jember Regency in 2007-2020 fluctuated, this was related to the large number of rainy days which was also influenced by rainfall.

The impact of global climate change is a change in rainfall patterns on a local or regional scale such as a shift in the trend of daily rainfall patterns and another factor that has a close relationship to affect rainy days is solar radiation [10]. Solar radiation activity affects the heat radiation captured by the atmosphere so that solar variability is the dominant factor that causes climate change in the form of daily cycles [19].

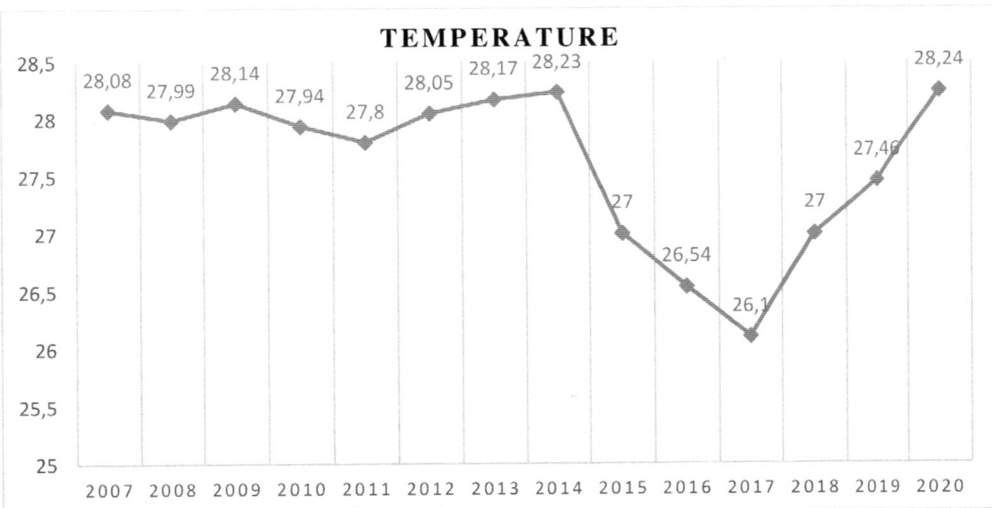

Figure 4. Temperature

Jember Regency has an average humidity of 74.28% with an average annual air temperature of 27.6 °C. In 2007 – 2014 the temperature tended to be stable then the air temperature graph decreased starting in 2015 until it reached the lowest point in 2017. The decrease in temperature was accompanied by an increase in humidity in Figure 5 then in the following year the humidity again decreased and there was an increase in the temperature graph. This shows that the higher the air temperature will be accompanied by a decrease in humidity.

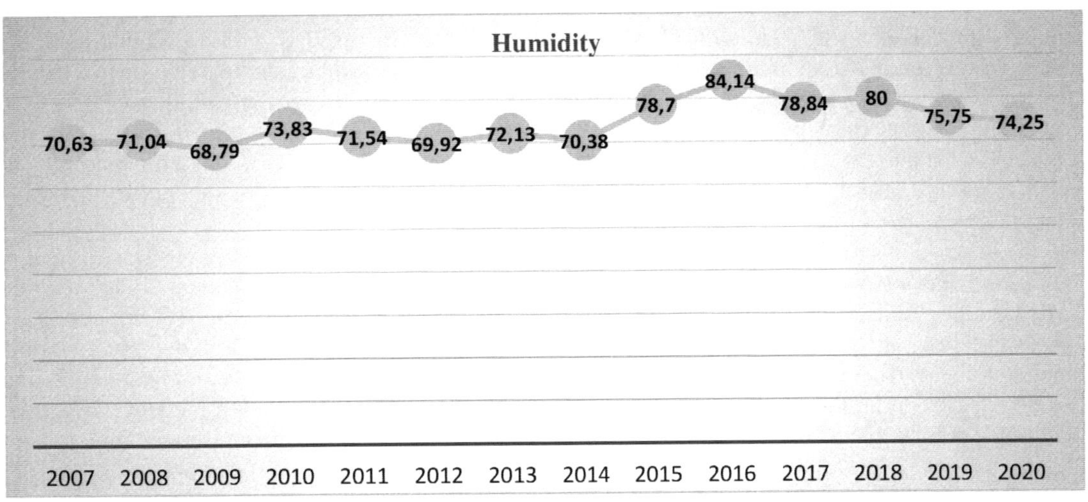

Figure 5. Humidity

3.2 Regression Modeling
3.2.1 Simultaneous Test
The hypotheses for simultaneously testing the significance of the regression parameters are:
H0 : β1 = β2 = … = β5 = 0
H1 : There is at least one βj ≠ 0; j = 1, 2, …, 5
Rejected H0 if p-value < α, for α = 0.05

Table 1. Analysis of Variance

Model	*df*	*Sum of square*	*Mean Square*	F-count	*P-value*
Regression	4	8733131893	2183282973	7,25	0,007
Residual	9	2708776757	300975195		
Total	13	11441908649			

Based on the results of the simultaneous test in Table 1, it shows that the P-value of 0.007 is less than 0.05 (α: 5%), or the F-count of 7.25 is greater than F table (3.63) so it is decided to reject H0, this means that all independent variables simultaneously have a significant effect on NO tobacco production in Jember Regency. So that the regression model or equation formed from this analysis is appropriate and meets the assumptions of the goodness of the regression model.

3.2.2 Partial Test
A partial test was used to determine the effect of each independent variable partially on Na-Oogst tobacco production in Jember Regency. The results of the partial test analysis are shown in table 2 as follows:

Table 2. Partial Test Result

Variable	Coefficient	SE Coefficient	T-Count	P-value	VIF
(Constant)	-362629	622081	-0,58	0,574	
Rainfall (mm)	0,1285	0,0937	1,37	0,203	1,48
Rainy Days	2,93	1,14	2,57	**0,030***	1,91
Temperature	16630	16360	1,02	0,336	5,57
Humidity	-1377	2413	-0,57	0,582	5,28

It can be seen in Table 2 that there is one partially significant. This is indicated by the P-value which is less than 0.05, which is in addition when compared to the T value of the statistical table (± 2.26), the t-value is greater than the T.table on the **rainy day variable**. So it can be concluded that there is an independent variable that partially significantly influences the production of Na-Oogst tobacco in Jember Regency, namely the rainy day variable. In line with the opinion of tobacco farmers in Temanggung, the current problems faced by tobacco farmers are weather and climate factor. According to farmers, the current weather is less favorable and inhibits the growth rate of tobacco plants, and even leads to crop failure. In addition, changes in weather and climate also trigger the spread of pests and plant diseases [20].

3.2.3 Regression Models
Based on the results of the regression analysis of the coefficient of determination (R^2) amount 76,33%, This means that the contribution of the independent variables, namely rainfall, rainy days, humidity, temperature, on Na-Oogst tobacco production in Jember Regency is 76.33% and the rest is influenced by other factors. The R^2 value that is close to 100% indicates that the regression model built is correct and can represent the actual situation. The regression model that is formed from is as follows:

$$Y = -362629 + 0.1285\,X1 + 2.93\,X2 + 16630\,X3 - 1377X4$$

Where Y is Production of Na-Oogst Tobacco (quintals); X1 is Rainfall (mm/year); X2 is Rainy days; X3 is Temperature (°C); X4 is Humidity (%)

The regression model shows that the constant or intercept is **-362,629**. This means that if all variables are constant or zero, then the production of Na-Oogst tobacco is reduced by **362,629** quintals. Then the regression coefficient on the rainfall variable (X1) is **0.1285**, which means it has a positive effect on Na-Oogst tobacco production. It can be assumed that if the rainfall (X1) increases by 1 unit, the increase in Na-Oogst tobacco production is **0.1285** quintals with the assumption that the other independent variables are constant. Then the regression coefficient on the variable number of rainy days (X2) is **2.93** which means it has a positive effect on Na-Oogst tobacco production. It can be assumed that if the number of rainy days (X2) increases by 1 unit, then tobacco production increases by **2.93** quintals with the assumption that the other independent variables are constant. While the regression coefficient on the temperature variable (X3) is **16,630** which means it has a positive effect on the production of Na-Oogst tobacco. It can be assumed that if the temperature (X3) increases by 1%, there will be an increase in tobacco production of **16,630** quintals with the assumption that the other independent variables are constant. The regression coefficient for the air humidity variable (X4) is – **1377**, which means it negative effect on NO tobacco production. It can be assumed that if the air temperature (X4) increases by 1°C, there will be a decrease in tobacco production of **1377** quintals with the assumption that the other independent variables are constant.

Although the regression equation has been formed, it must meet the classical assumptions in regression. The following are the results of the examination and testing of assumptions on the formed regression model [21].

1) Normality Test

The normality test was conducted to determine the residuals obtained from the difference between Y estimates and Y original data following a normal distribution. Test the normality of the data using the Kolmogorov Smirnov test with the following hypothesis testing:

H0: Data is normally distributed

H1: Data is not normally distributed

Level significant: 5 % (0,05)

The results of the normality test show the following results

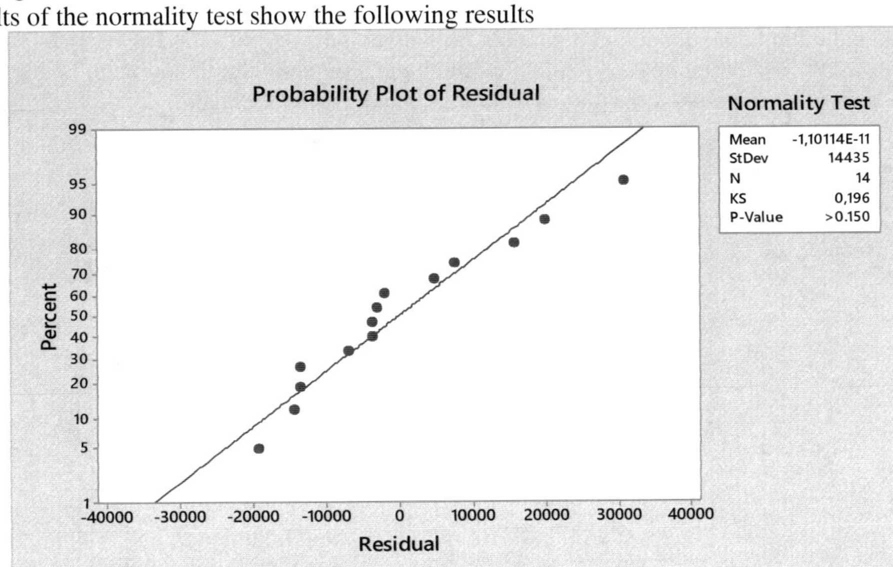

Figure 6. Normality Test

The results of the normality test can be seen from the magnitude of the p-value which is more than 0.05 and visually it can be seen that the residual points follow the normality line. This shows that the residuals meet the assumption of a normal distribution [22].

2) Multicollinearity Test

The second assumption that must be met is that there is no multicollinearity between the independent variables. One way to detect multicollinearity is by looking at the Variance Inflation Factor (VIF) if the VIF value > 10 indicates the presence of multicollinearity. Based on Table 2, it is known that the VIF value in all independent variables is < 10, so it can be concluded that there is no multicollinearity between independent variables in the regression model.

3) Autocorrelation Test

The third assumption that must be met is that there is no autocorrelation between residuals, or there is no relationship or influence from the first residual data to the next data residual. One of the ways to detect autocorrelation is the Durbin Watson test.

H0 : $\rho=0$ (autocorrelations)
H1 :$\rho\neq0$ (not autocorrelations)
Critical value:
Accepted H0, if dU<DW<(4-dU)

Durbin Watson Statistics: $DW = \frac{\sum_{t-2}^{n}(e_t - e_{t-1})}{\sum_{t=1}^{n} e_t^2}$

Table 3. Durbin Watson Test

R Square	Durbin-Watson
76,33 %	1,03805

Where Durbin-Watson Statistics = 1,0305, while from the Durbin Watson table with the number of variables 5 (p = 5), and the number of observations 14 (n = 14), it is obtained dL = 0.505 and dU = 2.296. Because the Durbin Watson statistical value is between dU and dL, it can be concluded that the Durbin Watson test results are not convincing. So that checking autocorrelation using other methods. One way to detect autocorrelation among residuals can be seen from ACF plotting.

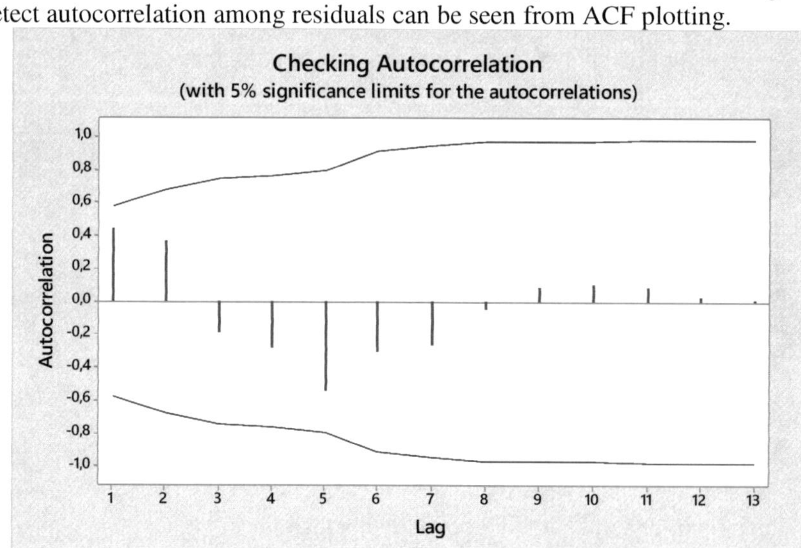

Figure 7. Autocorrelation Plot

Based on the results of the autocorrelation image, it can be seen that no lag that exceeds the red line, where the red line is the upper limit and the lower limit of the autocorrelation line. So it can be concluded that there is no autocorrelation between residues from the t data to the next t-1 data [23].

4) Heteroskedasticity Test

One way to detect the presence or absence of heteroscedasticity is to perform the Glejser test [24]. The Glejser test proposes to regress the absolute value of the residual on the independent variable. Probability results are said to be significant if the significance value is above the 5% confidence level. If the sig value in the Glejser test for each independent variable is <0.05, then there is heteroscedasticity or is called the unequal residual. If the sig value in the Glejser test for each independent variable is > 0.05, then there is no heteroscedasticity or it is called the same residual variance.

Analysis of Variance

Source	DF	Adj SS	Adj MS	F-Value	P-Value
Regression	4	252333261	63083315	0,86	0,522
Rainfall	1	222902560	222902560	3,05	0,115
Rainy days	1	73481335	73481335	1,00	0,343
Temperature	1	4652935	4652935	0,06	0,807
Humidity	1	511080	511080	0,01	0,935
Error	9	658663482	73184831		
Total	13	910996743			

Figure 8. Glejser Test

Based on the results of the Glejser test, it is known that the significance value (p-value) of all independent variables after being restored with absolute residual value is above 0.05 so that it can be concluded that the residual variance is the same or there is no heteroscedasticity. Based on the examination of the classical assumptions of the regression, all four have been fulfilled, so that the regression model obtained using the OLS approach is the Best Linear Unbiased Estimator (BLUE).

4 Conclusion

The mathematical model of the analysis of the effect of rainfall, number of rainy days, air temperature, and humidity on NO tobacco production in Jember Regency is as follows.

Y = -362629 + 0.1285 rainfall + 2.93 rainy days + 16630 temperature – 1377 humidity

The coefficient of determination is 76.33%, this shows the influence of the variables of rainfall, number of rainy days, temperature, and humidity resulting in fluctuations in Na-Oogst tobacco production in Jember Regency and 23.67% caused by other factors.

5 Acknowledgments

This research was supported by PNBP State Polytechnic of Jember 2021.

6 References

[1] Rachmawati 2018 Analisis Data Tembakau Jawa Timur 2018 (Surabaya: PT Sinar Murni Indoprinting.)

[2] Herminingsih H 2014 Hubungan adaptasi petani terhadap perubahan iklim dengan produktivitas tembakau pada lahan sawah dan tegalan di kabupaten jember *J. Sos. Ekon. Pertan.* **7** 22636

[3] Direktorat Jenderal Perkebunan 2018 *Statistik Perkebunan Indonesia* (Jakarta: Direktorat Jenderal Perkebunan)

[4] KUSUMANDARU W 2015 Analisis indeks kualitas tanah di lahan pertanian tembakau kasturi berdasarkan sifat kimianya dan hubungannya dengan produktivitas tembakau kasturi di Kabupaten Jember

[5] Hani E S, Soetriono S and Paramu H 2009 Pemodelan dan Strategi Competitiveness Agribisnis Tembakau Besuki Na-Oogst di Jawa Timur *JSEP (Journal Soc. Agric. Econ.* **3** 59–69

[6] Arifandi J A, Wardhono A and Indrawati Y 2018 *Panduan Praktik Budidaya Tembakau Besuki Na-Oogst* (Pustaka Abadi)

[7] Widiastuti D W Faktor-faktor yang Mempengaruhi Produksi Tembakau Besuki Na-Oogst Di Desa Jambe Arum Kecamatan Puger Kabupaten Jember

[8] Muktianto R T and Diartho H C 2018 Komoditas tembakau besuki Na-Oogst dalam perspektif pembangunan berkelanjutan Di Kabupaten Jember *Caraka Tani J. Sustain. Agric.* **33** 115–25

[9] Hari R, Tanti K and Titik I 2018 The main problems of na-oogst tobacco agribusiness in Jember *IOP Conference Series: Earth and Environmental Science* vol 207 (IOP Publishing) p 12012

[10] Herlina N, Azizah N and Pradiga E P 2020 Pengaruh Suhu dan Curah Hujan terhadap Produktivitas Tembakau (Nicotiana tabacum L.) di Kabupaten Malang *PLANTROPICA J. Agric. Sci.* **5** 52–63

[11] Soetriono, Evita Solihahani, Zulan Anisa F, Inayatin N, Susanti N and Zuniana Q 2014 *Agribisnis Tembakau Besuki Na- Oogst :*

[12] Sinaga E K, Matondang Z and Sitompul H 2019 *Statistika: Teori dan Aplikasi Pendidikan* (Yayasan Kita Menulis)

[13] Badan Pusat Statistik Kabupaten Jember 2020 Kabupaten Jember Dalam Angka (Jember)

[14] Gujarati D N 2016 Ekonometrika dasar

[15] Drapper N R and Smith H 1992 Analisis Regresi Terapan (Edisi Kedua) *Alih Bhs. oleh PT. Gramedia Pustaka Utama. Jakarta*

[16] Irawan D S and Rochayati S 2017 Proyeksi kebutuhan pupuk sektor pertanian melalui pendekatan sistem dinamis *Bogor, Balai Penelit. Tanah*

[17] Muksin D, Simbolon D, Wiyono E S and Sondita M F A 2020 Effect of global climate (ENSO) on regional climate (rainfall and air temperature) in the Morotai Island region *Adv. Environ. Sci.* **12** 97–109

[18] Meteorology Climatology and Geophysics Council (BMKG) 2020 El Nino and La Nina

[19] Syamsudin F I 2018 ANALISIS PENGARUH AKTIVITAS MATAHARI TERHADAP PERUBAHAN IKLIM *Prosiding SNPS (Seminar Nasional Pendidikan Sains)* pp 179–83

[20] Nurjani E, Harini R, Sekaranom A B and Mutaqqin A S 2020 Tobacco farmers Perspective towards increasing climate change risk on agriculture sector: a case study of Temanggung-Indonesia *IOP Conference Series: Earth and Environmental Science* vol 451 (IOP Publishing) p 12101

[21] Harlianingtyas I, Hartatie D, Salim A S and Supriyadi S 2019 MODELING OF THE FACTORS THAT INFLUENCE SUGAR PRODUCTION IN ASEMBAGUS SITUBONDO SUGAR FACTORIES *Proceeding of the International Conference on Food and Agriculture* vol 2

[22] Harlianingtyas I and Hartatie D 2021 Modeling of factors affecting the productivity of sugarcane in Jember Regency *IOP Conference Series: Earth and Environmental Science* vol 672 (IOP Publishing) p 12026

[23] Harlianingtyas I, Hartatie D and Salim A 2018 Modeling of rainfall and fertilization factor of sugarcane productivity in Asembagus sugar factory Situbondo *E&ES* **207** 12013

[24] Ghozali I 2013 Aplikasi Analisis Multivariate degan Program SPSS. Cetakan Keempat, Badan Penerbit Universitas Diponegoro, Semarang. Empirical Evidence on Canadians Firms. School of Accounting University of Central Florida

IOP Conf. Series: Earth and Environmental Science **980** (2022) 012006 doi:10.1088/1755-1315/980/1/012006

The Feasibility of Water Quality of Bedadung River for Inland Fishery Business Development and Sanitation Hygiene

AA Gemaputri[1*], IAA Pongoh[2*], OJ Anggraeni[2]

[1] Agro-industry Management, Politeknik NegeriJember, Jl. Mastrip PO Box 164 Jember 68101, Indonesia
[2] Agribusiness Management, Politeknik Negeri Jember, Jl. Mastrip PO Box 164 Jember 68101, Indonesia
[3] Agribusiness Management, Politeknik Negeri Jember, Jl. Mastrip PO Box 164 Jember 68101, Indonesia

ida_adha@polije.ac.id

Abstract. This research was conducted using a quantitative descriptive method for measuring water quality in the form of measuring chemical, physical and biological parameters of the waters at the research location, then using a qualitative descriptive method to describe the condition of the designated area in the watershed of the research location.
The research location is the Biting River, the Jompo River and the Besini River. The purpose of the study was to assess the feasibility of water quality at the research site for inland fishery businesses and the water needs for sanitation hygiene for the community. The results of the measurement of water quality parameters at the research location showed that the Biting river was in an unpolluted and fertile condition, while the Jompo River indicated that it was contaminated with *Oscillatoria* bacteria which is very dangerous for human health and aquatic organisms. The water quality of the Besini river shows that the waters are fertile, not polluted but not suitable for community hygiene and sanitation activities because it is brackish with an average water salinity of 25 ppt.

1. Introduction

Jember Regency is one of the regencies in the province of East Java which has a fairly large watershed, namely the Bedadung watershed with a length of 161 km and an area of 499.5 km2, the Bedadung watershed also has quite a number of tributaries that cross several rivers. Sub-districts in Jember Regency, the tributaries include the Biting River, Arjasa River, Baratan River, Jompo River, White River, Petung River and Besini River [1] . The utilization of the Bedadung tributary by the community around the river is very diverse, ranging from use for household activities (drinking water, bathing, washing and latrines), industrial activities and agricultural, livestock and fishery activities [2] .

A water has different water quality is determined by several factors, namely physics, chemistry, biology and interactions therein. Physical parameters commonly used to determine water quality include light, temperature, brightness and turbidity, water color, total solids, dissolved solids, suspended solids, and salinity [3] . Furthermore [4] said that the chemical characteristics of water greatly affect aquatic life, these characteristics are pH, salinity, dissolved CO2 and O2, as well as the content of toxic materials including organic materials such as ammonia, nitrite, heavy metals and synthetic chemicals. Microbiologically water quality is determined by the parameters of polluting, pathogenic and toxin-producing microbes. Sustainable management, exploitation and use of fresh water very important for people's lives. Preservation and protection good water quality is highly dependent on the conservation of biodiversity around freshwater sources, both in terms of sanitation and the environment [5] .

Information on water quality in a river water area as the largest source of fresh water plays a very important role in the management and development of inland aquaculture activities and their benefits for people's lives.

This study aims to study the quality of the waters of the Bedadung River as a development area for inland aquaculture, clean water sources and river ecotourism in Jember Regency, through several stages, namely: (1) Identifying water quality in the form of physical, chemical and biological parameters at the research site in a measurable manner in accordance with water quality standards for agriculture, fisheries, animal husbandry and sanitation hygiene, (2) Analyzing the results of identification and (3) Synchronization the results of the analysis with the results of interviews with the community around the research location

Based on the results of the identification of water quality and synchronization with the opinions of the community of river users in the research location, a blueprint was then prepared as a discourse to be used as study material in determining policies regarding the use of watersheds in Jember Regency.

2. Material and Method

Data collection in this study was carried out in two ways, namely primary data collection in the form of direct data collection at the research location and secondary data collection in the form of indirect data collection, namely through extracting data or information from the Marine and Fisheries Service of Jember Regency and the Jember Regency Environmental Service.

2.1 Research methods

Data collection in this study was carried out in three stages, namely:

a. Field observations; preliminary observations and research at the research location, namely the Biting River, the Jompo River and the Besini River
b. Primary data collection, namely water quality measurement in the form of observing physical, chemical and biological parameters at the research site
c. Secondary data collection obtained from the Department of Marine Affairs and Fisheries, the Environment Agency, and the Jember Regency Statistics Office, either through surveys or statistical compilations.

2.2 Water Quality Data Collection

Water quality data were collected from 3 research sites namely Biting River [8°06'40.3"S 113°46'06.7"E]; Jompo River [8°11'07.6"S 113°41'02.1"E]; and Besini River [8°21'58.8"S 113°27'54.0"E], with 2 observation stations at each location, interpolated at the observation station points for specific analysis. Determination of sampling points using the Inverse Distance Weight (IWD) method. Water quality parameters observed include:

Tabel 1. Parameters of Observation of Water Quality

No.	Parameters	Unit	Measurement Method Specification
Chemical			
1.	Dissolved Oxygen	mg/L	DO meter
2.	Phospat (PO4)	mg/L	Refraktometrik
3.	TDS	mg/L	TDS meter
4.	BOD	mg/L	Titrimetrik
5.	Nitrite	mg/L	IKM/5.4.13 K
6.	Nitrate	mg/L	IKM/5.4.13 K
7.	Ammonia	mg/L	IKM/5.4.13 K
8.	pH		pH meter
9.	Salinity	ppt	Handrefracto meter
Physics			
1	Temperature	^{o}C	Thermometer
2.	Water Depth	meter	Secondary data
Biology			
1.	Bacteria	cell/ml	Total count
2.	Plankton density	cell/ml	Direct counting (haemocytometer)

3. Result and Discussion

Bedadung tributaries used as research locations are the Biting River, the Jompo River and the Besini River, an overview of the research locations for the three tributaries can be seen in Figure 1.

a b c

Figure 1. Research Location (a) Biting River, (b) Jompo River, (c) Besini River

Observations of water quality at the 3 research locations showed characteristics and conditions of the waters that differed from one location to another, the results of observations of water quality chemically, physically and biologically at the 3 research locations can be seen in full in Table 2.

Tabel 2. Observation of Physical, Chemical and Biological Water Quality Parameters in Biting River, Jompo River and Besini River

Parameters	Parameter Items	Location of Observation		
		Biting River	Jompo River	Besini River
Physics	Temperature (^oC)	25,7	25,1	26,6
	Water Colour	Brownish green	Light gray & Brown	Brownish yellow
	Smell	No smell	Smells	Smells
	Depth	> 6 meters	> 6 meters	>10 meters
Chemical	Nitrite	0,041 mg/L	0,358 mg/L	0,091 mg/L
	Nitrate	9,2 mg/L	10,6 mg/L	7,8 mg/L
	Ammonia	0,022 mg/L	0,047 mg/L	0,006 mg/L
	BOD	1,93 mg/L	6,84 mg/L	3,30 mg/L
	Phospat	0,44 mg/L	0,34 mg/L	0,39 mg/L
	DO	28,5 mg/L	16 mg/L	12,6 mg/L
	pH	8,15	7,63	7,32
	TDS	208 mg/L	174 mg/L	727 mg/L
	Salinity	6 ppt	6 ppt	25 ppt
Biology	**Plankton Density:**			
	Cyclotella	200 cell/ml	-	-
	Licmophora	-	50 cell/ml	50 cell/ml
	Navicula	150 cell/ml	400 cell/ml	20 cell/ml
	Nitzchia	50 cell/ml	-	50 cell/ml
	Pleurosigma	5 cell/ml	5 cell/ml	-
	Chlorophyceae	500 cell/ml	-	-
	Bacteria Density :			
	Oscillatoria	-	50 unit/ml	-
	Macrozoobentos :			
	Brotia testudinaria	found	-	-
	Brotia costula	found	found	-
	Pomacea canaliculta	found	found	-
	Viviparus javanicus	found	found	-
	Anadara granosa	-	-	found
	Uca sp	-	-	found

Description: Biting rivers and Jompo rivers have fresh water, Besini river has brackish water

Unpolluted water is water that does not contain certain foreign materials in amounts exceeding the specified limits so that the water can be used normally for certain purposes, for example for drinking water, aquatic animal life (river and lake water), irrigation (agriculture), recreational and industrial purpose [6] , and [7] said that the water quality class defined depending on the physical, biological, and the chemical parameters intended for water determine its use water like; drinking water, water used for agriculture, or water used in industry. The use of river water for fishery, livestock and agricultural activities has been regulated through Government Regulation of the Republic of Indonesia Number 22 of 2021 concerning the Implementation of Environmental Protection and Management. Where in the regulation is stated the River Water Quality Standard and its designation as drinking water (class 1), in freshwater recreation and fishery activities (class 2), in livestock and agricultural activities (class 3), and crop irrigation (class 4).

The results of observations of water quality in Table 2 show that the water quality of the Biting River supports the life of the fish population, which is indicated by the measurement results of a pH value of 8.15, a DO value of 28.5 mg/L and an ammonia level of 0.022 mg/L. According to [8] and [3] , water that has a pH value between 7 to 9 can support the life of of aquatic organisms, and dissolved oxygen content in fresh water is between 15-25 mg/L and ammonia levels are less than 0.1 mg/L is good for fish growth and breeding.

The results of the measurement of biological parameters indicate that plankton from the *Chlorophyceae* species dominates the density of plankton in the Biting River. The presence of *Chlorophyceae* in waters is an indicator of fertile water conditions with good water quality. Biting river water also shows a phosphate content of 0.44 mg/L, where phosphate is an essential element for aquatic plants/algae and can affect water productivity as long as the content is in the water not to exceed 1 mg/L [3] . The results of observations of other biological parameters are the presence of macrobenthos as fertility and water pollution [9] . The macrobenthos found in the Biting River consisted of 4 species, namely *Brotia testudinaria, Brotia costula, Pomacea canaliculta*, and *Viviparus javanicus*. Species of *Brotia testudinaria* and *Brotia costula* are macrobenthos species that cannot be found in polluted waters (fresh water) so it can be said that the quality of Biting river water is in the unpolluted category.

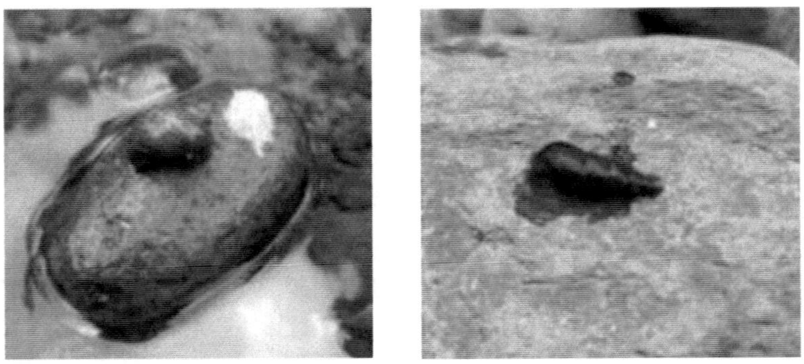

a b

Figure 2. Macrobenthos found at the research site (a) *Pomacea canaliculata* (b) *Brotia costula*

Observation of the water quality of the Jompo river shows that this location has a high organic matter content which is indicated by the highest BOD value compared to the other 2 research locations, namely 6.84 mg/l, BOD is the amount of oxygen needed by aerobic microorganisms to break down organic matter in water. and is a description of the high levels of organic matter and the abundance of aerobic microorganisms in the water [10] . Pollution of river water with organic pollutants has adverse impacts on the fish and affect the livelihood of many farmers. Communicating the findings of this study becomes necessary to mitigate pollution-related problems [11] . Another indicator that shows the high organic matter in the Jompo river is the presence of plankton of the Navicula species. The results of the TDS measurement in this river of 174 mg/L indicate that the solubility of inorganic materials is quite low and can be categorized as a standard of healthy water that is suitable for consumption, because according to WHO, water with a TDS value below 1000 mg/L is healthy water that is suitable for consumption, but based on the results of the measurement of nitrite (0.358 mg/L) which are high and the presence of plankton of the *Oscilatoria* type (the cyanobacteria group) which is toxic, then this river water can be categorized as not eligible the requirements for health (hygiene sanitation). High Nitrite content in waters can inhibit the process of binding oxygen by blood hemoglobin so that it is dangerous for the life of aquatic organisms [12] and the presence of *Oscilatoria* belonging to cyanobacteria in waters can endanger the life of aquatic organisms and human health because these bacteria are capable of producing nodularin toxin which is toxic, which causes damage to liver and kidney tissue through the food chain [13] .

The third research location is the Besini River which is located downstream of the Bedadung river which is approximately 1 km from the Indonesian Ocean so that the water in this river is influenced by the entry of sea water at high tide which causes the water to be a bit salty so it is categorized as brackish water with an average salinity of 25 ppt, and this river is also a route traffic of boats or fishing vessels that will sail to the sea or have just returned from the sea. The results of water quality observations show that the river water is brownish yellow which is an indicator of the height of plankton from the Diatom species (*Navicula, Limpthora and Nitzchia)*. The BOD value in this river is high, namely 3.30 mg/L which indicates that a lot of dissolved organic matter is oxidized in the waters, this can be because this river is a fishing vessel traffic lane and along this river are residential areas, where waste liquid and solid household dumped into this river. The high value of BOD is an indicator of pollution in the waters that can affect the decrease in dissolved oxygen in the water [14] . The dissolved oxygen (DO) in the Besini River also shows a low value, which is in accordance with the statement that the low DO indicates high demand for oxygen by the microorganisms [15] .

The results of the measurement of the TDS value in this river (727 mg/L) were higher than the other 2 research locations, this was influenced by the salt concentration which was indeed higher in this river. The concentration of TDS is determined by the concentration of salt, the higher the concentration of salt, the concentration of TDS will increase [16] . Based on observations of water quality, the Besini River can be said to have a fairly low level of water pollution, however this river is not suitable for community sanitation hygiene because the water is a bit salty (brackish) and not good for inland fishery business development because this river is a boat traffic lane or fishing boats.

4. Conclusion

Based on the identification of the physical, chemical and biological parameters of the waters of the Biting River, Jompo River and Besini River, it can be concluded that the Biting River can be said to be suitable for developing inland fisheries business and for community hygiene sanitation activities. The Jompo River can be said to be polluted and unsuitable as a source of water for the development of inland fishery businesses and for community sanitation hygiene, as well as the Besini River, although the level of water pollution is quite low, it is not suitable for community sanitation hygiene and for the development of inland fisheries business.

Acknowledgments

Thank you to the Politeknik Negeri Jember, especially the Center for Research and Community Service for facilitating research activities and ICOFA 2021 International seminars

References

[1] Dinas Kebersihan Dan Lingkungan Hidup 2007 *Kumpulan Data Status Lingkungan Hidup Daerah Kabupaten Jember Tahun 2007 Dinas Kebersihan Dan Lingkungan Hidup Kabupaten Jember Tahun 2007*

[2] Santoso B, Hendrijanto K, Rahmawati A and Jannah R 2010 Model Intervensi Pengelolaan Daerah Aliran Sungai (DAS) *Lap. Has. Penelit. Hibah Bersaing* 1–11

[3] Hanafie E 2003 *Telaah Kualitas Air Bagi Pengelolaan Sumberdaya dan Lingkungan Perairan* (Yogyakarta: Kanisius)

[4] Lesmana D S 2001 *Kualitas Air untuk Ikan Hias Air Tawar* (Jakarta: Penebar Swadaya)

[5] Arimoro F O, Abubakar M D, Obi-Iyeke G E and Keke U N 2021 Achieving sustainable river water quality for rural dwellers by prioritizing the conservation of macroinvertebrates biodiversity in two Afrotropical streams *Environ. Sustain. Indic.* **10** 100103

[6] Srikandi Fardiaz 1992 *Polusi Air dan Udara* (Yogyakarta: Kanisius)

[7] Sargaonkar A 2003 Development of an overall index of pollution for surface water based on a general classification scheme in Indian context *Environ. Monit. Assess.* **1** 43–67

[8] Pradana H A, Wahyuningsih S, Novita E, Humayro A and Purnomo B H 2019 Identifikasi Kualitas Air dan Beban Pencemaran Sungai Bedadung di Intake Instalasi Pengolahan Air PDAM Kabupaten Jember *J. Kesehat. Lingkung. Indones.* **18** 135

[9] Septiani B Y A, Jati W N and Zahida F 2015 Keanekaragaman Jenis Makrozoobentos sebagai Penentu Kualitas Air Sungai Mruwe Yogyakarta *J. Tenobiologi* 1–11

[10] Anrosana I A and Gemaputri A A 2018 Kajian Daya Dukung (Carrying Capacity) Lingkungan Perairan Pantai Pasir Putih Situbondo bagi Pengembangan Usaha Karamba Jaring Apung *J. Ilm. Inov.* **17** 73–9

[11] Mekuria D M, Kassegne A B and Asfaw S L 2021 Assessing pollution profiles along Little Akaki River receiving municipal and industrial wastewaters, Central Ethiopia: implications for environmental and public health safety *Heliyon* **7** e07526

[12] [12] Endang Dewi Masithah 2020 *Cyanophyta, Antagonisme Pembunuh dan Pionir Kehidupan* (Surabaya)

[13] Fahrur M, Makmur and Rachmansyah 2012 DINAMIKA KUALITAS AIR DAN HUBUNGAN KELIMPAHAN PLANKTON DENGAN KUALITAS AIR DI TAMBAK KECAMATAN BONTOA , KABUPATEN MAROS *Indoaqua* pp 881–94

[14] Uddin M J and Jeong Y K 2021 Urban river pollution in Bangladesh during last 40 years: potential public health and ecological risk, present policy, and future prospects toward smart water management *Heliyon* **7** e06107

[15] Effendi H, Romanto and Wardiatno Y 2015 Water Quality Status of Ciambulawung River, Banten Province, Based on Pollution Index and NSF-WQI *Procedia Environ. Sci.* **24** 228–37

[16] Khairunnas and Gusman M 2014 Analisis Pengaruh Parameter Konduktivitas, Resistivitas dan TDS Terhadap Salinitas Air Tanah Dangkal pada Kondisi Air *J. Bina Tambang* **3** 1751–60

IOP Conf. Series: Earth and Environmental Science **980** (2022) 012007 doi:10.1088/1755-1315/980/1/012007

Application of pulsed electric field in pasteurization of orange juice of siam cultivar: study on nutritional, physical, chemical properties, and total microorganism

MJ Wibowo[1], A Bakri[2], B Hariono[3], R Wijaya[4], A Brilliantina[5]

[1] Department of Technic, Politeknik Negeri Jember, Indonesia
[2,3,4,5]Departmen of Agriculture Technology, Politeknik Negeri Jember, Indonesia

Corresponding author : rizza.wijaya@polije.ac.id

Abstract. Orange juice, extracted from orange of siam cultivar, was pasteurized using a pulsed electric field (PEF) method. The aim of this research was to analyze the changes on nutritional, physical and chemical properties and total microbes from orange juice after pasteurization. The research was conducted from June until September 2021 located in Jember. The PEF treatment was carried out using treatment time variation for 5 minutes (1 cycle), 10 minutes (2 cycles) and 15 minutes (3 cycles). Several parameters of Vitamin C, density, pH, total soluble and total microbes were observed. The result showed that PEF treatment did not significantly change nutritional, physical and chemical properties and total microbes after pasteurization with PEF in compared with no treatment pasteurization. The best treatment was found in the variation of 10 minutes (2 cycles).

1.Introduction

Food processing is an important step to make food ready for consumption and to improve the quality and shelf life of food products [1]. However, what is often neglected in the food processing process is the degradation of the nutritional content of food ingredients [2]. The damage to the nutritional content of these foodstuffs is generally due to the processing of foodstuffs using a thermal process [3]. One of the newest and not so familiar technologies in Indonesia in the field of food processing without involving a thermal process is to use a Pulsed Electric Field (PEF).

The PEF process is based on the application of short pulses at high voltage to a food material placed between two electrodes. The process is very short ranging from one microsecond to one millisecond with short pulses. HPEF is widely applied to liquid materials, although recently the use of HPEF has expanded to solid materials such as the pretreatment process for essential oil leaves [4]. In addition, PEF technology can be used to increase the yield in fruit juice production, accelerate water transfer during drying operations, as well as increase the extraction of valuable compounds (such as antioxidants, dyes, or flavors) from the interior of the cells [5]. In addition, PEF can also extend the shelf life while maintaining the physical, chemical, and sensory properties of food [6]. The added value is that the purpose of this research is to know the characteristics of Siamese orange juice using the HPEF method.

1.1 Orange Juice

Citrus juice is a drink that is widely consumed compared to other fruit juices because of its taste as well as its health benefits related to the bioactive components contained in orange juice which act as antioxidants such as vitamin C and phenolic compounds. Recent studies have shown that 66% of the antioxidant capacity of orange juice is supported by vitamin C [7].

In the process of making orange juice on an industrial scale, the pasteurization process is generally carried out before being packaged. Pasteurization of orange juice is generally carried out at 80°C for 30 seconds. It aims to kill spoilage microbes, namely mold, and bacteria [8]. High Pulsed Electric Field (HPEF) is one of the electricity-based processing techniques. The application of short electrical pulses at high voltages allows the reduction of thermal processing effects, such as ohmic heating [13], and moderate electric fields. The HPEF method is a microbial inactivation technique without detrimental effects on the attributes of food products so that the nutritional content of food is maintained [9].

The application of PEF for the treatment of food commodities has been widely carried out, from the results of research the PEF method has been known to preserve food products without reducing their nutritional and sensory quality values [10]. Basically, the working principle of this sterilizer with high-voltage electric pulse field technology (High Pulsed Electric Field) is very simple, namely, liquid food media is subjected to an electric pulse field with a certain intensity generated from a voltage generator (coil). Various types of generators can be selected according to their needs, for example, Pearson Coil has been used to generate high voltage by. The Important aspects in HPEF technology are the high-voltage power generation unit, the design of the treatment chamber that is able to suppress the increase in temperature and the design of the electrode that is able to minimize the effects of electrolysis. High field intensity is achieved by storing large amounts of energy in the capacitor bank (a series capacitor) of the DC power supply which is then converted into high-voltage pulses. Energy requirements in PEF are more efficient than thermal pasteurization, especially when using a continuous system.

The main components in the electrical pulse-forming network are: 1) electrical energy generation. A high-voltage electrical energy generation unit capable of providing electrical energy at a specified voltage level up to about 40kV; 2) one or more banks of capacitors, inductors and/or resistors. Capacitors to store electrical energy. Capacitors are connected in parallel to increase the stored electrical energy. Inductors temporarily store magnetic energy thereby delaying the rise in electric current. Resistors are mainly represented by the charge of the process chamber and dissipate electrical energy. 3) one or more switches that transmit electrical energy to the electrodes and foodstuffs. The switches are connected in series and parallel for alternating high voltage and electric current. The pulse length ranges from less than a microsecond to a few milliseconds. 4) one or more process chambers with two electrodes between the food sample being flowed or placed. When the switch is closed, the capacitor discharges the stored energy into the process chamber and an electric pulse flows through the sample of food. The rate at which the voltage increases across the electrodes depends on the closing speed of the switch, the electric wire and the resistivity of the material. Electrical interconnection faults can occur due to using multiple process spaces in series or parallel. 5) Oscilloscope to measure the voltage that passes through the electrodes and displays an image of the shape of an electric pulse. The shape of the pulse depends on the network that forms the electric pulse. Some forms of electric pulses are exponential decay or square wave (both can be monopolar or bipolar), oscillatory pulse or instant reversal pulse, the commonly used pulse forms are exponential decay and square waveform [11].

2. Methods

The first stage is the optimization of the Semboro orange juice pasteurization process (Low Grade) using the HPEF method based on the time factor, duration of exposure with electrical pulses (Pulse Duration) and the voltage range used (Pulse interval).

2.1 Research Procedure

This research is planned until the year 2026, where researchers will divide into basic, applied, development research and finally get orange juice products that are safe for consumption. Researchers have published scientific articles from 2018-2020 related to HPEF technology for pasteurizing cow's milk, and on this basis this technology will be applied to orange juice based on local resources. In the basic research, a prototype of the HPEF method pasteurization device will be built with optimization of the HPEF dose and material contact time. In the following year, non-destructive sterilization technologies with multi-chamber will be developed. At the end of the research roadmap, orange juice products will be obtained that are safe for consumption and the creation of an orange juice agro-industry cluster based on local resources.

2.2 Research Location

This research was conducted at the TEFA Bakery Laboratory and the Food Analysis Laboratory of Politeknik Negeri Jember.

2.3 Tool and Material

The material used is Siamese orange juice obtained from the extraction of Siamese oranges without the addition of other ingredients. Siamese oranges were purchased from the Tanjung-Jember Market. Other materials are 95% alcohol, cotton and sterile bottles. The tools used are a series of PEF tools, high voltage voltmeter, pH meter, petri dish, tds meter, juice extractor, and cool box.

2.4 Research Methodology

The research was carried out from July to September 2021 at TEFA Canning Jember State Polytechnic. The method used is descriptive experimental method. Testing is done by varying the processing time. Processing time is 0, 5, 10, and 15 minutes. The volume of apple cider entered in the treatment chamber is 3 liters. For each processing time with PEF, 3 (three) replicates of samples will be taken which will then be analyzed for changes in nutritional value (vitamin C), chemical properties (pH and total dissolved solids) and the number of microbes (total microbes and coliforms) contained in the PEF will be calculated. Siamese orange juice processed with PEF. The high-voltage pulse generator used in the pasteurization of Siamese oranges with the PEF method consists of several blocks, namely the keypad block, microcontroller, display, flyback converter circuit, high-voltage transformer and chamber. The keypad is used to enter the high voltage setting and the time required for treatment.

The high voltage can be set from 20 kV to 40 kV and the length of treatment can be set in the range of 0 minutes to 15 minutes. The microcontroller serves to display the high voltage and treatment time that is set through the keypad. The flyback converter circuit will receive the output of the microcontroller in the form of square pulses that can be adjusted for the pulse width. The output of the flyback converter in the form of a voltage pulse will count the input voltage of the high-voltage transformer so that the output of the transformer will be in the form of a high-voltage pulse. High-voltage transformers can produce a maximum output of 40 kV. Constant frequency 20 kHz. All PEF components are assembled in a high-voltage generator box made of mica. The high-voltage pulses produced are directly fired into the chamber (the treatment area) which is coated on the outside by electrodes made of copper. The treatment chamber is made of 4 mm thick stainless steel plate which is safe for all food products. The treatment chamber is in the form of a four-legged cylinder with a volume of 1.7 liters equipped with an exhaust valve. To protect the entire pulsed electric field tool, the outside is provided with a barrier made of a clear mica box.

3. Experiment and Result

This study was conducted to determine the characteristics of orange juice with HPEF treatment. Analysis on chemical parameters such as Vitamin c, pH, Dissolved Solids, and Total Microbes were analyzed.

Vitamin C

The vitamin C content of Siamese orange juice before pasteurization is 6.205 mg/100gr. The highest level of vitamin C after pasteurization in the 5-minute treatment was 6.181 mg/100gr and the lowest in the 15-minute treatment was 6.091 mg/100gr. The graph of changes in vitamin C levels against treatment time is shown in Figure 1.

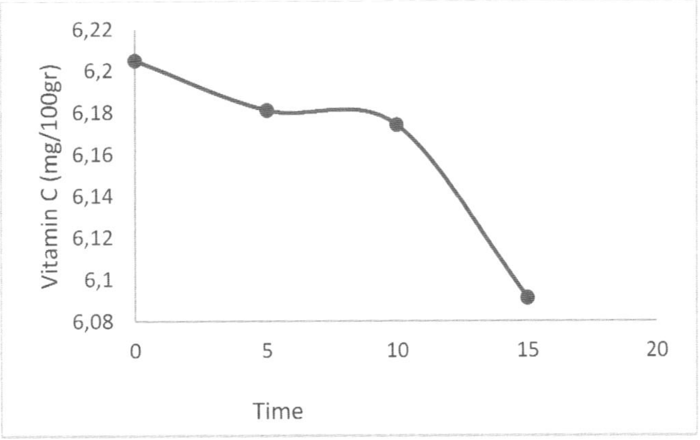

Figure 1. Vitamin C

Figure 1 it can be seen that the non-thermal pasteurization treatment with a high pulsed electric field (HPEF) will cause a decrease in vitamin C levels, but this change tends to be insignificant compared to without HPEF. According to [8] the processing of orange juice using heat will reduce the quality of the citrus juice produced due to oxidation which causes a reduction in the vitamin C content and loses the original taste and aroma of the orange. Meanwhile, when using HPEF technology, Siamese orange juice does not lose vitamin C, the original taste or aroma of the orange juice. This is also in accordance with the opinion [9] that the higher the temperature and the duration of heating, the greater the degradation of vitamin C. Oxidation of vitamin C (ascorbic acid) will convert ascorbic acid into L-dehydroascorbic acid which is chemically very labile and can undergo further changes to Ldiketogulonic acid which has no vitamin C activity anymore.

pH

The pH of Siamese orange juice before pasteurization was 4.64. The highest pH after pasteurization in the treatment for 10 minutes was 4.74 and the lowest in the treatment for 5 and 15 minutes was 4.4. Graph of changes in pH with treatment time is shown in Figure 2.

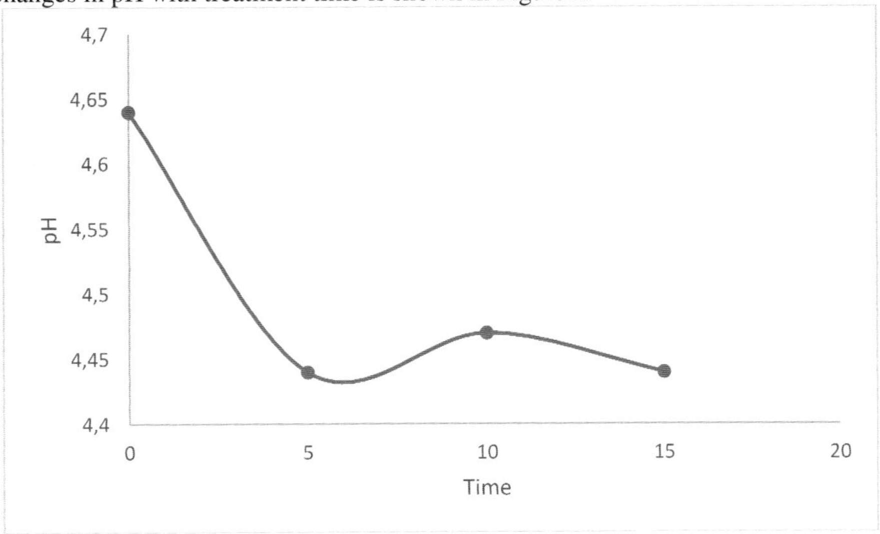

Figure 2. pH Value

In Figure 2 it can be seen that the pH of Siamese orange juice from non-thermal pasteurization using a pulsed electric field tends to decrease the pH of Siamese orange juice but the changes are not significant and are still in the average range with the pH before pasteurization. This is in line with research [10] that the pH of the mixture of orange juice and carrots treated with PEF did not change significantly. [11] also stated that HPEF treatment did not significantly change the pH of milk.

Dissolved Solid

The total soluble solids of Siamese orange juice before pasteurization was 776 ppm. The value of total soluble solids from orange juice after HPEF pasteurization tended not to change, except for the 15 minute treatment of 678 ppm. The graph of the change in total dissolved solids with respect to treatment time is shown in Figure 3.

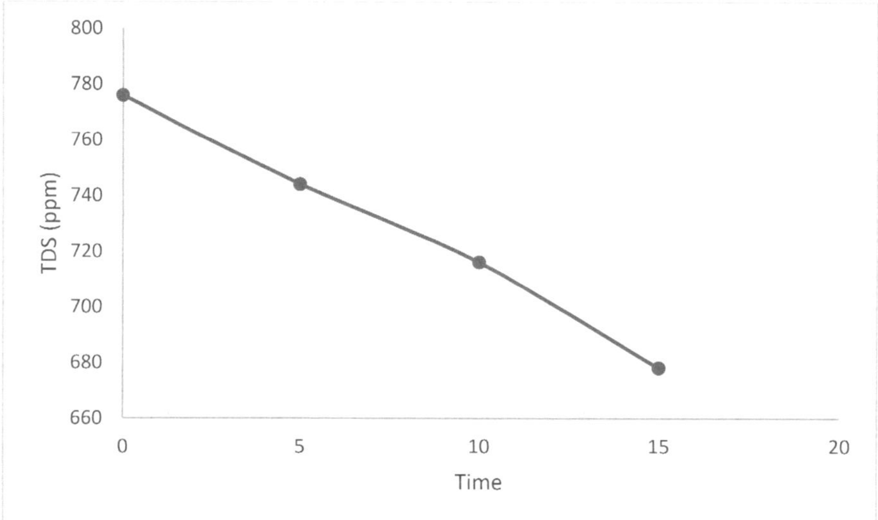

Figure 3. Changes in total dissolved solids with treatment time

The total value of soluble solids is the sum of the total content of all substances, both inorganic and organic, dissolved in food. The components in the fruit consist of water-soluble components, such as glucose, fructose sucrose, and water-soluble protein (pectin). The higher the dissolved solids number, the more cloudy the fruit juice, the more cloudy the drink made will reduce the level of consumer preference for the product [12]. Based on research results, HPEF technology can reduce the value of dissolved solids in a fruit juice. The decrease in the total dissolved solids value is thought to be because the content of the material contained in the Siamese orange juice is not degraded due to processing using HPEF, so that it can reduce the total soluble solids in the Siamese orange juice.

Total microbes

The total microbes of Siamese orange juice before pasteurization were 1.22x104 cfu/ml. The total of these microbes not in accordance with the Indonesian National Standard (SNI) where the maximum total microbes contained in fruit juice drinks is 2x102 cfu/ml. The highest total microbes after pasteurization in the 5 minute treatment was 5.09x103 cfu/ml and the lowest was 3.58x102 cfu/ml in the 15 minute treatment. The graph of the decrease in total microbes with time changes is shown in Figure 4.

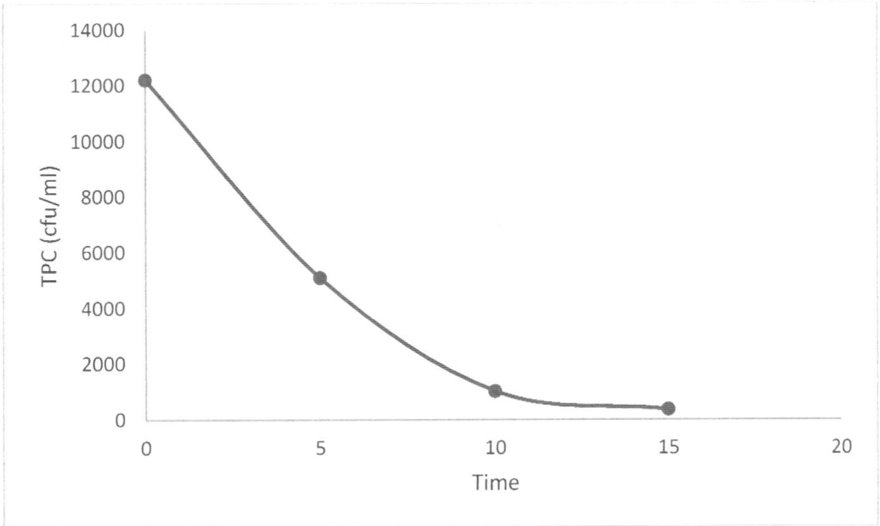

Figure 4. Decrease in total microbes with treatment time

The results of the analysis showed the presence of coliforms in the fresh Siamese orange juice. At the time after pasteurization, coliforms were still found in the treatment time up to 10 minutes, but at 15 minutes there were no coliforms. This is in accordance with the Indonesian National Standard (SNI) where the maximum permissible coliform limit is 20 APM/ml. No coliforms were found in the PEF pasteurized orange juice because the pH condition of the apple cider was quite acidic which ranged from 4.44-4.64. Under acidic conditions, coliforms are not able to live properly.

PEF is the latest non-thermal technology for liquid and semi-solid food processing systems. The process involves administering short pulses of 1 s to 100 s of a high voltage electric field between 20–80 kV/cm to a food material placed between two electrodes at room temperature [13]. Giving high-voltage pulses to food will damage the bacterial membrane causing the bacteria to die [14]. The possibility that arises from this phenomenon is the occurrence of abnormal metabolic activity or an increase in cell body metabolism that is too sharp so that it interferes with the work and physiological functions of cells. Electric shock with high voltage causes physical damage to cells. Microbial death due to high-voltage electric shock is thought to be influenced by damage to other cell structures, such as damage to the cell's cytoplasmic membrane. Although naturally the cytoplasmic membrane can be synthesized again but with high voltage, the damage in the form of holes in the outer membrane of the cell cannot be repaired, thus allowing the entry and exit of macromolecular compounds from the cell and cause death [15]. The length of treatment also affects the amount of microbial decline, where the longer the treatment time the microbial mortality is also getting bigger. It is suspected that the largest total microbes contained in orange juice have a high susceptibility to electric current, causing a high mortality rate [16]. Meanwhile, the remaining total living microbes were thought to have high resistance to electric shocks, so that the number of subsequent electric shock treatments was not so large in killing the remaining total microbes [17]. According to [18], spore death by electrocution treatment was limited by the strength of the electric field generated, the duration of the shock, the number of shocks applied and the cell size. Meanwhile, according to [19], electricity with high voltage for a long time has a greater effect in reducing the number of spores contained in the media.

4. Conclusion

The PEF treatment was carried out using treatment time variation for 0-15 minutes. Several parameters of Vitamin C, density, pH, total soluble and total microbes were observed. The result showed that PEF treatment did not significantly change nutritional, physical and chemical properties and total microbes after pasteurization with PEF in compared with no treatment pasteurization.

IOP Conf. Series: Earth and Environmental Science **980** (2022) 012007 doi:10.1088/1755-1315/980/1/012007

Acknowledgments
Our thank goes to Politeknik Negeri Jember, who has helped support for this research.

References

[1] Hermanuadi D Brilliantina A and Novitasari E K, 2021 Value added analysis of cassava "tape" supply chain in Bondowoso regency *IOP Conf. Ser. Earth Environ. Sci.* **672**, 1 p. 1–6.
[2] Yudiastuti S O N Wijaya R and Budiati T, 2021 The effect of ozonation time and contact time of edamame washing on color changes using the continuous type ozone washing method *IOP Conf. Ser. Earth Environ. Sci.* **672**, 1.
[3] Kosasih E A Zikri A and Dzaky M I, 2020 Effects of drying temperature, airflow, and cut segment on drying rate and activation energy of elephant cassava *Case Stud. Therm. Eng.* **19**, February p. 100633.
[4] Arnal Á J *et al.*, 2018 Implementation of PEF treatment at real-scale tomatoes processing considering LCA methodology as an innovation strategy in the agri-food sector *Sustain.* **10**, 4.
[5] Niu D *et al.*, 2020 Review of the application of pulsed electric fields (PEF) technology for food processing in China *Food Res. Int.* **137**, September p. 109715.
[6] Lebovka N I Shynkaryk N V. and Vorobiev E, 2007 Pulsed electric field enhanced drying of potato tissue *J. Food Eng.* **78**, 2 p. 606–613.
[7] Sundari D Almasyhuri A and Lamid A, 2015 Pengaruh Proses Pemasakan Terhadap Komposisi Zat Gizi Bahan Pangan Sumber Protein *Media Penelit. dan Pengemb. Kesehat.* **25**, 4 p. 235–242.
[8] Hariono B Djamila S Bahariawan A Wijaya R and Pambudi S, 2020 Energy study of edamame freezing process at PT Mitratani Dua Tujuh Jember *J. Phys. Conf. Ser.* **1450**, 1.
[9] Hariono B Wijaya R Kurnianto M F Sutrisno S Seminar K B and Brilliantina A, 2020 Quality of Goat's Milk Exposed Ultraviolet and High Pulsed Electric Field *IOP Conf. Ser. Earth Environ. Sci.* **411**, 1.
[10] Donsì F Ferrari G Fruilo M and Pataro G, 2011 Pulsed Electric Fields – assisted vinification *Procedia Food Sci.* **1** p. 780–785.
[11] Hariono B Kurnianto M F Bakri A Ardiansyah M and Wijaya R, 2018 Improvement of Sensory and Chemistry Quality of Fried Edamame by Freezing *IOP Conf. Ser. Earth Environ. Sci.* **207**, 1 p. 1–7.

IOP Conf. Series: Earth and Environmental Science **980** (2022) 012008 doi:10.1088/1755-1315/980/1/012008

Characterization of lignocellulolytic fungi isolated from cocoa rhizosphere under different cropping patterns

Intan Nirmalasari[1*], Dwi Andreas Santosa[1], Erwin Prastowo[2]

[1]Department of Environmental and Soil Biotechnology, Bogor Agricultural University, Bogor
[2]Indonesian Coffee and Cocoa Research Institute, Jember

*email : intan.nirmalasari186@gmail.com

Abstract. Cocoa plantations may provide a considerable amount of fungal biodiversity with high potential as soil organic matter biodecomposer agents. Therefore, the isolation and characterization of potential fungi is the first step to obtain the wide variety of lignocellulolytic enzymes. This study aimed at identifying potential lignocellulolytic fungi isolated from different type cocoa cropping pattern. The lignocellulose-degrading enzymes was determined by growing the isolated fungi on Guaiacol containing PDA media, and CMC media with additional of Red Congo solution. Their potential index (IP) was scored by measuring the diameter of the clear zone around the fungal colonies. Our identifications suggest the presence of both ligninolytic, and cellulolytic enzymes through a reddish, and clear zones produced around the colonies. Additionally, we identified two types of potential fungi with higher potential index both ligninolytic and cellulolytic, including strain fungal CSF#5 and CSF#15 which belongs to the group of fungi *Ascomycetes.*

1. Introduction

The diversity of shade vegetation types and specific microclimate characteristics in cocoa plantations provides a large amount of biodiversity, especially in the soil. Various types of fungi have diverse roles and functions and synergize with each other, one of which is as an agent of organic waste in cocoa plantations. One of the main obstacles in the processing of cocoa plant waste is the long decomposition time due to the structure of cocoa organic matter which is rich in lignocellulose. Lignocellulosic wastes represent about 60% of plant biomass and their disposal usually poses difficulties due to their volume and characteristic hard structure [1]. Biological biodegradation using microbes is considered the safest and most environmentally friendly [2].

In general, fungi have a better ability than bacteria in breaking down plant residues (hemicellulose, cellulose, and lignin) [3]. The growth of hyphae from fungi class *Basidiomycetes* and *Ascomycetes* (hyphal diameter 5–20 m) more easily penetrates the walls of tubular cells which are the main constituents of wood tissue accompanied by the release of enzymes that dissolve the cell walls of wood tissue (polymer complexes of lignin, cellulose and hemicellulose) [4]. Several enzymes involved in the breakdown of organic matter include -glucosidase, lignin peroxidase (LiP), manganese peroxidase (MnP), laccase and versatile peroxidase. These enzymes are produced by rot fungi such as *Pleurotus eryngii, P. ostreatus,* and *Bjekandera adusta* [5] and *Phanerochaete chrysosporium* [6] and *Trichoderma sp* [7].

The use of microbes, especially fungi as bioactivators in lignocellulosic degradation, is a technique that can be applied to increase the effectiveness of lignocellulosic waste degradation. Wood rot fungi and *Trichoderma sp.* can reduce the lignin and cellulose content in OPEFB by 20.83% and 33.77% [8]. Based on the description above, it can be seen that fungi play an important role in the process of biodegradation of lignocellulosic waste. Therefore, in this study, qualitative exploration and selection of potential lignocellulolytic fungi will be carried out on solid media. The results of the study can be used as a recommendation for the use of these fungal inoculants for decomposer agents in the treatment of lignocellulosic waste from cocoa plantations and other organic wastes.

2. Material and Methods

2.1. Isolation of fungi from cocoa plantation

Soil samples from 5 cocoa growing locations with different shade patterns were isolated using PDA media with the dilution method. A total of 10 g of soil samples were suspended in 90 ml of physiological solution, i.e. NaCl 0.85%, with 5 series of dilutions, namely 10-1 to 10-5. The last three dilution series, 10^{-3}, 10^{-4} and 10^{-5}, were taken 100 µl and poured into sterile petri dishes containing PDA media and then incubated at room temperature for 4-8 days [9][10]. After that, the total colonies that grew were calculated using the Total Plate Count method, then purification was carried out with the isolates were coded based on the type of fungus. Each selected fungal strain from the purification results was then selected for its potential in degrading lignin and cellulose.

2.2. Screening test of lignocellulolytic activity

The primary selection was performed using a qualitative method by measuring the ligninolytic and cellulolytic potency indices on petri dishes. Selection of ligninolytic potential was carried out using PDA media supplemented with Guaiacol reagent. The selection of cellulolytic potential using CMC + Congo red media.

2.3 Ligninolytic activity test

Fungal cultures aged ± 7 days were inoculated into PDA media, each of which was added with 0.05% Guaiacol [11][12], then incubated at room temperature for 3-7 days. A positive reaction on Guaiacol media is indicated by the presence of a reddish brown zone around the colony which is an indication of the production of the extracellular enzyme phenol oxidase (laccase) by oxidizing guaiacol [13]. The ligninolytic potency index can be obtained by the following formula:

$$\text{Lignolytic Potency Index} = \frac{Colored\ zone\ area}{Colony\ zone\ area}$$

2.4 Cellulolytic activity test

Fungal cultures were inoculated onto 1% Carboxymethyl Cellulose (CMC) media and incubated for 3-7 days at room temperature. Then the petri dish was soaked with 15 ml of 0.1% Congo red solution and let stand for 30 minutes, then rinsed twice using 15 ml of 1 N NaCl and let stand for 15 minutes [14]. A positive reaction was indicated by the presence of a clear zone around the colony which was an indication of cellulase enzyme activity.

$$\text{Cellulolytic Potency Index} = \frac{Clear\ Zone\ Diameter\ (mm) - Colony\ Diameter\ (mm)}{Colony\ Diameter\ (mm)}$$

From the two selections above, fungal isolates that showed positive results from each selection method were morphologically identified to determine the characteristics of fungi in more detail.

2.5 Identification of lignocellulolytic fungi

Identification was carried out morphologically, both macroscopically and microscopically. Macroscopically with the top and bottom surfaces of agar petri dishes by observing the shape, color, and diameter of the fungus. Microscopically, it was carried out using a microscope by observing the shape of the spores and the structure of the conidiospores and sporangiospores on the fungal body.

3. Result and Discussion

3.1. Lignocellulolytic fungal population in cocoa plantation

Based on the results of fungal isolation carried out on 5 soil samples from cocoa plantations with a series of 10^{-3}, 10^{-4} and 10^{-5} dilutions using PDA media, 23 types of fungal isolates were obtained with different morphological characteristics as shown in the table below (Table 1).

Table 1. Distribution of lignocellulolytic fungal species diversity in 5 different cacao cropping patterns

Isolate Code	Isolate Source					Zone	
	Soil-Lamtoro Shade	Soil-Lamtoro + Coconut Shade	Soil-Lamtoro + Palm Oil Shade	Soil-Lamtoro + Betel nut Shade	Soil-Paranet Shade	Lignolytic	Cellulolytic
CSF#1	-	+	-	-	-	-	+
CSF#2	-	+	-	-	-	-	+
CSF#3	-	+	-	-	-	-	+
CSF#4	+	+	+	+	+	-	+
CSF#5	-	+	-	-	-	+	+
CSF#6	+	+	+	-	-	-	+
CSF#7	+	-	-	-	-	-	+
CSF#8	+	-	-	-	+	-	+
CSF#9	+	+	-	-	+	-	-
CSF#10	-	-	+	-	-	+	+
CSF#11	-	-	+	-	-	-	-
CSF#12	+	-	-	-	-	-	+
CSF#13	+	-	+	-	-	-	+
CSF#14	-	-	+	-	-	-	-
CSF#15	+	+	+	-	-	+	+
CSF#16	+	-	-	-	-	-	+
CSF#17	+	-	-	-	-	-	-
CSF#18	+	+	+	-	-	-	-
CSF#19	+	-	-	-	+	-	-
CSF#20	+	-	-	-	+	-	-
CSF#21	+	+	+	-	+	+	+
CSF#22	-	-	-	+	-	-	+
CSF#23	+	-	-	-	+	-	+
Total Types of Fungi	**15**	**10**	**9**	**2**	**7**	**4**	**17**

CSF#1: Cocoa Soil Fungal No. 1…etc ; +: present; -:absent

The data on the diversity of 23 types of fungi from cocoa plantations in the table above describes the distribution pattern of locations and their ability to degrade lignin and cellulose. The cacao cropping pattern with lamtoro shading had the most diversity, there are 15 species compared to other shading patterns, while the lamtoro + betel nut shading patterns had the smallest diversity of fungi, namely 2 types of fungi. This is in accordance with the research of Tondok *et. al.*[15] that in cocoa (*Theobroma cacao*), the presence of shade and leaf age affected the diversity of endophytic fungi isolated from the site, but was not affected by the geographical location of the plant and the presence or absence of pathogen infection. As for the ability of fungi to degrade lignin and cellulase, there were 4 types of fungi that had ligninolytic abilities, namely CSF#5, CSF#10, CSF#15 and CSF#21. While fungi that have cellulolytic ability there are 16 types of fungi. Fungi that have ligninolytic abilities also have cellulolytic abilities, so that 4 types of fungi are potential that can degrade lignin and cellulose or lignocellulosic fungi.

3.2. Screening result of lignocellulolytic activity in solid medium

3.2.1. Screening result and enzyme potency index of ligninolytic activity

Pure cultures (single strain) aged ± 7 days from the isolated samples were selected using PDA media supplemented with 0.05% Guaiacol. The results showed that among the 23 isolates tested, there were 4 isolates that showed a color change reaction to reddish brown after being incubated for 3-7 days at room temperature.

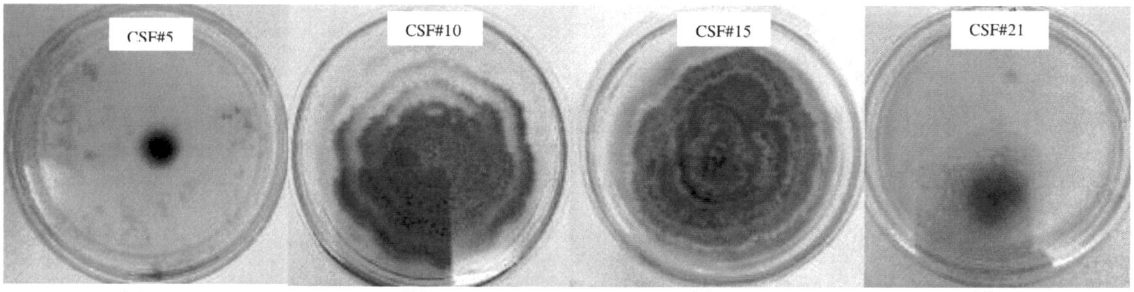

(a) **(b)** **(c)** **(d)**

Figure 1. Selection results of lignolytic fungal on PDA media supplemented with Guaiacol 0.05% after 7 days. (a) CSF#5; (b) CSF#10; (c) CSF#15; (c) CSF#21

As shown in Figure 1, there were differences between fungal colony growth, area and thickness of the reddish-brown zone produced after incubation for 7 days. The CSF#15 isolate produced a wider colored zone which occupied the entire petri dish, while the CSF#5 isolate produced a narrower colored zone but the color was more intense. The growth of CSF#5 and CSF#15 fungi colonies was positively correlated with the production of colored zones where the wider the colony growth, the more colored zones were produced. While the isolates of fungi CSF#10 and CSF#21, the growth of fungal colonies and the production of colored zones were negatively correlated, because the growth of colonies was faster than the production of colored zones. CSF#10 isolate had faster fungal colony growth and a wider and more concentrated color zone than CSF#21 isolate. The reddish-brown color was produced because the polymer oxidation was enhanced by guaiacol which started to appear in the media along with mycelium growth fungi on solid media as incubated [16].

The quantification test was carried out by measuring the colony growth area and the colored zone which was measured twice on the 3rd and 7th incubation days to measure the ligninolytic potential index of each fungal isolate. The lignolytic potency index indicates the lignolytic enzyme as shown in Figure 2 below.

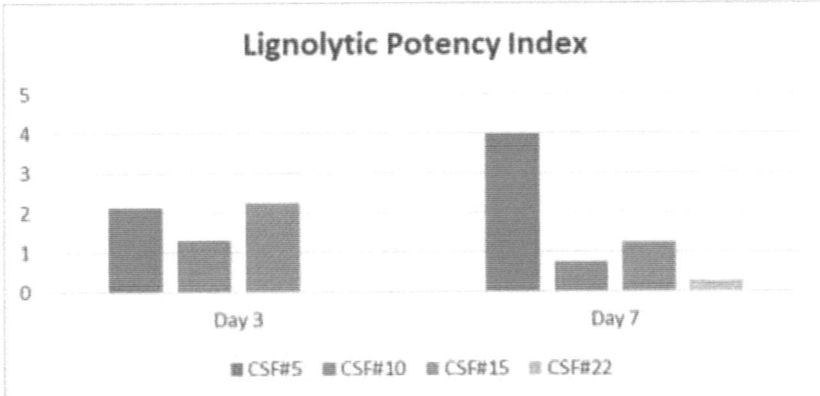

Figure 2. Index of lignolytic potential of isolate fungi CSF#5; CSF#10; CSF#15 and CSF#21
CSF: Cocoa Soil Fungal

As shown in figure 2, on the 3rd day of observation, isolate CSF#15 produced a higher potency index than the other three isolates, but isolate CSF#21 did not have a ligninolytic index because the colored zone had not been produced. On the 7th day of observation, the lignolytic potential index decreased in isolates CSF#10 and CSF#15, while isolates CSF#5 experienced a significant increase, and isolates CSF#21 began to form colored zones with the lowest ligninolytic potential index. than other isolates. Based on the results of the ligninolytic potency index measurement, isolate CSF#5 produced the largest potency index on the 3rd and 7th day of observation, followed by isolates CSF#15, CSF#10 and CSF#21. A strain is categorized as having a high ligninolytic potential index in the range of values (PI > 1,000), while a low ligninolytic potential index in the range of values (PI < 1,000) [14]. Fungi with a lower potency index require a longer adaptation to produce enzymes to be oxidized [17][18].

On the colony growth parameters, isolate CSF#5 had the slowest colony growth and isolate CSF#15 had the fastest colony growth compared to other isolates on the 3rd and 7th day of observation. This is because the production of the reddish-brown zone produced by isolate CSF#5 expanded until the 7th day compared to colony growth which tended to be slow, resulting in the highest ratio of ligninolytic potential index compared to other isolates. A higher potency index value indicates a higher ligninolytic enzyme activity and vice versa [19].

3.2.2. Screening and enzyme potency index of cellulolytic activity

Pure culture (single strain) aged ± 7 days from the isolated sample was selected using 1% CMC media then incubated for 7 days and soaked in 0.1% Congo red for 30 minutes to determine the cellulolytic ability. The results showed that among the 23 isolates tested, there were 17 isolates that produced a clear zone around the colony after being incubated for 7 days at room temperature.

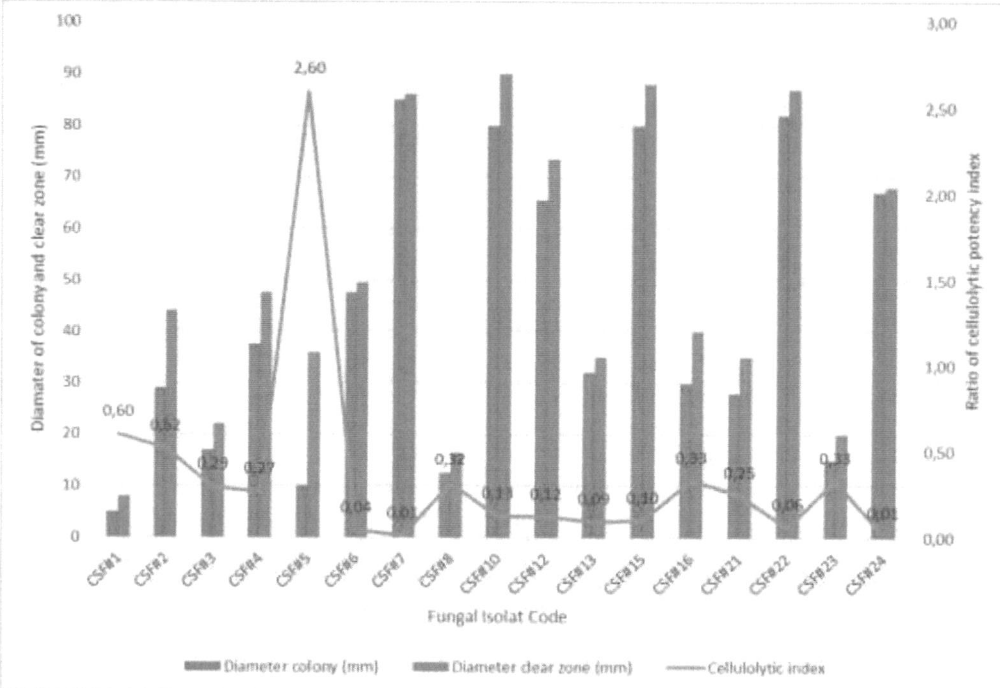

Figure 3. Relationship of fungal colony growth diameter, clear zone and cellulolytic potential index of 17 isolates of fungi on the 7th day of observation

As shown in table 3 and figure 3, the diameter of fungal colony growth, clear zone production and cellulolytic potency index were related to each other. The fastest colony growth diameter after 7 days of observation was found in isolates CSF#10, CSF#15 and CSF#21 and CSF#7, with values of 80 to 85 mm. This was positively correlated with the production of a wider clear zone than colony growth along with the length of incubation time, but resulted in a low ratio below 1. 50 mm after 7 days of incubation. However, there is a unique difference from isolate CSF#5 which has the highest cellulolytic potential index ratio of 2.60 with colony growth which is classified as low/slow, namely 10 mm on the 7th day of observation. CSF#5 isolate produced a clear zone that was relatively fast compared to colony growth, resulting in a higher ratio than other isolates. The low potency index is because fungi require a longer adaptation time to produce enzymes for oxidation. The difference in the cellulolytic activity index is thought to be due to the cellulase enzyme excreted by each fungal isolate with different potential to decompose substrates in growth media [20]. The greater the cellulase index in the isolate, the greater the cellulolytic activity produced [21].

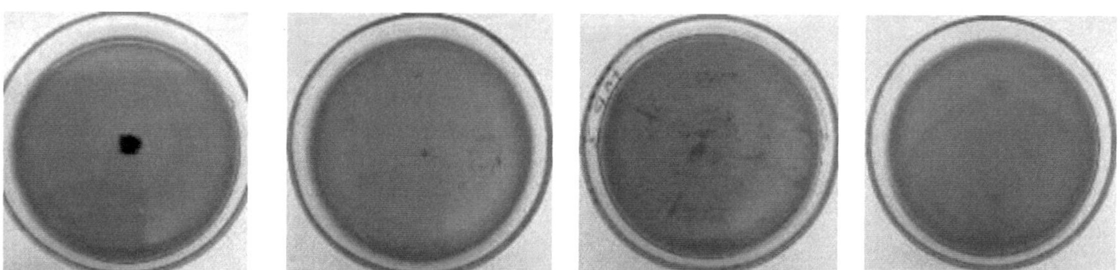

(a) (b) (c) (d)

Figure 4. Selection results of cellulolytic fungal on 1% CMC media after soaking in 0.1% Congo red on the 7th day of observation. (a) CSF#5; (b) CSF#10; (c) CSF#15; (c) CSF#21

Based on the results of each ligninolytic and cellulolytic selection, it shows that there are 4 types of fungi that have two abilities, both ligninolytic and cellulolytic or called lignocellulolytic. The largest ligninolytic and cellulolytic potency indices were found in CSF#5 isolated after the 7th day of incubation, but this isolate had a lower/slower colony growth than the other 3 lignocellulolytic isolates. While the fastest growth of fungal colonies was found in isolate CSF#15 both on the 3rd and 7th day of observation. This isolate had the highest ligninolytic potential index on the 3rd day of observation, but decreased on the 7th day along with colony growth. The isolates CSF#10 and CSF#21 had faster colony growth than isolates CSF#5, but had low lignocellulolytic abilities. Based on the qualitative test of the ligninolytic and cellulolytic abilities of each fungus accompanied by the speed of colony growth, it can be concluded that the most potent isolates were CSF#5 and CSF#15.

3.3. Identification of strain
Based on the morphology of isolates CSF#5 and CSF#15, it was known that the two isolates belonged to the Ascomycetes group strain due to the presence of insulated conidium and hyphae.

4. Conclusion
The results showed that 23 species of fungi were isolated from the soil at 5 cocoa planting locations with different types of shade. The results of the ligninolytic selection showed that there were 4 types of fungal strains that had the ability to produce ligninolytic enzymes, and there were 17 types of fungal strains that had the ability to produce cellulolytic enzymes. From the results of the two qualitative selections on solid media, it can be concluded that there are two potential isolates that have the highest ligninolytic and cellulolytic abilities, namely strains CSF#5 and CSF#15 which are thought to come from the *Ascomycetes* fungi group.

Acknowledgement
The author would like to thank the Indonesian Coffee and Cocoa Research Institute which has facilitated this research activity, both in the form of laboratories and funding costs for analysis services and meeting the needs of tools and materials.

References
[1] Rodrigues, C 2016 Lignocellulosic materials as raw material for obtaining biomolecules of commercial value. Biotechnology *J. Applied to Agroindustry* pp 283–314
[2] Lamid M, Nugroho TP, Chusniati S, Rochiman K 2011 Exploration of cellulolytic bacteria from beef cattle rumen fluid as inoculum material for agricultural waste *J Scientific Veterinary Medicine* **4** 37-42
[3] Alexander, M 1977 *Introduction to Soil Microbiology* (New York: John Wiley and Sons0
[4] Saraswati R, Santosa E dan Yuniarti E 2006 *Organic Fertilizers and Biological Fertilizers: Organisms that Transform Organic Materials* (Center for Agricultural Land Resources Research and Development Agency for Agricultural Research and Development)
[5] Lankinen, P 2004 Ligninolytic enzymes of the basidiomycetous fungi *Agaricus bisporus* and *Phlebia radiata* on lignocellulose-containing media. *Academic Dissertation in Microbiology*. http://www.u.arizona.edu/~leam/lankinen.pdf.
[6] Chen Y, Wang Y, Xu Z, Liu Y, Duan H 2019 Enhanced humification of maize straw and c anola residue during composting by inoculating Phanerochaete chrysosporium in the cooling period *Bioresources Technology* **293** 1-4
[7] Bohacz J, Kowalska TK 2020 Modification of post-industrial lignin by fungal strain of the genus *Trichoderma* isolated from different composting stages *J. Environmental Management* **266** 1-11
[8] Sudirman LI, Alhidayatullah dan Dharmaputra OS 2014 The ability of wood rot fungi isolates

JPA and Trichoderma Sp. S2-2 in degrading oil palm empty fruit bunches to produce cellulose *J. Plantation Tower* **8** 51-56

[9] Gautam SP, Bundela, PS, Pandey AK, Jamaludin, Awasthi MK, Sarsaiya S 2012 Diversity of cellulolytic microbes and the biodegradation of municipal solid waste by a potential strain. *Int. J. Microbiol.* IJOM(2012)325907

[10] Sukmawati 2018 Isolation of cellulolytic bacteria from banana peel waste *Biotropic Journal* **2** 46-52

[11] Gao H, Wang Y, Zhang W, Wang W and Mu Z 2011 Isolation, identification and application in lignin degradation of an ascomycete GHJ-4. *African J. Biotech.,* **10** 4166-4174

[12] Ranimol G, Venugopal T, Gopalakrishnan S, Sunkar S 2018 Production of laccase from *Trichoderma harzianum* and its application in dye decolourisation *J. Biocatalysis and Agricultural Biotechnology* **16** 400-404

[13] Mtui G, Masalu R. 2008 Extracellular enzymes from brown-rot. fungus *Laetiporus sulphureus* isolated from mangroves forests of coastal Tanzania *J. Scientific Research and Essay, 3* 154–161

[14] Kaur H, Kapoor S, Sharma S 2018 An efficient method for qualitative screening of ligninolytic enzyme potential of ganoderma lucidium.*Int J of Current Microbiology & Appl Sci* **7** 2442-2459

[15] Tondok EF, Sinaga MS, Widodo, Suhartono MT 2015 Diversity of Endophytic Fungus on Cocoa Fruit and Its Potential in Controlling Phytophthora Fruit Rot *Dissertation* (Bogor Agricultural Institute)

[16] Fijai A, Mulyaningsih Y, Sepwin NS, Aditya HB. 2019. Detection of ligninolytic capability of isolated fungi from decayed root and stem of oil palm tree. *J of Pysics* 1402 (2019) 033036

[17] Kaal E E, Field J A, Joyce T W 1995 Increasing ligninolytic enzyme activities in several white-rot basidiomycetes by nitrogen-sufficient media *J. Bioresource Technology* **53** 1339

[18] Keyser P, Kirk T K, Zeikus J G 1978 Ligninolytic enzyme system of Phanaerochaete chrysosporium: synthesized in the absence of lignin in response to nitrogen starvation *Journal of bacteriology* **135** 790-7

[19] Ang TN, Ngoh GC and May Chua AS 2011 A quantitative method for fungal ligninolytic enzyme screening studies *Asia-Pac. J. Chem. Eng.* **6** 589–595

[20] Sudiana, Rahayu, Imaduddin, dan Rahmansyah 2001 Cellulolitic Bacteria of Soil of Gunung Halimun Nasional Park *Berita Biologi* **5** 703-710

[21] Apun K, Jong BC dan Salleh MA. 2000. Screening and Isolation of A Cellulolytic and Amylolitic *Bacillus* from Sago Pith Waste. *J of Gen. Appl. Microbiol.* **46** 263 -267

Viability of biofertilizer bacteria *Rhizobium spp* based on household waste

I Erdiansyah[1]*, R Taufika[1], TW Widodo[1], Damanhuri[1], dan DM Jannah[1], and H Prayitno[2]

[1]Department of Agriculture Production, Politeknik Negeri Jember, Jl. Mastrip Po Box 164 Jember, Indonesia
[2]Department of Animal Science, Politeknik Negeri Jember, Jl. Mastrip Po Box 164 Jember, Indonesia

*E-mail: iqbal@polije.ac.id

Abstract. The research was carried out from May 2021 to August 2021. The research was carried out at the Jember State Polytechnic Plant Protection Laboratory. This study used four types of household waste media, namely rice washing water, rice bran water, soybean dregs water, and Yeast Mannitol Agar media as controls. Each treatment was repeated 5 times to obtain 20 experimental units. The data obtained were then analyzed qualitatively and quantitatively. Quantitative data analysis by calculating the number of colonies with TPC (total plate count) and T-test techniques, qualitative data obtained by observing bacterial purity and gram staining. The results of this study indicate that alternative media for bran water waste conventionally produced the highest bacterial population with a colony number of 6,80 x 10^{24} CFU and the lowest bacterial population came from rice alternative media with a colony number of 1.28 x 10^{24} CFU. Bacterial viability observed for 4 months between aseptic media and conventional media showed the results of the t-test with a count of 1.49 x 10^{-1} CFU. shows the results (ns) are not significantly different.

1. Intoduction

Viability is the ability to live a living thing which is indicated by the presence of a growth or biomass. Knowledge of the factors that affect the viability of a microbe is very important to know to treat a microbe. Preservation of *Rhizobium spp.* for long-term purposes is a crucial stage in studying microbiology to maintain and maintain its metabolic activity so that it does not change [1]. [2] states that, "viability is the possibility of a living being to be able to live". [3] Viability is the activity of life, or the possibility of life as indicated by the presence of growth in bacteria. *Rhizobium* is one of the bacteria that can bind nutrients in the form of N when in symbiosis with legume plants [4] in conducting a *Rhizobium* viability test requires a suitable medium as a place for growth and storage. A suitable medium for growth and storage must meet the requirements, namely the adequate intake of nutrients needed by bacteria [5].

YMB (Yeast Mannitol Broth) is a liquid medium that is very effective in the growth and storage of Rhizobium spp. for a long time because YMB (Yeast Mannitol Broth) is a specific medium for the growth of Rhizobium spp. bacteria, but the price of YMB (Yeast Mannitol Broth) instant media is very expensive and also very difficult to obtain among Indonesian farmers, therefore, it takes an alternative media that can grow *Rhizobium spp.* quickly, cheaply and have materials that are easily obtained and available among farmers. This study aims to obtain the best alternative liquid media to restrain the rate of bacterial metabolic activity so that viability or growth power can be maintained to maintain microbial isolates so that they have good growth power and high survival with minimum character changes.

2. Materials and Methods

The research was carried out from February to May 2021, the research was carried out at the Jember State Polytechnic Plant Protection Laboratory. This study used four types of household waste media, namely rice washing water, rice bran water, soybean dregs water, and Yeast Mannitol Agar media as controls. Each treatment was repeated 5 times to obtain 20 experimental units. The data obtained were then analyzed qualitatively and quantitatively. Quantitative data analysis by calculating the number of colonies with TPC (total plate count) and T-test techniques, qualitative data obtained by observing bacterial purity and gram staining.

2.1. Liquid Selective Media Production

YMB is a liquid medium that has specifications (selective media) in growing Rhizobium spp. The following is the composition of YMB (Yeast Mannitol Broth) media aseptically in 1000 ml of distilled water: 10 grams of Mannitol, 0.5 grams of K2HPO2, 0.4 grams of MgSO4.7H20, 0.2 grams of NaCl, 1 gram of Yeast extract. All ingredients were put into an Erlenmeyer containing 1000 ml of aqua dest. When all the ingredients have been mixed in aqua dest, the ingredients are stirred with a stirrer on a hot plate until homogeneous. After all the ingredients are completely dissolved. Erlenmeyer's mouth was closed with aluminum foil and rubber, then sterilized using an autoclave at 121oC with a pressure of 17.5 psi for 15 minutes.

2.2. Liquid Alternative Media Production

The ingredients used as alternative liquid media are rice, bean, and soybean, then rice, bran, tempeh is washed each in a separate container with a ratio of 1 kg of material: 3 liters of water, each jar contains 1 liter of alternative media. Soybean water, rice water, and bran water are each placed in an Erlenmeyer then added a solution of brown sugar as much as 10g/liter and shrimp paste 2.5/liter then filtered, then the ingredients are stirred with a stirrer on a hot plate until homogeneous. After all the ingredients are completely dissolved. Erlenmeyer's mouth was closed with aluminum foil and rubber then sterilized using an autoclave at 121oC with a pressure of 17.5 psi for 15 minutes.

2.3. Bacterial Inoculation

The isolate used was isolated [6] which had the highest adaptation rate of 6.88 x 1030 CFU/g of soil from Ajung District with a depth of 15 cm from the soil surface. Rhizobium spp. which will be inoculated into alternative media and selective media first cultured using a shaker for 4 days using YMB (Yeast Mannitol Broth) media. After 4 days, the inoculant culture was inoculated into a jar containing media.

2.4. Bacterial Purity and Preparation of YEMA + Congo Red Media Media

The materials that will be used in the manufacture of YEMA + Congo Red media are weighed using a digital Ohaus balance according to the composition of each ingredient. The materials used were 3 g Yeast Extract, 10 g Sucrose, 0.2 g MgSO4 (Magnesium Sulfate), 0.1 g NaCl, 0.5 g K2HPO4, 3 g CaCO3, 20 g Agar and 0.025 g Congo Red were weighed. using a digital Ohaus balance for each treatment. then the solution was homogenized using a hot plate stirrer equipped with a magnetic stirrer, then the pH was checked using a pH meter. The desired pH by Rhizobium is a pH of 6.8. after the solution is homogeneous and then cooked at a temperature of 1000C. After boiling, it was placed in an Erlenmeyer and autoclaved for 15 minutes at a temperature of 121 C and a pressure of 17.5 psi.

2 r

Gram staining is used to determine the bacteria *Rhizobium spp.* this is a type of gram-negative bacteria or gram-positive bacteria by first sterilizing the preparation by spraying it with 96% alcohol and then rinsing with distilled water, dividing the solution into 4 kinds, namely crystal violet solution A, B iod solution, 96% alcohol C solution, D safranin solution After that, take 1 ose of bacterial isolate and scratch it on the surface of the preparation until it is evenly distributed and then fixation is carried out. Take 1 drop of solution A (crystal violet) to add on the surface of the preparation that has been coated with bacteria and then let stand for 1 minute. Then the preparations were rinsed with distilled water and added 1 drop of B (iod) solution on the surface of the preparations and allowed to stand for 1 minute. The preparations were rinsed with aqua dest and then dripped with solution C (96% alcohol) and allowed to stand for 30 seconds, after that, added 1 drop of D solution (safranin) on the surface of the preparations and allowed to stand for 1 minute then the preparations were rinsed with distilled water and waited until dry.

3. Results and Discussion

Rhizobium spp. which has been stored at room temperature, viability was observed by calculating the number of bacterial colonies that grew. Calculations were carried out using the plate count method and dilution was carried out before being grown on YEMMA medium so that after incubation several colonies would be formed on the plate in countable amounts. The petri dish is used to count the number of bacterial colonies in a petri dish containing 30-300 colonies [7]. The effectiveness of *Rhizobium spp.* can survive if the microorganism can be active in a variety of different environmental conditions and survive in various forms. These bacteria must meet several criteria, including: (1) can be mass-produced; (2) remain stable and viable for a long time in storage conditions; (3) can survive if symbiotic with legume plants; and (4) when applied it has a beneficial effect on the host. After storage at room temperature according to treatment time and grown in YEMMA media, the number of colonies was calculated after 24 hours of incubation. Colony growth of Rhizobium spp. in the viability test can be seen in Figure 1. Based on Figure 1. it is known that different time treatments produce different variations in the number of populations. Changes in the number of populations are caused by the growth of bacteria in the media [8]. Based on the graph, there are 3 forms of growth rate, namely increasing growth rate, optimum growth rate, and decreasing growth rate.

Table 1. Bacterial Viability Rate for 4 Months

Treatment (Conventional/Aseptic)	Rhizobium Spp. Bacteria Population/Sample	The Average Population of Rhizobium Spp. / Samplel
R0 (YMB Conventional 1)	3,53X 10^{24} CFU/ml	3,44X 10^{24} CFU/ml
R0 (YMB Conventional 2)	3,58X 10^{24} CFU/ml	
R0 (YMB Conventional 3)	3,30X 10^{24} CFU/ml	
R0 (YMB Conventional 4)	3,34X 10^{24} CFU/ml	
R1 (Soybean Dregs Water 1)	5,00X 10^{24} CFU/ml	6,31X 10^{24} CFU/ml
R1 (Soybean Dregs Water 2)	6,81X 10^{24} CFU/ml	
R1 (Soybean Dregs Water 3)	6,94X 10^{24} CFU/ml	
R1 (Soybean Dregs Water 4)	6,48X 10^{24} CFU/ml	
R2 (Rice Washing Water 1)	1,05X 10^{24} CFU/ml	1,22X 10^{24} CFU/ml
R2 (Rice Washing Water 2)	1,45X 10^{24} CFU/ml	
R2 (Rice Washing Water 3)	1,28X 10^{24} CFU/ml	
R2 (Rice Washing Water 4)	1,10X 10^{24} CFU/ml	
R3 (Rice Bran Water 1)	6,53X 10^{24} CFU/ml	6,80X 10^{24} CFU/ml
R3 (Rice Bran Water 2)	6,93X 10^{24} CFU/ml	

R3 (Rice Bran Water 3)	6,75X 10^{24} CFU/ml	
R3 (Rice Bran Water 4)	6,99X 10^{24} CFU/ml	
S0 (YMB Aseptic 1)	3,00X 10^{24} CFU/ml	3,07X 10^{24} CFU/ml
S0 (YMB Aseptic 2)	3,26X 10^{24} CFU/ml	
S0 (YMB Aseptíc 3)	3,08X 10^{24} CFU/ml	
S0 (YMB Aseptic 4)	2,93X 10^{24} CFU/ml	
S1 (Soybean Dregs Water 1)	4,35X 10^{24} CFU/ml	5,34X 10^{24} CFU/ml
S1 (Soybean Dregs Water 2)	5,55X 10^{24} CFU/ml	
S1 (Soybean Dregs Water 3)	5,78X 10^{24} CFU/ml	
S1 (Soybean Dregs Water 4)	5,66X 10^{24} CFU/ml	
S2 (Rice Washing Water 1)	1,24X 10^{24} CFU/ml	1,28X 10^{24} CFU/ml
S2 (Rice Washing Water 2)	1,29X 10^{24} CFU/ml	
S2 (Rice Washing Water 3)	1,33X 10^{24} CFU/ml	
S2 (Rice Washing Water 4)	1,24X 10^{24} CFU/ml	
S3 (Rice Bran Water 1)	6,08X 10^{24} CFU/ml	6,59X 10^{24} CFU/ml
S3 (Rice Bran Water 2)	6,91X 10^{24} CFU/ml	
S3 (Rice Bran Water 3)	6,59X 10^{24} CFU/ml	
S3 (Rice Bran Water 4)	6,79X 10^{24} CFU/ml	

Note : CFU/ml = Jumlah koloni per unit dalam 1 ml media
6,80 X 0^{24} = The highest bacterial population came from aseptic bran alternative media
1,28 X 10^{24} = The lowest bacterial population comes from aseptic rice alternative media

The growth rate increased in two alternative media, namely wastewater bran, and soybean water waste with a storage time of 4 months, on alternative media for wastewater bran and soybeans, the results obtained were conventional bran water waste from the first month of 6,53 x 10^{24} CFU menjadi 6,99 x 10^{24} CFU in the fourth month so that it increased by 7%, alternative media for aseptic bran water waste in the first month was 6.08 x 10^{24} CFU to 6.79 x 10^{24} in the fourth month so that it increased by 13%, alternative media for soybean water waste conventional method in the first month 5,00 x 10^{24} CFU to 6,48 x 10^{24} CFU in the fourth month so that it increased by 36%, alternative media for aseptic soybean wastewater in the first month of 4,35 x 10^{24} CFU to 5,66 x 10^{24} CFU in the fourth month so that it increased by 30%.

According to [9], at the time of growth, the speed of division is high, the metabolic activity of cells is high, and the generation time is short. The high increase was due to the bacteria synthesizing substances contained in the bran and soybean media which could trigger the bacteria to secrete their cell metabolites for optimal cell growth. The high protein content in bran and soybean media can be used as a source of nutrition for bacteria during storage.

The optimum growth rate occurred in alternative media for conventional and aseptic rice with a storage time of 4 months, on alternative media for conventional rice water waste the results obtained were 1,05 x 10^{24} CFU to 1,10 x 10^{24} CFU in the fourth month so that it increased by 4%, the increase of only 4% is still in the category of stagnant growth so that the growth rate is optimum. In the alternative media of aseptic rice water waste, the results obtained were from the first month of 1,24 x 10^{24} CFU to 1,24 x 10^{24} CFU in the fourth month so that there was an increase of 0% so that in this alternative media the optimum bacterial growth did not decrease and increase in the fourth month but increased in the second and third months.

Optimum growth of bacteria is caused by the substrate in the bran carrier medium that is still able to provide nutrition or is still supportive for the life of the growing bacterial population. In line with

[10], that the climax point in the number of bacterial populations depends on the amount of food/nutrients, maximum cleavage speed, and high metabolic activity in synthesizing substances contained in the rice carrier medium. The high growth of bacteria in a medium indicates the high viability of bacteria in rice carrier media.

The growth rate decreased in conventional and aseptic YMB (Yeast Mannitol Broth) selective media with a storage time of 4 months, on conventional YMB (Yeast Mannitol Broth) selective media the results obtained in the first month were 3,53 x 10^{24} CFU to 3,34 x 10^{24} CFU in the fourth month so that it increased by 5.3%. On aseptic YMB (Yeast Mannitol Broth) selective media the results obtained were from the first month of 3,00 x 10^{24} CFU to 2,93 x 10^{24} CFU in the fourth month increasing by 2.5%. The population decline was due to a lack of growth factors such as a decrease in the supply of nutrients contained in YMB (Yeast Mannitol Broth) media.

According to [11], the decline in the number of populations in an organism is due to the reduced supply of food contained in a medium, with a reduced supply of limited food, the population of organisms will decrease drastically. This decrease in population is due to the formation of dead cells faster than the formation of cells. -new cells. In line with [10] that decreased growth rate can occur due to increased mortality or decreased reproduction of an organism or both can occur simultaneously. According to [9], such a situation would indicate the existence of competition among organisms for nutrients and space and was also caused by the materials released by cells into the medium.

According to [12], the length of storage time affects the growth and survival (viability) of bacteria with the occurrence of cell lysis which causes turbidity and the number of cells counted directly will decrease. According to [13] almost all the bacteria in their viability will begin to decrease when the storage time of bacteria exceeds the 3-month storage time limit.

The existence of three forms of bacterial growth rate indicates that there are factors that affect the growth of bacteria in a medium. According to [14] the factors that influence bacterial growth, in general, include storage time, nutrient availability, water density, temperature, dissolved oxygen, pH. Bacteria will do binary fission every 20 minutes and bacteria that do fission need nutrients for their division. If nutrition is inadequate, then growth will decline. Water is needed by bacteria because water plays a role in metabolic reactions in cells.

Bacteria generally grow and reproduce in media with high aw (water activity) (0.91), therefore a medium that is too concentrated causes cells to lack water and dies because the growth of the bacterial population is affected by the length of storage time. According to [15], population growth is a function of time (Nt = f(t)) meaning that bacterial population growth is strongly influenced by the length of time the bacteria live in a medium. So, if the bacteria's lifetime is not too long in a medium, the growth of the bacterial population will be small.

Based on Table 1 the results of the viability test by observing and counting the number of colonies from the eight media used in bacterial propagation, on media treated aseptically and conventionally showed that the highest mean bacterial population observed for 4 months was in conventional bran wastewater media (R3), which is 6,80 x 10^{24} CFU and the lowest bacterial population is found in aseptic rice water waste media (S3), which is 1,28 x 10^{24}CFU.

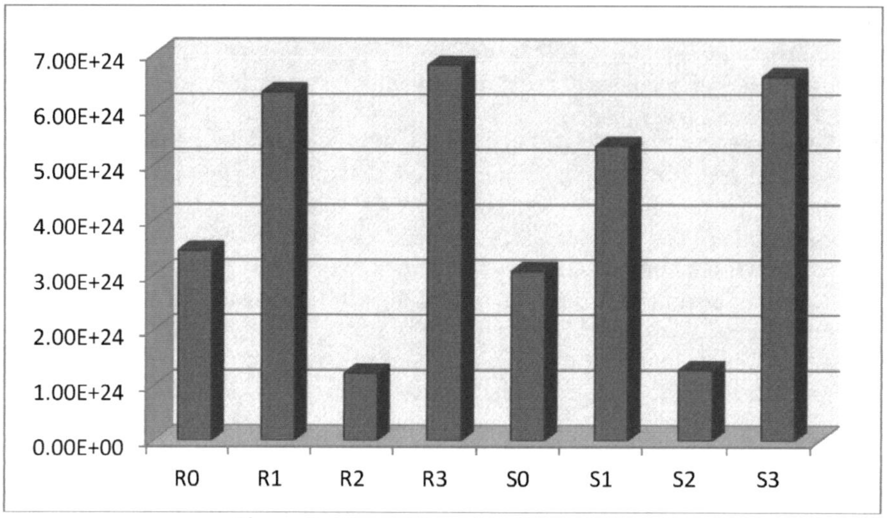

Figure 1. Average Bacterial Population

The growth of aseptic soybean liquid alternative media (S1) was not as high as the results of conventional soybean liquid alternative media (R1) which experienced a significant increase in population. The comparison between the first month to the second month is 36% of the total population in the first month 5,00 x 10^{24} CFU and in the second month 6,81 x 10^{24} CFU until it reaches the peak of the population in the third month which is 6,94 x 10^{24} CFU and back decreased in the fourth month, namely 66,48 x 10^{24} CFU, so that the average can be taken which is 6,31 x 10^{24} CFU so that the difference in results between alternative media for soybean water waste and the highest bacterial population is in alternative media for bran water waste which has a total population 6,80 x 10^{24} CFU is just a 7% difference.

Based on the growth pattern in Figure 1. then the growth of Rhizobium spp. The alternative media for soybean wastewater (R1) found third place after alternative media for aseptic (S4) and conventional (R4) bran, which experienced a growth rate of 36% from the first month of storage. This is because the bacteria experienced regrowth after experiencing a decline in the growth graph. [16] The ability of Rhizobium spp. bacteria to grow again is due to the presence of a source of nutrients originating from other bacterial cells that have died, so that the new nutrient source can be utilized by bacteria as a metabolic material for these bacterial cells. Bacterial bodies undergo decomposition so that they can be used as a source of energy for the remaining living bacteria to grow again.

Based on the results of the t-test analysis of the number of bacterial populations between the two treatments of aseptic media and conventional media during the viability test, which was carried out for 4 months, the results were 1.49 x 10^{-1} CFU. This shows that the effect of aseptic media and conventional media in bacterial propagation was not significantly different. The comparison between aseptic and conventional treatment media is said to be not significantly different considering that aseptic treatment is so complicated because it requires many installations such as electrical installations, hoses and requires expensive equipment such as autoclaves and ovens in the sterilization process which differs greatly from conventional treatment which is so simple. so that conventional treatment is easier to reach by farmers who will propagate *Rhizobium spp* [17].

4. Conclusion

Alternative media for bran water waste conventionally produced the highest bacterial population with a colony number of 6.80 x 10^{24} CFU and the lowest bacterial population came from aseptic rice alternative media with a colony number of 1.28 x 10^{24} CFU. Viability of bacteria observed for 4 months between aseptic media and conventional media showed the results of the t-test with t count 1.49 x 10^{-1} CFU showed the results (ns) were not significantly different.

IOP Conf. Series: Earth and Environmental Science **980** (2022) 012009 doi:10.1088/1755-1315/980/1/012009

Acknowledgment
We would like to thank the Head of The Research Center and Community Service who has funded this research through the State Polytechnic of Jember, Fiscal Year 2021 contract number 372/PL17.4/PG/2021

References
[1] Partanto dan Dahlan. 1991. *Kamus Ilmiah Popular.* Surabaya: Arloka
[2] Kamisa. 1997. *Kamus Lengkap Bahasa Indonesia.* Surabaya: Kartika
[3] Pritanti, L. 1995. *Uji Viabilitas Candida Tropicalis, Strain G XIII 2.A. Terhadap Senyawa Fenol Pada Medium Air Laut Sintetik.* Laporan Penelitian Magang Balibang Mikrobiologi. Bogor: LIPI
[4]. Sari, R. Dan Prayudyaningsih, R., 2015. *Rhizobium*: *Pemanfaatannya Sebagai Bakteri Penambat Nitrogen.* Balai Penelitian Kehutanan, Makassar. Info Teknis Eboni. Vol. 12 No.1, Juli 2015: 51 – 64
[5] Atlas, Ronald M. 2004. *Handbook of Microbiological.* Media fourth Edition Volume 1. United States Of America: CRC Press.
[6] Fajrin, V. N., Damanhuri, I.Erdiansyah. 2017. “Koleksi Dan Identifikasi Bakteri Penambat N Pada Pusat Lokasi Tanaman Kedelai Edamame (Glycine Max (L.) Merr.) Di Kabupaten Jember.” *Agriprima* 1(2):158–69.
[7] Hadioetomo, R.S. 1990. *Mikrobiologi Dasar Dalam Praktek.* Jakarta: PT Gramedia.
[8]. Valentina R.P, 2005. *Uji Viabilitas Mikroorganisme Efektif BIOEDU_UNS-I Pada Berbagai Media Carrier.* Surakarta: Skripsi S1 Pendidikan Biologi FKIP UNS.
[9] Mavitra Ellanvihara. 2005. *Pertumbuhan dan Aktivitas Eksoprotease Bacillus Licheniformis dan Bacillus Megaterium di Medium Ekstrak Limbah Padat Udang.* Skripsi S1 Biologi FMIPA UNS Surakarta.
[10]. Naughton dan Larry L. W. 1990. *Ekologi Umum.* Yogyakarta: Gadjah Mada University Press.
[11] Pelczar J.M, Chan E.S.C. 1986. *Dasar-dasar Mikrobiologi 1.* Jakarta: UI Press.
[12]. Darmawan Ari N. 2000. *Mikrobiologi Industri. www.tip.ugm.ac.id.* [12 Juni 2006].
[13] Widyastuti, 1995. *Proyek Penelitian dan Pengembangan Bioteknologi.* Pusat Penelitian Bioteknologi-LIPI. *www.dbriptek.lipi.go.id.* 12 Juni 2006.
[14] Gaman PM dan Sherrington KB. 1992. *Ilmu Pangan Pengantar Ilmu Pangan Nutrisi dan Mikrobiologi.* Yogyakarta: Gadjah Mada University Press.
[15] Robert, W. Poole. 2000. *An Introduction To Quantitative Ecology.* Cornell University.
[16] Erdiansyah, I., Sari, V. K., Pratama, A. W., & Wiharto, K. S. (2020). Utilization of Rhizobium spp as substitution agent of nitrogen chemical fertilizer on soybean cultivation. In *IOP Conference Series: Earth and Environmental Science* (Vol. 411, No. 1, p. 012065). IOP Publishing.
[17]. Erdiansyah, I. "Performance of Resistance of Rice Varieties Recommendation of Jember Regency to Brown Planthopper Pest (Nilaparvata lugens Stal.)." *IOP Conference Series: Earth and Environmental Science.* Vol. 207. No. 1. IOP Publishing, 2018.

Intercropping patterns effect of cardamom and cayenne pepper on population and intensity of whitefly pest's attacks (*Bemisia tabaci*)

I Erdiansyah [1*], E Eliyatiningsih [1], V K Sari [2], D Nurahmanto [3], and A H Prayitno [4]

[1] Department of Agriculture Production, Politeknik Negeri Jember, Jl. Mastrip Po Box 164 Jember, Indonesia
[2] Faculty of Agriculture, Universitas Jember, Jl. Kalimantan No. 37, Sumbersari, Jember 68121, Indonesia
[3] Faculty of Farmacy, Universitas Jember, Jl. Kalimantan No. 2, Sumbersari, Jember 68121, Indonesia
[4] Department of Animal Science, Politeknik Negeri Jember, Jl. Mastrip Po Box 164 Jember, Indonesia

* email: iqbal@polije.ac.id

Abstract. The study aimed to determine the population and intensity of attacks caused by whitefly pests on intercropping and monoculture cultivation areas. This research was conducted from June to August 2021 in Pace Village, Silo District, Jember Regency. The method used is to compare two intercropping and monoculture cultivation areas of medicinal plants cardamom - cayenne pepper. The first plot was cultivated in monoculture with pest control using an insecticide with the active ingredient Alfametrin 15 EC, while in the second plot, cayenne pepper - cardamom intercropped cultivation with a spacing of 1x2 meters and pest control without pesticides. A systematic random sampling of 50 plants in both cropping patterns on cultivated land. Observation of pest populations using insect nets, pitfall traps, and insect suction in the field. The data were analyzed using SPSS version 15.0. The data were tested for normality using Kosmorogrof Smirnov with non-parametric analysis. The results showed that the whitefly population in monoculture cayenne pepper cultivation had a higher average of 57.58 individuals, while in cayenne-cardamom intercropping cultivation, the population average was 51.00 and showed significantly different results. The intensity of whitefly attack showed significantly different results in monoculture cultivation of 45.58% while intercropping pattern of 50.10%.

1. Introduction

Maintenance of a cultivated plant by applying technical cultural control in the form of the use of an intercropping system that aims as a barrier in overcoming pests and diseases of cultivated plants, especially cayenne pepper. The production of cayenne pepper in Jember Regency 2019 based on Jember in Figures issued by the Central Statistics Agency [1] Jember reached 215,723 quintals when converted into kilograms, it was 21.5 million kilograms, but this amount was not sufficient for regional chili needs in Jember Regency. One of the problems in increasing the production of cayenne pepper, especially on a national scale, is the attack of pests and plant diseases.

Pest attacks can reduce the growth and yield of cayenne pepper plants by up to 75% and even result in crop failure if not controlled properly [2]. So far, the number of insects that have been identified and proven to be associated with cayenne pepper plants has reached 200 species, which are divided into 154 types of pests, 59 species of predators, and 44 species of parasitoid insects. The pests that most often attack cayenne pepper plants are whitefly or Bemisia tabaci. Until now, the most frequently used business is chemical control, the use of chemical insecticides used to control whitefly pests such as Karbosulfan and others. Chemical pesticides have not succeeded in controlling whitefly pests in cayenne pepper cultivation and some chemical pesticides have not succeeded in controlling whitefly pests, causing these pests to become resistant [3]. Therefore, an integrated pest control method is needed to reduce or limit the use of chemical pesticides to reduce adverse effects.

The concept of integrated pest management is a control technique and tactic that utilizes plant cultivation using technical culture, namely intercropping cultivation of cultivated plants as an alternative step to control whitefly pests (Bemisi tabaci) compared to monoculture cultivation. Increased activity of useful insects such as predators and pest parasitoids around intercropping plants [4]. Intercropping between chili peppers and herbal plants can suppress pest populations on cultivated land and can increase crop yields. The combination of techniques or tactics of optimal soybean pest control technology components is determined based on knowledge of appropriate information about soybean pests, ecosystems, and socioeconomics based on the IPM approach [5]. This study was conducted to utilize the intercropping pattern of cayenne pepper and cardamom on cultivated land, and another objective was to determine how effective the intercropping pattern was in suppressing the population and intensity of attack on cultivated cayenne pepper.

2. Materials and Methods

The research was carried out in June – August 2021 at the Plant Protection Laboratory of the Jember State Polytechnic and in Pace Village, Silo District, Jemer Regency, East Java Province. It is located at coordinates 11330 West Longitude – 114 East Longitude, 8LU-830LS. The research method used is non-parametric statistics based on the assumption of population distribution by comparing two chili cultivation areas that are intercropped with cardamom plants and chili pepper plants in monoculture.

2.1. Plant Cultivation Techniques

Land preparation is done by hoeing the soil to make it lose, then measuring the land as much as two beds, namely the first bed for intercropping treatment and the second bed for monoculture treatment, each stretch of land measuring 40 x 40 meters. Planting was done by planting two seeds per planting hole and one cardamom tuber next to chili plants with a size of 70 cm x 70 cm x 2 meters for intercropping treatment and cayenne pepper plants 70 cm x 70 cm in monoculture treatment.

2.2. Fertilizing and Laying Silver Mulch

Fertilization intercropping was carried out using NPK Mutiara. Fertilizer was applied twice at the age of 15 DAP and 35 DAP, while for cardamom plants it was done twice at the age of 25 DAP and 60 DAP using NPK Phonska fertilizer. Mulch is installed to protect plants from weeds and to maintain soil moisture after the beds are formed.

2.3. Observation of Whitefly Population

This is done by counting the number of whiteflies in the plant sample. Observations begin when the plants are two weeks old after planting, every two weeks until before harvest.

2.4. Observation of Whitefly Attack Intensity

The intensity of the whitefly attack was observed every two weeks when the plants were two weeks after planting. The calculation of the intensity of the whitefly pest attack is carried out by calculating the damage score of plant parts using the formula [6].

$$P = \frac{\Sigma(nxv)}{Z\ x\ N}\ x\ 100\% \qquad (1)$$

Information:
P = Intensity of attack/Damage
n = Number of samples observed for each category of damage
v = Score value for each category of damage
Z = The highest damage category score
N = Number of observed sample units

Observation of score values for plant leaf damage:
Score 0: No damage to plant leaves was observed
Score 1: There is less than 25% damage to the observed leaves
Score 2: There is damage between 25% - 50% on the observed leaves
Score 3: There is damage between 50% - 75% of the observed leaves
Score 4: There is more than 75% damage to the observed leaves

2.5. Insect Identification
Insects caught in insect nets and Yellow Pantrap were counted and identified in the plant protection laboratory of the Jember State Polytechnic.

2.6. Data Analysis
The data that has been obtained is processed statistically using the SPSS 15.0 application which includes a normality test using Kosmorogrof Smirnov, conducting a homogeneity test using One Way Anova and if there is a normal and homogeneous data distribution then it is continued with the Paired T-Test test, and if the distribution is not normal and not homogeneous or heterogeneous, then proceed with the Mann Whitney test.

3. Results and Discussion

3.1. Whitefly (Bemisia tabaci) Population
Observation of the whitefly population was carried out once a week for each sample of cayenne pepper on two lands with different planting systems.

Table 1. Whitefly Population

Treatments	Average (±SD)
Monoculture	$57{,}58{\pm}5{,}04^{a}$
Intercropping	$51{,}00{\pm}4{,}15^{b}$

Note: The data followed by different letters are significantly different according to the Mann Whitney test ($p<0.05$)

According table 1. to Mann Whitney, the cultivation technique between monoculture and intercropping shows that the whitefly population is significantly different, according to [7], stated that the use of cardamom plants can be a barrier to whitefly pests and become a place for natural enemies to live so that whitefly populations can survive. controlled naturally and tends to be low compared to monoculture cultivated land.

The whitefly has a high reproductive potential and quickly develops resistance to insecticides. Yield loss due to whitefly attack due to loss of nutrient fluid, extensive soot dew on soybean pods and leaves not filled [8]. The development of the whitefly population is influenced by the host and the microclimate. Research revealed that the personality value of whitefly on weeds at 25°C and 29°C was higher than in chili plants. The low population and intensity of whitefly attacks on soybeans are thought to be caused by local climatic conditions that do not support the development of whitefly pests. [9], stated that the factors that often trigger the development of pests are diversity in cultivation practices which include unequal planting time, seed quality, lack of water availability, soil fertility, and level of control [10].

3.2. Color Intensity of Whitefly Attack
The intensity of whitefly attack on the two treatments of intercropping system and monoculture of cayenne pepper showed significantly different results.

Table 2. Intensity of Whitefly Attack

Treatments	Average (±SD)
Monoculture	45,58±3,56[a]
Intercropping	50,10±2,03[b]

Note: The data followed by different letters are significantly different according to the Mann Whitney test ($p<0.05$)

The table 2 above shows that the intensity of the whitefly attack was not significantly different between monoculture and intercropping land. This difference can be seen from the maintenance of plants when applying pesticides on cultivated land wherein monoculture treatment the application of the pesticide Alfamethrin is carried out where the nature of this pesticide is as an antifeedant, which gives an effect that can prevent insects from eating plants that are applied to pesticides so that the intensity of whiteflyattacks can be controlled.

According to [11], The presence of natural enemies can suppress the presence of whitefly pests on cultivated land, especially ladybugs on intercropping land. Insects found in the intercropping area were whitefly, Trips tabaci, green ladybug, and pollinator butterfly. On cultivated land, natural predators of the whitefly were found, namely, the dome beetle (Menochilus sexmaculatus) which can prey on 200-400 whitefly nymphs, M. sexmaculata can prey on up to 51.5 whiteflies in 24 hours. In addition, the beetle Delphastus sp, sharp-eyed spider / Oxyopes was also found [12].

3.3. Cultivated Crop Yield
The weight of the sample of cayenne pepper on the two cultivated lands has a significant difference whereas cayenne pepper on monoculture land has a higher yield of 45.80 gr. while on intercropping land the yield is 1,424 g.

Table 3. Intensity of Whitefly Attack

Treatments	Average (±SD)
Monoculture	53,58±5,04[a]
Intercropping	40,40±4,15[b]

Note: The data followed by different letters are significantly different according to the Mann Whitney test ($p<0.05$).

The difference in yield between the two cropping patterns of intercropping and monoculture is caused by the light intensity reception. Table 3 In monoculture chili land, it received more sunlight than intercropping which experienced many etiolated plants because cardamom plants were higher than cayenne pepper.

The second factor is the level of damage to cayenne pepper plants on intercropping cultivation land which has lower yields due to attacks by other pests such as fruit flies [13].

4. Conclusion

The conclusions in this study are the population of whitefly on monoculture cultivated land has a higher average of 57.58 tails individuals while on intercropping cultivation land has a population average of 51.00 tails individuals with significantly different results, the intensity of whitefly attack showed significantly different results with a value of 45.58% in the monoculture treatment and 50.10% in the intercropping treatment, and the yield showed a significantly different value between monoculture treatments, namely 53.58 g and intercropping 40.40 g.

Acknowledgment

We would like to thank the Head of The Research Center and Community Service who has funded this research through the State Polytechnic of Jember, Fiscal Year 2021.

References

[1] Badan Pusat Statistik. 2020. Jember Dalam Angka. https://jemberkab.bps.go.id/publication/2020/05/20/c43569a520090bef8f9b8919/kabupaten-jember-dalam-angka-2020.html.

[2] Marwoto dan A. Inayati. 2011. Kutu Kebul: Hama Kedelai yang Pengendaliannya Kurang Mendapat Perhatian. Iptek Tanaman Pangan 6(1): 87–98.

[3] Inayati, A. dan Marwoto. 2012. Pengaruh Kombinasi Aplikasi Insektisida dan Varietas Unggul terhadap Intensitas Serangan Kutu Kebul dan Hasil Kedelai. J. Penelitian Pertanian Tanaman Pangan 31(1): 13–21.

[4] Subagyo, V.N.O. dan P. Hidayat. 2014. Neraca kehidupan Kutukebul Bemisia tabaci (Gennadius) (Hemiptera: Aleyrodidae) Pada Tanaman Cabai dan Gulma Babadotan pada Suhu 25 °C dan 29 °C. J. Entomologi Indonesia 11(1): 11–18

[5] Agastya I M I, Julianto R P D dan Marwoto. 2020. Review : Pengaruh Pemanasan Global Terhadap Intensitas Serangan Kutu Kebul (Bemisia Tabbaci Genn) Dan Cara Pengendaliannya Pada Tanaman Kedelai. Buana Sains Vol 20 No 1 : 99-110. http://repository.unitri.ac.id/1665/1/Jurnal%20Buana%20Sains%201.pdf.

[6] Digilib, its. 2003. Pengendalian Hama Tanaman. Available at :http://digilib.its.ac.id/public/ITS-Undergraduate-10642-Chapter1.pdf. Diakses 8 juni 2010.

[7] Marwoto, S. Hardaningsih, dan A. Taufik 2008. Hama, penyakit dan masalah hara pada tanaman kedelai. Identifikasi dan Pengendaliannya. Puslitbangtan. Bogor. 66 p.

[8] Berlinger, M. J. 1986. Host Plant Resistance to Bemisia tabaci. Agriculture, Ecosystems & Environment 17(1–2): 69–82.

[9] Marwoto, F.C. Indriani, A. Sulistyo, dan R.T. Hapsari. 2011. Diagnosis Ledakan Populasi Hama Kutu Kebul Bemissia tabaci pada Pertanaman Kedelai (Studi Kasus Faktor Penyebab Ledakan Populasi Kutu Kebul di KP Muneng MK 2009). Hlm. 277–288. Dalam A. Setyorini dan Marwoto: Kutu Kebul, Sistem Pengairan, dan Budidaya Kedelai 262.

[10] Sulistiyo Dwi Setyorini dan Marwoto. 2017. Perkembangan Populasi dan Serangan Kutu Kebul pada Kedelai dengan Sistem Pengairan dan Teknik Budidaya Berbeda. Prosiding. Balai Penelitian Tanaman Aneka Kacang dan Umb. Malang. https://balitkabi.litbang.pertanian.go.id/wp-content/uploads/2017/07/pros16_31.pdf.

[11] Widjono, Hermanto, M. Muchlish Adie, Y. Prayogo, Suharsono, Sholikin, A.A, Rahmianna, N. Nugrahaeni, N. Saleh, A. Kasno, Subandi, dan Marwoto (peny.). Prosiding Serninar Nasional Hasil Penelitian Tanaman Kacang-kacangan dan Umbi-umbian Tanggal 21 Desember 2009.

[12] Erdiansyah, I., Eliyatiningsih, E., Sari, V. K., & Nurahmanto, D. (2021, March). Utilization of javanese ginseng and citronella for insect diversity in Pace village, Jember regency. In *IOP*

Conference Series: Earth and Environmental Science (Vol. 672, No. 1, p. 012096). IOP Publishing.

[13] Erdiansyah, I., Damanhuri, F. N. U., & Azizah, P. Z. (2019). Isolasi dan Identifikasi Cendawan Mikoriza Arbuskular Pada Tiga Pusat Lokasi Tanaman Jagung di Kabupaten Jember. *Agriprima: Journal of Applied Agricultural Sciences*, *3*(2), 41-50.

IOP Conf. Series: Earth and Environmental Science **980** (2022) 012011 doi:10.1088/1755-1315/980/1/012011

The growth inducing endophyte: optimizing the non-host endophyte as growth promoter

Sugiyarto[1], C Trwidiarto[1], Supriyadi[1], R Firgiyanto[1*], H S Addy[2,3], A Wafa[2,3]

[1] Department of Agricultural Production, State Polytechnic of Jember, Jl. Mastrip, Jember, Indonesia
[2] Plant Protection Study Program Department of Plant Protection Faculty of Agriculture University of Jember, Indonesia, Jl. Kalimantan No 37 Jember, Indonesia
[3] Center of Development Center for Development of Advanced Science and Technology (CDAST), University of Jember, Indonesia

refa_firgiyanto@polije.ac.id

Abstract. The plants present a complex, spatially and temporally diverse ecological habitat. As living support of human, microorganism, microorganism and fungal. They are life in the host plant as endo, ecto, semi endo or other kind living mechanism. The symbiotic associations between microorganisms and plants are ancient and fundamental. Most of plant protection scientist develop the broad range of host in endphopytic microbes for the future purposes, such as biological control agents, growth inducer, the plant resistance inducer and soon. However, the development non true host of endophytic microbes are challenging. The research was done by isolating potential endophyte from three toga plant and then tested to spinach. The growing and plant pathogenic possibility test was done to eliminating the potential plant pathogen microbes. From 40 isolate, was detected more than 20 isolate have possibility to be a pathogen of Spinach, with different symptoms appearance level. The level of growth induced result also have different result among the potential endophyte. They appear on the growing time, the long of root and shoot number of spinach. The optimizing of growth promoter endophyte especially for non-host endophyte, need the extra path for convinced the isolated endophyte not to be potential plant pathogen.

1. Introduction

The plants present a complex, spatially and temporally diverse ecological habitat. The root and other part of plant became as living support system for macroscopic organisms, microbial communities and fungal. The symbiotic associations between microorganisms and plants are ancient and fundamental. Endophytic microbes are an intriguing group of organisms associated with various tissues and organs of terrestrial and some aquatic plants, and are the subject of increasing interest to mycologists, ecologists, and plant pathologists.

In general, "endophytic" infections are inconspicuous, the infected host tissues are at least transiently symptomless, and the microbial colonization can be demonstrated to be internal, either through histological means, by isolation from strongly surface disinfected tissue, or, most recently, through direct amplification of fungal nuclear DNA from colonized plant tissue. Infections of land plants by endophytes are ubiquitous, having been found throughout a broad range of host orders, families, and genera worldwide, and representing a diverse array of terrestrial and aquatic habitats [1].

Most of scientist develop the broad range of host in endphopytic microbes for the future purposes, such as biological control agents, growth inducer, the plant resistance inducer and soon. It's isolated from a sample plant and optimizing for other plant, not only in same genus but also for some long distance relationship plant. For axample, the fungal endophytic from aquatic legumes, *Rhizobium leguminosarum* bv. Trifolii from their aquatic legumes, was use on aplicate them into rice. As the result the inducer reaction was appear on the infected host [2].

The same pathways are used in this project. Fungi and bacterial endophite was isolate from three different family medicinal plants or in Indonesia called Toga. There was Gotu Kola or Pegagan (*Centella asiatica*), the cat whiskers plant or kumis kucing (*Orthosiphon aristatus*) and Drumstick tree or Kelor (*Moringa oleifera*). The fungal and bacterial endophyte from thet plant were tested on spinach (*Spinacia oleracea*).

2. Material and Methods

2.1. Microbial Endophytic Isolation

The isolation methodology of endophytic fungi from three different described plant performed by methods described by Braun *et al.* [3]. Hyphaltips were taken from initial isolates and transferred onto plates of potato dextrose agar (PDA, Hi Media India), Malt Ekstract Agar (MEA, Diftco Franklin Lakes, N.J.) and Tryptic Soy Agar (TSA, Merk), maintained at 18 8C or room temperature, and transferred every few months. Each isolates were selected for further morphological characterization based upon differences in gross colony morphology. The bacterial endophytic was isolate by maceration technique.

2.2. Endophytic Screening

For fungal morphological examination, isolates were cultured on PDA, and water agar. Cultures were incubated the 14 h light (dark) at 22-28°C. Following 2 weeks incubation, isolates were examined for colony morphology and morphological characteristics of the conidia and conidiophores. Conidia and conidiophores were taken approximately 2 mm inside the colony margin, mounted in water on a microscope slide, and observed with an Olympus BX61 microscope at 40x magnification.

The bacterial endhophyte was examine by bacterial colony examination in TS Agar medium. Than a microscope slide, and observed the colony shape and other morphological type. The first screening of microbial endophytic was done by seed pathogenicity test. By follow [4] methodology for seed health. The *S. oleracea* seed were tested individually on each isolated endophyte. The necrotic reaction in the test indicated the microbial endophytic have a possibility as plant pathogenic in spinach seedling. And for that microbes will not use for more purpose.

2.2. Growing test

The possibility of endophytic growth inducer on the Spinach was done by [5] modification methodology. The modification was done by adding the medium test for each endophytic microbes type. The bacterial endophytic was growth in LB medium for 48 hour and shake in 150 rpm at 28°C and incubated for 2 weeks. The surface sterilized seed adding before the shaker turn on. The seed planted on sandy soil on the tray. Each bacteria replicated three times with five plant unit per replication. The fungi was test by water agar seedling test. Each fungi test was replicated three times with five plant unit per replication at 28°C and incubated for 2 weeks 14 h light.

3. Result and Discussion

The plant endophytic life as an endosymbiont. It mostly from bacterium and fungus. That lives within a host on specific host part or cell, without causing apparent disease. According to Ahmad *et al* [6] It live asymptomatically inside plant tissue and found the most plant kingdom. The symbiont of endophyte and plant are mostly mutualism, where each species has a net benefit. For example fungi Endophytic fungi receive shelter and nutrition from host plants and in return provide great advantages to the host.

However, according to the [7], the evolution of endophytic microbes was start as normal plant pathogen, attacks their host and absorbed the nutrition from their host just like other plant pathogens.

Moreover, [8] informed that endophytic-host plant associations are generally treated separately from parasitic, pathogenic and saprophytic interactions and are viewed as mutualistic associations. However, endophyte–host interactions are based on mutual exploitation. Benefits to the partners are rarely symmetric and conflicting selection forces are likely to destabilize them.

The mutualistic associations of endophytic and host plant appear not only at one time. They develop their symbiotic interaction for long time, from geneteration to generation, to specific plant to specific microbe's species. To specific microbes to specific plant part. Some of them can more specifically to the strain, race, or pathovar. It fact make the development of non-true host endophytic microbes are challenging. Need to extra purposes for development of endophytic microbes from other plant host to other plant genus or species.

The same result and theory was showed in this project. According to research, the necrotic reaction (Table 1) and seedling symptom disease (Figure 1) was appear on the spinach body part. The appearance of necrotic in the spinach part indicated the interaction e.g infection process was started. According to [9] the fungal endophytic behaviour changed due to abiotic and biotic factor. By that factor the effector and elicitor pathways on plant was started. Moreover, [10] reported that the abiotic factor like high light triggers endophyte pathogenicity, while low light supports endosymbiotic development. The pathogenicity under high light resulted from light-induced production of H2O2 by the fungus, triggering hypersensitivity, cell death, and tissue necrosis.

Table 1. Plant Pathogenic Posibility of Isolated Endophytic

Isolate Code	Necrotic Reaction on Spinach	Necrotic Clearance Level	Appearance Level
B.C.R1	-	-	-
B.C.R2	-	-	-
B.C.R3	-	-	-
B.C.R4	-	-	-
B.C.R5	-	-	-
B.C.R6	-	-	-
B.C.R7	-	-	-
B.C.R8	-	-	-
B.C.S1	+	5	Clear Spot
B.C.S2	+	1	Ambiguous
B.C.S3	-	-	-
B.C.L1	+	1	Ambiguous
F.C.L1	+	3	Mostly Clear
F.C.R1	-	-	-
F.C.R1	-	-	-
F.C.S1	-	-	-
F.C.L1	-	-	-
F.C.L2	-	-	-
B.K.R1	-	-	-

B.K.R2	-	-	-
B.K.R3	-	-	-
B.K.R4	+	1	Ambiguous
B.K.R5	-	-	-
B.K.R6	-	-	-
B.K.R7	-	-	-
B.K.R8	-	-	-
B.K.S1	-	-	-
B.K.L1	-	-	-
B.K.L2	-	-	-
F.K.R1			
F.K.R2	+	1	Ambiguous
F.K.R3	-	-	-
F.K.L1	-	-	-
B.M.R1	-	-	-
B.M.R2	+	1	Ambiguous
B.M.R3	-	-	-
B.M.R4	-	-	-
B.M.R5	-	-	-
B.M.L1	+	1	Dim
B.M.L2	+	1	Dim
F.M.R1	-	-	-
F.M.R2	-	-	-
F.M.L1	+	5	Mostly Clear
F.M.L2	+	3	Dim
F.M.L4	+	4	Mostly Clear
F.M.L3	+	3	Mostly Clear

The clearance level of necrosis in the spinach indicated two think. The microbial endophyte was triggers a hypersensitive response visible as local necrotic lesions of the host. And the second indication was the necrosis on the development non-host endophyte should be treated as independent features in the evaluation system. The clearance level of necrosis on the spinach indicate the conflicting selection forces by the plant host was undergo. According to the [11], most plant pathogens carry genes encoding plant cell wall-degrading enzymes. However, nonphytopathogens may possess glycoside hydrolase other than cellulase/hemi-cellulase (or cell wall degradation hydrolases). Presence of this enzyme in numerous endophytes is consistent with its possible role in the diversity of sugar utilization that might be a useful component of a competent endophyte.

The growth promoter is the one of several symbiotic reactions of the microbial endophyte and the true-new host. According to Gomez *et al* [12], plant growth-promoting endophyte may be used to reduce the need for chemical fertilizers, but their efficiently be affected by several factor. The utilization of endophytes can be protected the plant host from the rhizospheric competitors and extreme environmental conditions, could overcome those problems and successfully promote the crops under field conditions.

The plant host species, the microbe type, species, race and mechanism are the most factors to influence the growth of plant. Not all endophytic microbes have mechanism to boost the growth of the host. The growth promoter is just one of symbiotic mechanism among plant host and endophytes. However, the deep screening is the fist pathway to the develop the endophytic. Its also the most important path on the development non host endophytic microbes. While, the visible necrotic symptom is the most part on evaluation system during the endophytic sreening.

Figure 1. The growth promoter reaction compared to control (Green Box) and Seedling Symptoms (Red box) in Seedling Test compared to control (yellow box)

4. Conclusion

The growing and plant pathogenic possibility test was done to eliminating the potential plant pathogen microbes. From 40 isolate, was detected more than 20 isolate have possibility to be a pathogen of Spinach, with different symptoms appearance level. The level of growth induced result also have different result among the potential endophyte. They appear on the growing time, the long of root and shoot number of spinach. The optimizing of growth promoter endophyte especially for non-host endophyte, need the extra path for convinced the isolated endophyte not to be potential plant pathogen.

Acknowledgment

To the Research and Social Service Center of State Polytechnic of Jember by The Applied Research (Penelitian Terapan) Grand on 2021 with The contrack number: 23/PL17.4/PG/2021

References

[1] Stone, Jeffrey K, Charles W B, and James F White Jr 2000 Microbial Endophytes 17–44

[2] Yanni, Youssef G, Rizk R Y, Corich V, Squartini A, Ninke K, Philip-Hollingsworth S, Orgambide G, De Bruijn F, Stoltzfus J, and Buckley D 1997. "Natural Endophytic Association between Rhizobium Leguminosarum Bv. Trifolii and Rice Roots and Assessment of Its Potential to Promote Rice Growth." Pp. 99–114 in Opportunities for biological nitrogen fixation in rice and

other non-legumes. Springer
[3] Braun, Karen, Romero J, Liddell C, and Creamer R 2003 Mycological Research 107 980–88
[4] Dube, Praxedis, Struik P C, and Ngadze E 2018 African Journal of Agricultural Research 13 753–70
[5] Zou, Tengyue, Huang C, Wu P, Ge L, and Xu Y 2020 Plants 9 490
[6] Ahmad R Z, Ameen F, Khalid R, Alghuthaymi M A, Alsalmi R, and Li C 2019 Sustainable Agriculture Research 8 66-72
[7] Wilson D 1995 Endophyte: the evolution of a term, and clarification of its use and definition Oikos 274-276
[8] Saikkonen K, Wäli P, Helander M, and Faeth S H 2004 Trends in plant science 9 275-280
[9] Mengistu A A 2020 Endophytes: Colonization, behaviour, and their role in defense mechanism. International Journal of Microbiology https://doi.org/10.1155/2020/6927219
[10] Alvarez-Loayza, J. F. White Jr., M. S. Torres et al., 2011. PLoS One, vol. 6, 2011
[11] Ali S, Duan J, Charles T C, and Glick B R. 2014. Journal of6eoretical Biology 343 193–198
[12] Jiménez-Gómez A, Saati-Santamaría Z, Kostovcik M, Rivas R, Velázquez E., Mateos P F, Menendez E, and García-Fraile P 2020 Agronomy 10 1788 https://doi.org/10.3390/agronomy10111788

IOP Conf. Series: Earth and Environmental Science **980** (2022) 012012 doi:10.1088/1755-1315/980/1/012012

The effect of Biofertilizer Application and Shoot Pruning on the Production of chili plants on sandy media

M Z Sukri[1], H F Rohman[1] and R Firgiyanto[1*]

[1] Department of Agricultural Production, State Polytechnic of Jember Mastrip Street PO BOX 164 Jember, Indonesia

*refa_firgiyanto@polije.ac.id

Abstract. Red chili as one of the important horticultural crops cultivated commercially, this is because in addition to chili has a fairly complete nutritional content, it also has a high economic value which is widely used both for household consumption and for the food industry. Therefore, it is necessary to apply intensification technology to increase chili production. These technologies include the application of Plant Growth Promoting Rhizobacteria (PGPR) to increase soil fertility and the application of shoot pruning to increase branching of chili plants. The purpose of this study was to determine the effect of PGPR and Trichoderma sp. And the treatment of shoot pruning on the production of red chili plants.This research was conducted at the Jember State Polytechnic Greenhouse, Sumbersari District, Jember Regency with an altitude of ± 89 masl. This study used a complete randomized block design (CRD) factorial consisting of 2 factors. The first factor was the provision of biological agents consisting of 4 levels of treatment. The second factor was the time for cutting shoots, consisting of 3 levels of treatment. The observation variables is productivity of chili. The results showed that branch pruning and biological agent treatment did not significantly increase the production of chili plants. However, there is a tendency for better treatment with pruning and without biological agents.

1. Introduction

Puger is the center of sandy land agribusiness in Jember Regency, Chili is a potential commodity to be developed in Puger because it has wide adaptability, short life and has high economic value with good market opportunities [1] [2] so this plant is considered very potential to be developed and cultivated by the Puger community.

Sandy soil in Puger District belongs to the Udipsamments soil type with characteristics characterized by sandy soil texture, very fast soil drainage, very shallow to moderate effective depth, very low salinity, low CEC, acidic to neutral pH, very low organic C, Total N is very low and phosphate is moderate [3] [4]. To increase tolerance and adaptability of plants to stress, including climate change, it is necessary to optimize the use of coastal land, one of which is the use of biological agents.

Biological agent is any organism that includes all species or varieties of insects, nematodes, protozoa, bacteria, fungi, viruses, microplasms, and other organisms which in all stages of development can be used as pest control in a process of plant cultivation and processing of agricultural products [5]. Biological agents that function as soil fertility enhancers are PGPR and *Trichoderma* sp.

PGPR can directly produce growth regulators and increase nutrient uptake by plants. Indirectly, rhizobacteria are associated with the production of metabolites such as antibiotics and siderophores, which can function to reduce the growth of phytopathogens [6].

Trichoderma sp. is one of the useful microorganisms and is a harmless symbiotic fungus, even mutually beneficial between soil-borne fungi and plant roots. Trichoderma sp. Acts as a root growth stimulant, because Trichoderma sp. Has the ability to increase plant growth hormones such as auxins and cytokinins [7]. In addition to functioning to improve the root structure of plants, *Trichoderma* sp. Also has a role as an antagonist fungus against pathogens. Trichoderma sp. can inhibit the growth of several disease-causing fungi in plants, including Rigidiforus lignosus, Fusarium oxysporum, Rizoctonia solani, Sclerotium rolfsi [8]

Increased production of chili plants can also be done through pruning which aims to streamline plant growth and development in a more productive direction. Another goal is to increase the efficiency of nutrient use [9]. Therefore, it is necessary to conduct research on the provision of biological agents and shoot pruning to find out how they affect the productivity of chili plants on sandy media.

2. Material and Methods

2.1. Material

Chili seeds, organic fertilizers, anogranics, mulch, PGPR, *Trichoderma sp.* Ruler, calipers, scissors, colling box, calipers, stationery, sprinkler, cameras, wires, hoes, buckets, analytical scales, gauges and knives.

2.2. Methods

This study used a complete randomized block design (CRD)

Factor I : Pruning times :

- P0 : Without Pruning
- P1 : Pruning 14 Day After Planting
- P2 : Pruning 21 Day After Planting

Factor II : Provision Biological Agents

- H0 : Control
- H1 : PGPR
- H2 : *Trichoderma sp*
- H3 : PGPR + *Trichoderma sp*

3. Result and Discussion

The results showed that branch pruning and biological agent treatment did not significantly increase the production of chili plants (Table 1). However, there is a tendency for better treatment with pruning and without biological agents (Figures 1 and 2).

Table 1. Lists of various types of PGPR enrichment and pruning on plant height and number of chili branches

Source of Diversity	Production
P	ns
H	ns
P x H	ns

Noted: P = Pruning, H = PGPR enriched, * = Significant, ** = Very Significant, ns = Non Significant
WAP = Week After Planting

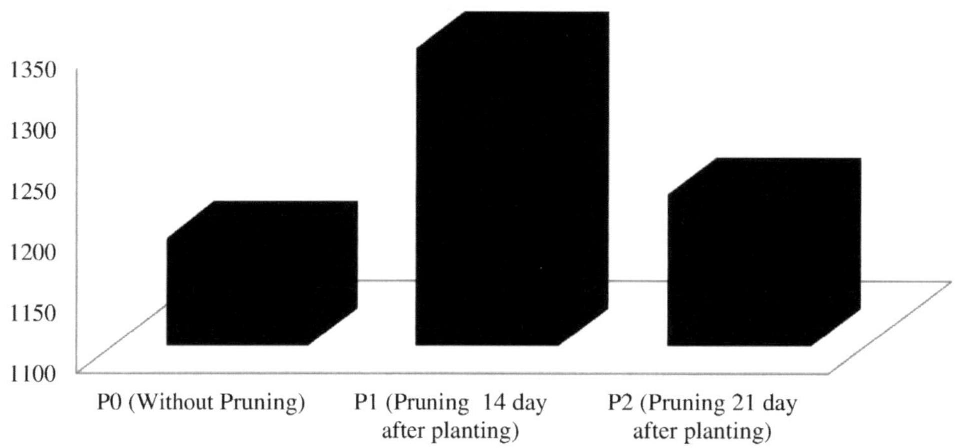

Figure 1. Effect of pruning on chili production

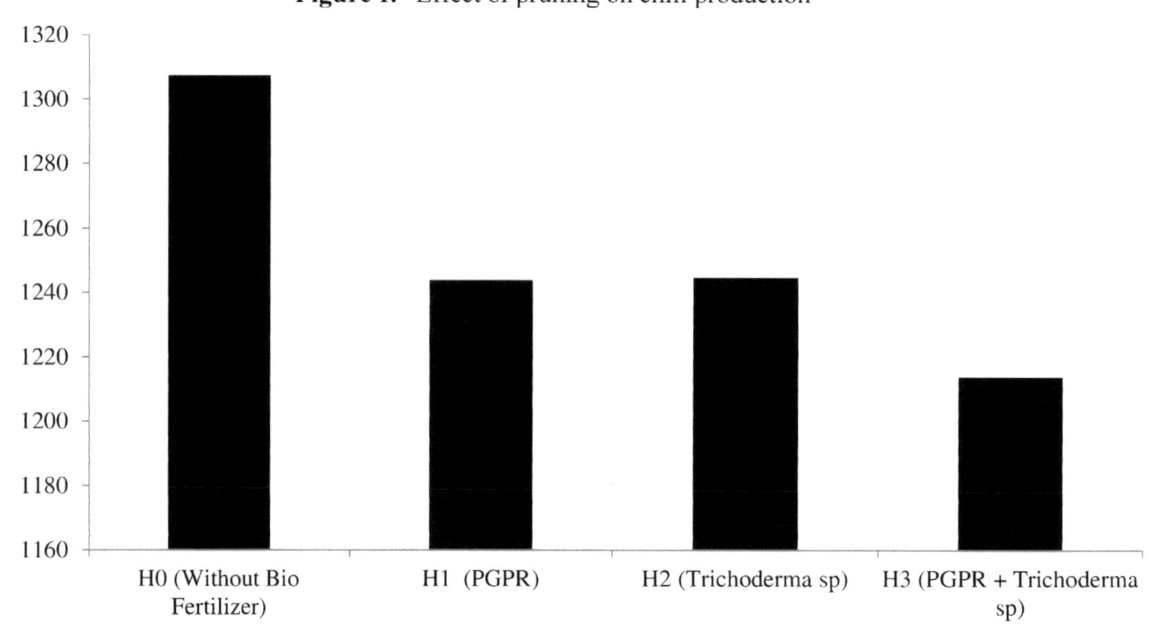

Figure 2. Effect of biological agents on chili production

The phenomenon of growth in the shoot or canopy is more dominant than the lateral or other parts which are influenced by the hormone auxin, causing the growth of lateral/axillary shoots to become dormant and stunted. These shoots will start their growth after the primary apical shoot meristem/principal becomes a permanent organ, such as a flower or flowering meristem [10]. This has an impact on the low production per plant. Therefore, to stimulate the growth of lateral shoots, the influence of apical dominance needs to be inhibited. According to [11] one way in which restricted lateral shoot growth can be increased is the removal of the apical portion of the shoot by cutting or pruning. Therefore, there is a tendency for better production, presumably because chili plants that have been pruned vegetatively become inhibited so that the intake of photosynthetic products becomes focused on the needs of the generative process of chili plants.

Application of biofertilizers such as PGPR enriched with Trichorderma sp. is part of an effort to increase soil fertility. PGPR which has been enriched with Trichoderma sp. is expected to play a role in spurring growth plants and play a role in direct or indirect mechanisms through control disease to maintain plant productivity. The direct mechanism occurs through nitrogen fixation, phosphate solubilization, and production of siderophores, phytohormones, and 1-aminocyclopropane- 1-carboxylate deaminase, while the indirect mechanism is through the production of antibiotics,hydrogen cyanide (HCN), and siderophores; ecological niche competition (environment growth/ecological niche), and induction of systemic resistance. The results that do not significantly affect the application of biofertilizers are thought to be because the successful use of beneficial living organisms in agriculture is not only influenced by the quantity of cells present in the inoculants, but is also influenced by energy sources, application of inoculants, environmental factors (temperature, rainfall). and methods of storing the product before use. This is also in accordance with the opinion of [12] that the life activities of soil organisms are strongly influenced by climate, soil and vegetation factors. In addition, the use of doses that have not met the needs is also suspected to cause the effect given to plants is not optimal because the number of microorganisms is not enough to significantly increase the productivity of the planting media which has an impact on plant growth.

4. Conclusion

The conclusion of this study is that pruning and application of biological fertilizers and their interactions have not been able to increase the production of chili plants. Therefore, it is necessary to do further on these two studies, because there is a tendency for better results to be found.

Acknowledgment

Authors wishing to acknowledge assistance or encouragement from State Polytechnic of Jember, because this research was carried out on the research grant of PNBP State Polytechnic of Jember Budget Year in 2020

References

[1] Surtadi C A, and Wirasti 2017 J Pengkajian dan Pengembangan Teknologi Pertanian **20** 125-139
[2] Supriadi D R, Susila A D, and Sulistyono E 2018 J. Hort. Indonesia **9** 38-46
[3] Sukri M Z, Firgiyanto R, Sari V K, Basuki. 2019 J Penelitian Pertanian Terapan **19**142-146
[4] Irawanata M H Y 2017 Evaluasi kesesuaian lahan pasir pantai untuk Pengembangan tanaman wijen (Sesamum Indicum L.) di Kecamatan Puger Kabupaten Jember [Skripsi] Fakultas Pertanian (Yogyakarta Indonesia: Universitas Muhammadiyah Yogyakarta)
[5] Sopialena 2018 Pengendali hayati dengan memberdayakan mikroba Mulawarman University Press Samarinda 98 hal
[6] Kloepper J W, Ryu C M, and Zhang S 2004 Phytopatology **94** 1259-1266
[7] Syahri 2011 Potensi Pemanfaatan Cendawan Trichoderma Spp. sebagai Agens Pengendali Penyakit Tanaman Di Lahan Rawa Lebak Balai pengkajian teknologi pertanian (BPTP) Sumatera selatan
[8] Herlina L, and Pramesti D 2009 Penggunaan Kompos Aktif Trichoderma sp. dalam Meningkatkan Pertumbuhan Tanaman Cabai Semarang (ID) Universitas Negeri Semarang
[9] Tony H 2003 Berkebun Hidroponik Secara Murah Penebar Swadaya Jakarta 96 hlm.
[10] Shimizu S S and Mori H, 2001 *J. Plant Physiology* **127**.
[11] Wilkins Malcolm B 1984 *Advanced Plant Physiology*. Pitman Publishing Inc. London

[12] Hakim N M, Nyapka Y, Lubis A M, Nugroho S G, Rusdi S M, Hong G, H Hailey 1986 *Dasar-dasar Ilmu Tanah* Universitas Lampung Press Lampung

Chitosan application for maintaining the growth of lettuce (*Lactuca sativa*) under drought condition

S Nurliana [1*], **S Fachriza**[1], **N M Hemelda** [1,2] **and R Yuniati** [1,2*]

[1] Department of Biology, Faculty of Mathematics and Natural Sciences Universitas Indonesia, 16424, Depok, Indonesia
[2] Research Group of Metabolomics and Chemical Ecology, Department of Biology, Faculty of Mathematics and Natural Sciences, Universitas Indonesia

* ratnayuniati@sci.ui.ac.id

Abstract. Drought stress is one of the most common abiotic stress in agriculture. The use of antitranspirant agents to reduce the effect of drought stress on crops has been considered as a potential method. The application of chitosan as an antitranspirant agent has been reported to be effective in several crops. This study was aimed to analyze the effect of chitosan foliar application for maintaining the growth of lettuce under drought conditions. Three concentrations of chitosan (0.2 g/L; 0.4 g/L; 0.6 g/L) were tested in this study. The parameter observed were the height of the lettuce, number of leaves, leaf length and the width, root length, as well as fresh and dry weight. The drought condition without chitosan application reduced the growth of lettuce in all parameters. Chitosan application at 0.2 g/L was able to reduce the effect of drought stress and maintain the growth of lettuce. However, higher chitosan concentration applications (0.4 g/L and 0.6 g/L) were observed to reduce the growth of lettuce. Based on this study, chitosan was reported to reduce the effect of drought stress in lettuce. As chitosan affects the transpiration process by stomatal closure, a higher concentration of chitosan may also affect the photosynthesis process leading to growth inhibition.

1. Introduction

Drought stress is one of the most common abiotic stress in agriculture. In general, drought stress occurs when the water availability on the soil is fewer, and the atmospheric condition causes continuous water loss due to transpiration or evaporation [1]. The main cause of drought is climate change. As many as 51.2% of Indonesia's territory in 2020 experienced a dry season marked by reduced rainy days and a low amount of rainfall measured on the surface [2],[3]. Most areas in East Nusa Tenggara, West Nusa Tenggara, Bali and Java have experienced consecutive days without precipitation ranging from 20-60 days [2]. Meanwhile, in 2021 it is also predicted that several regions in Indonesia have the potential to experience meteorological drought in the classification of alert (days without precipitation >31 days) to alert (days without precipitation >61 days) with an estimated chance of precipitation (<20 mm/10 days) [4].

Constant drought stress can be harmful to the farmer. A decrease in crop yield because of low water content in soil may lead to decreased photosynthesis as well as biomass production in plants is decreased [5]. Drought conditions would encourage stomatal closure to reduce water loss [6]. However, the closure of stomata may also cause the gas exchange hindering so that CO_2 concentrations decrease [7],[8]. In addition, drought stress can also reduce the synthesis of photosynthetic enzymes [8]. Decreased CO_2

concentrations and reduced photosynthetic enzymes cause the process of photosynthesis of plants to be inhibited [8], even within a certain period, which may result in the death of the plant [1].

The response of plants to drought conditions varies, depending on the severity and duration of the drought as well as the plant species and its growth stage [9]. Lettuce is one of the plants that are susceptible to drought [10]. That matter is caused by the fact that lettuce has a short rooting system [1], [10],[11]. In addition, lettuce cultivation carried out by farmers is generally done conventionally using soil, which causes lettuce to become susceptible to drought caused by the availability of water in the soil are fewer. Moreover, the uncertain dry season in Indonesia worsened the drought conditions, which further reduced the availability of water in the ground. Thus, the results of lettuce yield and productivity can decrease [12].

Lettuce requires a high water content (>95% of the field capacity of the soil) to support its growth [13]. Lettuce is experiencing mild drought if the need for lettuce encounters 75-90% of the field capacity [14] and severe drought if the field capacity of the soil is <50% [15]. When lettuce is exposed to drought, lettuce will be wilted, growth will decrease, and leaves will turn yellow [16]. Based on Sayyari et al (2013) research, water deficits of 30% to 40% drastically reduce fresh weight and lettuce crop production and may increase oxidative damage [5].

Over the past few years, various cultivation techniques have been applied to overcome the water deficit in crops [17]. One strategy to reduce the impact of water deficit is by using antitranspirant compounds that could potentially induce drought tolerance in plants [18]. One of the antitranspirant compounds that can be used is chitosan [19]. Some examples of other antitranspirant compounds often used are magnesium carbonate, abscisic acid, salicylic acid, phenylmercuric acetate, and kaolin [17]. Chitosan will act as an antitranspirant because it can induce the closure of stomata in plants so that plants can reduce water loss in their bodies [17].

Chitosan can induce the closure of the stomata by creating a signal to the chloroplasts so that they can form H_2O_2 and NO as the second messenger responsible for improving the synthesis of abscisic acid (ABA) [20]. In addition, chitosan can act as an elicitor because chitosan can protect plants from various oxidative stress by inducing different enzymes and secondary metabolites in the biosynthetic or signaling pathway [21]. Chitosan can induce phytoalexin, polyphenolic, lignin, and flavonoid to increase plant immunity [22]. Chitosan is also biocompatible, biodegradable, non-toxic, and environmentally friendly and is frequently used as a substitute for pesticides because it can act as an antimicrobial agent in plants [23], [24]. Application of chitosan can be used in various ways such as foliar application, seed coating, seedling root dipping, enrichment soil, and addition of chitosan as a supplement in plant tissues [25]. A commonly used method of chitosan application is a foliar application [26]. Foliar application using chitosan is most widely used because, generally, water loss in plants occurs through stomata [26].

The application of chitosan to overcome drought stress has been reported in some crops. Spraying leaves using chitosan is said to reduce the transpiration rate of pepper plants (*Capsicum* sp.) by 26–43% without a reduction in crop biomass [27]. In addition, a foliar application using chitosan can also improve the parameters of growth, crop yield, and quality of cowpea (*Vigna unguiculata* (L) Walp) under drought conditions [28].

Chitosan application on lettuce plants has also been reported. Xu and Mou (2018) said that the application of chitosan on soil could improve growth parameters, phytochemical efficiency, and gas exchange in lettuce [29]. Meanwhile, Kurzawińska (2007) also reported that seed and root dipping in chitosan solution could increase germination and reduce pathogen attacks on lettuce [30]. However, neither study was conducted in drought conditions. Therefore, research to evaluate the effectiveness of chitosan in facing the impact of drought on lettuce plants needs to be done.

The beneficial effect of chitosan depends on the concentration, the method of application, environmental conditions, and the growth status of the plant [21]. Foliar application using chitosan with a concentration variation of (0.1–0.4 g/L) was reported to improve growth and physiology parameters and the quality of the fruit of the pepper plant [27], common bean [31], and basil [21]. Based on previous research results, the concentration of chitosan was small, so that in this study, a narrower range of chitosan concentrations will be used (0.2; 0.4; and 0.6 g/L). This study aimed to analyze the effect of chitosan foliar application with concentration variations of (0.2; 0.4; and 0.6 g/L) for maintaining the growth of lettuce (*Lactuca sativa*) under drought conditions.

2. Material and Methods

2.1. Material
The materials used in the study were plastic tray [72x72], polybag size 20x25 cm (diameter x height), black plastic (90x120 cm), a mixture of soil: goat manure: sand (1:1:1), lettuce seeds [cv. Grand Rapid], chitosan powder, 0.5% acetic acid, 5% NaOCl, aquades, water, kitchen towel, and envelope.

2.2. Methods
The experiment was conducted in the greenhouse of the Department of Biology, Faculty of Mathematics and Natural Sciences, Universitas Indonesia, from November 2020 to January 2021. The research was conducted experimentally using a Completely Random Design consisting of 5 treatments. There were five treatments in this study: positive control, negative control, chitosan 0.2 g/L, chitosan 0.4 g/L, and chitosan 0.6 g/L. Each treatment consists of 5 plants, so that the total number of plants for five treatments were 25 plants.

The planting medium used was soil with a mixture of sand and goat manure by comparison (1:1:1). Lettuce seeds used were sterilized first by soaking the seeds in a 5% NaOCl solution for 5 minutes and then rinsed with water five times [32]. Seedlings that already had 4-5 leaves were transplanted to polybags [33]. Before being given drought treatment, lettuce was still watered with regular irrigation (100% field capacity) in the morning and afternoon for up to 2 weeks as an acclimatization stage [34]. The chitosan solution was made by dissolving the chitosan powder in 1000 mL of 0.5% acetic acid. This study used chitosan levels consisting of (0.2, 0.4, and 0.6 g/L).

Drought treatment was carried out after the acclimatization period was complete (2 weeks after the transplant). Negative control, chitosan 0.2 g/L, chitosan 0.4 g/L, and chitosan 0.6 g/L groups that received drought treatment were watered with a frequency of once a day in the morning with 75% field capacity [35]. Meanwhile, the treatment of positive control, which does not experience drought, remains watered with a frequency of twice a day in the morning and afternoon with 100% field capacity [35]. During the experiment, the application of chitosan was used two times over 30 days. Approximately 20 ml per plant of these solutions were sprayed at the surface of the upper and lower lettuce leaves using a hand sprayer [27].

The criteria for lettuce that can be harvested was when lettuce reached the age of 2-3 months (35-60 days) after transplantation [36]. Harvesting was done in the morning by removing all parts of lettuce and its roots [36]. The parameter observed were divided into 2 aspects, environmental and growth parameters. Environmental parameters observed were temperature and humidity. Meanwhile, growth parameters observed included the plant height, leaves number, leaf length, and width, root length as well as fresh and dry weight of the plant.

2.3. Statistical analysis
The result data were tested with a one-way variant analysis test (ANOVA) with a significance of 95% to determine the effect of chitosan application on the growth of lettuce plants. All statistical analysis was done using Statistical Product and Service Solution (SPSS) ver. 24.

3. Result and Discussion

3.1. Environmental Parameters
Environmental parameters measured included temperature and humidity. The measurement results of environmental parameters were presented in the form of an average table during planting to harvesting (Table 1). The average temperature in greenhouse tends to be high beyond the optimal limit of lettuce (25±2°C) [37], especially during the day (33.38±3.36°C) (Table 1). In contrast to temperature, the

highest humidity occurs in the morning (72±10.08%) (Table 1). Along with the increase in temperature, the humidity decreases. From the result, we know that the average temperature of a greenhouse was not ideal for supporting lettuce growth. That condition worsened the dry treatment for lettuce growth.

Table 1. Environmental Parameters in the Greenhouse

Time	**Temperature (°C)**	**Humidity (%)**
Morning	29.69±2.12	72±10.08
Noon	33.38±3.36	60±11.62
Afternoon	32.08±2.97	62±14.09

3.2. Growth Parameters

Growth parameters measured included the plant height, leaves number, leaf length, and width, root length as well as fresh and dry weight of the plant. The positive control group showed better growth on all parameters than the negative control group (Table 2). This matter indicates that drought stress can reduce the parameters of lettuce growth. However, the positive control group tends to be not differed from the negative control group (Table 2). During the experiment, the temperature inside the greenhouse was higher than the suitable temperature required for lettuce to grow optimally. Thus, we assume that lettuce in the positive control group also suffers from drought stress like negative control.

Based on this research, the application of chitosan 0.2 g/L was able to increase the height of lettuce, the leaves number, leaf length and width, root length, as well as fresh and dry weight of the lettuce than the negative control (Table 2). Meanwhile, chitosan 0.4 and 0.6 g/L was only able to increase the height of lettuce compared to the negative control (Table 2). The number of the leaves, the leaves size, root length, and produce biomass on chitosan 0.4 and 0.6 g/L was tended to be not differed from the negative control group (Table 2).

Table 2. Growth Parameters

Treatment	**Plant height (cm)**	**Leaves Number**	**The Leaf Length (cm)**	**The Leaf Width (cm)**	**Root Length (cm)**	**Fresh Weight (g)**	**Dry Weight (g)**
Positive Control	8.1±1.64	7.4±1.52	9.05±1.56	4.1±0.58	8.9±1.95	3.79±1.41	0.17±0.05
Negative Control	7.02±1.18	6.6±1.14	8.95±1.08	4.05±0.48	8±3.02	2.08±1.05	0.11±0.06
Chitosan 0.2 g/L	11.83±4.37	7.25±2.50	9.88±2.26	5.06±1.33	9.13±2.95	4.02±3.06	0.19±0.05
Chitosan 0.4 g/L	9.3±2.77	5±1.41	8.2±2.83	4.6±1.52	7.3±2.36	2.31±2.36	0.12±0.13
Chitosan 0.6 g/l	9.6±4.14	4±2.07	7.55±2.78	3.45±1.64	6.3±1.48	1.96±1.51	0.09±0.07

Lettuce treated with a chitosan treatment can grow higher under drought conditions compared to the negative control group. Uthairatanakij et al (2007) reported chitosan application can increase plant growth by increasing the synthesis of auxin hormone through the independent pathway of tryptophan [38]. According to Saharan and Pal (2016), the growth increase is due to the activation of the expression of the AX 11 gene, a gene that plays a role in signaling the synthesis of auxin hormone [39]. However, the mechanism of increasing the synthesis of these hormones is not yet known [40].

Based on the result, the application of chitosan 0.2 g/L increased the leaf number, leaf length and width, and root length of the lettuce than chitosan 0.4 and 0.6 g/L treatment although not significantly. This is supported by Abu-Muerifah (2013) that chitosan 0.2 g/L increased the leaf number and leaf area of common bean (*Phaseolus vulgaris* L.) under drought stress than chitosan 0.4 g/L [31]. Shehzad et al

(2020) also report that the application of chitosan 0.2 g/L increased the root system development and improved the water absorption capability in sunflowers (*Helianthus annuus* L.) [41].

Drought stress causes increased transpiration rates, which cause the water content of the cell to fall below optimum. Physiological processes are not to be running optimally because the water content in cells decrease so that plants will reduce biomass production and reduce growth [5]. However, based on this result the application of chitosan 0.2 g/L was able to increase the fresh and dry weight of the plant under drought conditions. Abu-Muerifah (2013) also reports similarly that chitosan 0.2 g/L can increase the yield and quality of the common bean [31].

Chitosan as an antitranspirant is known to decrease transpiration rate by inducing the closure of stomata in plants. Bittelli et al. (2001) reported that spraying leaves with chitosan could reduce the concentration of potassium ions in guard cells [27]. If the concentration of potassium ions in guard cells is low, then the stomata will close and vice versa [42].

In addition to the decrease in potassium ions, the reduction in transpiration rate is also caused by the abscisic acid (ABA) hormone [20]. Chitosan can bind to the cell membrane through interactions between positively charged amine groups of chitosan with negatively charged phospholipids [30]. The interaction induces secondary messenger signals in cells so that in chloroplasts, hydrogen peroxide (H_2O_2) is formed via the octadecanoid pathway and forms NO [20]. The emergence of H_2O_2 compounds stimulates the accumulation of Reactive Oxygen Species (ROS) and the synthesis of ABA hormones [20]. Meanwhile, NO creates phosphoric acid through the phospholipase D (PLD), phospholipase C (PLC), and diacylglycerol kinase (DGK) pathways [20]. Phosphoric acid plays a role in increasing the work of ABA by inhibiting the ABI1 gene, which is a negative regulator of ABA [20]. Therefore, ABA can induce the closure of stomata in plants.

In addition to providing a positive effect on plant growth, the application of chitosan can also have a negative impact. Davenport et al (1969) stated that giving chitosan with too high concentrations can cause damage to the surface of the leaves [26]. Giving chitosan at concentrations that are too high can also decrease plant productivity. If the concentration of chitosan given is too high, carbon assimilation in plants decreases because water absorption is suppressed [19]. Based on this result, if the stomata close for a long time may lead to a reduction in photosynthetic activity that plays an essential role in plant growth [42]. That matter causes concentrations of chitosan 0.4, and 0.6 g/L cannot increase lettuce growth compared to chitosan 0.2 g/l. Therefore, the concentration of chitosan plays an essential role in determining the effectiveness of chitosan.

Chitosan 0. 2 g/L was confirmed to increase lettuce growth even if not significantly. Similarly, it was also reported by Abu-Muerifah that 0.2 g/L of chitosan showed the best growth in the common bean [31]. Meanwhile, higher chitosan concentration (0.4 and 0.6 g/L) was confirmed to reduce the effectiveness of chitosan. This was evidenced by all the observed parameters, lettuce growth with chitosan treatment of 0.4 and 0.6 g/L was lower than chitosan 0.2 g/L and did not even present a significant difference when compared to the negative controls.

Based on the previous study, the application of chitosan 0.4 g/L reportedly did not have a significant effect on sunflower (*Helianthus annuus* L.) [41] and sage (*Salvia officinalis* L.) [44]. However, the application of chitosan 0.4 g/L and 0.6 g/L can increase the growth parameter in basil (*Ocimum basilicum*) [45] and stevia (*Stevia rebaudiana* Bertoni) [46] significantly. That study shows that the application of chitosan 0.4 g/L and 0.6 g/L still has a significant effect on plant growth. This indicates that optimal concentrations may vary depending on plant species and chitosan concentration.

In addition to chitosan concentration and plant species, the influence of environmental conditions also allows for different results. Temperature conditions in the greenhouse were higher than the suitable temperature required for lettuce to grow optimally [33],[36]. Higher temperatures can increase the evaporation and transpiration net in the plant [47].

Based on this research, the concentration of chitosan that showed the best results for maintaining the growth of lettuce plants was chitosan 0.2 g/L.

4. Conclusion

Chitosan application at 0.2 g/L was able to reduce the effect of drought stress and maintain the growth of lettuce. However, higher chitosan concentration applications (0.4 and 0.6 g/L) were observed to reduce the growth of lettuce.

5. Acknowledgment

The authors would like to thank Universitas Indonesia for funding this research through PUTI Grant with contract number NKB-3546/UN2.RST/HKP.05.00/2020.

References

[1] Jaleel C A, Manivannan P, Wahid A, Farooq M, Somasundaram R and Panneerselvam R 2009 *International Journal of Agriculture and Biology* **11** 100-105
[2] BMKGOTIM (Badan Meteorologi Klimatologi dan Geofisika Stasiun Meteorologi H.Asan Kota Waringin Timur) 2020 Perkembangan musim kemarau tahun 2020 [online]. Available at:https://bmkgkotim.info/2020/06/27/perkembangan-musim-kemarau-tahun-2020/ [Accessed 26 July 2021]
[3] Thirafi, H 2020 Perkembangan musim kemarau 2020 [online]. Available at:https://www.bmkg.go.id/press-release/?p=perkembangan-musim-kemarau-tahun-2020&tag=press release&lang=ID [Accessed 26 July 2021]
[4] Prasetyaningtyas, K 2021 Analisis Dinamika Atmosfer Dasarian II Juli 2021 [online]. Availabe at:https://www.bmkg.go.id/iklim/dinamika-atmosfir.bmkg [Accessed 26 July 2021]
[5] Sayyari M, Ghavami M, Ghanbari F and Kordi S 2013 *International Journal of Agriculture and Crop Sciences* **5**(17) 1951–1957
[6] Anjum S A, Xie X, Wang L, Saleem M F, Man C and Lei W 2011 *African Journal of Agricultural Research* **6**(9) 2026–2032
[7] Farooq M, Wahid A, Kobayashi N, Fujita D and Basra S M A 2009 *Agronomy for Sustainable Development* **29** 185–212
[8] Huseynova I M, Rustamova S M, Aliyeva D R, Babayev H G and Aliyev J A 2016 ''Photosynthesis, antioxidant protection, and drought tolerance in plants'', in Drought Stress Tolerance in Plants Vol 1 ed Hossain M A, Wani S H, Bhattacharjee S, Burritt D J and Tran L P (Basel: Springer International Publishing) pp 349–378
[9] Demirevska K, Zasheva D, Dimitrov R, Simova-Stoilova L, Stamenova M and Feller U 2009 *Acta Physiologiae Plantarum* **31** 1129–1138
[10] Kiran S 2019 *Notulae Botanicae Horti Agrobtanici Cluj-Napoca* **47**(2) 352–358
[11] Jackson L E 1995 *Plant Cell and Environment* **18**(8) 885–894
[12] Tamura Y, Mori T, Nakabayashi R, Kobayashi M, Saito K, Okazaki S, Wang N and Kusano M 2018 *Frontiers in Plant Science* **9**(665) 1–11
[13] Finch H J S, Samuel A M and Lane G P F 2014 *Lockhart & Wiseman's Crop Husbandry Including Grassland* 396–430
[14] Paim B T, Crizel R L, Tatiane S J, Rodrigues J V, Rombaldi C V and Galli V 2020 *Scientia Horticulturae* **272** 109578
[15] Furn-Wei L, Lin K, Wu C, Chang Y, Lin K & Wu C 2019 *American Society of Horticultural Science* **55**(1) 89–95
[16] Cabillo C M 2019 *International Journal of Scientific & Engineering Research* **10**(12) 928–935
[17] Pandey P P, Sharma R and Neelkanthe S S 2017 *Plants Archive* **17**(2) 1146–1156
[18] Guleria V and Shweta 2020 *International Journal of Current Microbiology and Applied Sciences* **9**(5) 1671–1678
[19] Iriti M, Picchi V, Rossoni M, Gomarasca S, Ludwig N, Gargano M, and Faoro F 2009 *Environmental and Experimental Botany* **66** 493—500
[20] Pichyangkura R and Chadchawan S 2015 *Scientia Horticulturae* **196** 49–65
[21] Pirbalouti A G, Malekpoor F, Salimi A and Golparvard A 2017 *Scientia Horticulturae* **217** 14–122

[22] Malerba M and Cerana R 2016 *International Journal of Molecular Sciences* **17**(7) 996
[23] Darvill A, Augur C, Bergmann C, Carlson R W, Cheong J J, Eberhard S and Albersheim P 1992 *Glycobiology* **2**(3) 181–198
[24] Ahmed K B M, Khan M M A, Siddiqui H and Jahan A 2020 *Carbohydrate Polymers* **227** 115331
[25] Dzung N A, Khanh V T P and Dzung T T 2011 *Carbohydrate Polymers* **84** 751–755
[26] Davenport D C, Hagan R M and Martin P E 1969 *California Agriculture* **4** 16
[27] Bittelli M, Flurya M, Campbell G S and Nichols E J 2001 *Agricultural and Forest Meteorology* **107** 167–175
[28] Farouk S and Amany A R 2012 *Egyptian Journal of Biology* **14** 14–26
[29] Xu C and Mou B 2018 *HortTechnology* **28**(4) 476–480
[30] Kurzawińska H 2007 *Polish Chitin Society Monograph XII* 173–178
[31] Abu-Muriefah S S 2013 *International Research Journal of Agricultural Science and Soil Science* **3**(6) 192–199
[32] Yap Q C and Teo S S 2019 *Agriculture and Food Sciences Research* **6**(2) 203–210
[33] Zhang G L, Ma L, Beuchat L R, Erickson M C, Phelan V H and Doyle M P 2009 *Journal of Food Protection* **72**(12) 2471–2475
[34] Zha L, Liu W, Zhang Y, Zhou C and Shao M 2019 *Frontiers in Plant Science* **10** 1440
[35] Molina-Montenegro M A, Zurita-Silva A, and Oses R 2011 *Ciencia e Investigación Agraria* **38**(1) 65–74
[36] Dosem I R, Astuti Y Th M and Santoso T N B 2018 *Jurnal Agromast* **3**(1) 1–11.
[37] Gao M, Liu Y, Dong Y and Song Z 2020 *Journal of Hazardous Materials* **401**
[38] Uthairatanakij A, da Silva J A T and Obsuwan K 2007 *Orchid Science Biotechnology* **1** 1–5
[39] Saharan V and Pal 2016 *Chitosan based nanomaterials in plant growth and protection* (India: Springer) pp x+55
[40] Trzcińska M A, Bogusiewicz A, Szkop M and Drozdowski S 2015 *Forests* **6** 3165–3176
[41] Shehzad M A, Nawaz F, Ahmad F, Ahmad N and Masood S 2020 *Ecotoxicology and Environmental Safety* **187** 109841
[42] Willmer C M and Pricker M 1996 *Stomata* 2nd ed (London: Chapman & Hall) pp 315
[43] Holding D R and Streich A M 2013 *Extension Circular* **1268** 1–10
[44] Vosoughi N, Gomarian M, Pirbalouti A G, Khaghani S and Malekpoor F 2018 *Industrial Crops and Products* **117** 366–374
[45] Gerami M, Majidian P, Ghorbanpour A and Alipour Z 2020 *Physiology and Molecular Biology of Plants* **26**(5) 965—974
[46] Maleekpoor F, Pirbalouti A G and Salimi A 2016 *Research on Crops* **17**(2) 354–359
[47] Chandra P, Wunnava A, Verma P, Chandra A and Sharma K R 2021 *Pedosphere* **31**(3) 496–509

IOP Conf. Series: Earth and Environmental Science **980** (2022) 012014 doi:10.1088/1755-1315/980/1/012014

Effectivity of inorganic fertilizer NPK (15-15-6) to growth and yield of lowland rice (*Oryza sativa* L.) on alfisol soil

E Fidiyawati [1], E Latifah [1], A Krismawati [1], D Sihombing [1], D Setyorini [1], A Bakar [1], and Sugiono [1*]

[1] AIAT East Java, Jl Raya Karangploso Km 4 Malang, Indonesia

* email: bptpsugiono@gmail.com

Abstract. This study was conducted to determine the effectiveness of NPK (15-15-6) fertilizer on the growth and yield of lowland rice. It was carried out on irrigated rice fields in the second season (March-June), on Alfisol soil (25 m asl). RCBD experimental design with 8 doses of fertilizer treatment and 3 replications. Treatment: a) without fertilizer, b) local recommendation of NPK (15-15-15) 300 + Urea 200 kg/ha, c) NPK (15-15-6) 200 + Urea 0 kg/ha, d) NPK (15-15-6) 300 + Urea 0 kg/ha, e) NPK (15-15-6) 400 + Urea 0 kg/ha, f) NPK (15-15-6) 200 + Urea 200 kg/ha, g) NPK (15-15-6) 300 + Urea 200 kg/ha, and h) NPK (15-15-6) 400 + Urea 100 kg/ha. The results showed that the application of NPK (15-15-6) 300 + 200 urea kg/ha had a significant effect on plant height, productive tillers and yield. The yield was 7,448 t/ha and the RAE *(Relative Agronomic Effectiveness)* value was 106.46, while the recommended fertilizer yield was 7.20 t/ha and the RAE value was 100. NPK fertilizer (15-15-6) dose of 300 + 200 Urea kg/ha can recomedate as an alternative NPK fertilizer in lowland rice.

1. Introduction

Efforts to forwarded rice self-sufficiency in Indonesia are still confront hindrance. This constraint is caused by rice production just notably concentrated in Java, the high rate of population growth and rivalry for land use [1]. National rice production from the last two years showed an increase of 57.16 million tons in 2012 [2]. In 2013, the rice production raised to 59.88 million tons [2]. One of several endeavor to increase growth and production is intensification through technological improvements, these include: the use of high-yielding varieties, the use of labeled seeds and fertilization to sufficient nutrient requirements [3]. NPK significantly increased grain yields [4] and effectively raised rise production [5].

Various attempts that can be made to overcome the lack of nutrients are the provision of inorganic fertilizers which are usually in the form of single fertilizers such as Urea, TSP/SP-36 and KCL. But in practice, most farmers only apply fertilizer in the form of nitrogen, so it is feared that it can lead to nutrient deficiency in the soil. To cope with this problem, it is necessary to apply a balanced fertilization policy. The concept of balanced fertilization is that plants are not only fertilized with nitrogen and phosphate, but need to be fertilized with other nutrients according to plant requirements and their availability in the soil [6]. Ways to increase domestic production can be done by extensification and intensification of agriculture, especially fertilization which is one of the methods to increase yields, especially fertilization with NPK compound fertilizers [7]. In principle, the overall nutrient balance or fertility must be such that it can promote optimal and normal plant growth [8]. According [9] 50% organic fertilizer (total N 4%) and 50% inorganic fertilizer led to an increase in NPK availability and rice yield during treatment of 100% inorganic fertilizer, reducing the use of inorganic fertilizers and conducing to sustainable agriculture. While fertilizer subsidies are decreasing, research results [10] that fertilizer subsidies have a significant effect on the paddy industry.

In its development, balanced fertilization is generally applied, so that the efficiency of fertilization is low [11]. In general, farmers only provide macronutrients, because they directly affect the quantity of the harvest. If this continues for a long period of time, there will be an imbalance of nutrients in the soil, it will be exacerbated if the organic matter content is also lower in agricultural lands.

Productivity that tends to slop recently is closely related to the nutrient imbalance in paddy fields where the average organic matter is less than 2%. Because few farmers are aware of using organic fertilizers for their paddy fields. With the application of NPK fertilizer and the application of organic fertilizer, it is expected to improve the nutrient adequacy for rice plants, so that crop yields will increase.

The purpose of the study was to determine the effectiveness of NPK (15-15-6) fertilizer on the growth and production of lowland rice plants.

2. Materials and Methods

The location of the research on the effectiveness of NPK fertilizer (15-15-6), in Nglebur Village, Kedungpring, Lamongan, on second season in March - September 2018, in irrigated rice fields. The Inpari 30 rice variety is a new superior rice that has been planted in many locations. NPK (15-15-6) and NPK (15-15-15) inorganic fertilizers, urea fertilizers, and organic fertilizers.

Field research method, cropping pattern in one year paddy-paddy-secondary crops, experimental plots were arranged based on Randomized Block Design with 3 replications, there were 8 treatments. The application of organic fertilizer at a dose of 2 tons per hectare, in the last tillage before leveling the fertilizer evenly distributed. Treatment plots were carried out between 3 days after being leveled or adjusted to soil conditions so that it was easy to plot by making plots according to treatment and replication. Move planting age 21 days, planting jajar legowo 2:1, the number of seeds 2-3 plants per planting hole. Application of NPK fertilizer (15-15-6) and comparison of NPK (15-15-15) at the age of 10 days after planting (days after planting), Urea fertilizer application 3 times, the first and second time application of 0.5% NPK fertilizer, the last application at the age of 35 dap. Optimal pest and disease control, taking into account the principles of IPM to obtain optimum results.

Table 1. NPK (15-15-6) Fertilizer Effectiveness Test Treatment.

Treatment	Treatment Kinds and Dosage of Fertilizer (kg/ha)			Nutrient content (kg/ha)		
	Urea	NPK (15-15-15)	NPK (15-15-6)	N	P_2O_5	K_2O
A	0	0	0	0	0	0
B	200	300	0	137	45	45
C	0	0	200	30	30	12
D	0	0	300	45	45	18
E	0	0	400	60	60	24
F	200	0	200	120	30	12
G	200	0	300	135	45	18
H	100	0	400	105	60	24

Information: Treatment A was control without fertilization to calculate RAE. Treatment B fertilizer recommendation.

The results of soil analysis at the test site are: pH neutral, P_2O_5: very high, K available: low and Mg content: very high, low C-organic and CEC is high and the texture is classified as clay. Observations were made on the components of growth and yield, namely: (a) plant height at the age of 50 days after harvest and before harvest, (b) number of productive tillers. Observation of yields: (a) panicle length, number of filled grain, percentage of void grain, weight of 1,000 grains; and (b) yield - DGH (Dry Grain Harvest).

Observational data were analyzed using the F test followed by DMRT at a level of 5%. The effectiveness of NPK Fertilizer (15-15-6) was measured using the RAE method *(Relative Agronomic Effectiveness)* to standard fertilizers. As a standard fertilizer, the recommended dose is 200 kg/ha Urea + 300 kg/ha NPK (15-15-15). The RAE formula is as follows:

$$RAE = \frac{Yield\ on\ tested\ fertilizers - Yield\ on\ control}{Yield\ on\ standard\ fertilizers - Yield\ on\ control}\ X\ 100\ \%$$

3. Results and Discussion

The research treatment of NPK fertilizer (15-15-6) with the addition of organic fertilizer 2t/ha, showed an effect on plant height, at the age of 50 days after harvest and before harvest. At the age of 50 DAP showed a proportion of NPK and Urea fertilizers in comparison to NPK (15-15-6) and NPK fertilizers (15-15-15). Plant height before harvest application of NPK (15-15-6) 300 kg and Urea 200 kg/ha, showed optimal growth and had a significant effect. According [12] soil analysis showed that the increase in soil fertility was indicated by an increase in C-organic content from 0.89% to 2.43%.

Table 2. Plant height and number of tillers effectiveness of NPK Fertilizer (15-15-6).

Treatment	Plant Height		Tillers Number	
	50 (dap)	Before harvest	50 (dap)	Before harvest
A	67.27 c	88.50 f	7.36 c	8.16 f
B	87.57 a	108.90 b	12.53 a	15.73 a
C	70.57 bc	96.76 e	9.30 b	10.26 e
D	71.13 bc	99.47 d	9.30 b	11.46 de
E	73.46 b	103.13 c	9.56 b	12.13 cd
F	88.06 a	109.93 ab	11.50 a	14.16 ab
G	89.63 a	111.90 a	12.967 a	14.56 ab
H	85.26 a	110.03 ab	12.06 a	13.76 bc
CV	2.98	1.45	7.81	7.56

Information: The line numbers followed by the same letter were not significantly different based on the DMRT follow-up test at the 0.05 level.

The number of tillers in the NPK (15-15-6) treatment with a combination of Urea fertilization and the comparison of NPK (15-15-15) aged 50 days after planting, were not significantly different. The number of productive tillers combination of NPK (15-15-6) + Urea 200 kg/ha and the comparison of NPK (15-15-15) had a significant effect on the number of productive tillers and had a significant effect on the number of tillers aged 50 DAP and productive tillers. According [13] NPK fertilization is able to increase the maximum number of tillers and the highest productive tillers in rice.

The number of tillers is strongly influenced by the availability of nitrogen and phosphorus nutrients in the soil, although the number of tillers that grow not all produce panicles [14]. Tillers begin to form at the age of 10 days and reach a maximum at the age of 50-60 days after planting [15]. After reaching the maximum, the number of tillers decreases. According [16] stated on wheat that there was a very significant difference in the number of effective tillers, plant height, grain content, panicle length, number of grain, straw and harvest index. Based on the results of [13] it was shown that the dose of site-specific NPK fertilizer for the growth of swamp rice plants was a dose of 150% NPK and a dose of 10 Mg ha^{-1} compost straw. This is indicated by the maximum number of tillers and the highest productive tillers in rice plants.

The number of filled grains, the effectiveness test of NPK fertilizer (15-15-6) showed a significant effect on the number of grains per panicle, the number of filled grains was not significantly different from NPK (15-15-6) 400 + Ura 100, NPK (15-15-6) 6) 200 + Urea 200 kg/ha and comparator NPK (15-15-15) 300 + 200 kg/ha. The percentage of empty grain had no significant effect on all NPK fertilizer treatments (15-15-6).

Table 3. Void Grain Percentage, Yield t/ha and RAE value of NPK Fertilizer (15-15-6)

Treatment	Void Grain Percentage	Weight 1000 Grains (gram)	Yield t/ha (DGH)	RAE Value
A	4.51 a	27.30 c	3.33 e	-
B	5.86 a	28.56 ab	7.20 ab	100
C	4.22 a	27.76 bc	4.41 d	28,17
D	4.68 a	27.76 bc	4.70 d	35,4
E	6.56 a	28.23 ab	5.46 c	55,04
F	6.19 a	28.20 ab	7.12 ab	98,12
G	5.62 a	28.86 a	7.44 a	106,46
H	5.51 a	28.60 ab	6.71 b	87,34
CV	26.33	1.61	4.82	

Weight of 1000 grains and real yield, the highest weight was indicated by NPK (15-15-6) 300 + Urea 200 kg/ha treatment. The highest weight was not significantly different in the NPK (15-15-6) 400 kg/ha treatment and the ratio treatment for NPK (15-15-15) 300 + Urea 200.

4. Yield

The yield of the effectiveness test of NPK fertilizer (15-15-6) has a significant effect on the dry grain harvest (DGH). The highest yield was shown in the application of NPK 300 + Urea 200 kg/ha, production of 7.45 t/ha. The results were not significantly different at the dose of NPK (15-15-6) 200 + Urea 200 kg/ha and the comparison of NPK (15-15-15) 300 + Urea 200 kg/ha from the results of various studies showing that the average NPK nutrient transported by superior varieties of inbred rice were 17.5 kg N, 3 kg P and 17 kg K for each tonne of grain and straw produced [17]. [18] Showed the NPK nutrient requirement for hybrid rice 19.1; 3.2; and 1.8 kg/t grain. In general, the more grain and straw produced (to a certain extent), the higher the NPK nutrients absorbed by the plant. According [19] stated that the yield of red rice with a dose of NPK fertilization consisted of urea: 150, SP-36: 100, and KCL: 70 kg/ha and according [20], NPK and organic fertilizers increase rice farming yields.

To determine the value of the effectiveness of the fertilizer used "*Relative Agronomic Effectiveness*" value (RAE) to fertilizer recommendations. As a recommended fertilizer, the recommended dose of NPK (15-15-6) is used, which is used as the basis for the RDKK preparation (Rencana Detail Kebutuhan Kelompok), i.e. 300 kg Phonska + 200 kg Urea. The RAE value of the use of NPK fertilizer (15-15-6) in the treatment of NPK 15-15-6 (300 + 200 Urea/ha) was 106.46 suitable for application as a substitute for NPK fertilizer in rice plants.

5. Conclusion

Application of NPK fertilizer (15-15-6) 300 + 200 Urea kg/ha and comparison of NPK (15-15-15) 300 + 200 kg/ha, Inpari 30 rice varieties in plant height and number of productive tillers were not significantly different, production 7,448 t/ha above the recommended fertilizer production of 7.20.t/ha. The highest RAE value in the treatment (NPK (15-15-6) 300+200 Urea/ha) was 106.46. So that NPK fertilizer (15-15-6) is suitable for application as a substitute for NPK fertilizer in rice plants. Research results [21] the use of fertilizers with the right dose will greatly increase crop yields.

6. Acknowledgement

We express our gratitude to PT. Saraswati as a fertilizer producer for their cooperation, Head of AIAT East Java for direction in carrying out activities. The Sumadi Field Officer and all the implementing teams and all parties assisted in the completion of the research. Donald Sihombing as the supervisor of Scientific Writing.

References

[1] Nataatmadja, H 1988 Perkembangan Produksi dan Kebijaksanaan Pemerintah dalam Produksi Beras. Bogor: Pusat Penelitian dan Pengembangan Tanaman Pangan
[2] BPS-Statistics Indonesia 2014 Statistik Padi dan Palawija Tahun 2012 dan Ramalan Kondisi 2013 http://www.beritaresmistatistik.com
[3] Suparyono, Suprihanto and Sudir 2001 Pemanfaatan Benih Sehat dan Mikroorganisme Terbawa Benih Sebagai Komponen Utama PHT Beberapa Penyakit Penting Tanaman Padi.
[4] Miah M A, Gaihre Y K, Hunter G, Singh U and Hossain S A 2016 Fertilizer Deep Placement Increases Rice Production: Evidence from Farmers' Fields in Southern Bangladesh J. Agron. 108(2) 805–812
[5] Mi W, Sun Y, Xia S, Zhao H, Mi W and Brookes P C 2018 Geoderma effect of inorganic fertilizers with organic amendments on soil chemical properties and rice yield in a low-productivity paddy soil Geoderma 320 23–29
[6] Fagi A M and Makarin A K 1990 Pelestarian Swasembada Beras: Peluang dan Tantangan. Risalah Rapat Kerja Hasil dan Program Penelitian Tanaman Pangan
[7] Baharsyah S, Kasryno F and Darmawan D H 1988 Padi Buku I Bogor
[8] Rauf A W, Syamsudin T and Sihombing S R 2000 Peranan Pupuk NPK pada Tanaman Padi Departemen Pertanian
[9] Moe K, Htwe A Z, Thi T, Thu P and Kajihara Y 2019 Effects on NPK status, growth, dry Matter and yield of rice (Oryza sativa) by organic fertilizers applied in field condition J. Agric. 9(109) 1-15
[10] Ramli N N, Shamsudin M N, Mohamed Z and Radam A 2012 The impact of fertilizer subsidy on Malaysia Paddy / rice industry using a system dynamics approach J.Sci 2(3) 213–219
[11] Adiningsih J S and Soepartini M 1995 Pengelolaan Pupuk pada Sistim Usahatani Lahan Sawah. Makalah pada apresiasi metodologi pengkajian sistem usahatani berbasis padi dengan wawasan agribisnis Bogor 7-9 September 1995 p. 76
[12] Lasmini S A, Kusuma Z, Santoso M and Abadi A L 2015 Application of organic and inorganic fertilizer improving the quantity and quality of shallot yield on dry land Int. J. Sci. Technol. Res. 4(4) 4–7
[13] Aksani D, Budianta D and Hermawan A 2018 Determination of site-specific NPK fertilizer rates for rice grown on Tidal Lowland J. unila 23(1) 19–25
[14] Yoshida S 1981 Fundamentals of rice crop Los Banos Philippines: International Rice Reseach Institute
[15] Vergara B S 1995 Plant growth and Development Rice Production Manual Philippines: International Rice Reseach Institute

[16] Datta R, Prasad L, Subedi R and Prasad B 2020 Heliyon Assessment of productivity and profitabilty of wheat using Nutrient Expert ® -Wheat model in Jhapa district of Nepal Heliyon 6(6) e04144
[17] Dobermann A and Fairhurst T 2000 Rice: Nutrient disorder and nutrient management Philippines: International Rice Reseach Institute
[18] Mahajan B R, Pandey R N, Datta S C, Kumar D, Sahoo R N and Parsad R 2014 Fertilizer nitrogen, phosphorus and sulphur prescription for aromatic hybrid rice (Oryza sativa L.) using targeted yield approach in Proceeding of the National Academy of Science 84(3) 537-547
[19] Masni Z and Masli ME 2019 Yield performance and nutrient uptake of red rice variety (MRM16) at different NPK fertilizer rates International Journal of Agronomy 2019(2) 1-6
[20] Lanya I and Subadiyasa N N 2017 Original research article mineral fertlizer as an alternative fertilizer in Int. J. Biosci. Biotechnol. 4(2) 99–109

The effect of marinade concentrations of different local herbs and spices on the hedonic test of super native chicken breast

H Oktafa[1], AH Prayitno[2*], HT Handayani[3], and DL Rukmi[2]

[1]Department of Health, Politeknik Negeri Jember, Jl. Mastrip Po Box 164 Jember, Indonesia
[2]Department of Animal Science, Politeknik Negeri Jember, Jl. Mastrip Po Box 164 Jember, Indonesia
[3]Department of Agriculture, Politeknik Negeri Jember, Jl. Mastrip Po Box 164 Jember, Indonesia

*E-mail: agushp@polije.ac.id

Abstract. This study was conducted the effect of marinade concentrations of different local herbs and spices on the hedonic test of super native chicken breast. The research material consisted of super native chicken breast, chili powder, sugar, salt, pepper, garlic, cinnamon, palm oil, lime juice, monosodium glutamate, and ginger. The treatment levels of marinade concentration were 10, 20, and 30% of the weight of meat. The hedonic test of the super native chicken marinade was carried out by 40 untrained panelists. The hedonic test data of super native chicken breast marinated were analyzed by non-parametric analysis through the Hedonic Kruskal-Wallis test and if there was a significant difference ($P<0.05$) then further tested with Duncan's New Multiple Range Test. The results showed that the marinade concentration level of different local herbs and spices had a significant effect on acceptability, but did not affect on color, aroma, taste, texture, juiciness, and tenderness of super native chicken breast marinated. The marinade concentration of local herbs and spices at the level of 20% was the best treatment with highest score hedonic of 4.38 so that it was preferred by the panelists when compared to other treatments.

1. Introduction

Super native chicken is a cross between a native rooster and a layer chicken. Super native chicken is very popular for consumption by the people of Indonesia because it is known that it tastes better, is lower in fat, cholesterol, and takes a shorter time of about 45 to 60 days to be ready to be harvested. Native chicken takes a longer time than super native chicken, about 75 to 80 days to be harvested. Super native chicken meat as a source of Indonesian animal germplasm continues to be developed into ready-to-cook products. Ready to cook (RTC) products that are currently trending are marinated chicken meat products. Ready-to-cook products are becoming popular in the community because there are restrictions on the Covid-19 pandemic, making it easier for the cooking process. Marinating technology is the process of soaking meat in a marinade. The marinade is a seasoned solution that functions as a meat marinade, usually used to improve the taste, impression of juiciness, and tenderness of meat after cooking [1,2]. Marinade ingredients that can be used are local herbs and spices which are widely available in Indonesia as a source of local potential.

Local herbs and spices such as chili powder, sugar, salt, pepper, garlic, cinnamon, palm oil, lime juice, and ginger have been extensively studied separately for use as marinades and flavorings in meat processing. Meat marinade technology can also improve food safety and added value to the product [3]. The principle of meat marinade technology is soaking meat in a marinade containing certain ingredients so that slowly passive transport of the marinade material into the meat by osmosis occurs [1]. The increase in the taste and tenderness of the meat due to the marinating process is caused by the increase in the water holding capacity of the meat [4]. The application of marinade technology on broiler meat can improve the taste, tenderness, and acceptability of the panelists [5]. Local herbs and spices have the potential to be used together to be used as marinade ingredients for super native chicken. Breast meat is one of the most preferred parts of meat by some consumers [6]. So far, no one has evaluated the application of local herbs and spices marinade technology for super native chicken breast. The hedonic test is generally carried out to evaluate the panelists' preference for processed meat products [7]. Therefore, this study was conducted to know the effect of marinating local herbs and spices with different concentrations on the hedonic test of super native chicken breast.

2. Materials and Methods

2.1. Materials
The material used in this study consisted of native chicken breast, chili powder, sugar, salt, pepper, garlic, cinnamon, palm oil, lime juice, ginger, and monosodium glutamate (MSG).

2.2. Methods
This research consists of several processes. This research was started from marinade preparation, meat preparation, marinating process, cooking, and hedonic test.

2.2.1. Preparation of meat. The breast super native chicken used in this study came from super native chickens (unisex) obtained from local farms in Jember.

2.2.2. Preparation of marinade. The marinade formulation used in this study used local herbs and spices which included: 12% chili powder, 12.25% sugar, 8% salt, 1% pepper, 0.25% cinnamon, 17% garlic, 25% palm oil, 10% lime juice, 12% MSG, and 2.5% ginger.

2.2.3. Marination process. The chicken breast mixed with marinade with treatment concentration levels, namely: 10, 20, and 30% of the meat total weight. The chicken breast marination until it is evenly distributed. The chicken breast was then vacuum-packed with time marination for 60 minutes.

2.2.4. Cooking process. The chicken breast of marinated was cooked by steaming for 45 minutes and after the cooked meat it is then cooled and then hedonic tested by untrained panelists.

2.3. Hedonic Test
The super native chicken breast which has been marinated with different levels of marinade concentration was tested hedonic by 40 untrained panelists [8]. The parameters observed were: color, flavor, taste, texture, tenderness, juiciness, and acceptability [7]. The hedonic scale used is 1 (dislike extremely), 2 (dislike), 3 (rather like), 4 (like), and 5 (like extremely).

2.4. *Statistic Analysis*
The data of hedonic test were analyzed by analysis non-parametric test of Hedonic Kruskal-Wallis and and if there was a significant difference ($P<0.05$) then further tested with Duncan's New Multiple Range Test [9].

3. Results and Discussion

The hedonic test is one of the sensory characteristics of meat products that is widely tested to determine the level of preference of the panelists for the meat products being tested [10]. The results of the hedonic test that have been carried out by the panelists on super native chicken breast cooked with local herbs and spices with different concentration levels were presented in Table 1.

Table 1. Hedonic test results of super native chicken breast marinated with herbs and spices with different levels of marinade concentration

Variable	Marinade concentration level		
	10%	20%	30%
Color^{ns}	3.40	3.60	3.73
Flavor^{ns}	3.93	3.70	3.83
Taste^{ns}	3.95	3.88	3.73
Texture^{ns}	3.80	4.05	3.88
Tenderness^{ns}	3.93	3.98	3.85
Juiceness^{ns}	3.73	3.95	3.95
Acceptability	3.93^{a}	4.38^{b}	4.15^{ab}

^{ab}Different superscripts at the same row indicate significant differences ($P<0.05$)

3.1. Color

The results showed that the concentration level of marinade of different local spices and herbs had no significant effect ($P>0.05$) on the color of the marinated breast meat. The color score of the marinated breast meat ranged from 3.40 to 3.73, i.e. from rather like to like. The level of preference of the panelists on the color of breast meat that has been marinated with the marinade of local herbs and spices at different concentration levels did not change. This could be because the panelists had difficulty distinguishing the color of the marinated breast meat with local herbs and spices at different concentration levels visually by eye. Perception of meat color can influence panelists' decisions in choosing meat and its processed products [11]. Visually, the panelists are thought to have the same perception of all colors of super chicken breast that have been marinated with different levels of marinade concentration of local herbs and spices so that the assessment of the color score of the marinated meat does not differ much for each treatment..

3.2. Flavor

The results showed that the concentration level of marinade of different local spices and herbs had no significant effect ($P>0.05$) on the flavor of the marinated breast meat. The aroma score of the marinated breast meat ranged from 3.70 to 3.93, i.e. from rather like to like. The level of panelists' preference for the aroma of breast meat that has been marinated with the marinade of local herbs and spices at different concentration levels did not change. This could be because the panelists had difficulty distinguishing the flavor of the marinated breast meat from the marinade of local herbs and spices at different concentration levels by smelling it. The flavor of the meat was produced from a combination of various components that stimulate the olfactory receptors in the nasal passages [11], while the flavor-forming compounds derived from local herbs and spices as a marinade for super native chicken breast with its concentration level are thought to have almost the same flavor after being smelled by panelists so that based on these perceptions, the panelists' assessment of the flavor score of the marinated meat did not differ much for each treatment.

3.3. Taste

The taste is a sensory quality of meat-related to the sense of taste [7]. The results showed that the concentration level of marinade of different local herbs and spices had no significant effect ($P>0.05$) on the taste of the marinated breast meat. The taste score of the marinated breast meat ranged from 3.73 to 3.95, which was from rather like to like. The level of preference of the panelists on the taste of breast meat that has been marinated with the marinade of local herbs and spices at different concentration levels did not change. This could be because the panelists had difficulty distinguishing the taste of the marinated breast meat from the marinade of local herbs and spices at different levels of concentration by the way they were tasted. The taste of the meat was produced from a combination of various components that stimulate the taste receptors on the tongue [11], while the specific flavoring compounds derived from local herbs and spices as a marinade for super native chicken breast with the level of concentration are thought to have almost the same taste after being tasted by the panelists taste buds so that based on these perceptions, the panelists' assessment of the taste score of the marinated meat did not differ much for each treatment.

3.4. *Texture*
The texture is a sensory property of meat that is related to the level of smoothness of the meat[7]. The results showed that the concentration level of marinade of different local herbs and spices had no significant effect (P>0.05) on the texture of the marinated breast meat. The texture score of the marinated breast meat ranged from 3.80 to 4.05, i.e. from rather like to like. The level of preference of panelists on the texture of breast meat that has been marinated with the marinade of local herbs and spices at different concentration levels did not change. This could be because the panelists had difficulty distinguishing the texture of the marinated breast meat from the marinade of local herbs and spices at different concentration levels when chewing the meat. Panelists are suspected of having almost the same perception of the texture of the meat used, namely the breast which is known to have a smooth texture [6] so that based on this perception, the panelist's assessment of the texture score of the marinated super native chicken breast is not much different for each treatment.

3.5. *Tenderness*
The tenderness is the main parameter in determining the quality of meat that is sensory tested [7]. The results showed that the concentration level of marinade of different local spices and herbs had no significant effect (P>0.05) on the tenderness of the marinated breast meat. The tenderness score of the marinated breast meat ranged from 3.85 to 3.98, ie from rather like to like. The panelists' preference for tenderness of breast meat that had been marinated with the marinade of local herbs and spices at different concentration levels did not change. This could be because the panelists had difficulty distinguishing the tenderness of the marinated breast meat from the marinade of local herbs and spices at different concentration levels when biting the meat. Panelists are thought to have almost the same perception of the tenderness of the meat used, namely the breast which is known to be tender [6] so that based on this perception, the panelist's assessment of the tenderness score of marinated super native chicken breast is not much different for each treatment.

3.6. *Juiciness*
The juiciness is a sensory property related to the level of wetness of the meat [7]. The results showed that the concentration level of marinade of different local spices and herbs had no effect (P>0.05) on the juiciness of the marinated breast meat. The juiciness score of the marinated breast meat ranged from 3.73 to 3.95, ie from rather like to like. The level of panelists' preference for the juiciness of breast meat that had been marinated with the marinade of local herbs and spices at different concentration levels did not change. This could be because the panelists had difficulty distinguishing the juiciness level of the marinated breast meat from the marinade of local herbs and spices at different concentration levels when chewing the meat. Panelists are thought to have almost the same perception of the level of wetness of the meat, so based on this perception, the panelist's assessment of the juiciness score of marinated super native chicken breast is not much different for each treatment.

3.7. *Acceptabiliy*
The acceptability is part of the sensory properties of the meat to the level of acceptance of the panelists to all the sensory properties of the meat tested [7]. The results showed that the concentration level of

marinade of different local spices and herbs had a significant effect ($P<0.05$) on the acceptability of marinated breast meat. The color score of the marinated breast meat ranged from 3.93 to 4.38, from like to like extremely. The level of preference of panelists on the acceptability of breast meat that has been marinated with the marinade of local herbs and spices at the concentration level has changed. The acceptability scores starting from the highest were obtained from breast meat that had been marinated with marinade concentration levels of 20%, 30%, and 10%, namely 4.38; 4.15; and 3.93 respectively. This could be due to the panelists as a whole that the marinade concentration level of 20% was the right concentration for marinating super native chicken breast.

4. Conclusion

The results showed that the marinade concentration level of different local herbs and spices had a significant effect on acceptability, but did not affect on color, flavor, taste, texture, juiciness, and tenderness of super native chicken breast marinated. The marinade concentration of local herbs and spices at the level of 20% was the best treatment with highest score hedonic of 4.38 so that it was preferred by the panelists when compared to other treatments.

Acknowledgment

The author thanked profusely to the P3M of Politeknik Negeri Jember for providing research funding through PNPB 2021 sources so that this research could be carried out.

References

[1] Brooks C 2011 Marinating of Beef for Enhancement *Beef Res.*

[2] Yusop S M, O'Sullivan M G and Kerry J P 2011 *Marinating and enhancement of the nutritional content of processed meat products* (Woodhead Publishing Limited)

[3] Nurwantoro, Bintoro V P, Legowo A M and Purnomoadi A 2012 Pengolahan daging dengan sistem marinasi untuk meningkatkan keamanan pangan dan nilai tambah *Wartazoa* **22** 72–8

[4] Carroll C D, Alvarado C Z, Brashears M M, Thompson L D and Boyce J 2007 Marination of turkey breast fillets to control the growth of Listeria monocytogenes and improve meat quality in deli loaves *Poult. Sci.* **86** 150–5

[5] Birk T, Grønlund A C, Christensen B B, Knøchel S, Lohse K and Rosenquist H 2010 Effect of organic acids and marination ingredients on the survival of Campylobacter jejuni on meat *J. Food Prot.* **73** 258–65

[6] Resnawati H 2005 Karakteristik karkas dan preferensi konsumen terhadap daging dada ayam yang diberi ransum mengandung cacing tanah (Lumbricus rubellus) *Prosiding Seminar Nasional Teknologi Inovatif Pascapanen untuk Pengembangan Industri Berbasisi Pertanian* (Bogor: Balai Besar Penelitian dan Pengembangan Pascapanen Pertanian) pp 424–31

[7] Prayitno A H, Suryanto E and Zuprizal 2010 Kualitas fisik dan sensoris daging ayam broiler yang diberi pakan dengan penambahan ampas virgin coconut oil (VCO) *Bul. Peternak.* **34** 55–63

[8] Setyaningsih D, Apriyantono A and Sari M P 2010 *Analisa Sensori Industri Pangan dan Agro* (Bogor: IPB Press)

[9] Riadi E 2014 *Metode Statistika: Parametrik & Non-Parametrik* (Tangerang: Pustaka Mandiri)

[10] Prasetyo B and Prayitno A H 2021 The sensory characteristics of fortified beef sausage with duck eggshell *The 3rd International Conference On Food and Agriculture* (Jember: IOP Conference Series: Earth and Environmental Science) pp 1–6

[11] Syamsir E 2011 Karakteristik mutu daging *Kulinologi Indones.* 1–7

The fortification effect of duck eggshell nano-calcium on the physical quality of beef sausage

AH Prayitno[1*], DL Rukmi[1], A Widiyawati[2], and B Prasetyo[1]

[1]Department of Animal Science, Politeknik Negeri Jember, Jl. Mastrip Po Box 164 Jember, Indonesia
[2]Department of Health, Politeknik Negeri Jember, Jl. Mastrip Po Box 164 Jember, Indonesia

*E-mail: agushp@polije.ac.id

Abstract. This research was aimed the fortification effect of duck eggshell nano-calcium on the physical quality of beef sausage. The materials include beef, duck eggshell nano-calcium, tapioca, soy protein isolate, palm oil, salt, pepper, garlic, onion, shallot, coriander, nutmeg, sugar, frankfurter, sodium tripolyphosphate, monosodium glutamate, and ice. The treatments for fortification of duck eggshell nano-calcium were 0; 0.15; 0.3; 0.45; and 0.6% of the total dough. The parameters tested were pH value, water holding capacity, tenderness, and water activity. The physical quality data were analyzed by analysis of variance and if there was a significant difference ($P<0.05$) then further tested with Duncan's New Multiple Range Test. The results showed that the fortification of duck eggshell nano-calcium had a significant effect ($P<0.05$) on the pH value, water holding capacity, tenderness, and water activity of beef sausage. Fortification of duck eggshell nano calcium up to 0.6% can increase the value of pH, tenderness, water activity, and decrease water holding capacity.

1. Introduction

Duck eggshells are one of the wastes from duck farming which are abundant with a high calcium content. The utilization of by-products from duck eggshells [1] can be used as a source of dietary calcium. Duck eggshells contain about 94-97% calcium carbonate [2]. The calcium content of eggshells in the form of flour is as much as 50.75% [1]. The function and economic value of duck eggshells can be increased by the application of nanotechnology to form nanoparticles. High energy ball milling is nanotechnology that can be applied to change the size of eggshell particles into nano. The process of applying nanotechnology with high energy ball milling can change the particle size of duck eggshells from 13,229 nm to 347 nm in the form of calcium oxide [3] with a calcium content of around 54.36-59.27% [4]. Particles in nanosize have the advantage of being easy to dissolve and high absorption in the intestine so that they are more efficient when compared to micro-sized calcium [5].

The performance of nano-sized particles is better due to an increase in a surface area [6]. One type of metal oxide that has been widely applied is nano calcium oxide (NCaO). Nano calcium oxide can be used as an antibacterial agent [7], catalyst [8], food additive [9–13], drug delivery system [14] which can increase absorption [15] so that it can be absorbed almost 100% by the body [16]. Nano calcium duck eggshell as a source of food is very good compared to other sources of calcium as a component that is functional and has a positive impact on body health. The results of research conducted by the International Osteoporosis Foundation (IOF) show that 1 in 4 Indonesian women with an age range of 50 to 80 years has a risk of osteoporosis so that Indonesian women have a risk of osteoporosis 4 times higher than men at risk of osteoporosis [17]. Men up to 55 years of age and after 55 years of age have a higher risk of osteoporosis than women. Low intake of calcium into the body causes the risk of osteoporosis [18].

One of the things that the body needs to meet bone health is calcium as a macromineral. Fulfillment of body calcium obtained from broccoli, spinach, soybeans, milk, and dairy products. Fulfillment of body calcium can also be met in other ways such as consuming calcium in the form of supplements. Whereas there is another way that is safer and healthier, namely by consuming food products fortified with calcium. Eggshell nano-calcium as a natural source of calcium has been developed as a fortified material for functional food products. Fortified sausage is one of the most popular forms of food in Indonesia. Sausage is one of the comminution products of meat which is in the process of being made with or without the addition of other food ingredients that are inserted into the sausage casing with or without a cooking process [19]. Sausage fortified with eggshell calcium can be used as a functional food [20]. Sausage fortified with eggshell calcium has been studied [21] and has the potential to prevent osteoporosis for those who consume it [22]. So far, there has been no supporting research on the effect of nano-calcium fortification on the physical quality of sausages. The physical quality of sausage is one of the variables used to evaluate the quality of sausage products that are tested objectively. Therefore, this study was conducted to determine the effect of duck eggshells nano-calcium fortification of on the physical quality of beef sausage and its potential as a functional food.

2. Materials and Methods

2.1. Materials

The materials used in this study consisted of beef, duck eggshell nano-calcium, tapioca, palm oil, salt, garlic, pepper, monosodium glutamate (MSG), sodium tripolyphosphate (STPP), soybean protein isolate, frankfurter, ice, and collagen casing.

2.2. Methods

This research consists of several steps. This research started from the preparation of duck eggshell nano-calcium, sausage processing, physical quality test, and statistical analysis.

2.2.1. Preparation of duck eggshell nano-calcium. Eggshell nano-calcium are made using high energy ball milling [3]. Duck eggshells were soaked in hot water for 10 minutes, the eggshell membranes were cleaned, dried at 105°C for 12 hours, then mashed using a sample mill. Eggshell flour was calcined at 1000°C for 2 hours to produce calcium oxide. Eggshell calcium oxide was further processed using high energy ball milling for 60 minutes to produce nano calcium oxide powder.

2.2.2. Preparation of ingredients. The sausage dough formulations that will be used in this study are beef (50%), tapioca (16.5%), soybean protein isolate (2.5%), palm oil (10.5%), salt (1.2 %), STPP (0.5%), MSG (1%), pepper (0.2%), garlic (1.2%), shallot (2%), onion (2%), coriander (0.2%), nutmeg (0.2%), sugar (0.5%), frankfurter (1%), and ice (10.5%).

2.2.3. Sausage processing. The level of fortification of eggshell nano-calcium in this study were: 0; 0.15; 0.3; 0.45; and 0.6% of the total sausage dough. Beef cut into small pieces, cleaned of connective tissue, then ground. Ground meat and oil were mixed, added salt, sodium tripolyphosphate, eggshell nano-calcium treatment, and half of the ice. All spices are ground, and oil is added to the dough, soy protein isolate, tapioca, and the remaining half of the ice are mixed until smooth. Sausage dough is inserted into the collagen casing. Raw sausage is then boiled for 45 minutes at a temperature of 60-70°C and after being cooked the sausage is cooled and then tested for physical quality.

2.3. *Physical Quality Test*
The fortified beef sausage with duck eggshell nano-calcium tested physical quality, namely the pH value [23], water holding capacity [24], tenderness [25], and water activity [26].

2.4. *Statistic Analysis*
The data physical quality were analyzed by analysis of variance using completely randomized design and if there was significantly different (P<0.05), then tested further by the Duncan's New Multiple Range Test [27].

3. Results and Discussion
The physical quality is a parameter of meat quality consisting of pH value, water holding capacity, tenderness, and water activity of meat products that were tested objectively [28]. The results of the physical quality test of sausage fortified with duck eggshell nano-calcium were presented in Table 1.

Table 1. The results of the physical quality test of sausage fortified with duck eggshell nano-calcium

Variable	Fortification Level				
	0%	0,15%	0,30%	0,45%	0,60%
Value of pH	6.38^{a}	6.68^{a}	7.38^{b}	8.43^{c}	8.93^{d}
Water holding capacity (%)	93.96^{a}	92.70^{b}	85.80^{c}	79.30^{d}	74.41^{e}
Tenderness (mm/g/ 5 second)	8.30^{ab}	8.50^{bc}	8.45^{b}	8.00^{a}	8.90^{c}
Water activity	0.68^{ab}	0.78^{ab}	0.78^{ab}	0.75^{ab}	0.83^{b}

^{abcde}Different superscripts at the same row indicate significant differences (P<0.05)

3.1. *Value of pH*
The results showed that different levels of fortified duck eggshell nano-calcium had a significant effect (P<0.05) on the pH value of beef sausage. The pH value of beef sausage produced ranged from 6.38-8.93. The higher the level of fortified duck eggshell nano-calcium, the pH value of beef sausage increased. This is following the research of Suryanto et al. [12] that fortification of eggshell calcium can increase the pH value of processed meat products. Eggshell nano-calcium fortified in a beef sausage in the form of calcium oxide which decomposes into Ca^{2+} and O^{2-} [1]. Ca^{2+} is alkaline and O^{2-} is acidic so oxygen will evaporate into gas and beef sausage fortified with duck eggshell nano-calcium has higher levels of alkaline calcium, which causes the pH value of beef sausage to increase.

3.2. *Water holding capacity*
The results showed that different levels of fortified duck eggshell nano-calcium had a significant effect (P<0.05) on the water holding capacity of beef sausage. The water holding capacity of beef sausage produced ranges from 74.41-93-96%. The higher the level of fortified duck eggshell nano-calcium, the water holding capacity of beef sausage decreases. The decreasing water holding capacity of sausages is followed by the increasing pH value. The water holding capacity of sausages is strongly influenced by the pH value [1]. Increasing the pH value causes changes in the protein content of meat [29] and with the presence of fortified duck eggshell nano-calcium in sausage processing, the ability of meat proteins to bind water decreases as a result, the water holding capacity of sausages decreases.

3.3. *Tenderness*
The results showed that different levels of fortified duck eggshell nano-calcium had a significant effect (P<0.05) on the tenderness of beef sausage. The tenderness value of beef sausage produced ranges from 8.00-8.90 mm/kg/5 second. The higher the level of eggshell nano-calcium duck fortification, the tenderness value of beef sausage increased. The increase in the value of sausage tenderness indicates that the resulting sausage becomes more tender. This can be influenced by the water holding capacity of the sausage which is decreasing so that the texture of the sausage is getting more open as a result the sausage becomes more tender. In addition, the increasing pH value of sausages affects the binding

capacity of sausage water (Table 1). This is in accordance with the research of Prayitno et al. [30] that tenderness has a relationship with the pH value and water holding capacity

3.4. Water activity

The results showed that different levels of fortified duck eggshell nano-calcium had a significant effect ($P<0.05$) on the water activity of beef sausage. The water activity value of beef sausage produced ranged from 0.68-0.83. The higher the level of fortified duck eggshell nano-calcium, the water activity value of beef sausage increased. The increase in water activity indicates that the free water content in the sausage is getting higher. This can be influenced by the decreasing water holding capacity so that the ability of meat protein to bind the free water content decreases as a result the water activity value of sausages increases. In addition, the increase in water activity can also be affected by the meat processing process. This is in accordance with Novasina [31] that the water activity value of meat products decreases during the processing where the water activity value of sausages ranges from 0.97-0.96 so that the sausages fortified with duck eggshell nano-calcium in this research still meet the standards for water activity value.

4. Conclusion

The results showed that the fortification of duck eggshell nano-calcium had a significant effect on the pH value, water holding capacity, tenderness, and water activity of beef sausage. Fortification of duck eggshell nano-calcium up to 0.6% can increase the value of pH, tenderness, water activity, and decrease water holding capacity.

Acknowledgment

The author thanked profusely to the P3M of Politeknik Negeri Jember for providing research funding through PNPB 2021 sources so that this research could be carried out.

References

[1] Prayitno A H, Suryanto E and Rusman 2016 Pengaruh fortifikasi nanopartikel kalsium laktat kerabang telur terhadap sifat kimia dan fisik bakso ayam *Bul. Peternak.* **40** 40–7
[2] Nurlaela A, Dewi S U, Dahlan K and Soejoko D S 2014 Pemanfaatan limbah cangkang telur ayam dan bebek sebagai sumber kalsium untuk sintesis mineral tulang *J. Pendidik. Fis. Indones.* **10** 81–5
[3] Prayitno A H and Sutirtoadi A 2019 *Karakteristik Nano Kalsium Alami Berbagai Jenis Kerabang Unggas* (Jember: Politeknik Negeri Jember)
[4] Prayitno A, Prasetyo B and Sutirtoadi A 2020 Synthesis and characteristics of nano calcium oxide from duck eggshells by precipitation method *IOP Conf. Ser. Earth Environ. Sci.* **411** 1–6
[5] Poulain N and Nakache E 1998 Nanoparticles from vesicles polymerization. II. Evaluation of their encapsulation capacity *J. Polym. Sci. A Polym. Chem.* **36** 3035–43
[6] Habte L, Shiferaw N, Mulatu D, Thenepalli T, Chilakala R and Ahn J W 2019 Synthesis of nano-calcium oxide from waste eggshell by sol-gel method *Sustain.* **11** 1–10
[7] Roy A, Gauri S S, Bhattacharya M and Bhattacharya J 2013 Antimicrobial activity of CaO nanoparticles *J. Biomed. Nanotechnol.* **9** 1570–8
[8] Gopalappa H, Yogendra K, Mahadevan K M and Madhusudhana N 2012 A comparative study on the solar photocatalytic degradation of Brilliant Red azo dye by CaO and CaMgO2 nanoparticles *Int. J. Sci. Res.* **1** 91–5
[9] Prasetyo B and Prayitno A H 2021 The sensory characteristics of fortified beef sausage with duck eggshell *The 3rd International Conference On Food and Agriculture* (Jember: IOP Conference Series: Earth and Environmental Science) pp 1–6
[10] Prasetyo B and Prayitno A H 2020 *Aplikasi Nano Kalsium Alami Kerabang Telur Pada Sosis dan Potensinya Sebagai Pangan Fungsional* (Jember)
[11] Suryanto E, Setiyono, Rusman and Prayitno A H 2014 Firmness and microstructure properties

of chicken meatball fortified with eggshell calcium powder *Sustainable Livestock Production in The Prespective of Food Security, Policy, Genetic Resources, and Climate Change* (Yogyakarta: The Asian-Australasian Association of Animal Production Societies) pp 1280–183

[12] Suryanto E, Setiyono, Rusman and Prayitno A H 2014 Chemical composition, cooking, physical and sensorial properties of chicken meatball fortified with eggshell calcium powder *XIVth European Poultry Conference* (Stavanger: World's Poultry Science Association) pp 1–3

[13] Prayitno A H, Suryanto E, Rusman, Setiyono, Jamhari and Utami R 2019 Pengaruh fortifikasi kalsium dan nanopartikel kalsium laktat kerabang telur terhadap sifat sensoris bakso ayam *Prosiding Seminar Nasional Teknologi Peternakan dan Veteriner* (Jember: Pusat Penelitian dan Pengembangan Peternakan) pp 725–32

[14] Balaganesh A S, Sengodan R, Ranjithkumar R and Chandarshekar B 2018 Synthesis and characterization of porous calcium oxide nanoparticles (CaO NPS) *Int. J. Innov. Technol. Explor. Eng.* 2278–3075

[15] Malsch N H 2005 *Biomedical nanotechnology* (Boca Raton (US): CRC Press)

[16] Suptijah P 2009 Sumber Nano Kalsium Hewan Perairan *101 Inovasi Indonesia* (Jakarta: Kementerian Riset dan Teknologi)

[17] Kementerian Kesehatan 2005 *Data dan Kondisi Penyakit Osteoporosis di Indonesia* (Jakarta: Pusat Data dan Informasi Kementerian Kesehatan RI)

[18] Jahari A B and Prihatini S 2014 Risiko Osteoporosis Di Indonesia *Gizi Indones.* **30** 1–11

[19] SNI 2015 *Sosis Daging* (Jakarta: Badan Standardisasi Nasional)

[20] Astawan M 2011 Pangan Fungsional Untuk Kesehatan Yang Optimal *Kompas*

[21] Daengprok W, Garnjanagoonchorn W and Mine Y 2002 Fermented pork sausage fortified with commercial or hen eggshell calcium lactate *Meat Sci.* **62** 199–204

[22] Murota I, Baba T, Shimada M and Sato N 2010 Safety evaluation of excessive intake of calcium-fortified fish sausage to prevent osteoporosis in humans *J. Japanese Soc. Food Sci. Technol.* **57** 163–70

[23] AOAC 2019 *Official Methods of Analysis of the Association of Analytical Chemists* (Washington DC: Association of Official Analytical Chemist)

[24] Luckose F and Pandey M C 2014 Combined Effect of Non Meat Proteins and Different Binders on Low Salt Poultry Meat Systems *Int. J. Adv. Res.* **2** 413–24

[25] Dhana I G N A O and Wikandari P R 2019 Pengaruh konsentrasi enzim protease dari isolat Lactobacillus plantarum B1765 terhadap keempukan daging *UNESA J. Chem.* **8** 33–7

[26] Petracci M, Laghi L, Rocculi P, Rimini S, Panarese V, Cremonini M A and Cavani C 2012 The use of sodium bicarbonate for marination of broiler breast meat *Poult. Sci.* **91** 526–34

[27] Riadi E 2014 *Metode Statistika: Parametrik & Non-Parametrik* (Tangerang: Pustaka Mandiri)

[28] Sujarwanta R O, Jamhari, Suryanto E, Yuliatmo R and Prayitno A H 2019 Physicochemical and sensory characteristics of chicken nugget with curcuma (Curcuma zanthorrhiza) flour fortification *IOP Conference Series: Earth and Environmental Science* vol 387

[29] Knipe L 2003 EMULSIFIERS | Phosphates as Meat Emulsion Stabilizers *Encyclopedia of Food Sciences and Nutrition* (Cambridge: Academic Press) pp 2077–80

[30] Prayitno A H, Suryanto E and Zuprizal 2010 Kualitas fisik dan sensoris daging ayam broiler yang diberi pakan dengan penambahan ampas virgin coconut oil (VCO) *Bul. Peternak.* **34** 55–63

[31] Novasina 2008 *The influence of water activity (aw) in meat products* (Lavalle: EQUINLAB S.R.L.)

Potential of tamarind seeds (*Tamarindus indica* L.) as prebiotics on the growth of lactic acid bacteria

M. M. D. Utami[1], A. C. Dewi[1], and N. Ningsih[1]

[1]Department of Animal Science, Politeknik Negeri Jember, Jl. Mastrip PO Box 164 Jember 68121, Indonesia

*E-mail: merry.mdu@polije.ac.id

Abstract. The purpose of this study was to examine the potential of tamarind seeds as a prebiotic by conducting tamarind seeds with a temperature and incubation time. Prebiotic testing was carried out using *Lactobacillus casei* grown on media with the addition of tamarind seeds. The research used a completely randomized design. The temperature was divided into two levels (60 and 70°C) and the incubation time is divided into three levels (120, 180, and 240 minutes). The treatments were T1 (control), T2 (60°C, 120 minutes), T3 (60°C, 180 minutes), T4 (60°C, 240 minutes), T5 (70°C 120 minutes), T6 (70°C, 180 minutes), and T7 (70°C, 240 minutes). Each experiment was repeated twice. Data were analyzed by analysis of variance, if there was a significant effect then continued with Duncan test. Based on the results of this study: tamarind seeds which were incubated at a temperature of 60°C for 240 minutes increased the number of colonies significantly ($P<0.05$) compared to the incubation time below, incubation tamarind seeds at 70°C increased the number of bacterial colonies significantly ($P<0.05$) in all-time ranges, however the optimal time 180 minutes. Tamarind seeds have potential as prebiotics, all the parameters showed the ability of tamarind seeds as a prebiotic.

1. Introduction

Microflora imbalance in the digestive tract due to colonization of pathogenic bacteria interferes with livestock growth. Probiotic bacteria can improve the balance of microflora in the digestive tract. The main effect of probiotics in poultry is to improve growth performance, meat quality, immune response, and affect not only morphology but also gut microbiota.

Prebiotics are carbohydrates that are not digested and not absorbed by monogastric livestock, such as chickens. Several studies reported that prebiotics modulates the growth of probiotic bacteria such as Lactobacillus and Bifidobacteria by modifying the balance of intestinal microflora [1]. The main prebiotic mechanism of action is the modulation of the gut microbiota by selectively providing food for beneficial groups of probiotic bacteria [2] and reducing colonization of unwanted intestinal pathogenic bacteria, thereby increasing the integrity of the intestinal mucosa [3]. The mechanism of prebiotic activity affects the composition of bacteria in the gut and enhances immunity to chicken [4] [5].

The general requirements of prebiotics cannot be absorbed when passing through the upper gastrointestinal tract, resistant to acidic pH, stimulating the growth of non-pathogenic bacteria, and modulating the host defense system [6]. The prebiotics tried in chickens are oligosaccharides such as fructooligosaccharides, inulin [7], galactooligosaccharides [8], mannaoligosachharides [9], lactose [2], and xylooligosaccharides [10].

Some commercial prebiotics includes yeast cells [11] [12] *Saccharomyces cerevisiae* or yeast cultures showed a prebiotic effect [13]. In addition to the presence of commercial prebiotics from Saccharomyces, some plants contain polysaccharides such as oligosaccharides, inulin, which have potential as prebiotics.

Tamarind seeds (*Tamarindus indica* L), contain quite large polysaccharides around 50%-60% [14]. Fructooligosaccharides are fructose polymers that are undigested in the intestines of poultry [15] [16]. Oligosaccharides and polysaccharides undigested by non-ruminant digestive enzymes. Tamarind seeds are known to contain a large number of polysaccharides, but it is not clear whether tamarind seeds can be used for prebiotics. Therefore, this study was conducted to study the effect of tamarind seed powder on the potential prebiotics and to determine the potential of tamarind seeds as prebiotics, it is necessary to research tamarind seeds as prebiotics by observing the growth of probiotic bacteria given tamarind seeds by calculation of bacteria with Total Plate Count (TPC).

2. Materials and Methods

2.1. Preparation of tamarind seeds powder

The tamarind seeds were mashed with a blender until the size was 60 mesh, then incubated in the divided water bath at 60 and 75°C. Each sample with a different temperature was then incubated for 120, 180, and 240 minutes. Furthermore, each sample of tamarind seeds with temperature and time determination is taken 2% by dissolving in aqua dest. Proximate analysis of tamarind seeds was carried out in the laboratory of chemistry, food biochemistry, and agricultural products at the University of Jember. Proximates composition of *Tamarindus indica* (Table 1.)

Table 1. Proximates Composition of *Tamarindus indica*

Sample	Moisture (%)	Crude Protein (%)	Crude Fat (%)	Ash (%)	Carbohydrates (%)
simplo	9.73	15,61	3,26	2,70	68,71
duplo	9,28	15,60	3,31	2,62	69,19
mean	9.51	15.61	3.29	2.66	68.45

2.2 Preparation of Lactobacillus casei

The microbial culture was *Lactobacillus casei* FNCC 0090 obtained from the Center for Food and Nutrition Studies, Gadjah Mada University. The bacterial growth medium was MRSA. Dilution using (plating method). A total of 1 ml *Lactobacillus casei* placed in a test tube containing 9 ml of Sodium Chloride solution 0.9%, then diluted by taking 1 ml put in a test tube containing 9 ml of Sodium Chloride solution 0.9% so that a 10^{-1} dilution is obtained, then from the 10^{-1} dilution tube 1 ml is taken and put into the test tube containing 9 ml of Sodium Chloride 0.9% solution, so that it can be diluted 10^{-2}. The procedures were handled continuously until the level of dilution is 10^{-8}.

2.3 Treatments and trial procedure

The research used a completely randomized design. The temperature was divided into two levels (60 and 70°C) and the incubation time is divided into three levels (120, 180, and 240 minutes). The treatments were T1 (control, 27°C), T2 (60°C, 120 minutes), T3 (60°C, 180 minutes), T4 (60°C, 240 minutes), T5 (70°C 120 minutes), T6 (70°C, 180 minutes), and T7 (70°C, 240 minutes). Each experiment was repeated twice.

A total of 1 ml of each dilution treatment was put into sterile Petri dishes containing agar medium by pouring method. The percentage of tamarind seeds is divided into control (without tamarind) and 2% (dissolved in 100 ml aquadest). The pour plate was then incubated at 36°C for 24 hours. The number of colonies grown on the media was calculated using the plate count method according to [17].

2.4 Data Analysis

This study used a completely randomized design. Statistical Product of Service Solution software was used to establish the validity of the models based on analysis of variance (ANOVA), if there was a significant effect then continued with the Duncan test.

3. Results and Discussion

Prebiotics are not digested or absorbed in the gastrointestinal tract, besides providing a food source for probiotic bacteria, such as Lactobacillus, reducing the attachment of pathogenic bacteria including Salmonella, and promoting the gut microbiota. Prebiotics contain polysaccharides that inhibit the binding of pathogenic bacteria to the intestinal mucosa [18] and modulate the immune system by eliminating pathogenic bacteria from the intestinal tract [19].

Tamarind seeds are indicated to contain antinutrients in the form of tannins, free phenols, and oligosaccharides, so tamarind seeds must be processed first to be utilized [20] essential oils, and some natural polymers such as starch [14] a source of phytochemicals [21] [22] and consisting of phenolic antioxidants [23]. Tamarind seed extract exhibited potential antioxidant and anti-microbial activity [21].

One of the limitations on the use of tamarind seeds is the presence of tannins. Tannins bind proteins to form complex bonds so that these proteins are difficult to digest by protease enzymes. Tannins are compounds that can inhibit microbial growth by inhibiting the work of enzymes such as cellulose, pectinase, oxidative peroxide, and others [24]. According to [25] the phenols present in tannin compounds in high concentrations can be toxic to bacteria. The heating process can reduce the content of tannins, phytic acid, and trypsin inhibitors which are anti-nutritional substances [26]. The steaming treatment reduced the tannin content of tamarind seeds, the tamarind seeds without treatment contain 0.2849% tannin, 10 minutes of steaming reduces the tannin value to 0.2683%. 0.2663%. This showed that the steaming treatment of tamarind seeds can reduce the value of tannins contained in tamarind seeds. Tannins and phenols are water-soluble compounds. Tannins can be removed by soaking, heat treatment, or cooking [27].

The colonies number of *Lactobacillus casei* with different temperatures and time incubation of tamarind seeds are shown in Table 2.

Table 2. The colonies number of *Lactobacillus casei* (cfu/mL)

Treatments	Replications		Means
	1	2	
T1 (control)	2.85×10^7	1.94×10^7	2.39×10^{7a}
T2	1.0×10^7	1.0×10^7	1.0×10^{7a}
T3	2.45×10^7	1.86×10^7	2.15×10^{7a}
T4	387.54×10^7	356.85×10^7	372.19×10^{7b}
T5	572.47×10^7	536.61×10^7	554.54×10^{7c}
T6	917.33×10^7	917.16×10^7	917.25×10^{7e}
T7	671.87×10^7	656.93×10^7	664.40×10^{7d}

$^{a, b, c, d, e}$ Different superscripts in the same column show significant differences ($P<0.05$)

Heating tamarind seeds at 60°C for 120 minutes (T2) and 180 minutes (T3) have not shown inactivates anti-bacterial activity, from the number of bacterial colonies on T2 was the lowest of all treatments and controls, as well as on T3. Flavonoids are the largest group of tannin compounds, phenolic compounds have effective properties to inhibit the growth of viruses, bacteria, and fungi. The mechanism of flavonoid compounds in inhibiting fungal growth is by damaging the cell wall of Candida albicans which consists of lipids and amino acids. Tannin compounds with their ability to form protein complexes and damage cell membranes by denaturing protein bonds in cell membranes, resulting in cell membrane lysis [28]. However, at 240 minutes of incubation, the number of bacterial colonies increased significantly ($P<0.05$) compared to 180 minutes (T3) and 120 minutes (T2). The incubation time of 240 min is estimated to have damaged tannin in tamarind seeds, thereby reducing the inhibition of bacterial growth.

Lactobacillus casei is a lactic acid bacteria [29]. The growth and metabolic activity of lactic acid bacteria are strongly influenced by the biochemical and biophysical environment [30]. The biochemical environment is a bacterial growth medium that must provide nutrients according to the requirements of bacteria [31] [32]. A large amount of research has studied the relationship between nutritional requirements and metabolic activities of lactic acid bacteria [30].

Tamarind seeds generally contain high organic matter [33] in the form of carbohydrates, proteins, and fats [34]. Based on the results of the proximate analysis of tamarind seeds in this study, the carbohydrate content was 68.45% and the protein content was 15.61%. In addition [35] reported that tamarind seeds contain essential amino acids, such as isoleucine, leucine, lysine, methionine, phenylalanine, valine, and are a source of essential fatty acids, as well as minerals, especially calcium, phosphorus, and potassium.

The main metabolic activity of lactic acid bacteria is breaking down carbohydrates for energy [31] [32] [36] and breaking down proteins, lipids, and other compounds also essential for normal growth [30]. Tamarind seeds have phytic acid that reduces the bioavailability of certain minerals, interferes with protein utilization due to the formation of phytate protein complexes and phytate-protein minerals [37] [38]. Heating tamarind seeds can reduce the cyanogen content, reduce trypsin inhibitor activity and reduce the formation of phytate protein. Heat treatment can increase the nutritional value by destroying trypsin inhibitors and increasing the digestibility of proteins and amino acids, fats, and carbohydrates contained in feed ingredients [39]. Trypsin inhibitor activity can be reduced through the steaming process [20] so that food substances can be used for bacterial growth, this can be seen from colonies on heating 70°C (T5, T6, and T7) was significantly higher ($P<0.05$) than 60°C heating (T2, T3, and T4). The optimum incubation time at 70°C is 180 minutes, this is indicated by the number of colonies T6 higher significantly ($P<0.05$)compared to the incubation time of 120 minutes (T5) and 240 minutes (T7), but the incubation time of 240 (T7) minutes provides better conditions for bacterial growth than the incubation time of 120 minutes (T5).

4. Conclusion

Based on the results of this study:

1. Incubation of tamarind seeds at a temperature of 60°C for 240 minutes increased the number of colonies significantly compared to the incubation time below
2. Bacterial colonies using 2% tamarind seeds which were incubated at 70°C increased the number of bacterial colonies in the all-time ranges.
3. Tamarind seeds have potential as prebiotics with incubation at 70°C for 180 minutes.

Acknowledgments

The authors express deep appreciation to Politeknik Negeri Jember. This research was funded by Politeknik Negeri Jember through the PNBP scheme in 2021.

References

[1] Abudabos A M, Alyemni A H, Dafalla Y M and Khan R U 2017 *J. Appl. Anim. Res.* 538–542
[2] Hajati H and Rezaei M. 2010 *International Journal of Poultry Science*. 298-304
[3] Iji PA and Tivey D R 1998 *World's Poultry Science Journal*. **54** 129-143

[4] Bozkurt M, Aysul N, Küçükyilmaz K, Aypak S, Ege G, Catli A U, Akşit H, Çöven F, Seyrek K and Çınar M 2014 *Poultry Science*. **93** 389-399
[5] Dhama K, Verma V, Sawant P M, Tiwari R, Vaid R K and Chauhan R S 2011 *Journal of Immunology and Immunopathology*. **13** 1-19
[6] Patterson J A and Burkholder K M 2003 *Poultry Science*. **82** 627-631
[7] Adhikari P A and Kim W K 2017 *Annals of Animal Science*. **17** 949
[8] Jung S J, Houde R, Baurhoo B, Zhao X and Lee B H 2008 *Poultry Science* **87** 1694-1699
[9] Pourabedin M and Zhao X 2015 *FEMS Microbiology Letters*. **362** 122
[10] Aachary A A and Prapulla S G 2009 *Bioresource Technology* **100** 991-995
[11] Perricone M, Bevilacqua A, Corbo M R and Sinigaglia M 2014 *Food Microbiology*. **38** 26-35
[12] Santin E, Maiorka A, Macari M, Grecco M, Sanchez J C, Okada T M and Myasaka A M 2001 *Journal of Applied Poultry Research*. **10** 236-244
[13] Roto S M, Rubinelli P M and Ricke S C 2015 *Frontiers in Veterinary Science*. **2** 28
[14] Rao Y, Saideswara and Mary M K 2012 Handbook of Herbs and Spices *Woodhead Publishing* 512-533
[15] Roberfroid M B and Delzenne N M 1998 *Annual Review of Nutrition*. **18** 117-143
[16] Adhikari P, Lee C H, Cosby D E, Cox N A and Kim W K 2019 *Poultry Science*. **98** 1235-1242
[17] Kirchman D, Sigda J, Kapuscinski R and Mitchell R 1982 *Applied and Environmental Microbiology*. **44** 376-382
[18] Gaggìa, F, Mattarelli P and Biavati B 2010 *International Journal of Food Microbiology* **141** S15-S28
[19] Alloui M N, Szczurek W and Swiatkiewicz S 2013. *Annals of Animal Science*. **13** 17
[20] Pugalenthi M, Vadivel V, Gurumoorthi P and Janardhanan K 2004 *Tropical and Subtropical Agroecosystems*. **4** 107-123
[21] Tsuda T, Watanabe M., Ohshima K, Yamamoto A, Kawakishi S and Osawa T 1994 *Journal of Agricultural and Food Chemistry*. **42** 2671-2674
[22] Muyonga J H, Andabati B and Ssepuuya G 2014 *Food Science & Nutrition*. **2** 9-16
[23] Aengwanich W, Suttajit M, Srikhun T and Boonsorn T 2009 *International Journal of Poultry Science*. **8** 749-751
[24] Sinsabaugh R L., Antibus R K and Linkins A E 1991 *Agriculture, Ecosystems & Environment*. **34** 43-54
[25] Krastanov A, Alexieva Z and Yemendzhiev H 2013 *Engineering in Life Sciences*. **13** 76-87
[26] Manohar B, Ramakrishna P and Udayasankar K 1991 *Journal of Food Engineering*. 13 241-258
[27] Singh U 1988 *Plant Foods for Human Nutrition*. 38 251-261
[28] Abbas T F, Abbas M F and Lafta A J 2020 *Annals of Tropical Medicine and Public Health*. **23** 23-1146
[29] Yoon K Y, Woodams E E and Hang Y D 2006 *Bioresource Technology*. **97** 1427-1430
[30] Saeed A H and Salam A I. 2013 *Food and Nutrition Sciences*. **1**.
[31] Sanchez S and A. L. Demain 2008 *Microbialogy and Biotechnology*. **1** 283-319
[32] Hoefnagel M H N, Starrenburg M J C, Martens D E, Hugenholtz J, Kleerebezem M, Swam I I V, Bongers R, Westerhoff H V and Snoep J L 2002 *Microbiology*. **148** 1003-1013
[33] Yusuf A A, Mofio B M and Ahmed A B 2007 *Science World Journal* **2** 1
[34] Krithika V and Sri S R 2014 *International Journal of Research in Science*. **1** 117-123
[35] Bagula M and Arya S S 1998. *Seed*. **70**-75
[36] König H and Fröhlich J 2009 *Biology of Microorganisms on Grapes, in Must and in Wine, Springer* 3-29
[37] Reddy N R, Sathe S K and Salunkhe D K 1982. *Advances in Food Research*. **28** 1-92
[38] Vijayakumari K, Siddhuraju P, Pugalenthi M and Janardhanan K 1998 *Food Chemistry*. **63** 259-264
[39] Flavin D F 1982 *Veterinary and Human Toxicology*. **24** 25-28

IOP Conf. Series: Earth and Environmental Science **980** (2022) 012018 doi:10.1088/1755-1315/980/1/012018

Early production performance of crossing chickens raised on indoor or with outdoor access

R T Hertamawati *[1], B Prasetyo and U Suryadi

[1] Animal Husbandry Department, Politeknik Negeri Jember, Jl. Mastrip 164 Jember, East Java, Indonesia

*E-mail : rosa_trihertamawati@polije.ac.id

Abstract. An experiment was conducted to determine the effect of the rearing system and feed on early growth performance of crossing chicken (male local chicken x female Isa Brown final stock laying chicken). One hundred 1-d-old chicks were randomly assigned to 2 factor treatments, the first factor was rearing cages (indoor and outdoor access) and the second factor was feed ration (20% and 17% crude protein). In the indoor treatment, the chickens were housed in floor pens (5 birds/m2). In the outdoor access treatment, the chickens were housed in a similar indoor house; in addition, they also had an outdoor grass paddock (5 birds/m2). All birds were provided ad libitum water and were raised from 14 to 35 days of age. The results showed that early production performance (average daily gain, feed intake and feed conversion ratio) were different among treatment (P>0.05) at 28 and 35 days. The early growth performance of crossing chicken between Bangkok chicken and layer chicken have good performance in the indoor or outdoor cages with nutritional content of 20% crude protein and energy metabolize 2900 kcal/kg

1. Introduction

The rearing of local chickens has not shown optimum productivity, due to local chickens have relatively slower growth and lower egg production compared to purebred chickens. The slow growth of local chickens was caused by the low quality of their genetics [1] [2]. Efforts should be made to increase the productivity of these local chickens.

The performance of local chickens can be improved by improving feed, maintaining and improving genetic quality through the crossing process[3][4]. One of the local chickens that are commonly crossed is the Bangkok chicken, actually most of this chicken was crossing between pure male Bangkok with Kampung chicken [5]. Bangkok chickens are generally kept as fighting cock, because they have a larger and stockier morphology than native chickens in general. Besides being used as fighting chickens because they have high body weight gain, these chickens can be used to produce local broilers through crosses with broilers or layers. Free-range chickens are able to reach a weight of 0.85 kg within 2 months of rearing compared to native chickens which only reach a weight of 0.50 kg [6].

Improving the performance of local chickens which includes body weight gain, feed consumption, mortality and feed conversion is not enough just to be crossed, but other factors also need to be considered, for example, feed and environmental factors, especially the cage used to raise the crossbred chickens. Improving the growth performance of Bangkok chickens can be done through improving the quantity and quality of the diet [7]. The previous studies showed that the protein requirement for crossbreed Bangkok chicken was 20% and the Metabolic Energy content in the feed was 2,800 Kcal/Kg during its growth period [8]. Local chickens in the growth period require 17% protein and metabolic energy 3200 kcal/kg feed [9]. The protein and energy requirements for native chickens lower than broilers. There is a difference between the protein requirements of cross-bred chickens and local chickens with relatively expensive feed prices and limited availability, it is necessary to determine the optimal protein content for the growth of chicken crosses.

On the other hands housing factor also determines the growth of livestock. According to [10], chicken coops based on the type of wall or ventilation are divided into closed cages (indoor) and open cages (outdoor). The use of the cage system also affects feed consumption, the use of a wider cage makes chickens require a lot of energy to mobilize, so that feed consumption also increases [11]. The purpose of this study was to evaluate the effect of cages and feed on the performance of Bangkok chickens and broilers.

2. Material and methods

2.1. Animal and treatments

A total of 100 crossbred one day old chicks (DOC) derived from natural mating between male Bangkok chickens and female chickens (Isa Brown) with a sex ratio of 1:10. The DOC were housed in a brooder battery cage made of wire with a density of 20 birds per cage until the age of 2 weeks. After 2 weeks of age divided into 2 treatment factors, the first factor is the raising cage with indoor or outdoor access (A1 = outdoor and A2 = indoor access) with a cage size of $1m^2$ for 10 chickens and the second factor is diet, containing crude protein (CP) content 20 % (B1) and 17% (B2) with metabolic energy of 2900 kcal/kg. One unit of cage was used as the treatment unit.

All chicks were given Newcastle and IBD vaccines, antibiotics, minerals and vitamins as needed. The ingredients were used in diets are corn, palm kernel, fish meal, rice bran, oil and premix. The diets were prepared in crumble.

2.2. Performance data collection

SPSS 26 software was used to analyze the studied traits of the average body weight gain, feed intake and feed conversion ratio at 21 days to 35 days old of treatment, assuming the Completely Random Factorial Design. The test of LSD was conducted to diagnosing the significant differences between the means of the levels of each factor.

3. Result and discussion

3.1. Feed intake

The result of the present study showed that there was a significant effect of cage to feed intake in the 28 to 35 days. The average feed intake for each treatment is showed in Table 1. with an average range of 113.84 g/head to 211.30 g/head at the age of 21 days to 35 days, it average is lower than the results of research by [5]. Feed intake in indoor cages was higher than outdoor cages at the age of 28 and 35 days, chickens reared outdoors consumed less feed due to initial stress and adjustment to the environment. In accordance with the results of research by [11] that different types of cages have a very significant effect on feed intake, and feed intake is influenced by environmental temperature, chicken health, cages, feed containers, nutrient content in feed and stress that occurs [12]. The temperature in the outdoor cage is higher so that stress can occur which will affect the decrease in feed consumption resulting in a decrease in body weight [13].

The protein content of feed does not affect feed consumption because the energy content of the treatment feed is the same according to the opinion that the high consumption of chicken feed is influenced by the energy content of the feed, crude fiber, density of feed type or density and crude fat.

Table 1. Mean of feed intake at the growth period of crossbred chickens (g/head/week)

Days	Cage (A)	Feeding (B)		Average	P value		
		B1	B2		A	B	AxB
		-----g/head-----					
21 d	A1 (outdoor)	115.45±18.9	112.24±19.3	113.84±18.1			
	A2 (indoor)	124.28±33.2	121.76±30.3	123.00±30.0	0.45	0.81	0.98
	Average	119.86±25.9	117.00±24.5				
28 d	A1 (outdoor)	157.22±0.5	146.78±13.3	152.0±10.9^{a}			
	A2 (indoor)	168.28±7.6	164.78±6.6	167.03±7.1^{b}	0.01**	0.73	0.46
	Average	163.25±8.8	155.78±13.7				
35 d	A1 (outdoor)	192.22±22.9	170.94±27.1	181.57±26.2^{a}			
	A2 (indoor)	207.69±11.6	204.92±15.9	211.30±13.6^{b}	0.005**	0.45	0.14
	Average	199.93±18.9	192.93±31.2				

Notes: ** The different superscript letter within a column showed the significantly different (P<0.05) BNJ 5%

3.2. Body weight gain

Weight gain is an important factor to consider in observing chicken performance. The result of this study showed that there was a significant effect of diet to body weight gain from 21 to 35 days, but the outdoor or indoor access had no effect on body weight gain, according research [13]. However, in other studies, the weight gain with the free-range treatment were lower than with the indoor treatment [14].

Table 2. Mean of body weight gain at the growth period of crossbred chickens (g/head/week)

Days	Cage (A)	Feeding (B)		Average	P value		
		B1	B2		A	B	AxB
		-----g/head-----					
21 d	A1 (outdoor)	68.11±7.05	44.16±6.53	56.13±14.2			
	A2 (indoor)	67.25±2.91	44.65±5.86	55.95±12.7	0.94	0.001**	0.799
	Average	67.68±5.11^{a}	44.51±5.85^{b}				
28 d	A1 (outdoor)	97.53±24.6	52.89±16.64	75.21±30.75			
	A2 (indoor)	108.20±22.21	60.83±14.34	84.51±30.56	0.311	0.001**	0.880
	Average	102.86±22.79	56.86±15.23				
35 d	A1 (outdoor)	122.90±16.15	57.25±19.43	90.07±38.48			
	A2 (indoor)	124.56±16.47	67.11±11.45	95.83±33.09	0.436	0.001**	0.578
	Average	123.73±15.40	62.18±15.91				

Notes: ** The different superscript letter within a column showed the significantly different (P<0.05) BNJ 5%

The growth of crossing increasing until 35 days due to in the growth phase showed on Figure 1. The weight gain of crossing chicken increase or grow rapidly at the age of 1 - 2 months and decreasing at the age of 2 - 3 months. This is in line with the opinion of [15] which states the growth of chicken body weight will be fast and reach a specific weight at a young age.

3.3 Feed efficiency

Feed Conversion Ratio (FCR) in crossing chickens was significantly affected by diet, while the cage indoor or author factor does not have a significant effect, however, in other studies feed conversion ratio with the free-range treatment were lower than in the indoor treatment [14]. Dietary 20% crude protein gave a better conversion rate ($p<0.05$) in the range of 1.63 when compared to the feed containing 17% crude protein. It can be seen that the longer it takes for chickens to reach a certain body weight, give lower efficiency to use of feed.

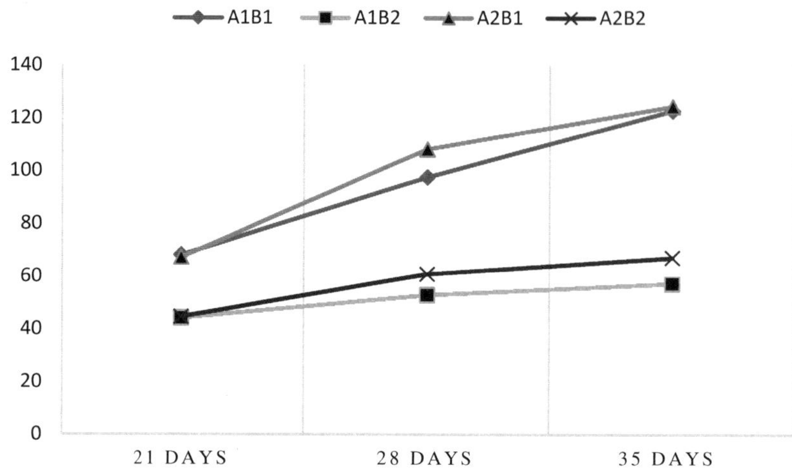

Figure 1. Body weight gain (g/head)

Table 3. Mean of feed conversion ratio (FCR) at the growth period of crossbred chickens

Days	Cage (A)	Feeding (B)		Average	P value		
		B1	B2		A	B	AxB
21 d	A1 (outdoor)	1.70±0.30	2.59±0.65	2.15±0.67			
	A2 (indoor)	1.86±0.65	2.81±0.94	2.34±0.88	0.529	0.006*	0.921
	Average	1.78±0,43	2.70±0.77				
28 d	A1 (outdoor)	1.68±0.37	3.03±1.08	2.35±1.04			
	A2 (indoor)	1.63±0.39	2.86±0.84	2.25±0.89	0.739	0.01**	0.872
	Average	1.66±0.36	2.94±0.91				
35 d	A1 (outdoor)	1.59±0.31	3.25±1.05	2.42±1.14			
	A2 (indoor)	1.69±0.28	3.31±0.79	2.50±1.02	0.792	0.001**	0.945
	Average	1.64±0.28	2.46±1.05				

Notes: ** The different superscript letter within a column showed the significantly different (P<0.05) BNJ 5%

4. Conclusion

It could be concluded that the performance of early growth performance of crossing chicken between Bangkok chicken and layer chicken have good performance in the indoor or outdoor cages with nutritional content of 20% crude protein and energy metabolize 2900 kcal/kg.

5. Acknowledgement

This research was supported by PNBP applied research grant project from Politeknik Negeri Jember.

References

[1] Nataamijaya A G 2016 Karakteristik dan Produktivitas Ayam Kedu Hitam *Bul. Plasma Nutfah* **14** 85

[2] Habiburahman R, Darwati S and Sumantri C 2018 Pola Pertumbuhan Ayam Silangan Pelung Sentul Kampung Ras Pedaging (IPB D-1) G4 Umur 1-12 Minggu *J. Ilmu Produksi dan Teknol. Has. Peternak.* **6** 81–9

[3] Lapihu Y L, Telupere F M S and Sutedjo H 2019 Kajian Fenotip dan Genetik Performa Pertumbuhan dari Persilangan Ayam Lokal dengan Ayam Ras Petelur Isa Brown *J. Sain*

Peternak. Indones. **14** 298–305
[4] Bintang H R dan I A K 2010 Produktivitas Ayam Lokal *Lokakarya Nas. Inov. Teknol. Pengemb. Ayam Lokal* 121–6
[5] Kholik A, Sujana E and Setiawan I 2016 Performa Ayam Hasil Persilangan Pejantan Bangkok Dengan Betina Ras Petelur Strain Lohman *Jurnal.Unpad.Ac.Id*
[6] Rahayu H. S. I, Darwati S and Mu'iz A 2019 Morfometrik Ayam Broiler dengan Pemeliharaan Intensif dan Akses Free Range di Daerah Tropis *J. Ilmu Produksi dan Teknol. Has. Peternak.* **7** 75–80
[7] Mokodongan A R, Nangoy F, Leke J R and Poli Z 2017 Penampilan Pertumbuhan Ayam Bangkok Starter Yang Diberi Pakan Dengan Level Protein Berbeda *Zootec* **37** 426
[8] J.F. Singarimbun, L. D. Mahfud dan E S and Fakultas 2013 Effect of Feeding With Different Protein Levels On The Carcass Quality of Crossbred of Bangkok and Arabic Chicken **2** 10
[9] Iskandar S 2012 Optimalisasi Protein Dan Energi Ransum Untuk Meningkatkan Produksi Daging Ayam Lokal *Pengemb. Inov. Pertan.* **5** 96–107
[10] Fanatico A C, Pillai P B, Emmert J L, Gbur E E, Meullenet J F and Owens C M 2007 Sensory attributes of slow- and fast-growing chicken genotypes raised indoors or with outdoor access *Poult. Sci.* **86** 2441–9
[11] Setiawati T, Afnan R and Ulupi N 2016 Performa Produksi dan Kualitas Telur Ayam Petelur pada Sistem Litter dan Cage dengan Suhu Kandang Berbeda *J. Ilmu Produksi dan Teknol. Has. Peternak.* **4** 197–203
[12] Uzer F, Iriyanti N and Roesdiyanto R 2013 Penggunaan Pakan Fungsional Dalam Ransum Terhadap Konsumsi Pakan Dan Pertambahan Bobot Badan Ayam Broiler *J. Ilm. Peternak.* **1** 282–8
[13] Chen X, Jiang W, Tan H Z, Xu G F, Zhang X B, Wei S and Wang X Q 2013 Effects of outdoor access on growth performance, carcass composition, and meat characteristics of broiler chickens *Poult. Sci.* **92** 435–43
[14] Castellini C, Mugnai C, Moscati L, Mattioli S, Amato M G, Mancinelli A C and Dal Bosco A 2016 Adaptation to organic rearing system of eight different chicken genotypes: Behaviour, welfare and performance *Ital. J. Anim. Sci.* **15** 37–46
[15] Depison D, Puteri N I and Gushairiyanto G 2020 Growth Patterns, Body Weight, and Morphometric of KUB Chicken, Sentul Chicken and Arab Chicken *Bul. Peternak.* **44** 67–72

Effect of Low Grade Vegetable Soybean on Performance of Days Old Broiler Chicken

S Nusantoro[1], M Andriani[1], Nurkholis[1]*

[1]Department of Animal Sciences, State Polytechnic of Jember

*nurkholis@polije.ac.id

Abstract. This experiment aimed to examine the effect of low grade vegetable soybean (LGS) via drinking water on performance of days old chicken. Two hundreds day old boiler chicken (unsexed) were randomly assigned into completely randomized design, using 4 treatments (Non= 0% LGS, Vit=vitamin, LG5=5% LGS, and LG10=10%LGS solution) and 5 replication. Data were collected from day-1 until day-7. LGS significantly ($p<0.05$) affected final body weight and weight gain of broiler chicken not feed conversion ratio ($p>0.05$). The results showed that LGS of 10% can be applied via drinking water to improve performance during initial phase of broiler chicken rearing period.

1. Introduction

The use of phytochemicals appear to be widely applied in many sectors, including livestock and broiler management. It has positive impact not only on performance but also on the health status of animals. The banned of antibiotic inclusion in feed in many country has been pushing some actors, such as researchers, producers, and farmers to look for alternative ways in its replacement. Phytochemicals, i.e. substances derived from secondary metabolites of plants, are considered to be safety and nature friendly for meat consumers.

Amongst phytochemicals, flavonoids groups (for example genistein and daidzein of isoflavone) are well studied. Some legumes and fruits such as soybean, alfalfa, apple, and *Citrus Limon* are source of flavonoids. In human, flavonoids were known as compounds that act as anticancer [1]. Polyphenolic compounds showed biological activities and processes that are dysregulated during cancer development[2]. In poultry, isoflavones play a role in reproductive performance and carcass traits. For example supplementation of 10 mg/kg daidzein in feed increased laying rate of broiler breeder. Also, commercial broiler chicken given grade level of soy isoflavone showed decreasing abdominal fat [3]

Vast majority research of the utilisation of flavonoids in poultry (for example [2]–[4]) were delivered through feed either as single supplement or combination with other substance. From practical point of view, this might be difficult for farmer due to its low level of applications need accuracy in weighting. Thus this experiment aimed to examine the effect of low grade vegetable soybean given through drinking water on performance of days old broiler chicken.

2. Material & Method

2.1. Animal and rearing management

This experiment was conducted in Politeknik Negeri Jember during August through October 2021. The amount of 200 day old broiler (DOB) having initial body weight (IBW) 40.10 ± 0.81 g were purchased from local poultry shop. The bird were kept in opened housing system following breeder management guide.

2.2. Experimental setup

DOB were randomly assigned in completely randomized design. Chicken were given 4 treatment i.e. Non= 0% LGS, Vit=vitamin, LG5=5% LGS, and LG10=10%LGS solution. Each treatment was replicated 4 times. LGS solutions were prepared by grinding low grade edamame (from local producer) and then mixed with water accordingly. Treatments were delivered as broiler drink. During experiment, feed (Table 1) was given ad libitum

Tabel 1 Formulation of experimental feed given to broiler chicken for 7 days.

No	Ingredient	Percentage
1	Maize	60.91
2	Soybean meal	24.85
3	Meat and bone meal	7.00
4	Corn gluten meal	3.00
5	Palm oil	2.00
6	$CaCO_3$	0.64
7	NaCl	0.15
8	Premix	1.00
9	Methionine	0.24
10	Lysine	0.21
Total		100
Calculated composition:		
Metabolisable energy (kcal/kg)		3,118.88
Crude protein (%)		22.25

2.3. Performance calculation

Feed intake (FI) = feed given subtracted by residue of feed; Body weight gain (BWG) = final body weight at day 7 (FBW) subtracted by IBW; Feed conversion ration = FI divided by BAG.

2.4. Data analysis

All data were analysed using one way anova of SPSS 22 for Mac. All data were ensured having variance homogeneity and normally distributed. FBW and BWG were tested using Tukey HSD post hoc.

3. Results & Discussion

Performance of broiler during 7 days is presented in Table 2. Treatments affected FBW and BAG ($P<0.05$). Booth have the same trend, in which broiler given LG10 showed the the highest value compare others. IBW, FBW, and FCR is not affected by treatment ($P>0.05$). Data also showed that the use of low grade vegetable soybean (LG5) was comparable to Vitamin or even better in LG10. This finding might be better in minimising management cost. Similar FCR in all treatments could be addressed to relatively the same feed intake. LGS in the present experiment was expected to increase broiler performance by increasing ability of chicken to face stress in open housing system.

IOP Conf. Series: Earth and Environmental Science **980** (2022) 012019 doi:10.1088/1755-1315/980/1/012019

Table 2. Performance of broiler chicken until 7 days

Parameters	Non	Vit	LGS5	LGS10	P value
IBW (g/bird)	39.67	40.27	39.13	40.60	0.25
FBW[1] (g/bird)	117.97^{a}	123.77^{ab}	118.90^{a}	130.93^{b}	0.02
BWG[1] (g)	78.30^{a}	83.50^{ab}	79.77^{a}	90.33^{b}	0.03
FI (g/bird)	140.30	141.10	140.10	140.33	0.99
FCR (g feed/ g BAG)	1.80	1.69	1.76	1.56	0.32

[1]Means value in a row with unlike letters were significantly different ($P<0.05$)

Improvement of FBW and BWG found in this experiment confirmed the results of previous workers who used flavonoid as supplement for chicken [5] [6]. One of mode of actions regarding the role of phytochemicals is reducing oxidative stress which may allow chickin to show better performance. The value of FCR is comparable than common days old broiler chicken performance. There were no significant difference in FCR due to the amount of FI was also similar but BWG was better in chicken given LGS10.

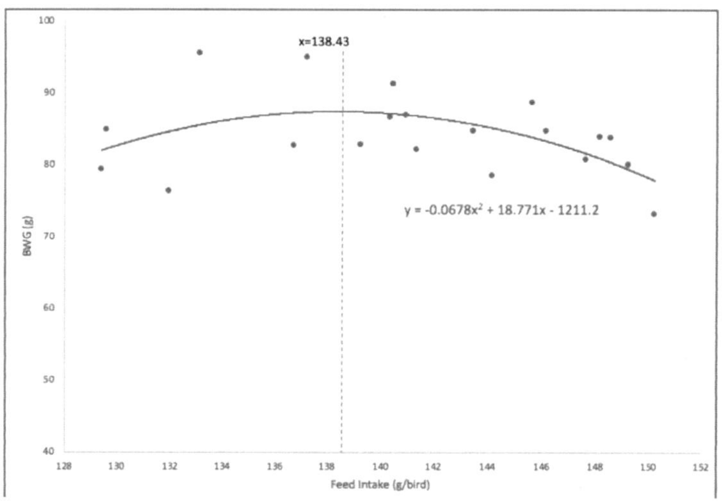

Figure 1. Second order polynomial between feed intake and body weight gain during 7 days.

Instead of linear, the relationship between FI and BAG is quadratic (Figure 1), in which after chicken reach as much as 148.43 g of feed consumption, the growth is declining. Factors affecting FI are rearing temperature, age of bird, and genetic traits. In this case our data showed (unpublished) that temperatur appeared to affect feed intake as we used opened housing system.

4.Conlusion

LGS significantly ($p<0.05$) affected final body weight and weight gain of broiler chicken not feed conversion ratio ($p>0.05$). The results showed that LGS can be applied via drinking water to improve performance during initial phase of broiler chicken rearing period.

Acknowledgement
Authors acknowledge State Polytechnic of Jember, providing research grant in the scheme of Penelitian Dasar PNBP 2021 managed by Pusat Penelitian dan Pengabdian Masyarakat.

References
[1] Thakur M Singh K and Khedkar R, 2020, Phytochemicals, in *Functional and Preservative Properties of Phytochemicals*, (Elsevier), p. 341–361.
[2] Sharma N Dobhal M Joshi Y and Chahar M, 2011 Flavonoids: A versatile source of anticancer drugs *Pharmacogn. Rev.* **5**, 9 p. 1.
[3] Payne R L Bidner T D Southern L L and Mcmillin K W, Aug. 2001 Dietary Effects of Soy Isoflavones on Growth and Carcass Traits Of Commercial Broilers *Poult. Sci.* **80**, 8 p. 1201–1207.
[4] Azzam M M *et al.*, May 2019 Effect of Soybean Isoflavones on Growth Performance, Immune Function, and Viral Protein 5 mRNA Expression in Broiler Chickens Challenged with Infectious Bursal Disease Virus *Animals* **9**, 5 p. 247.
[5] . M Y M . M S R . M K I and . M M M, Jul. 2004 Effects of Saponin and L-Carnitine on the Performance and Reproductive Fitness of Male Broiler *Int. J. Poult. Sci.* **3**, 8 p. 530–533.
[6] Ma J S *et al.*, Nov. 2015 Effects of flavones of sea buckthorn fruits on growth performance, carcass quality, fat deposition and lipometabolism for broilers *Poult. Sci.* **94**, 11 p. 2641–2649.

Fermented meat and digestive tract of snail as amino acid supplements for functional feed of native chickens

U Suryadi, R T Hertamawati, S Imam*

Department of Animal Science, Politeknik Negeri Jember, Mastrip Street PO BOX 164, Jember 68101, East Java, Indonesia

*Email: shokhirul_imam@polije.ac.id

Abstract. Sources of amino acids and lactic acid bacteria have been developed from the hydrolysis of snail meat and its digestive tract through a fermentation process using a consortium of microorganisms present in cow's milk whey, which is fermented by microorganisms from rice washing water with internal microorganisms from snails. The fermentation process used the Submerged Fermentation (SmF) method. The fermentation process was carried out using Lactic Acid Bacteria (LAB) inoculants contained in rice washing water. This product was researched and developed with the aim of being a feed supplement to reduce the use of fish meal or other protein sources such as concentrates in the preparation of feed for poultry. Product development studies were carried out on the LAB content in rice washing water and fermented snail meat and its digestive tract, as well as their amino acid content. The results of the study showed that at the dilution of rice washing water, approximately 3 different colonies of LAB were found, the dilution of 10^{-1} to 10^{-4} could not be counted but at the dilution of 10^{-5}, 50 colonies of LAB were obtained with an amount of 5×10^7 CFU/ml. The fermented snail meat and its digestive tract contained approximately 4 different LAB colonies and found 3×10^4 CFU/ml at a dilution of 10^{-3}, and contained 6 types of amino acids, namely; alanine, glycine, cysteine, arginine, lysine, and proline. The conclusion of this research is that fermented snail meat and digestive tract have potential as an amino acid supplement for functional feed for native chickens.

1. Introduction

Snails are wild animals as a source of local feed ingredients whose use is often neglected, and their existence is often exterminated because snails cause huge losses to the agricultural industry. Snails are natural enemies for farmers that can cause damage to plants. This snail has a high reproductive capacity so that it has a detrimental impact on the destruction of fields [1].

The population of wild snails is currently decreasing in their natural habitat, and some edible snail species are overexploited, so that the snails are threatened with extinction. therefore there is an increasing interest in snail cultivation [2]. However, many difficulties occur in snails because as agricultural pests, animals that are very vulnerable due to climate change and infection and their physiology has not been adequately explored [3].

Snails are very important ecologically and economically, and are used as food and medicine. Snails possess adaptive immunity, snails possess and develop a series of innate defense and immune mechanisms to respond to invading pathogenic bacteria and more importantly establish and maintain a beneficial symbiotic microbiota in the gut [4]. Snails physically and chemically externally and internally have defenses such as gastric mucus and gastric acid fluid [5]. lubricating mucus that covers the body and has a role as antimicrobial activity [6].

Snails are a rich source of protein containing large amounts of essential amino acids, calcium and a small amount of fat [7]. Snails contain nutrients that can be used as a source of protein, besides that they also contain microbiota in their intestines. Raw snails contain 81.96% water with crude protein content of 11.3 g/100 g, ash 2.56 g/100 g, fiber 0.18 g/100 g, and fat 6.78 g/100g on a dry basis, and contains significant amounts of minerals and amino acids Asp, Ser, Glu, Gly, His, Arg, Thr, Ala, Pro, Tyr, Val, Lys, Ile, Leu, Phe, Cys, Met, Trp [8].

Total bacterial content ranged from 1.00-1.50 x 10^8 cfu/g, *Coliform* count ranged from 1.68-2.20 x 10^7 cfu/g, *Salmonella/Shigella* count ranged from 5.2-8.2 x 10^7cfu/g, the number of lactic acid bacteria ranged from 1.03-1.30 x 10^8 cfu/g and the number of fungi ranged from 7.3 x 10^7 - 1.00 x 10^8 cfu/g. Microorganisms that can be isolated are *Bacillus subtilis*, *Staphylococcus aureus, Lactobacillus spp.*, *Escherichia coli, Micrococcus luteus* and *Bacillus cereus* while the fungal isolates are *Aspergillus terrus, Aspergillus fumigatus*, *Absidia sp., Fusarium oxysporum, Eurotium sp.* and *Aspergillus flavus* [9].

In the intestinal tract of snails, *Lactobacillus* was found which has the potential as a specific probiotic for itself. Intestinal commensal *lactobacilli* strains interact with the gastrointestinal mucosa of snails and enhance cellular immune responses such as: chemotaxis and phagocytosis, as well as the antimicrobial activity of hemolymph factors [3].

The technological innovation that can be used to increase the nutritional value and digestibility of feed ingredients is the fermentation process. Fermentation is the oldest technique in food preservation, besides being low in cost, fermentation can also produce products that have better nutritional value than the original ingredients [10]. Fermentation is all kinds of metabolic processes with the help of enzymes from microbes (microorganisms) to carry out oxidation, reduction, hydrolysis and other chemical reactions, resulting in chemical changes in an organic substrate by producing certain products [11].

Fermented foods are more nutritious than unfermented foods [12], this is due to the enzymatic processes produced by microorganisms through catabolic and anabolic processes, breaking down complex compounds, and synthesizing vitamin complexes and other growth factors [13]. Fermentation is a natural way to increase vitamins, essential amino acids, anti-nutrients, protein, food appearance, taste and aroma enhanced. Fermentation also helps in reducing the energy required for cooking as well as making safer products [14,15].

Taking into account the potential for the existence of snails in epidemiology, the content of nutritional elements, and microbiological elements, then with a touch of biotechnology application of snail meat fermentation and its digestive tract it is possible to process it into amino acid supplements. This study aims to examine the potential of snail meat and it's digestive track as a amino acid supplement for functional feed of native chickens.

2. Materials and Methods

2.1. Fermentation of snail meat and its digestive tract

Rice washing water is heated to a temperature of 63°C, then cooled and cured for 3 days. Rice washing water is used to ferment cow's milk for 5 days to form whey (cloudy liquid) and curd (clumps). The resulting whey is mixed with sugar water and then used to ferment the snail meat and its digestive tract in a 1:1:1 proportion. Fermentation was carried out for 3 weeks so that the pH of the liquid with an acid aroma was obtained, the color was cloudy brown, pH 3.71 and a temperature of 28°C. This liquid is used as a source of internal inoculants of microorganisms.

2.2. Isolation of lactic acid bacteria

Snail samples that have been fermented with several media as much as 25 ml were suspended in 225 ml of 0.85% NaCl aseptically (considered a dilution of 10^{-1}). Then a series of dilutions were carried out in stages from 10^{-2} to 10^{-5}. A total of 1 ml of suspension from each dilution was grown on GYP (Glucose

Yeast Extract Peptone) agar media (glucose 10 g, yeast extract 10 g, peptone 5 g, beef extract 2 g, Sodium Acetate 2 g, 0.5 ml salt solution), 10 g of tween 80 and 12 g of agar in one liter of distilled water) to which 5 g of $CaCO_3$ was added, then incubated at 37°C for 3 days. The isolation method was carried out by sprinkling the spread plate method. Lactic acid bacteria was indicated by the formation of a clear zone around the colony. Colonies considered as lactic acid bacteria were sampled using a sterile ose needle and inoculated on GYP media. $CaCO_3$ agar to obtain pure colonies. Pure colonies obtained were re-grown on GYP media. $CaCO_3$ agar slanted as stock for further tests.

2.3. Counting Total Bacteria
Total plate count is all colonies growing on NA media. The number of bacterial colonies counted on petri dishes is between 25-250 colonies. After that the amount obtained is multiplied by the dilution. The method of calculating TPC is: The number of bacteria in a petri dish x 1/ dilution factor.

2.4. Amino acid analysis
Amino acids were analyzed using High Performance Liquid Chromatography (HPLC). The principle of this amino acid analysis is that amino acids from proteins are liberated by hydrolysis with 6N HCl. The hydrolyzate was dissolved in sodium citrate buffer and each amino acid was separated using HPLC. Prior to the hydrolysis process, protein extraction was carried out using the Kjeldahl method.

3. Results and Discussion
Isolation and characterization of bacteria in food products is important to determine and differentiate the beneficial or detrimental effects of the microbiota in a particular sample. Isolation in the form of natural microbial communities can be exploited in one of the media with various bioprocesses to encourage the conversion process by selecting certain microbial species. One of the potential media that can be used as a source of inoculants is rice washing water. Rice water washing contains starch as a source of microbial food, especially Lactic Acid Bacteria (LAB).

The results of observations of LAB colonies from rice washing water were found about 3 different colonies. The morphological and physiological characteristics of the three LAB colonies used in this experiment are presented in Table 1. The existence of 3 types of LAB colonies contained in rice washing water, this may consist of 3 different types, according to [16], microorganisms grow together, both in natural communities and in artificial or synthetic cultures. The results of the 10^{-1} to 10^{-4} dilutions cannot be calculated, but the 10^{-5} dilutions obtained the number of 50 colonies with a total of 5 x 10^7 CFU/ml (Table 2.).

Table 1. Morphological and physiological characteristics of Lactic Acid Bacteria colonies from rice washing water.

Colony	Characteristics				
	Form colony	Edge colony	Colony surface	Top color	Bottom color
A1	Circular	Entire	Convex	Milky white	Creamy white
A2	Circular	Entire	Convex	Yellow	Yellow
A3	Circular	Entire	Convex	Creamy white	Cream

LAB is a group of bacteria that produce lactic acid as a result of carbohydrate fermentation in the substrate [17]. LAB is a large group of bacteria with the ability to ferment sugars into lactic acid. Lactic acid bacteria (LAB) is a type of gram-positive bacteria that uses carbohydrates as the only or main carbon source [18]. Fermentation microorganisms mainly involve LAB. such as *Enterococcus, Streptococcus, Leuconostoc, Lactobacillus,* and *Pediococcus* [19]. The fermentation process also increases the activity of microbial enzymes because it provides an acidic environment at a temperature of 22-25°C [19]. Lactic acid fermentation is an intracellular anaerobic enzymatic process to convert sugar into lactic acid, which is one approach to obtain the energy needed for cell life processes under anaerobic conditions [19–22].

Table 2. Results of Isolation of LAB and TPC from rice washing water.

Sample	Different types of LAB	TPC results					Results CFU/ml				
		Dilution									
		10^{-1}	10^{-2}	10^{-3}	10^{-4}	10^{-5}	10^{-1}	10^{-2}	10^{-3}	10^{-4}	10^{-5}
Rice washing water	3	CNC	CNC	CNC	CNC	50	CNC	CNC	CNC	CNC	5×10^7

Note: CNC = Cannot be calculated

The presence of LAB content from the soaking of rice washing water is possible that these bacteria can degrade snail meat through fermentation. The presence of microbial diversity implies that qualitatively there is an opportunity to convert substrates and and produce naturally produced secondary metabolites [16]. The results of the isolation of lactic acid bakeria colonies from meat and digestive tract of fermented snails found approximately 4 different LAB colonies and at a dilution of 10^{-3} found 3×10^4 CFU/ml.

Table 3. Observations of LAB colonies from meat and digestive tract fermented snail.

Colony	Characteristics				
	Colony form	Colony edge	Colony surface	Top color	Bottom color
AB1	Circular	Entire	Convex	Milky white	Cream
AB2	Circular	Entire	Flat	Clear white	Cream
AB3	Circular	Entire	Raised with concave beveled edge	Milky white	Cream
AB4	Circular	Entire	Convex	Cream	Yellow cream

Table 4. Results of isolation of bacteria and TPC LAB from meat and digestive tract of snails fermented.

Sample	Different types of LAB	TPC results					CFU/ml results				
		Dilution									
		10^{-1}	10^{-2}	10^{-3}	10^{-4}	10^{-5}	10^{-1}	10^{-2}	10^{-3}	10^{-4}	10^{-5}
Rice washing water + Snail	4	65	16	3	-	-	65×10^2	16×10^3	3×10^4	-	-

Note: - = Cannot be calculated

All fermented foods and beverages were identified to contain LAB which is the most important dominant microbiota that contributes to beneficial effects in fermented foods/drinks [19]. *Lactobacillus, Lactococcus, Streptococcus, Pediococcus, Leuconostoc*, and *Enterococcus* are some of the facultative/obligate anaerobic gram-positive and acidogenic (lactic acid) bacteria used to metabolize saccharides with different degrees of efficiency into alcohols, lactic acid, lipids, amino acids, and compounds. Aliphatic [23]. Enzymes such as amylase, proteinase, mannase, catalase, cellulose, are generally produced from the fermentation of microorganisms, especially *Bacillus* [24].

The presence of LAB from the soaking of rice washing water, it is possible that these bacteria can degrade snail meat through fermentation. Snail meat fermentation process can cause snail meat protein to be hydrolyzed into amino acids through the services of microorganisms that produce protease enzymes. The results of the analysis of the amino acid content of fermented snail meat and its digestive tract obtained 6 types of amino acids (Table 5).

Table 5. Amino acid content of fermented snail meat and its digestive tract.

No	Types of amino acids	Amino acid level (%)
1	Glycine	0.226
2	Alanine	0.189
3	Lysine	0.176
4	Arginine	0.121
5	Cystein	0.203
6	Proline	0.185

During the fermentation process, enzymes produced by microorganisms initially degrade the macronutrients of the substrate [15], The three main pathways by which LAB are involved in the manufacture of fermented foods and the development of their taste, are (a) glycolysis (fermentation of sugars), (b) lipolysis (fat degradation) and (c) proteolysis (protein degradation) [23]. Protein degradation will result in increased protein solubility and the availability of several micronutrients and essential amino acids [25]. Sharma et al., [26] microorganism activity plays an important role in food fermentation by showing changes in chemical and physical properties of food. As a fermentation strain, LAB must have several important metabolic characteristics, such as the ability to produce acid and aroma, the ability to hydrolyze protein, the ability to produce viscous exopolysaccharides and the ability to inhibit bacteria [27].

Fermentation techniques are used either to produce large numbers of microbial cells or to produce extracellular microbial products (e.g. lactic acid), enzymes, amino acids, vitamins and other pharmaceutical compounds [28]. There are various roles of LAB in fermented foods such as to produce better taste, aroma, and texture, increase nutritional value, safety, and some functional compounds [29].

Fermentation processes in foods often cause changes in nutritional and biochemical qualities relative to the starting material. Fermented foods consist of a very complex ecosystem consisting of: enzymes from raw materials that interact with the metabolic activity of fermenting microorganisms. Fermented microorganisms provide a unique approach to food stability through physical and biochemical changes in fermented foods. These fermented foods can benefit consumers compared to simple foods in terms of antioxidants, peptide production, organoleptic and probiotic properties, and antimicrobial activity. It also helps in anti-nutrient levels and toxin levels. The quality and number of microbial communities in fermented foods varies based on the manufacturing process and storage/durability conditions. This review contributes to current research on biochemical changes during food fermentation. The focus is on changes in biochemical compounds that determine the characteristics of the final fermented food product from the original food source [26].

Several strains of *Lactaseibacillus casei*, *Lacticaseibacillus rhamnosus*, and *Streptococcus thermophilus* contribute to the production of branched-chain amino acid derivatives and aromatic amino acid derivatives [30]. The final product of hydrolysis is free amino acids, which are rich sources of nitrogen. Enzymatic breakdown of protein is very important for LAB, because it provides free amino acids needed for protein synthesis and LAB growth itself, and allows a significant reduction in energy expenditure in amino acid synthesis [31]. The proteolytic system of lactic acid bacteria consists of proteinases, peptidases, and specific transport proteins. Proteinases break down casein into peptides, then peptidases (intracellularly) degrade the peptides into amino acids and smaller peptides [31].

Under the influence of proteases in meat, some muscle protein macromolecules; disruption of disulfide bonds; and an increase in the reactivity of the sulfhydryl, hydroxyl, carboxyl, and amino groups. During this process, sarcoplasmic proteins are hydrolyzed to a significant degree, and depending on enzyme specificity, myofibrillar and connective tissue proteins are partially hydrolyzed. Meat under the influence of proteases, the degree of hydration is increased, and the digestibility of protein is increased; that is, more meat is available for proteolytic digestive enzymes. Due to this process, the content of solutes and free amino acids and small peptides is increased, which improves the palatability of the meat. Proteolytic enzymes contribute to flavor enhancement and are protective factors against many unfavorable microorganisms [31].

The degradation of meat proteins into peptides is catalyzed by proteolytic enzymes present in LAB, and the peptides are then further hydrolyzed by exopeptidase and endopeptidase into small peptides and amino acids. LAB has only a weak proteolytic action on fermented meat myofibrillar proteins [32,33]. However, some *Lb. casei, Lb. plantarum, Lb. curvatus and Lb. active sakei* strains contribute to the hydrolysis of sarcoplasmic proteins [34] and the subsequent decomposition of peptides into amino acids.

Some of the activities of *Lb. sake, Lb. curvatus* and *Lb. plantarum* strain has leucine and valine amino-peptidase, which contribute to protein and peptide catabolism yielding free amino acids, precursors of flavor compounds in the final product [35]. The production of high-quality fermented milk products depends on the proteolytic system of the starter bacteria, because the peptidases and amino acids formed have a direct impact on taste or serve as flavor precursors in these products [36].

The proteolysis process can not only be carried out chemically and or the addition of commercial enzymes to the substrate can also use certain microbial strains that secrete proteases to hydrolyze substrate proteins [37]. *Bacillus* species are involved in the process of enzymatic hydrolysis of proteins as producers of peptides and amino acids, which are claimed to have health benefits [38]. The physicochemical properties of protein hydrolysates depend on the protein substrate, the specificity of the enzymes used for proteolysis, and the hydrolysis conditions [39]. Amino acid profile is the most important nutritional parameter (protein quality and digestibility) of a protein source food.

4. Conclusion

Fermentation of snail meat and its digestive tract produced 4 different LAB colonies with 6 types of amino acid content, namely; Alanine, Glycine, Cysteine, Arginine, Lysine, and Proline. Fermented snail meat and its digestive tract have potential as an amino acid supplement for functional chicken feed.

5. Acknowledgments

The author would like to express his gratitude to the Directorate of Research and Community Service, Deputy for Research and Development Strengthening, Ministry of Research and Technology / National Research and Innovation Agency, who has funded this research with contract number 753/PL17.4/PG/2021.

References

[1] Adeline Sneha J, Khamis N H, Chakravarthi R, Harun M and Yeow Y K 2018 Acoustic identification of snail for pest control in agriculture *J. Adv. Res. Dyn. Control Syst.* **10** 2349–56

[2] Parlapani F F, Neofitou C and Boziaris I S 2014 Microbiological quality of raw and processed wild and cultured edible snails. *J. Sci. Food Agric.* **94** 768–72

[3] Dushku E, Ioannou A, Staikou A and Yiangou M 2019 Probiotic properties and immunomodulatory activity of gastrointestinal tract commensal bacterial strains isolated from the edible farmed snail Cornu aspersum maxima. *Fish Shellfish Immunol.* **92** 792–801

[4] Nyholm S V and Graf J 2012 Knowing your friends: invertebrate innate immunity fosters beneficial bacterial symbioses *Nat. Rev. Microbiol.* **10** 815–27

[5] Charrier M and Brune A 2003 The gut microenvironment of helicid snails (Gastropoda:

Pulmonata): In-situ profiles of pH, oxygen, and hydrogen determined by microsensors *Can. J. Zool.* **81** 928–35

[6] Etim L, Aleruchi C and Obande G 2016 Antibacterial Properties of Snail Mucus on Bacteria Isolated from Patients with Wound Infection *Br. Microbiol. Res. J.* **11** 1–9

[7] Milinsk M C, Padre R das G, Hayashi C, de Oliveira C C, Visentainer J V, de Souza N E and Matsushita M 2006 Effects of feed protein and lipid contents on fatty acid profile of snail (Helix aspersa maxima) meat *J. Food Compos. Anal.* **19** 212–6

[8] Mohd Irwani Hafiz Sahid*1, Khalisanni Khalid1 S N A M 1Malaysian 2012 Nutritional quality changes after canning of Achatina fulica Mohd *Nov. Int. J. Anal. Inov.* **1** 16–20

[9] Adegoke A A, Adebayo-Tayo, Bukola C, Inyang, Comfort U, Aiyegoro, Olayinka A, Komolafe and Amos O 2010 Snails as meat source: Epidemiological and nutritional perspectives *J. Microbiol. Antimicrob.* **2** 1–5

[10] Phadke G G, Elavarasan K and Shamasundar B A 2014 Angiotensin-I converting enzyme (ace) inhibitory activity and antioxidant activity of fermented fish product ngari as influenced by fermentation period *Int. J. Pharma Bio Sci.*

[11] Azizah N, Al-Barrii A N and Mulyani S 2012 Pengaruh lama fermentasi terhadap kadar alkohol, ph, dan produksi gas pada proses fermentasi bioetanol dari whey dengan substitusi kulit nanas *J. Apl. Teknol. Pangan* **1** 72–7

[12] Hasan M, Sultan M Z and Mar-E-Um M 2014 Significance of Fermented Food in Nutrition and Food Science *J. Sci. Res.* **6** 373–86

[13] Kennedy D O 2016 B Vitamins and the Brain: Mechanisms, Dose and Efficacy--A Review *Nutrients* **8** 68

[14] Nkhata S G, Ayua E, Kamau E H and Shingiro J-B 2018 Fermentation and germination improve nutritional value of cereals and legumes through activation of endogenous enzymes *Food Sci. Nutr.* **6** 2446–58

[15] Xiang H, Sun-Waterhouse D, Waterhouse G I, Cui C and Ruan Z 2019 Fermentation-enabled wellness foods: A fresh perspective *Food Sci. Hum. Well.* **8** 203–243

[16] Canon F, Nidelet T, Guédon E, Thierry A and Gagnaire V 2020 Understanding the mechanisms of positive microbial interactions that benefit lactic acid bacteria co-cultures. Review article *Front. Microbiol.* **11** 2088

[17] Agrawal N and Prakashi A 2013 Isolation of lactic acid bacteria from fermented milk products and their antimicrobial activity against Staphylococcus aureus *J. Food Saf.* **15** 39–42

[18] George F, Daniel C, Thomas M, Singer E, Guilbaud A, Tessier F J, Revol-Junelles A-M, Borges F and Foligné B 2018 Occurrence and Dynamism of Lactic Acid Bacteria in Distinct Ecological Niches: A Multifaceted Functional Health Perspective *Front. Microbiol.* **9** 2899

[19] Mokoena M P, Chelule P K and Gqaleni N 2005 Reduction of fumonisin B1 and zearalenone by lactic acid bacteria in fermented maize meal. *J. Food Prot.* **68** 2095–9

[20] Janiszewska-Turak E, Kołakowska W, Pobiega K and Gramza-Michałowska A 2021 Influence of drying type of selected fermented vegetables pomace on the natural colorants and concentration of lactic acid bacteria *Appl. Sci.* **11** 1–13

[21] Cirlini M, Ricci A, Galaverna G and Lazzi C 2020 Application of lactic acid fermentation to elderberry juice: Changes in acidic and glucidic fractions. *LWT* **118** 108779

[22] Wu C, Li T, Qi J, Jiang T, Xu H and Lei H 2020 Effects of lactic acid fermentation-based biotransformation on phenolic profiles, antioxidant capacity and flavor volatiles of apple juice *LWT* **122** 109064

[23] Bintsis T 2018 Review. Lactic acid bacteria as starter cultures: An update in their metabolism and genetics *AIMS Microbiol.* **4** 665–684

[24] Chettri R and Tamang J P 2015 Naturally fermented ethnic soybean foods of India *J. Ethn. Foods* **2** 8–17

[25] Rollán G C, Gerez C L and LeBlanc J G 2019 Lactic Fermentation as a Strategy to Improve the Nutritional and Functional Values of Pseudocereals. *Front. Nutr.* **6** 98

[26] Sharma R, Garg P, Kumar P, Bhatia S K and Kulshrestha S 2020 Review microbial fermentation and its role in quality improvement of fermented Foods *Fermentation* **6** 106

[27] Wang Y, Wu J, Lu M, Shao Z, Hungwe M, Wang J, Bai X, Xie J, Wang Y and Geng W 2021 Metabolism characteristics of lactic acid bacteria and the expanding applications in food industry *Front. Bioeng. Biotechnol.* **9** 612285

[28] Bajagai Y S, Klieve A V., Dart P J and Bryden W L 2016 *Probiotics in animal nutrition – Production, impact and regulation* (Rome: Food and Agriculture Organization)

[29] Yudianti N F, Yanti R, Cahyanto M N, Rahayu E S and Utami T 2020 Isolation and Characterization of Lactic Acid Bacteria from Legume Soaking Water of Tempeh Productions *Digit. Press Life Sci.* **2** 00003

[30] Bancalari E, Sardaro M L S, Levante A, Marseglia A, Caligiani A, Lazzi C, Neviani E and Gatti M 2017 An integrated strategy to discover Lactobacillus casei group strains for their potential use as aromatic starters *Food Res. Int.* **100** 682–90

[31] Kieliszek M, Pobiega K, Piwowarek K and Kot A M 2021 Characteristics of the Proteolytic Enzymes Produced by Lactic Acid Bacteria *Molecules* **26** 1858

[32] Ammor M S and Mayo B 2006 Selection criteria for lactic acid bacteria to be used as functional starter cultures in dry sausage production: An update *Meat Sci.* **76** 138–146

[33] Li K Y 2004 Fermentation: Principles and Microorganisms *Handbook of Food and Beverage Fermentation Technology* (New York: Marcel Dekker Inc.) pp 594–608

[34] Christensen J E, Dudley E G, Pederson J A and Steele J L 1999 Peptidases and amino acid catabolism in lactic acid bacteria *Antonie van Leeuwenhoek, Int. J. Gen. Mol. Microbiol.* **76** 217–46

[35] Fox P F and Wallace J M 1997 Formation of flavor compounds in cheese *Adv. Appl. Microbiol.* **45** 17–85

[36] Akabanda F, Owusu-Kwarteng J, Tano-Debrah K, Parkouda C and Jespersen L 2014 The Use of Lactic Acid Bacteria Starter Culture in the Production of *Nunu*, a Spontaneously Fermented Milk Product in Ghana ed R G Moreira *Int. J. Food Sci.* **2014** 721067

[37] Hajfathalian M, Ghelichi S, García-Moreno P J, Moltke Sørensen A-D and Jacobsen C 2018 Peptides: Production, bioactivity, functionality, and applications *Crit. Rev. Food Sci. Nutr.* **58** 3097–129

[38] Nagai T and Tamang J P 2010 Fermented soybeans and non-soybeans legume foods *Fermented Foods and Beverages of the World* ed J P Tamang and K Kailasapathy (New york: CRC Press) pp 191–224

[39] Sila A, Sayari N, Balti R, Martinez-Alvarez O, Nedjar-Arroume N, Moncef N and Bougatef A 2014 Biochemical and antioxidant properties of peptidic fraction of carotenoproteins generated from shrimp by-products by enzymatic hydrolysis *Food Chem.* **148** 445–52

Bio-emulsifier supplementation from *Pseudomonas putida* that cultivated on medium containing waste palm oil as feed additive on poultry

M Nurhamzah[1], and D Pantaya[1]*
[1] Poultry Agribusiness Study Program, Department of Animal Science, Politeknik Negeri Jember, Indonesia

*e-mail: dadik_pantaya@polije.ac.id

Abstract. The purpose of this study to determine effect of bio-emulsifier from *Pseudomonas putida* as feed supplement on the performance of broiler chickens. The experiment was divided in two phases. In phase 1: test of oil bubbles formation in water using bio-emulsifier, and phase 2 : *in vivo* test bio-emulsifier in animal. 200 one-day-old male and female broiler strain Cobb were used. On d 14, chicks were separated by sex and placed in 40 pens, sex separate (100 males or 100 females) and randomly assigned 4 treatments and 5 replicates of 5 birds each for 14 d (21d-35 d). The diets were based of corn, concentrate and rice bran meal and supplemented with: 0, 0.5,1 and 1.5 g/kg bio-emulsifier. The observed parameters were feed intake (FI), average daily gain (ADG), and feed conversion ratio (FCR). Result of experiment 1: as expected, supplementation of bio-emulsifier was increased small oil bubbles formation in medium oil in water, whereas in experiment 2 showed FI, ADG and FCR unaffected by dietary supplementation with bio-emulsifier on broiler at finisher phase (week 4 and 5) (P>0.05) and did not decrease on broiler performance. The results give new information that bio-emulsifier from *Pseudomonas putida* could potentially as a new alternative feed additive in the future on broiler performance.

1. Introduction

Energy in animal feed is the main nutrient in addition to protein, fat and minerals that are needed by livestock. The cause of in-efficient of nutrient metabolism was the lack energy availability in the feed. Lack of energy consumption caused disturbances in poultry production. Therefore, energy consumption must be able to meet the standard needs of broiler chickens 2800-3200 Kcal/kg [1-3] . Corn has been recognized as the main source of energy in poultry, the use of this material ranges from 50-60 percent, but that amount is still needed for other energy sources to meet energy requirement. Among the ingredients in feed, oil contains energy and essential fatty acids [4]. Oil contains higher calory than carbohydrates which is responsible for providing increased energy density [5].

The addition of oil in the digestive tract of poultry is difficult to digest because it cannot mix with water. The low fat absorption capacity, especially in young cattle, due to incomplete physiological functions and low lipase enzyme production [6]. Fat to be absorbed is hydrolyzed by lipase enzymes and bile acid contribute to helping the formation of micelles before they can be absorbed in the cell membrane [7]. To increase the ability of micelle formation, it is necessary to add an emulsifying agent, namely an emulsifier. The addition of an emulsifier can increase feed digestibility and body weight of starter phase chickens [7]. The ability to break down fat in the digestive tract will stimulate easier absorption of fat in cell membranes [8]. Micelles are small globules that contain polar compounds on their surface, so that they will easily penetrate the un stirrer water part of the hydrophilic cell membrane, thus facilitating the process of simple fat absorption [9].

Emulsifiers that function as surfactants are mostly made from petroleum materials which are increasing competitive with fuel requirements, for this reason, surfactants from organic materials that do not cause contamination, are easily degraded and are effective in a certain temperature and pH *Pseudomonas putida* is one type of microbe that can produce biosurfactants for emulsifiers [10]. The

IOP Conf. Series: Earth and Environmental Science **980** (2022) 012021 doi:10.1088/1755-1315/980/1/012021

reaction between glucose and fatty acid or triglycerides as a substrates of *Pseudomonas putida* will produce a type of glycolipid surfactant [11]. This reaction can be produced when the bacterial growth medium contains fatty acids and glucose. Fatty acids can come from waste palm oil and glucose from molasses materials. The use of this bio-emulsifier is expected to increase the formation of micelles so as to facilitate the absorption of fat in the cell membrane so that it can increase the energy produced so as to improve performance in poultry. The objective of this present study at evaluating the effect of addition a bio-emulsifier as feed supplement on the performance of broiler chickens.

2. Materials and Methods

2.1. Production of bio-Emulsifier

The wild type of *Pseudomonas putida* was originated from Feed technology laboratory, Politeknik Negeri Jember. *Pseudomonas putida* was grown in the cultivated in the medium that containing: molasse 360 g/L, peptone 50 g/L, Potato Dextrose Broth (PDB 50 g/L mixed to 1 g/L, 1.5% waste palm oil, ammonium sulfate (Sigma Aldrich, Germany) and distillated water. The media was sterilized using an autoclave at temperature of 121 °C, 15 min and cooled for 2 hours. The media contained in the test tube was then planted with *Pseudomonas putida* bacteria and incubated at 37 °C ,200 x g for 24 h in a shaker incubator. After 24 hours, medium were transferred to volume 1000 ml Erlenmeyer and shake with the same procedure as before. Cell biomass was centrifuged at 4 °C, 9000 x g for 10 min. The harvested then dried in the oven during 48 h 40 °C and used for *in vivo* trial.

2.2. Bio-emulsifier test oil in water

The test was divided into 4 treatments, P0, control P1, 200 µl P2 300 µl, and P3 400 µl. The precipitate bio-emulsifier that obtained previously was mixed with 100 ml of distilled water and, shake for 1 min at 27 °C, then put in sterilized the petri dish (90 mm). Introduced 10 ml palm oil and observed sum of small bubbles formation during 15 minutes in petri dish.

Table 1. Treatment ration formulation

	(%)			
Parameter	P0	P1	P2	P3
Feed ingredients				
Corn	58	58	58	58
Concentrate	30	30	30	30
Rice Bran	10	10	10	10
Palm oil	1	1	1	1
Premix[1]	1	1	1	1
Bio emulsifier) g/kg feed*	0	0.5	1	1.5
Total %	100	100	100	100
Chemical composition				
Protein (%)	19.8	19.8	19.8	19.8
Fat (%)	5.45	5.45	5.45	5.45
Crude Fiber (%)	4.21	4.21	4.21	4.21
ME (Metabolism energy) (Kcal/kg)	2.929	2.929	2.929	2.929
Ca (Calcium) (%)	1.35	1.35	1.35	1.35
P (Phospor) (%)	0.66	0.66	0.66	0.66

[1]Premik each kg contains: vitamin A, 8000 IU; vitamin D, 2100 IU; vitamin E, 12 IU; vitamin B1 , 0.5 mg; vitamin B6, niacin 1 mg, 15 mg, pantothenic acid, 12.5 mg.

2.3 Animals and diets

200 one-day-old male and female broiler strain Cobb were used. On d 14, chicks were separated by sex and placed in 40 pens, sex separate (100 males or 100 females). There were consisted 2 dietaries: starter phase (1 to 21 d), finisher (21 to 35 d). Chickens individual body weight and feed intake were measured at 21-28 d and 28-35 d to evaluate bird performance. The diets were based of corn, concentrate and rice bran meal and supplemented with: P0, 0 (control), P1: 0.5, P2: 1, and P3 : 1.5 g/kg bio emulsifier from *Pseudomonas putida*. Feed nutrient was referred in the basal diet met that suggested by NRC (1994). The composition of the feed that will be used for the trial is shown in **Table 1.**

2.4. Data collection

Data collection starts from the beginning to the end of maintenance. The data taken included feed intake (FI), average daily gain (ADG) and feed conversion ratio (FCR) was taken once a week.

2.5. Statistical Analysis

Data were submitted to analysis of variance using the ANOVA of Minitab software. Means were compared by the Tukey test

3. Results and Discussion

3.1 Bio-emulsifier oil in water test

The result In the bubble test formation from experiment 1 is presented in **Table 2** and showed that P3 treatment with the addition of a bio emulsifier caused highest number of bubbles compared to P2, P1. formation of the number of bubbles indicates the occurrence of a fraction of oil in water (O in W). The oil fraction shows or indicates the occurrence of an emulsification process. *Pseudomonas sp.* produced biosurfactants that are surface active, characterized by a decrease in the surface tension of the liquid medium [12]. The biosurfactant produced by *Pseudomonas spp.* is a low molecular weight biosurfactant was identified as the amphiphilic surface-active glycolipids (rhamnolipids) that consisting of hydrophobic and hydrophilic molecules [13].

Table 2. The number of small bubbles after introduction bio-emulsifier (unit) in medium oil in water

Treatment	Small bubbles formation
P0	475±65.05^{a}
P1	602±9.90^{b}
P2	595±130.00^{b}
P3	690±67.00^{c}

According Gumel [14] *Pseudomonas putida* can growth in medium fatty acid C8:0 and C18: 1 as carbon and ammonium as nitrogen source that found in waste palm oil effluent.

3.2. Bio-emulsifier test in Broiler performance

During the observation period in the *in vivo* study, the chickens were in good condition and did not appear to have health problems and showed normal livestock behavior. Chickens did not display a number clinical symptom. Analysis of variances was showed no significant effect on feed consumption among treatments (P>0.05). The coefficient variation for feed intake were slightly higher for the chickens reared with sexes separated with males have larger variation than female. The absence of differences between treatments is thought to be due to the relatively the same energy content. According Massuquetto [15] the energy content in feed will affect the amount of feed consumed. Chickens will stop eating when the energy consumed is sufficient for chicken activities. The addition of bio emulsifier is expected to reduce feed consumption. This is in accordance with the opinion Guerreiro [16] which states that the addition of an emulsifier does not cause a change in performance.

Table 3. Broiler performance was fed treated by glycolipid as bio emulsifier produced from *Pseudomonas putida*

Parameter	P0	P1	P2	P3	SEM	P value
Cummulative Week 4-5						
Male						
Feed intake	1635,50	1567,50	1483,69	1456,26	390	0.754
ADG	873,89	808,92	797,49	833,12	90.74	0.711
FCR	1,87	1,94	1,86	1,75	0.18	0.289
Female						
Feed intake	1491,95	1382,85	1308,96	1345,18	131	0.187
ADG	770,55	709,51	728,59	714,76	153.1	0.959
FCR	1,94	1,95	1,80	1,88	0.22	0.337

SEM = Standard Error Mean

Data from ADG and FCR showed that the addition of bio-emulsifier had no significant effect (P>0.05) for both males and females. The increase in body weight gain is expected to be achieved by the addition of a bio-emulsifier. Bio-emulsifier agent is expected to increase fat absorption, because it can increase the hydrolyze of fat globules. There was no significant effect, this was thought to be caused by the product application carried out finisher period at week 4 (21-28 d). According Polin and Hussein [17] to in this period chickens produced lipase at a high concentration compared to chickens aged 42 days. The ability of fat absorption is sensitive in young broilers, where lipase production is not balanced with fat absorption capacity. [18] . The addition of an emulsifier in 7-day-old chicks can reduce oil use by 20%, when compared to 14 d and 21 d. the bio-emulsifier ability was not affected due to lipase activity at concentrations that were still able to increase fat absorption, so that the bio-emulsifier effect was not effectively effective. significantly in suppressing FCR which is related to the efficiency of production performance on broiler.

4. Conclusion
The results give new information that bio-emulsifier from *Pseudomonas putida* could potentially as a new alternative feed additive in the future for broiler performance.

5. Acknowledgment
The author expresses his deepest gratitude to PT. Indofood Sukses Makmur Tbk. for the research funding assistance provided in the Indofood Riset Nugraha (IRN) program 2019/2020.

References
[1]. Aftab, U., World's Poultry Science Journal, 2019. **75**(4): p. 507-514.
[2]. Pantaya, D., et al., IOP Conference Series: Earth and Environmental Science, 2020. **411**: p. 012040.
[3]. Wulandari, S. and T.M. Syahniar, IOP Conference Series: Earth and Environmental Science, 2018. **207**: p. 012034.
[4]. Kaur, N., V. Chugh, and A.K. Gupta, Journal of food science and technology, 2014. **51**(10): p. 2289-2303.
[5]. Antunes, M., et al., Semina: Ciências Agrárias, 2016. **37**: p. 429.
[6]. Meng, X., B.A. Slominski, and W. Guenter, Poult Sci, 2004. **83**(10): p. 1718-27.
[7]. Iqbal, J. and M.M. Hussain, American journal of physiology. Endocrinology and metabolism, 2009. **296**(6): p. E1183-E1194.
[8]. Werner A, K.F., Verkade HJ. *Fat Absorption and Lipid Metabolism in Cholestasis.* . In: Madame

Curie Bioscience Database [Internet] 2000-2013.
[9]. Pantaya, D., et al., IOP Conference Series: Earth and Environmental Science, 2020. **411**: p. 012041.
[10]. Fenibo, E.O., et al., Microorganisms, 2019. **7**(11): p. 581.
[11]. Siebenhaller, S., J. Grüninger, and C. Syldatk, U.T. Bornscheuer, Editor. 2018, AOCS Press. p. 293-313.
[12]. Tuleva, B.K., G.R. Ivanov, and N.E. Christova, Z Naturforsch C J Biosci, 2002. **57**(3-4): p. 356-60.
[13]. Haloi, S., et al., 3 Biotech, 2020. **10**(3): p. 120-120.
[14]. Gumel, A.M., M.S.M. Annuar, and T. Heidelberg, Industrial Microbiology • Braz. J. Microbiol. 45 (2) 2014.
[15]. Massuquetto, A., et al., Animal, 2020. **14**(6): p. 1139-1146.
[16]. Guerreiro Neto, A.e.a., Brazilian Journal of Poultry Science [online], 2011. **13. Number : 2** p. 119-125.
[17]. Polin, D. and T.H. Hussein, Poultry Science, 1982. **61**: p. 1697-1707.
[18]. Krogdahl, A., The Journal of nutrition, 1985. **115**: p. 675-85.

Carica papaya seed meal in diet can reduce egg quail cholesterol without reduce egg quality

R Rahmasari, R T Hertamawati*, N Ningsih, S Imam, U Suryadi and B A Nugraha

Animal Science Department, Politeknik Negeri Jember, PO BOX 164, Jember, Jawa Timur, Indonesia

E-mail: rosa_trihertamawati@polije.ac.id

Abstract. The purpose of this study was to determine the effect of papaya seed meal on the quality of quail eggs and to determine the best dose of quail egg quality. Using an experimental method with a completely randomized design (CRD), with 4 treatments and 6 replications, each replication consisted of 7 quails. The treatments were P0 (control feed), P1 (0.3% papaya seed meal in feed), P2 (0.6% papaya seed meal in feed) and P3 (1.2% papaya seed meal in feed). The parameters observed were egg white index, egg yolk index, egg yolk color, shell thickness, and egg cholesterol. The data was analysed by Analysis of Variance (ANOVA). The results showed that the addition of papaya seed meal in the feed had not a significant effect ($p<0,05$) on egg white index, on egg yolk index and shell thickness, but decreased egg yolk color. However addition papaya seed meal had significant effect ($p<0,05$) reduces egg yolk cholesterol. In conclusion, addition of papaya seed meal (*Carica papaya* L.) up to 1.2% reduce egg yolk cholesterol without effected the quality of quail eggs.

1. Introduction

Human body need good nutrition for their growth and health. One of the popular foods that contain good nutrition is eggs. Eggs that are frequently comsumed is quail eggs which are consumed as a protein source. Quail eggs production increase every year, the growth 0.46% from 2018 to 2019 [1]. The average consumption of quail eggs per capita per year in Indonesia increased from 7.8 to 9.2 from 2016 to 2017 [2]. Quail eggs have a complete nutrient, however quail eggs contain high cholesterol and that make a reason some people don't want to consume quail eggs. [3] Cholesterol content in quail eggs around 12.26 mg/g. [4] report that cholesterol level in quail egg around 844 mg/100g and thats number quit higher than cholesterol in chicken egg 372 mg/100 g. Food with high cholesterol level not favored by people who suffer with desease cause by cholesterol such as cardiovascular desease, high blood pressure and obesity [5]. Some research have been done to reduce cholesterol content on quail eggs, such as with diet treatment. Feed rich in vitamins, minerals, or phenolic compound such as tanin, saponin, and another compound can effect to egg cholesterol levels. One of the potential ingredients can be used on decreasing egg cholesterol levels without affected egg quality is papaya seeds.

Papaya is one of the fruit in Indonesia with high production. [6] papaya production in Indonesia about 1.016.388 tons at 2020. Some part of papaya fruit such as seeds and leaves are not widely used for human food. The amount of papaya seeds is about 16% from the whole papaya fruit and is considered as a waste [7]. [8] [9] Papaya seeds meal are rich with nutriens, such as 21.8% crude protein, 6.7% ash, 26.2% crude fiber, crude fat 28.3%, metabolism energy 3570 kcal/kg, 681.4 mg/100 g Ca, 2115.7 mg/100 g P and 423.5 mg/100 g Mg. Several experiments regarding papaya seeds meal for feed have been done.

Adding papaya seeds meal to quail feed 0.25%, 0.5% and 1% was unaffected egg weight and feed conversion [9]. Another research report that 0.5% papaya seed meal on broiler breeder diet showed that weigh of eggs and DOC weigh are heavier than control diet ($p<0.05$) [10]. However, data about affect of papaya seeds meal on quail still limited. This research is presents to evaluation of papaya seeds meal on quail diets on egg interior quality and egg cholesterol content.

2. Materials and Methods

2.1. Preparation of papaya seed meal

Papaya seed were collected from local papayas farmer around Jember district. Papaya seeds then were washed and sun dried for 4-5 days. The dried papaya seed were ground to powder.

2.2. Animal and diet treatments

The experiment used 168 of 75-days-old quail were reared for eight weeks. This experiment was conducted in a completely randomized design (CRD) with four treatments and six replications with seven quails in each . The diet treatments were P0: 0% papaya seed meal (control), P1: diet contain 0.25% papaya seed meal, P2: diet contain 0.5% papaya seed meal and P4: diet contain 1% papaya seed meal. The ingredients were used on diets are corn, rice bran, fish meal, soybean meal, oil, $CaCO_3$, premix and commercial feed. The diets were prepare in crumble. Diet and drinking water given ad libitum. Feed formulation and nutrien content are shown in **Table 1**.

Table 1. Feed formulations and nutrient contents of diet trearments

	Treatments			
	P0	P1	P2	P3
Ingredients (%)				
Corn	39.92	40.41	40.08	40.57
Rice bran	5.59	5.41	5.37	5.35
Soybean meal	19.33	19.57	19.41	19.65
Fish meal	6.45	6.53	6.63	5.90
Oil	3.63	3.67	3.64	3.69
Papaya seed meal	0	0.25	0.5	1
$CaCO_3$	5.32	5.39	5.34	5.41
Commercial feed	19.35	18.37	19.03	18.03
Premix	0.40	0.40	0.40	0.40
Nutrients content				
Gross energy (kcal/kg)	2964.84	2966.46	2966.69	2966.81
Crude protein (%)*	22.83	22.63	22.73	22.43
Crude fat (%)	6.16	6.20	6.19	6.24
Crude fiber (%)	3.67	3.67	3.77	3.87
Ca (%)	3.07	3.07	3.05	3.03
P (%)	0.64	0.63	0.63	0.60

*The analysis result from Laboratorium Teknologi Pakan, Politeknik Negeri Jember. P0: diet without papaya seed meal (control), P1: diet containing 0.25% papaya seed meal, P2: diet containing 0.5% papaya seed meal, and P3: diet containing 1% papaya seed meal.

2.3. Interior egg quality data collection

Variable of interior egg quality data in this research were egg white index, yolk index, yolk color, shell thickness and cholesterol content. egg white index, yolk index, yolk color, shell thickness was measured once a week during treatment. Cholesterol content was measured once time in the end of treatment. Data were to analysis of variance (ANOVA), if any indicated a signifinant effect, the analysis was continued with a post-hoc test by Duncan's multiple range test using SPSS software.

3. Results and Discussion

The averages data of egg white index, yolk index, yolk color and shell thickness of quail egg during treatment are shown in **Table 2.** Adding papaya seed meal in the diets (0.25; 0.5 and 1%) not significantly affected on egg white index, yolk index, and shell thickness ($p>0.05$) compared to the control. There was no difference in egg white index, yolk index and shell thickness means that no negative effect of papaya seed meal on the interior egg quality. Another factor was possibly because nutrients contents such as energy, protein and mineral in the diet treatments had the same value between control diet and diet treatments [9]. However, papaya seed meal was significantly affected yolk color ($p<0.05$). Yolk color was decreased by adding papaya seed meal. Score of yolk color on control diet treatment was higher than yolk color on diet with papaya seed meal ($p<0.05$). [11] some factors that influenced yolk color are xanthophylls pigment, zeaxanthin, lutein and synthetic pigments.

.

Table 2. The averages data of egg white index, yolk index, yolk color and shell thickness of quail egg during treatment

Treatments	P0	P1	P2	P3
Egg white index	0.13a	0.14ab	0.14ab	0.14ab
Yolk index	0.47	0.46	0.46	0.47
Yolk color	4.79c	3.87b	3.48a	3.29a
Shell thickness	0.03	0.03	0.03	0.03

P0: diet without papaya seed meal (control), P1: diet containing 025% papaya seed meal, P2: diet containing 0.5% papaya seed meal, and P3: diet containing 1% papaya seed meal. Means in the same row with superscripts differ significantly ($P<0.05$).

Table 3. Cholesterol content of yolk quail egg after treatment

Treatments	Cholesterol content (mg/100g)
P0	670.619c
P1	574.884b
P2	424.694a
P3	406.516a

P0: diet without papaya seed meal (control), P1: diet containing 0.25% papaya seed meal, P2: diet containing 0.5% papaya seed meal, and P3: diet containing 1% papaya seed meal. Means in the same row with superscripts differ significantly ($P<0.05$). Analyzed at Laboratorium Biokimia Nutrisi, Departemen Nutrisi dan Makanan Ternak, Fakultas Peternakan, Universitas Gajah Mada.

Cholesterol content of yolk quail egg after treatment are shown in **Table 3.** The result showed that adding papaya seed meal up to 1% on diet had significant effect on reducing egg yolk cholesterol levels ($p<0.05$). The same result was reported that total cholesterol and SGPT levels decreased in white rats suffering from hypercholesterolemia. The decrease in total cholesterol was around 13.39% and SGPT levels 31.4% ($p<0.05$) [12]. In addition high nutrient content, papaya seeds meal contain some antinutrient and antioxidant properties. Antinutrient compositions on papaya seed meals were phenols 28.32 mg/100g, hydrogen cyanide 1.87 µg/g, alkaloids 1.07% and oxalates 11.08 mg/100g [13]. The antinutrient may cause decreasing cholesterol level. Fat metabolism is inhibited due to inhibition of lipase enzyme secretion may cause by the addition 1.42 g papaya seeds meal on diet [14]. Obstructed fat metabolism is thought directly inhibit cholesterol synthesized. [15] Yolk precursors are synthesized in the liver and transported by plasma to ovary. Yolk cholesterol content is primarily dependent on the cholesterol content (triglyceride-lipoprotein). Inhibition of cholesterol synthesis can reduce the rate of lipoprotein synthesis and secretion which possibility cause reductions in egg yolk cholesterol levels.

4. Conslusion

The results showed that adding papaya seed meal on quail diet has not significant effect on egg white index, on egg yolk index and shell thickness, but decreased egg yolk color. However addition of papaya seed meal (*Carica papaya* L.) up to 1.2% reduce egg yolk quail cholesterol without effected the quality of quail eggs.

5. Acknowledgments

This research supported by PNBP 2020 grant project from Politeknik Negeri Jember

References

[1] Direktorat Jenderal Peternakan 2019 Produksi Telur Puyuh Menurut Provinsi
[2] Direktorat Jenderal Peternakan dan Kesehatan 2019 Statistika Peternakan dan Kesehatan Hewan
[3] Ali M A and Abd El-Aziz A A 2019 *Minia University* **5** (3): 39-56
[4] USDA (United States Department of Agriculture) 2018 Food composition data base. https://ndb.nal.usda.gov/ndb/foods/show/01140
[5] Wardah, J Rahmahani and T Sopandi 2016 *Asian J.of Agric. Res.* **10** (2): 114-125
[6] Badan Pusat Statistika 2020 Produksi Tanaman Buah-buahan
[7] Sugiharto S 2020 *Veterinary World* **13** (8): 1613-1619
[8] Maisarah A M, Asmah R and Fauziah O 2014 *J. Tissue Sci. Eng* **5** (1): 2-7
[9] Rahmasari R, Hertamawati R T, Rafli A K and Nugraha B A 2021 IOP Conf. Ser.: Earth Environ. Sci. **672** 012040
[10] Dassidi N, Oumbortime N, Kouame Y A E, Kokou V, Decuypere E, Gbeassor M, Onagbesan O and Tona K 2020 Int. J. of Poult. Sci. **19** (1): 1-9
[11] Lesson S and Summers J D 2005 Commercial Poultry Nutrition (Nottingham University Press) pp. 198-205.
[12] Saputri L O, Satriyasa B K dan Yasa W P S 2017 *Warmadewa Medical Journal* **2** (1): 1-10
[13] Cemaluk E A C 2017 *App. Sci. Report.* **17** (3): 75-81
[14] Salim A N, Sumardianto dan Amalia U 2018 *JPHPI* **21** (2): 188-198
[15] Naber E C 1976 *Poultry Science* **55** (1): 14-30

IOP Conf. Series: Earth and Environmental Science **980** (2022) 012023 doi:10.1088/1755-1315/980/1/012023

Early growing performance of local male weaning lambs fed edamame pod-based rations with and without exogenous enzyme additives

E Kustiawan[1], S Nusantoro[1], M Andriani[1], and TM Syahniar[1*]

[1]Department of Animal Science, Politeknik Negeri Jember, Mastrip Street PO BOX 164, Jember Regency 68101, East Java, Indonesia

*email: mahiseta@polije.ac.id

Abstract. This study aimed to evaluate the early growing performance of local male weaning lambs fed edamame pod-based rations with and without exogenous enzyme additives. Nine local male weaning lambs with initial weight about 16.88±2.33 kg/head were randomly assigned into into 3 levels treatment using an experimental completely randomized design (CRD). Every single level treatment was repeated 3 times. The levels treatment was 0% (control), 0.14% (w/w), and 0.28% (w/w) of exogenous enzyme addition into edamame pod-based rations. The lambs were maintained in individual pen which fed and given drinking water ad libitum. The parameters observed were nutrients consumption, liveweight gain, and gain to feed ratio during the first couple week maintenance with 2 weeks adaptation period. Data obtained were analysis using analysis of variance and further tested by DMRT for different result. The result presented that nutrient consumption, daily gain and gain to feed ratio were not affected ($P>0.05$) by various levels of exogenous enzyme additions. Thereby, it was concluded that the addition of exogenous enzyme up to 0.28% w/w potentially improved the nutrient consumptions and productivity performance of local male weaning lambs.

1. Introduction

Glycine max (L) Merill or edamame has been industrially produced and exported by Jember Regency. It also produced industrial waste around 10-15 tons of edamame pod waste every day. The potential of edamame pod waste (EPW) as an alternative local feed has been applied to beef and dairy cattle. However, this potential needs to be explored in order to optimize the use of EPW as sheep feed to support one of the Corporate Social Responsibility (CSR) programs for small-scale sheep breeders especially around Jember Regency.

The EPW was as a source of fiber with a crude fiber (CF) content of 33.93%; NDF 63.83%; ADF 43.10%; and 5.05% lignin but contain higher protein (13.53%; [1], unpublished) than elephant grass (9.7% protein and 36.1% fiber). There was an obstacle to EPW usage which it will potentially reduce nutrients digestibility [2], especially for sheep and even more for weaning lambs. The ability of digestive system to digest low quality forages on small ruminants was considered inadequate and showed lower capability than large ruminants [3]. Several solution explorations have been carried out to overcome this constraint for the EPW utilization. It was fermentation or silage [4,5] or created as complete feed [6], but both had not shown optimal results.

Based on constraints, this study faced how to improve the nutritional quality of edamame pod-based rations for local male weaning lambs so that it can be effectively and efficiently applied to present optimal early growing performance. This study approach was the application of mix-exogenous enzyme additives (fibrolytic, cellulolytic or mixture) in a total mixed ration (TMR) based edamame pod waste. TMR was designed to avoid selective rejection of unpalatable portions, to ensure uniformity distribution and nutrients intake, then to optimize feed efficiency. This study aimed to evaluate the early growing performance of local male weaning lambs fed edamame pod-based rations with and without exogenous enzyme additives.

2. Methods

2.1. Experimental design

This study was carried out in the Sheep Pen of the Department of Animal Science and in the Laboratory of Feed Technology at Politeknik Negeri Jember. Nine local male weaning lambs with initial weight about 16.88±2.33 kg/head were randomly assigned into 3 levels treatment using an experimental completely randomized design (CRD). Every single level treatment was repeated 3 times. The levels treatment was 0% (control), 0.14% w/w or 1.4 g/kg, and 0.28% w/w or 2.8 g/kg of exogenous enzyme addition into edamame pod-based rations. The edamame pod-based rations contained 89.08% DM, 91.69% OM, 71.4% TDN and 12.6% protein and the exogenous enzyme added was produced by Bioproton, PTY LTD, Australia. The exogenus enzyme was a multi-fibrolytic enzyme that contained cellulase, xylanase, β-glucanase, protease, α-amilase, hemicellulase, pectinase, phytase, lipase, hemicellulase, acid phytase and acid phosphatase.

The lambs were adapted to the individual pen and kept comfortably conditions for 2 weeks. They fed edamame pod-based rations gradually by reducing the provision of forages until the feed offered reached 100% edamame pod-based rations and given drinking water ad libitum. The lambs have been maintaining for two months but the data published here for early couple weeks observation due to this study was ongoing. The parameters observed were nutrient consumption such as dry matter intake (DMI), organic matter intake (OMI), total digestible nutrient (TDN) and protein intake; liveweight gain (kg/d); and gain to feed ratio during first couple week maintenance with 2 weeks adapation period.

2.2. Statistical analysis

Data obtained were analysis using analysis of variance (ANOVA) and further tested by Duncan's Multiple Range Test (DMRT) for the different result. It was conducted using SPSS 22.

3. Results and Discussions

The results of parameters for nutrient consumption and performances of local male weaning lambs fed edamame pod-based rations with or without exogenous enzyme addition were presented in Table 1. Nutrient consumption included dry matter, organic matter, total digestible nutrient (TDN) and protein intake were not affected (P>0.05) by various levels of exogenous enzyme additions. They were ranged 474.5 g/d, 435.1 g/d, 380.4 g/d, and 67.1 g/d for DMI, OMI, TDNI, and CPI, respectively.

The level addition of exogenous enzyme in the edamame pod-based rations as a dry feeds has not been effective to show significant result on nutrient consumption of local male weaning lambs. It was similar with Beauchemin et al. who informed that exogenous enzymes are more effective when applied to high moisture feeds (such as silages) compared to dry feeds [7]. The direct action of exogenous enzymes before feed consumed cause the release of reducing sugars arising from partial solubilization of cell wall components [8]. Then, an increase of releasing on reducing sugar will increase the availability of carbohydrates in the rumen thereby shortening the lag time needed for microbial colonization and can also enhance rapid microbial attachment and growth [9]. Additionally, exogenous enzymes increase hydrolytic activity before incubation of feed in the rumen then also help rumen microbes to raise microbial access into cell walls and increase the rate of fiber degradation during ruminal fermentation [10].

Table 1. Nutrient consumption and performances of local male weaning lambs fed edamame pod-based rations with or without exogenous enzyme addition

Parameters	Exogenous enzyme addition (w/w)			P-value
	0%	0.14%	0.28%	
DMI (g/d)	406.5	479.5	537.6	p>0.05
OMI (g/d)	372.7	439.7	493.0	p>0.05
TDNI (g/d)	325.8	384.3	430.9	p>0.05
CPI (g/d)	57.5	67.8	76.0	p>0.05
Total gain (kg)	3.1	4.4	5.6	p>0.05
Daily gain (g/d)	223.8	316.7	402.4	p>0.05
Gain to feed	0.5	0.7	0.8	p>0.05

DMI = dry matter intake, OMI = organic matter intake, TDNI = total digestible nutrient intake, CPI = crude protein intake

ADG and gain to feed ratio were tend to enhance from 223.8 to 402.4 g/d and 0.5 to 0.8, respectively. It was equivalent to the trend of tend to increase on nutrient consumption, specifically TDN consumption. TDN determined feed energy and its consumption described the reference for amount of energy consumed that will be used to perform muscle movements and growth. The more TDN content contained and consumed, so that the more energy obtained by livestock. It dependent on the physiological status of livestock to be used as a reference in fattening programs [11]. Nevertheless, this result was similar with previous study that the enzyme addition of low or high doses to edamame waste or oat straw in lambs fed was not affected in the average daily gain [6,12], whereas it was inconsistent with the previous study which state that the addition of exogenous enzyme improved sheep performances [13]. Meale et al. (2014) reported that the application of exogenous enzyme preparations to the diets of small ruminants has little impact on nutrient intake and production [14].

4. Conclusion

According to the results, it concluded that the presence or absent of exogenous enzyme addition presented similar results on nutrient consumption, daily gain and gain to feed ratio. However, the addition of exogenous enzyme up to 0.28% w/w potentially improved the nutrient consumptions and productivity performance of local male weaning lambs. Further observations that elaborate nutrient digestibilities and ruminal fermentations need to be conducted.

5. Acknowledgement

Authors acknowledge P3M Politeknik Negeri Jember for PNBP Research Grant 2021 contract number 13/PL17.4/PG/2021.

References

[1] Marzuki A and Kustiawan E 2020 (Jember)
[2] Baurhoo B, Ruiz-Feria C A and Zhao X 2008 *Anim. Feed Sci. Technol.* **144** 175–84
[3] Soto-Navarro S A, Lopez R, Sankey C, Capitan B M, Holland B P, Balstad L A and Krehbiel C R 2014 *J. Anim. Sci.* **92** 1621–9
[4] Herring E M, Kegley E B, Coffey K, Gadberry S, Powell J G and Pruden M C 2020 *J. Anim. Sci.* **98** 74–74
[5] Syahniar T M and Subagja H 2018 *J. Peternakan, Politek. Negeri Jember* 87–91

[6] Marzuki A, Syahniar T M, Andriani M, Nusantoro S and Kustiawan E 2021 *Proceeding International Conference of Food and Agriculture* pp 103–6

[7] Beauchemin K A, Colombatto D, Morgavi D P and Yang W Z 2003 *J. Anim. Sci.* **81** E37-E47

[8] Hassan A, Gado H, Anele U Y, Berasain M A M and Salem A Z M 2020 *Anim. Biotechnol.* **31** 365–72

[9] Dean D B, Adesogan A T, Krueger N and Littell R C 2005 *J. Dairy Sci.* **88** 994–1003

[10] Puastuti W, Haryati T and Yulistiani D 2015 *Prosiding Seminar Nasional Teknologi Peternakan dan Veteriner* pp 106–14

[11] Sihotang A, Sudrajat D and Dinansih E 2012 *J. Pertan.* **3** 78–90

[12] Bueno A L, Martinez G M, Garcia P H, Garcia J M and Perez F P 2013 *Anim. Nutr. Feed Technol.* **13** 355–62

[13] Estrada-Angulo A, Arteaga-Wences Y J, Castro-Pérez B I, Urias-Estrada J D, Gaxiola-Camacho S, Angulo-Montoya C, Ponce-Barraza E, Barreras A, Corona L, Zinn R A and Leyva-Morales J B 2021 *Animals* **11** 2390

[14] Meale S J, Beauchemin K A, Hristov A N, Chaves A V and McAllister T A 2014 *J. Anim. Sci.* **92** 427–42

The potential solvent for tannin extraction as a feed additive made of coffee husk (*Coffea canephora*) using Soxhlet Method

S B Kusuma[1], S Wulandari[1*], R A Nurfitriani[1], and A Awaludin[1]
[1]Department of Animal Science, Politeknik Negeri Jember, PO BOX 164 Jember, Indonesia.
*E-mail: suci@polije.ac.id

Abstract. The purpose of this research was to determine the best of solvent for extracting tannins as a feed additive from coffee husks using the soxhlet method. The method used was quantitative with a completely randomized design experimental. The coffee husk used was a type of Robusta taken from the Kaliwates Region, Jember Regency, Indonesia. The material was the waste from the grinding of coffee cherries, namely the pericarp (parchment, pectin layer, pulp, and outer husk). This research consisted of 3 treatments (P1: Methanol, P2: Ethanol, P3: Acetone), each treatment consisted of 3 replications. The variables were the tannin extracts and residue of tannin extracts. The results showed that different types of tannin solvents had a significant effect on the tannin extracts yield and residue of tannin extracts ($P<0.05$). Methanol and ethanol solvents produced higher tannin than acetone solvents: 0.5392%, 0.6120%, and 0.2457% respectively. Coffee husk with ethanol solvent gave the least residue compared to methanol and acetone: 61.64%, 66.89%, and 85.26% respectively. The conclusion of this research was Ethanol is a solvent that produces the highest total tannin (0.6120%) and the lowest residue (61.64 ± 0.6688%) from coffee husks compared to methanol and acetone. The result of this research can be used as a consideration for the tannins extraction of other plantation waste.

1. Introduction

The availability of coffee waste in Jember Regency is quite abundant, which is 11,863 ton in 2017[1]. The husk composition of coffee is quite large, it is about 43.6% of the total fresh coffee fruit [2]. Coffee husk is one type of waste that has a quite high tannin. Tannin is nontoxic organic compounds that is classified as polyphenols commonly obtained from plant extracts. Tannin is beneficial as antioxidants, food additives, and drugs[3].

Tannins are phenolic compounds derived from secondary plant metabolism and develop complex bonds with proteins [4]. Physically, when dissolved into water, tannins will form colloids and give a sour and bitter taste. When tannins are mixed in alkaloids and gelatin, it will become sediment. Tannin is shiny yellow or white amorphous powder, almost colorless, with distinctive smell and taste [3]. Chemically, tannins are complex compounds in the form of polyphenol mixtures that are difficult to separate so that they are difficult to crystallize.

Tannin compounds can preserve proteins during the fermentation process, reduce ammonia levels, and increase the nutrient of silage products[5] Tannin is also known for its benefits as antioxidants, antidiarrhea, antibacterial and astringents. Tannins have strong bonds with protein molecules and have the ability to precipitate proteins.

For ruminants, tannin can be either advantageous or disadvantageous, depending on the amount consumed by the animal, its type and chemical structure as well as the composition of the rest of the feed, especially the concentration of CP feed. In temperate forage, in low to moderate concentrations

(<50 g / kg DM), it is beneficial in terms of increasing protein utilization without negatively affecting feed intake and digestion of nutrients. This shows that coffee husk waste can be used as a potential material for improving the quality of feed nutrients.

Tannins can be extracted using a single solvent or solvent mixture. The solvents that can be for tannin extraction are methanol, ethanol, or acetone solutions, and also ethyl acetate [6]. Extraction using Soxhlet with liquid solvents is one of the best methods used in separating bioactive compounds from nature. This is because the heating that occurs during the extraction process can increase the ability to extract compounds that are not soluble at room temperature, so that the activity of pulling compound is more maximal [7]. Therefore, tannins have an important role as feed additives, so this research was conducted to observe the extraction of tannins from coffee husk waste using different solvents. The purpose of the research was to determine the best solvent for extracting tannins as feed additives from coffee husks using the Soxhlet method.

2. Materials and Methods

2.1 Making coffee husk powder

Robusta coffee husk waste (from the husk coffee plantation, Jember, East Java Province, Indonesia). Dry coffee husk (<15%) water content is smoothed using the Whiley *mill* (2 mm screen). The results of the analysis of the Feed Technology Laboratory, Politeknik Negeri Jember (2021) showed that the coffee husk used contained the following nutrients: BK (91.0%); PK (11.3%); LK (1.6%); SK (16.6%).

2.2 Extraction of coffee husk waste tannins using Soxhlet method

Twenty gram of coffee husk powder wrapped in Whatman filter paper, is extracted with organic solvents according to treatment (methanol (99.8%), ethanol (70%), acetone (70%) as much as 375 ml. The extraction process uses a Soxhlet tool with a temperature of 90°C or characterized by the spinning of the solution inside the pumpkin. Extraction ends until the solution is clear. Filtrate extraction method Soxhlet inserted into rotary vacuum *evaporator* (60°C) until obtained concentrated extract and dried into the oven (60°C) [7].

2.3 Qualitative analysis of tannins.

The concentrated extract obtained is taken as much as 10 g and added with 3-4 drops of ferrite chloride solution ($FeCl3$). The formation of bluish black indicates the presence of phenols (tannins) [8]. Observed parameters: 1) Results of coffee husk extract and tannin yield; 2) residue from extraction

Tannin yield (X%)

$$X\% = \frac{The\ amount\ of\ tanin\ produced}{The\ amount\ of\ coffee\ husk}\ x\ 100\%$$

2.4 Data Analysis

Data obtained from the extraction of Soxhlet coffee husk tannins analyzed variance with Complete Random Design (CRD). Each treatment is repeated three times, significant results are followed by Duncan's Multiple Range Test (DMRT).

3. Results and discussion

The tannin content of coffee husk by soxhlet method using methanol and ethanol as solvents were significantly different. Both of them produced quite high tannins compared to tannin extracts using acetone as a solvent (Table 1). From these results showed that the results of tannin extract were influenced by the kind of solvent. The highest content of coffee husk tannin was obtained from ethanol solvent by 0.612%, then the methanol solvent was 0.5392%, and the smallest one was the acetone solvent by 0.2457%.

Table 1. Tannin content of coffee husk using the soxhlet method on different types of solvents

Number	Solvent Types	Coffee Husk Extraction Results (%)	Total Tannin (%)
1	Methanol	29.53 ± 1.6399[a]	0.5392[b]
2	Ethanol	28.17 ± 1.2687[a]	0.6120[a]
3	Acetone	6.33 ± 0.5118[b]	0.2457[c]

[a,b,c] = Means were significantly different in each column ($P<0.05$)

The effectiveness of the extraction of a compound by a solvent depends largely on the solubility of the compound, that a compound will dissolve in a solvent of the same polarity. Tannins are polar compounds that will dissolve in polar solvents. Polar solvents include methanol, ethanol, and acetone. The dielectric constant measures the solvent's ability to reduce the strength of the electric field around the charged particles. Organic solvents based on their dielectric constants can be divided into two: polar and non-polar solvents. The dielectric constant is expressed as the resisting force between two electrically charged particles in a molecule. The higher the dielectric constant, the solvent is more polar. The dielectric constants of methanol, ethanol and acetone each have a dielectric value of 33.64; 25.16; and 21.13 respectively [9].

The results of coffee husk extract with organic solvents showed that coffee husk tannin extract could be done using polar solvents with a range of dielectric constants around 25.16 to 33.64 which found using methanol and ethanol solvents. Different solvents showed significant different of tannin production (total tannin). The largest content of coffee husk tannins is obtained from ethanol solvents by 0.612%; then methanol solvent was 0.5392%; and the smallest one was acetone solvent by 0.2457% (Table 1). The high total tannins of coffee husk extracts using ethanol showed that ethanol was the best solvents. This was happened because the acquisition of extraction products is based on the similarity of polarizing properties to solvents. It was also stated by Mandal [10] that ethanol heats up faster so that it can improve the extraction process. This was seen during the soxhlet process, ethanol solvents boiled first from other solvents indicated by the rotation of the solution in the soxhlet vessel. The type of solvent produced a different extraction residue, the largest being acetone at 85.26 ± 0.3921 followed by methanol a 66.89 ± 1.8619 and ethanol at 61.64 ± 0.6688 (Table 2).

Table 2. Residual levels of coffee husk using the soxhlet method in different alcohol solutions

Number	Solvent type	Residue (%)
1	Methanol	66.89 ± 1.8619[b]
2	Ethanol	61.64 ± 0.6688[c]
3	Acetone	85.26 ± 0.3921[a]

[a,b,c] = Means were significantly different in each column ($P<0.05$)

An extraction with ethanol solvents, it produced the highest tannins so that it would produce the lowest residues (Table 2). This condition was because the coffee husk that could be extracted (as tannins) was the highest. In contrast, extraction using acetone solvent produced the lowest amount of tannin with the highest residue. Polar compound not only releases tannins from the coffee husk, but also dissolves other components that are polar such as carbohydrates and minerals [6], so that the extracted tannin contains different types of impurities. Das [11] mentions that the selection of the right type of solvent and temperature can release carbohydrates from tannins so that more tannins are produced, which is indicated by fewer residues. Because of extraction process, the extracted tannin contains different types of impurities including minerals, stilbenes and sugars.

4. Conclusion

Ethanol is a solvent that produces the highest total tannin (0.6120%) and the lowest residue (61.64 ± 0.6688%) from coffee husks compared to methanol and acetone.

5. Acknowledgements
The authors would like to thank P3M Politeknik Negeri Jember for funding the research conducted through the 2021 PNBP funding source, number: 349/PL17.4/PG/2021.

References
[1] BPS, "Produksi Perkebunan Kopi Menurut Kabupaten/Kota di Jawa Timur tahun 2006 - 2017," 2018.
[2] D. F. Flórez-Delgado, "Effect of the fermentation time on the nutritional quality of Coffea arabica L. pulp silage," *Cienc. Tecnol. Agropecu.*, vol. 21, no. 2, 2020.
[3] N. J. N. Nnaji, C. O. B. Okoye, N. O. Obi-Egbedi, M. A. Ezeokonkwo, dan J. U. Ani, "Spectroscopic characterization of red onion skin tannin and it's use as alternative aluminium corrosion inhibitor in hydrochloric acid solutions," *Int. J. Electrochem. Sci.*, vol. 8, no. 2, hal. 1735–1758, 2013.
[4] T. V. C. Nascimento *et al.*, "Effects of condensed tannin-amended cassava silage blend diets on feeding behavior, digestibility, nitrogen balance, milk yield and milk composition in dairy goats," *Animal*, vol. 15, no. 1, hal. 100015, 2021.
[5] M. M. Lorenz, T. Eriksson, dan P. Udén, "Effect of wilting, silage additive, PEG treatment and tannin content on the distribution of N between different fractions after ensiling of three different sainfoin (Onobrychis viciifolia) varieties," *Grass Forage Sci.*, vol. 65, no. 2, hal. 175–184, 2010.
[6] M. Verdiana, I. W. R. Widarta, I. D. Gede, dan M. Permana, "Gelombang Ultrasonik Terhadap Aktivitas Antioksidan Ekstrak Kulit Buah Lemon (Citrus limon (Linn .) Burm F .)," *J. Ilmu dan Teknol. Pangan*, vol. 7, no. 4, hal. 213–222, 2018.
[7] E. P. Mokoginta, M. Revolta, J. Runtuwene, dan F. Wehantouw, "1. Pengaruh Metode Ekstraksi Terhadap Aktivitas Penangkal Radikal Bebas Ekstrak Metanol Kulit Biji Pinang Yaki (Areca Vestiaria Giseke)," *Pharmacon*, vol. 2, no. 4, hal. 109–113, 2013.
[8] P. Tiwari, B. Kumar, M. Kaur, G. Kaur, dan H. Kaur, "Phytochemical screening and Extraction: A Review," *Int. Pharm. Sci.*, vol. 1, no. 1, hal. 98–106, 2011.
[9] M. Mohsen-Nia, H. Amiri, dan B. Jazi, "Dielectric constants of water, methanol, ethanol, butanol and acetone: Measurement and computational study," *J. Solution Chem.*, vol. 39, no. 5, hal. 701–708, 2010.
[10] V. Mandal, Y. Mohan, dan S. Hemalatha, "PhcogRev-1-1-7.pdf," vol. 1, no. 1, 2007.
[11] A. K. Das, M. N. Islam, M. O. Faruk, M. Ashaduzzaman, dan R. Dungani, "Review on tannins: Extraction processes, applications and possibilities," *South African J. Bot.*, vol. 135, no. September, hal. 58–70, 2020.

IOP Conf. Series: Earth and Environmental Science **980** (2022) 012025 doi:10.1088/1755-1315/980/1/012025

Fermentation of soybean meal-hydrolysates as the medium that treated by papain enzyme with *Saccharomyces cerevisiae* for biomass production

D. Pantaya, [1], D. Pamungkas [2], S. Wulandari [1], M. M. D. Utami[1],
[1]Department of Animal Science, Politeknik Negeri Jember, Jl. Mastrip PO Box 164 Jember 68121, Indonesia
[2]Beef Cattle Research Institute, Grati, Pasuruan, East Java, Indonesia.

E-mail: dadik_pantaya@polije.ac.id

Abstract. The purpose of this study was to determine the effect of soybean meal (SBM) media processing using the papain enzyme on the amino acid content and its effect on yeast biomass production. The study was divided into 2 experimental phases, first test: effect of papain enzyme hydrolysis test on amino acid in soybean meal that treated by boiling and non-boiling medium, second test: yeast biomass production was cultivated using soybean meal hydrolysates as medium with 3 treatments, namely control (- soybean meal), with soybean meal (-enzyme) and soybean meal (+enzyme). Data were analyzed using a completely randomized design with ANOVA and post hoc analysis using Tukey's multiple range test. From the analysis, it was found that the addition of papain enzymes supplemented on soybean meal medium was affected on crude protein and amino acids composition (L-Arginine, L-proline, L-methionine, and L-phenylalanine). Enzyme treatment on SBM as substrate was increased the yeast biomass compared to control (P<0.01). In conclusion, enzyme treatment on soybean meal was improved the fermentation performance of yeast biomass production

1. Introduction

Yeast contains specific bioactive nutrients that have been widely used over the last two decades for feed supplements [1, 2]. The use of yeast in ruminants can play a role in stimulating the population of fiber-degrading microbes in the rumen, increasing the proportion of propionate [3], reducing lactic acid content, and regulating rumen pH [4]. One of the yeasts commonly used to produce single-cell protein biomass is Saccharomyces cerevisiae [4]. Cell wall biomass contains balanced amino acids, carbohydrate compounds, and several types of vitamins (especially vitamin B) [5].

During fermentation, yeast converts simple compounds into fermentation products such as ethanol, sugar, and biomass. The formation of yeast protein comes from soluble peptone compounds. Peptone is a protein hydrolysate that serves as a nitrogen source. The reaction resulting from protein breakdown by enzymes can increase the production of soluble proteins that yeast can use for cell wall protein synthesis [6, 7]. Lack of this compound will affect the fermentation process, inhibit yeast growth and reduce biomass production.

Some materials that contain high protein can be used for the production of peptone, among others, from animal and vegetable proteins. Soybean meal is one of the ingredients that contains very complete protein and amino acids. Optimization of the nutrient content of yeast media is carried out through pre-treatment to produce hydrolysates products that can increase the availability of nutrients for yeast growth. Media pre-treatment can be done in various ways, including physically, chemically, and enzymatically. Chemical hydrolysis of the substrate has the disadvantage that the product is not specific and requires more energy in the pre-treatment process [8]. While enzymatic hydrolysis has advantages such as not requiring high temperatures [9], this method can completely break down amino acids in the form of hydrolysates products. Papain enzyme is a type of enzyme capable of hydrolyzing proteins in amino acid groups and amide bonds

[10]. To evaluate the performance of the enzyme, it can be done by comparing it with physical pretreatment processing through heating, then testing the use of hydrolysates products as a yeast growth medium. The purpose of this study was to determine the effect of processing soybean meal using the papain enzyme on the composition of the amino acid content and its effect on yeast biomass production.

2. Materials and Methods

2.1. Microorganism

The microorganism used in this study was a *Saccharomyces cerevisiae* from the Feed Technology collection (Politeknik Negeri Jember, Indonesia) that was isolated from soil. The culture was stored in medium agar slants under freezing conditions (-18°C).

2.2. Enzyme preparation

Papain enzyme used in this experiment were provided by local commercial product with enzyme activity: 125.2 TU/mg.

2.3 Experiment 1

Soybean meal was milled with a mesh size of 1 mm using a sample mill (IKA-Werke M20, Germany) to obtain a small and uniform soybean meal, then analyzed the chemical composition of the AOAC method [11]. The test was carried out with the following treatments: P1: Boiled soybean meal + papain enzyme 500 U, P2: Non-enzyme boiled soybean meal, P3: Non-boiled soybean meal + enzyme, P4: Non-enzyme non-boiled soybean meal . Weigh 5 g of ground soybean meal using a 100 ml beaker glass, then add 25 ml of the solution according to the treatment (P1-P4). The sample was then heated for 1 hour at a temperature of 70°C. Weigh the tube before centrifugation, then centrifuged at 10,000 rpm for 10 min. Next, weigh the weight of the precipitate and supernatant and record the weight and volume. The protein fraction was analyzed for protein in the precipitate using the Kjeldahl method, and amino acid test in the supernatant solution using (Liquid chromatography-mass spectrometry) (LC-MS) amino acid content compared with standard amino acids L-Arginine, L- proline, L-methionine, and L-phenylalanine (Sigma-Aldrich). The amino acid fraction of the supernatant was analyzed according to the method [12]. One ml of the solution was homogenized using an ultra turra blender and then stored at -20°C in a polyethylene tube. One ml was put in a centrifuge tube and vortexed for 2 min and centrifuged 5,000 g, 5 min at -5°C. The supernatant was then transferred to a clean vial, then filtered using a 0.45 m nylon syringe before being injected into LC-MS.

Table 1. Fragment characteristic and selected ion monitored by LC MS in amino acid analysis

Amino acid	Fragmen ion, m/z	Selected Ion, m/z
L- Arginin	175,129	175
L- Proline	116,70	116
L-Methionine	150, 133, 104	150
L-Phenylalanine	166, 149, 120	166

2.4.Experiment 2

The test was carried out with the following treatments P1: Medium made from standard peptone, P2: Medium made from non-enzyme soybean meal, P3: Medium made from soybean meal + enzyme. The medium used for this test has the same composition as the composition of the liquid medium with 1.5% peptone. Inoculation was carried out on 10 ml liquid media and incubated for 24 hours with 100 x g shaking, then transferred to 100 ml media in a 250 ml Erlenmeyer tube and incubated for 48 hours. Yeast product collection was obtained by centrifugation at 10,000 x g for 5 min and dried in a dryer at 40°C. The number of yeast colonies was calculated by the total plate count (TPC) method at a temperature of 30°C with several dilutions. The amount of yeast that grew was calculated using a digital colony counter (Intech, Germany) as a quantitative growth value. The composition of the media for production consists of molasses, PDB (potatoes dextrose broth) as a carbon source [10], soybean meal, ammonium sulfate (NH3SO4), and a mixture of several minerals (mix minerals).

2.5 Statistic analysis
The Data were analyzed using a completely randomized design with ANOVA. The Minitab software package was used for static analysis (Minitab Inc, USA). The statistic models With the following models $Y_{ij}=\mu+\alpha_i+ \varepsilon_{ij}$. Where Y=dependent variables, μ: overall mean, α_i: the effect of treatment, ε_{ij}: random error. The post hoc Tukey's multiple ranges tests calculated and probability value less 0.05 were considered significant.

3. Results and Discussion

The results of the analysis showed that the protein content of soybean meal without enzymes was on average 40-41%, while with the addition of enzymes it was reduced by 34-35%. The difference is probably due to the reduced protein content in the precipitate into dissolved protein. The reduced protein content in the precipitate was caused by the increase in the hydrolysis process of the protein thereby reducing the protein fraction on the sediment. A previous study by Pantaya et al [13] reported that enzyme treatment could increase the dissolved protein content on incubation for 4 hours at 60°C on soybean meal. Papain enzymes including proteolysis enzymes can break down protein fractions into simple components. This simulation study proves that enzyme treatment can increase the soluble protein fraction.

There is no significant effect on the boiling treatment of the substrate, this is because boiling at a temperature of 60°C has not affected protein binding so it does not affect the released amino acid content. According to Jiang et al [14], the peak point of cooking loss usually occurs at a temperature of 90-100°C, so that to increase dissolved protein, an increase in heating or boiling temperature is required. This study shows that the use of papain enzymes is more effective for protein hydrolysis in producing amino acid compared to boiling system. The use of enzymes works at low temperatures and does not require high temperatures. Enzymes can hydrolyze protein compounds in specific chains. The advantages of the enzymatic process are that it does not require high temperatures, the hydrolysis process takes place specifically, and can conserve all amino acids.

The profile of the enzyme hydrolysis results as shown in Figure 2. The amino acid content of L-Arginine released was the highest compared to L-methionine, L-phenylalanine, and L-proline, which was 58 ppm. Enzyme treatment in a higher concentration of amino acids than without the enzyme. The papain enzyme is effectively used in hydrolyzing the crude protein fraction into amino acids. Crude protein is a complex compound composed of amino acids. The detected amino acid concentration indicates the profile of the amino acid content in soybean meals. The concentration indicates the specific character of the amino acids contained in a protein. The concentration of each amino acid varies in protein depending on its solubility. This increase in amino acids indicates the availability of amino acids that could be used for yeast growth media.

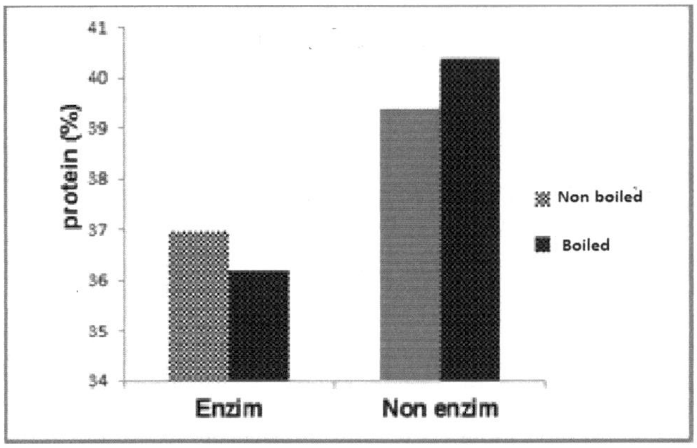

Figure 1. Protein profile in soybean meal residue (n=3)

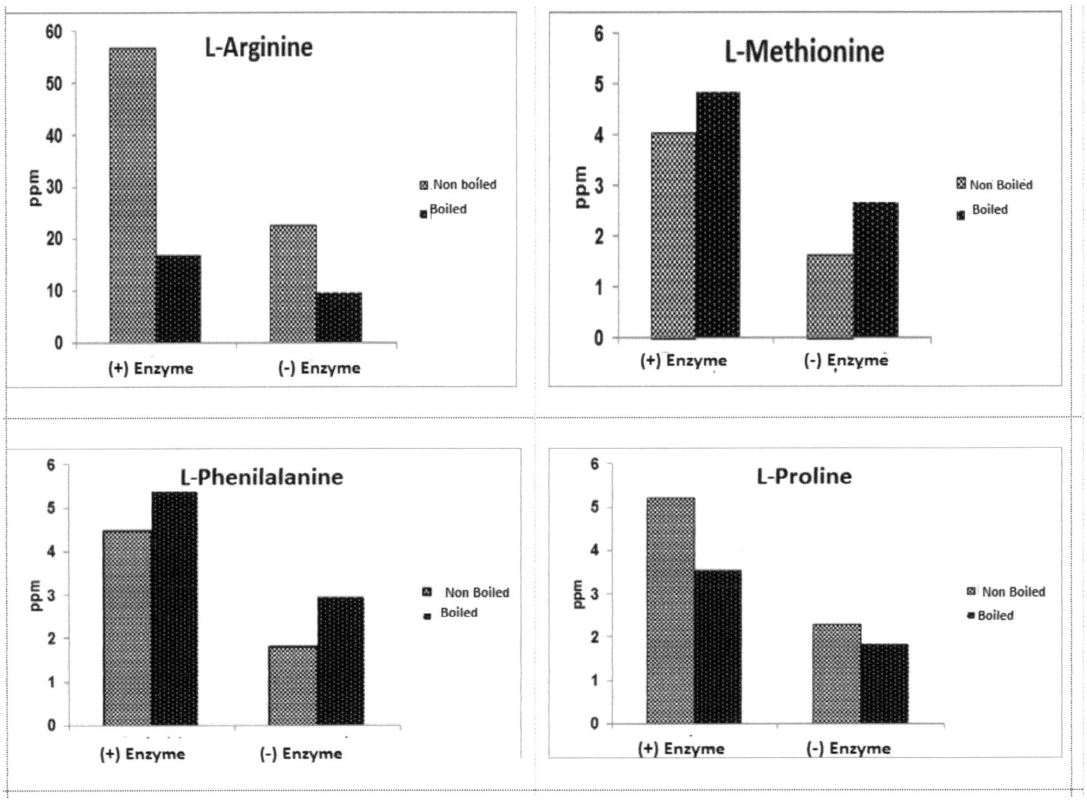

Figure 2. Influence of heating and papain enzyme of amino acid profile in the substrate based soybean meal (n=3)

Yeast grows on all mediums and produces biomass. The results showed the characteristics of yeast biomass production using soybean meal peptone as a medium (Table 2). The statistical analysis results demonstrated that the addition of enzymes had a significant effect on biomass production and Colny forming Unit (CFU/g) (P<0.05). In this study, the hydrolyzed media substrate can increase yeast biomass. These results showed that a positive correlation between amino acid concentration and biomass production.

Amino acids are components that play a role in the formation of proteins, value of concentrations of amino acids cause sufficient availability of cell wall formation. According to the opinion of Saenz [15] which states that amino acids are used as materials for protein synthesis, thus affecting biomass production. The amino acid content detected in the substrate was included in the essential amino acids, namely L-phenylalanine, L-methionine, while the conditionally essential amino acids were L-arginine and L-proline [6]. The transfer of amino acids in plasma through two systems, namely NCR (nitrogen catabolic repression) which regulates the permeases process, and NCI (nitrogen catabolic inactivation) which regulates the import of unfavorable nitrogen sources. The increase in biomass production is due to the increased use of nitrogen and amino acids [16] for anabolic processes [17]. The results demonstrated that L-penylalanine, L-methionine, L arginine and L proline in medium have a positive effect on biomass production.

Table 2. Yeast biomass production using a papain enzyme treatment in soybean meal culture (n=4)

Parameters	Biomass production g/100 ml medium	Dry matter of Biomass (g)	CFU Log10/g
Control	6.30 ± 0.12 [a]	1.53±0.06 [a]	9.07±0.09 [a]
SBM (-) Enzyme	6.31 ± 0.02 [a]	1.60±0.10 [a]	9.22±0.06 [a]
SBM (+) Enzyme	7.10 ± 0.06 [b]	2.00±0.06 [b]	11.20±0.08 [b]

CFU: Colony-forming unit
SBM: Soybean Meal

From these results, it is obvious that enzymatic treatment can increase dissolved protein so that it has a positive effect on increasing biomass production. The use of enzymes can increase the concentration of amino acids in the supernatant and can increase the production of yeast biomass.

4. Conclusion

The use of enzymes in the preparation of the medium resulted in ideal yeast biomass production. Enzymatic treatment using papain could increase the amino acid content resulting from protein hydrolysis of the media and had a positive effect on increasing yeast biomass production.

5. Acknowledgments

This research was funded by INSINAS program 2016, the Ministry of Research and Technology.

References

[1] Pantaya, D. and M.M.D. Utami, IOP Conference Series: Earth and Environmental Science, 2018. **207**: p. 012033.
[2] Wulandari, S. and T.M. Syahniar, IOP Conference Series: Earth and Environmental Science, 2018. **207**: p. 012034.
[3] Lettat, A., et al., BMC Microbiol, 2012. **12**: p. 142.
[4] Pantaya, D., et al., J Dairy Sci, 2016. **99**(12): p. 9759-9767.
[5] Jouany, J.P., et al., Anim Feed Sci Tech, 1998. **75**(1): p. 1-13.
[6] Berlowska, J., et al., Journal of the Institute of Brewing, 2017. **123**(3): p. 396-401.

[7] Øverland, M., et al., Aquaculture, 2013. **s 402–403**: p. 1–7.
[8] Aslanzadeh, S., et al., N. Qureshi, D.B. Hodge, and A.A. Vertès, Editors. 2014, Elsevier: Amsterdam. p. 3-36.
[9] Wang, T. and X. Lü, X. Lü, Editor. 2021, Woodhead Publishing. p. 137-159.
[10] Tapal, A. and P.K. Tiku, M. Kuddus, Editor. 2019, Academic Press. p. 471-481.
[11] AOAC. Series. 2005, Maryland, USA.7. H,.
[12] Ozcan, S. and H.Z. Senyuva, J Chromatogr A, 2006. **1135**(2): p. 179-85.
[13] Pantaya, D., et al., Seminar Nasional Hasil Penelitian dan Pengabdian Masyarakat, ISBN 978-602-14917-2-0, 2016.
[14] Jiang, Q., et al., International Journal of Food Properties, 2018. **21**: p. 2110-2120.
[15] Sáenz, D.A., M.S. Chianelli, and C.A. Stella, Journal of Amino Acids, 2014. **2014**: p. 283962.
[16] Taccari, M., et al., Bioresour Technol, 2012. **110**: p. 488-95.
[17] Schnierda, T., et al., Letters in Applied Microbiology, 2014. **58**(5): p. 478-485.

The effect of extorted wet edamame pod liquid waste in chicks performance

M Andriani[1*], N Muhamad[1], Sabran[2], TM Syahniar[1], and PR Ratri[3]

[1]Departement of Animal Science, Politeknik Negeri Jember, Jember 68122, Indonesia
[2]Departement of Health Information Management, Politeknik Negeri Jember, Jember 68122, Indonesia
[3]Departement of Clinical Nutrition, Politeknik Negeri Jember, Jember 68122, Indonesia

*Email: andriani@polije.ac.id

Abstract. This study aimed to determine the effect of liquid waste from extorted wet edamame pod (LEP) as a natural feed additive added to drinking water in chicks during rearing in a tropical environment. The study was designed using a completely randomized design (CRD) with 3 levels of LEP and 1 control treatment. Each level treatment replicated 4 times with 10 chickens for each experimental unit. The treatment levels were L0=control (using commercial additive based on the product dosage); L1=2.5% v/v LEP; L2 = 5%v/v LEP; and L3=7.5% v/v LEP as additive that added to drinking water. The variables observed broiler performance in the first week starter period which included water intake and feed intake, body weight gain and FCR. The statistical analysis used analysis of variance (ANOVA) then followed by Duncan's Multiple Range Test (DMRT) when showed significant result. The results presented that the usage of 2.5% v/v LEP increased water intake but presented similar effect on feed intake and the performance of first week broiler maintenance in the tropical environment. In conclusion, the usage of LEP up to 7.5% in broiler drinking water is safe to apply as a natural feed additive without reducing broiler performance.

1. Introduction

The poultry farming industry continues to increase along with the increasing public awareness in consuming animal protein, one of which is broiler farming. Broiler chickens are animals that are prone to stress or heat stress, especially at high environmental temperatures and humidity. Maintenance of broiler chickens in a tropical environment, where the fluctuating environmental temperature has an impact on broilers that tend to experience heat stress. Broiler chickens that experience stress can lead to an increase in free radicals and disrupt the hormonal balance in the body. These conditions reduce the availability of vitamin C as an antioxidant that plays a role in suppressing the formation of free radicals. Free radicals increase under stress conditions, as a result, if they do not get additional antioxidants from the outside, they can cause cell damage [1]. Damage to organs and cells caused by free radicals can ultimately reduce broiler productivity. This condition will have an effect on decreasing feed consumption, growth rate and affect feed efficiency, feed quality, and livestock immunity [2].

These problems are essential and require the role of feed additives to reduce the risk of heat stress and disease in poultry, including broilers. The use of synthetic additives is important to reduce oxidative stress in order to trigger the growth rate. However, there are concerns about chemical residues caused

IOP Conf. Series: Earth and Environmental Science **980** (2022) 012026 doi:10.1088/1755-1315/980/1/012026

by synthetic additives, which can settle in chicken meat and organs and will trigger metabolic disorders and be harmful to humans who consume them. Therefore, for the sake of human health factors, it is necessary to find an alternative substitute for synthetic additives in the form of innovative natural feed additives so that the health of livestock and humans is maintained.

Edamame has a fairly high nutritional value, namely 582 Kcal, 11.4 g protein; carbohydrates 7.4 g; fat 6.6 g; vitamin A 100 mg; B1 0.27 mg; B2 0.14 mg; B3 1 mg; and vitamin C 27%; and minerals such as phosphorus 140 mg; calcium 70 mg; iron 1.7 mg; and potassium 140 mg in 100 g of edamame [3], [4] and contains nine essential amino acids that the body needs, contains no cholesterol and little saturated fat and is rich in fiber, vitamins C and B, calcium, iron or magnesium, and fatty acids. folate [5]. Soybeans and soy products are reported to contain high amounts of isoflavones [6]. Phytochemical compounds that have the potential to show antioxidant activity in edamame pod waste water in the form of LEP as a natural feed additive for poultry, especially in days old chicks broiler drinking water, have not been studied before.

2. Metodology

2.1. Design of experiment

The study was conducted for 1 month in the Poultry Cage, Politeknik Negeri Jember Animal Science Department. Experiment used 160 days old chick (DOC) broiler of *Lohmann unsexing* whit the rate of body weight is about 40 g. The poultry cage in completed *thermohigrometer*, the cage temperatures are between 26 – 32 ^{o}C, feed and drinking water were offered *ad libitum*.

Experimental design used was completely randomized design (CRD) with four treatments and three replicates consisting teen DOC per replicate. Edamame pod (LEP) is made from rejected edamame which is taken from the skin and then rinsed and then boiled (ratio 1:5 [**7**]), wait for 5 minutes to boil, then filtered to take the water which will become LEP. The commercial additive (as additive syntetic) composition was shown in Table 1.

Tabel 1. Nutrient content of commercial addititive

Nutrients	Amount/kg
Vitamin A	6.000.000 IU
Vitamin D3	1.200.000 IU
Vitamin E	2.500 IU
Vitamin K3	3 g
Vitamin B1	2 g
Vitamin B2	3 g
Vitamin B6	1 g
Vitamin B12	2 mg
Vitamin C	20 g
Nicotina acid	15 g
Calcium D pantothenic	5 g
Electrolit such as Na, K, Ca, and Mg	750 g

Source: PT.Medion (2021)

The days old chicks subjected to treatments drinking water with or without water LEP (v/v) which is consisted of L0= control (commercial additives used dosage product); L1= drinking water + 2,5% LEP; L2= drianking water + 5% LEP; dan L3= drinking water + 7,5% LEP as using feed additive nature. Diets used in mesh, respectively as shown in Tabel 2

Tabel 2. Feed formulation and nutrient content

Ingredients	%
Corn	60.91
Soybean meal	24.85
Meat Bone Meal	7.00
Corn Gluten Meal	3.00
Palm Oil	2.00
CaCO3	0.64
Salt	0.15
Premix	1.00
DL-Methionine	0.24
Lysine	0.21
Total	100.00
Nutritions*	
EM (kkal kg^{-1})	3118.88
Crude Protein (%)	22.25
Calcium (%)	0.95
Phosphorus (%)	0.42

**base on calculation*

2.2. Variables of experiment

The variables measured were daily water intake, daily feed intake, body weight gain and feed conversion ratio calculated at the last day of experiment.

2.3. Statistical Analysis Data

Data of first week broiler (daily water intake, daily feed intake and body weight gain, and feed conversion ratio (FCR), were statistical analysis using analysis of variance ANOVA and any mean differences were further tested using analysis of Duncan's Multiple Range Test (DMRT). It was conducted using SPSS 22.

3. Result and discussion

Efforts to stimulate the growth and improve performance of chicks are to increase protein and antioxidant intake. Protein quality is determined, among other things, by the completeness and balance of essential amino acids. Essential amino acids cannot be synthesized in the body so their needs must be provided in the food and drink they consume. In the other hands, the administration of antioxidant supplements can increase humoral immunity in broiler chickens subjected to environmental stress. Increased production needs to be done with additional nutritional intake, one of which is by providing additional supplements (amino acids and antioxidant agent). [8] [9] The high amount of natural antioxidant and protein in edamame is expected to affect the quality of chickens seen from parameters such as: body weight gain, water intake, feed intake, and FCR value.

As our study results showed day old chick given to 2.5% LEP has a very significant effect on drinking water consumption, but at the level of 0%, 5% to 7% LEP, it does not show a significant effect (Table.3). The water intake ratio in this study was an average of 52.22±0.96. This shows that the higher the LEP level, the better the water intake efficiency.

Table 3. Performance of broiler as affected by LEP addition into drinking water

Treatment	Water Intake (ml/DOC)	Feed Intake (g/DOC/day)	Body weight gain g/DOC/day	FCR Feed:gain, g/g)
L0 (commercial additive)	52.41±2.48^{a}	19.81±1.39	12.03±1.00	1.65±0.16
L1(2,5% LEP)	55.78±1.19^{b}	20.55±0.89	12.41±0.74	1.66±0.13
L2 (5% LEP)	50.48±2.77^{a}	20.40±0.91	11.73±0.73	1.70±0.10
L3 (7,5% LEP)	50.21±1.52^{a}	19.34±0.95	12.04±0.76	1.65±0.09

Note: Means the same column with difference superscript differ significantly (P<0.05)

Although at Level 2.5% LEP was significantly different, it did not show a significant difference in feed intake, body weight gain, and FCR. This is in line with the research of Utami and Hertamawati at 2019 which showed that dietary edamame soybean isoflavone treatment did not affect feed intake, weight gain, final weight, and feed conversion of broilers chickens. [10] This results may occurred because the amount of antioxidants contained in the edamame skin waste with concentration 2,5%; 5%, and 7,5% that is not adequate enough to increase body weight gain. In addition to doses that may be less than optimal, other factors that also need to be considered in the growth of chickens are genetics, disease, temperature, cage sanitation, ventilation, treatment, and cage management. Feeding factors, feed travel in the digestive tract, physical form of feed and nutritional composition of feed also affect the value of feed intake, weight gain, final weight, and feed conversion.[11] In this study, the feed given in each treatment was the same.

4. Conclusion

The addition of liquid waste of extorted wet edamame pod into drinking water could significantly affect the water intake of broiler chicks but not significantly affect the feed intake, body weight gain, and FCR.

5. Acknowledgement

Authors acknowledgement P3M Politeknik Negeri Jember for PNBP Research Grant 2021, contract number 379/PL17.4/PG/2021.

References

[1] Surai, P. F. 2007. Natural Antioxidants in Poultry Nutrition: New Developments. 16th European Symposium on Poultry Nutrition. Pp. 669-676.
[2] Al-Fataftah, A.R.A. and Z. H. M. Abu-Dieyeh. 2007. Effect of cronic heat stress on broiler performance in Jordan. Int. J. Poult. Sci.6: 64-70.
[3] Johnson D., Wang, S., and Suzuki, A. 1999. Edamame vegetable soybean for Colorado. In Janick, J. (Ed.). Perspectives on New Crops and New Uses.
[4] Nguyen, V.Q. 2001. Edamame (Vegetable Green Soybean). In The Rural Industrial pege: 49-56. http://attar.ncut.org/attar-pub/Edamame.html.
[5] Pusdatin. 2014. Kedelai Jember Tembus Pasar Internasional. http://setkab.go.id/kedelai-jembertembus-pasarinternasional/.
[6] Kim EH, Kim SL, Kim SH, Chung IM. 2012. Comparison of isoflavones and anthocyanins in soybean (Glycine max (L.) Merill) seeds of different planting dates. J Agric Food Chem 60: 10196-202. DOI:10. 1021/jf3031259.
[7] Astuti R D, Wahyono F, Mangisah I, 2021. Pengaruh pemberian aditif cair buah naga merah (*Hylocereus polyrhizus*) terhadap kecernaan protein, energi metabolis dan produksi telur burung puyuh. Jurnal Ilmu-Ilmu Peternakan 25 (3): 80- 87.
[8] Sutomo B., Ihsan M N, Hamiyanti A A.2015. Pengaruh pemberian feed supplement viterna pada air minum terhadap penampilan ayam pedaging. J. Ternak Tropika. 16: 25-29.
[9] Aengwanich W, P. Sridama, Y. Phasuk, T. Vongpralab, P. Pakdee, S. Katawatin and S. Simaraks. 2003. Effects of ascorbic acid on cell mediated, humoral immune response and pathophysiology.

IOP Conf. Series: Earth and Environmental Science **980** (2022) 012026 doi:10.1088/1755-1315/980/1/012026

[10] Utami M.M.D & Hertamawati R.T. 2019. Dietary edamame soybean isoflavon concentrate on improving carcass quality of broilers. Second International Conference on Food and agriculture. 1—6.
[11] Lacy, M. and L. R. Vest. 2000. Improving Feed Convertion in Broiler: A Guide for Growers. Springer Science and Business Media Inc, New York.

The effect of phytobiotics ginger (*Zingiber officinale Rosc.*) extract in drinking water on hematological characters of KUB chickens

N. Ningsih, Gayuh Syaikhullah, M. Adhyatma, Zainur Rofiqi, M. F. Al Huzaini

Animal Science Department, Politeknik Negeri Jember, Jember, Indonesia

Corresponding author: niatiningsih@polije.ac.id

Abstract: This study was carried out to investigate the effect of phytobiotics ginger (*Zingiber officinale Rosc.*) in drinking water on haematological characters of Kampung Unggul Balitbangtan (KUB) chickens. A total of one hundred KUB chickens fourteen days old were assigned into 4 treatments with 5 replicates (5 birds in each replicate). The experimental treatments were control without red ginger extract (T0), drinking water with 0.5% ginger extract (T1), 1.0 % ginger extract (T2), and 1.5 % ginger extract (T3). The basal diet was yellow corn, BR2 poultry concentrate produced by PT Wonokoyo, rice brand, and vitamin premix, which contains 17.07% CP, 3097.79 kcal/kg ME, 0.81% Ca, and 0.56% Pav. Variables evaluated were haematological character of KUB chicken. The results were analyzed using ANOVA in a completely randomized design using Statistical Package for Social Science or SPSS. Duncan multiple range test (DMRT) were used to separate mean of data when p-value differ significantly ($P<0,05$). Results showed that suplementation ginger extract in the drinking water did not significantly effect haematological character of KUB chicken..

Keywords: Haematological Character; KUB Chicken; Phytobiotics; Red Ginger;

1. Introduction

Indonesia has at least 34 kinds of native chicken with different morphological characteristics. Native chicken is a potential resource as a provider of animal protein demand, approximately 12.33% of the total national consumption of chicken was contributed with native chickens [1]. KUB (Kampung Unggul Balitbangtan) chicken is an improved Indonesia native chicken breed by the Indonesian Research Institute for Animal Production (IRIAP). KUB chicken was launched as a commercial chicken by Agricultural Minister No. 274/Kpts/SR.120/2/2014 (Direktorat Jenderal Peternakan dan Kesehatan Hewan 2014). KUB chicken has the potential genetic to produce meat and egg with savory taste, delicious and in demand by the public. The KUB-1, the first generation of KUB produces eggs up to 50% hen day production (HDP) or about 180 eggs/hen/year [2].

Extreme climate change in Indonesia directly affects the physiology, productivity, and healthy of KUB chicken. Phytobiotics are used for increased animal health during extreme climate change. Phytobiotics are secondary plant metabolites, they function as antioxidants, digestive enhancers, nutraceuticals, and health promoting substances. They are used as growth promoters in feed, especially for monogastric [3]. Phytobitics have an activity to modify the gut microflora by reducing the number of pathogenic organisms. Phytobiotics lead to health benefits, promoting growth and improving meat and egg quality in poultry production [4]. One phytobiotic is ginger that can be used as a potential alternative for growth promotor. *Zingiber officinale* Rosc, the Zingiberaceae family, commonly recognized as ginger, is a monocotyledonous herb and popularly known as a feed additive that has various pharmacological characters [5]. Ginger has been used as an herbal medicine with zingerone, shogaol, and gingerol compounds [6]. The medicinal value of ginger is as antioxidant, antimicrobial, anti-inflammatory, anti-hypercholesterolemic, anti-hyperglycemic, and antispasmodic potential that arises from the oleoresins which are present in it [7].

Hematological characters are indicators of the physiological status of animals and are used in the diagnosis of animal health. Hematological studies are of ecological and physiological interest in helping to understand the relationship of blood characteristics to the environment. Animals with good blood composition are likely to show good performance. The examination of blood allows investigating the presence of several metabolites and other constituents in the body of animals and it plays a vital role in the physiological, nutrition, and pathological status of an organism [8]. The purpose of this study was to evaluate the effect of phytobiotic ginger to improve physiological KUB chicken concerning hematological characters.

2. Material and Methode

2.1 Animal, Diets, and Experimental Design

The study was conducted at Bagorejo Village, Gumukmas District, Jember, while the hematological character analysis was carried out at the New Prosenda Jember Laboratory. The number of 100 fifteen days old KUB chickens were divided into four treatments and five replicates with 5 birds in each replicate pen. Each pen was equipped with a feeder and a water trough. The basal diets were formulated based on yellow corn, soybean meal, and poultry concentrate which contained 17,07% crude protein (CP), 3097,79 Kcal/kg metabolizable energy (ME), 0.81% Calcium (Ca), and 0.56% available Phosphorus (Pav). Formulation and chemical compositions for the basal diets are shown in Table 1. The experimental treatments were consisted of control without red ginger extract (T0), drinking water with 0.5% red ginger extract (T1), 1.0 % red ginger extract (T2), and 1.5 % red ginger extract (T3).

Tabel 1. Ingredients and nutrient composition of the basal diet

Ingredients	Chemical compositions				
	Proportion (%)	CP (%)	ME (Kcal/kg)	Ca (%)	Pav (%)
Yellow corn	56.57	5.03	1866.81	0.01	0.13
Rice brand	17.17	9.98	735.28	0.79	0.21
Poultry concentrate	26.26	2.06	495.70	0,01	0.22
Total	100.00	17.07	3097.79	.0,81	0.56

2..2 Blood collection

Each replicate of birds was reared in individual pens. A bird on day 35 of age was selected randomly from each replicate and around 2 ml blood samples were collected from each bird through the brachial vein, using sterile needles and syringes. The blood samples were put in properly labeled and sterilized anticoagulant (EDTA) tubes, then used for hematological analysis. The blood tubes were placed at room temperature for 6 h and incubated overnight in the refrigerator at 4oC to obtain serum. The serum samples were kept at -20°C before biochemical analysis.

2.3 Evaluation of hematological character

The hematological character evaluated were hemoglobin, erythrocytes, hematocrit, concentration, mean cell hemoglobin (MCH), mean cell volume (MCV), mean cell hemoglobin concentration (MCHC), leucocyte, and thrombosit of KUB chicken, using hematological Analyzer Sysmex XP- 100 (Japan).

2.4 Ginger extract

The ginger extract was produced by washing the ginger rhizome with sterile distilled water, then drying it. The dried ginger rhizome was cut into small pieces, then weighed as much as 100 grams, and mixed with 100 ml of sterile distilled water (weight/volume 1: 1), then blended. The filtrate was then filtered using sterile gauze to obtain ginger extract (Mujim, 2010).

2.5 *Statistical Analysis*

Data of hematological character of KUB chicken were expressed as mean ± STdev. The data were statistically analyzed by one-way ANOVA with a completely randomized design, using Statistical Package for Social Science or SPSS (SPSS GmbH, Munich, Germany). Duncan's multiple range test (DMRT) was used to separate data with significant differences. All indication of significance was based on a probability of less than 5%.

3. Results and Discusion

The results for a hematological character with phytobiotics ginger (Zingiber officinale Rosc.) extract in drinking water are presented in Table 2. The data showed that treatment had no significant effect (P value > 0.05) on hemoglobin, erythrocytes, hematocrit, MCV, MCH, MCHC, Leucocyte, and trombosite of KUB chicken. The results were similar to [9] that showed the blood profile had not been affected by heat pressure. The metabolic processes with heat stress in the body have not had an impact on the blood profile of Indonesia native chicken.

Tabel 2. The data of hematological character with phytobiotics ginger extract in drinking water

Item	P0	P1	P2	P3	P value
Hemoglobin (g/dl)	13,80±2,55	11,64±3,50	10,42±3,96	11,38±3,48	0,475
Erytrocytes (x10^6)	2,39±0,39	2,06±0,29	1,86±0,32	2,0±0,26	0,537
Hematocrit (%)	29,72±2,37	25,54±3,27	22,26±3,86	24,62±3,22	0,448
MCV (fl)	122,62±2,85	125,28±2,31	118,32±1,71	123,42±2,64	0,200
MCH	56,84±1,12	56,80±0,55	55,62±0,71	56,66±1,06	0,743
MCHC (g/dl)	46,42±0,35	45,40±1,03	46,96±0,44	45,94±0,65	0,175
Leucocyt (x 10^3/μl)	59,36±4,14	50,74±2,63	44,16±1,02	48,65±5,70	0,411
Thrombosit (Plt)	12,40±2,93	7,40±0,93	8,80±0,86	5,80±1,02	0,72

The results of the present research were not in agreement with [10] who reported that supplemented broiler feed with ginger powder improved blood chemistry and hematological parameters in birds. The hematological character was sensitive to ginger-based diets, and these values increased with increasing dietary ginger intake. The different results in this study were due to the process of extracting phytochemical components from ginger not optimally so a few bioactive were obtained that can be used as phytobiotic for KUB chicken.

No significant effect from this research was due to the chickens undergoing heat stress for a long time and leading to tolerating and adapting to the existing environmental temperature. The bird will maintain their homeostasis to regulate their physiological response to heat stress [11]. The physiological homeostasis of chicken will keep its body temperature within normal temperature with release excess body temperature into the environment [12]. An imbalance between metabolic heat production was causes changes in livestock behavior and the function of several body organs, besides that it will indirectly affect the increased hormone corticosterone which has an impact on decreasing hemoglobin levels. Heat stress in poultry chickens will cause changes in hemoglobin concentration, erythrocyte count, and hematocrit levels in the body.

The hematological character with phytobiotics ginger (Zingiber officinale Rosc.) extract in drinking water blood acts as a pathological reflector of the status of animals exposed to xenobiotics and other extreme conditions [7]. Laboratory tests on the blood are vital tools that help detect any deviation from normal health status in animals. The examination of blood allows investigating the presence of several metabolites and other constituents in the body of animals and also play vital roles in the physiological, nutritional, and pathological status of an. Changes in hematological parameters are often used to determine the status of the body and stresses due to environmental, nutritional, and pathological factors. The normal values of hematological character obtained in the present study indicated that phytobiotics ginger extract added to chicken water had healthy blood composition are likely to show good performance [13].

4. Conclusion

Suplementation ginger extract in the drinking water did not significantly effect haematological character of KUB chicken.

5. Anklowlegdment

This research was fully supported by PNBP 2021 Politeknik Negeri Jember with contract number 384/PL17.4/PG/2021.

6. Refference

[1] Agustini, N., T. B. Julianto, I. P. C. P. Adnyana, and L. G. S. Astiti 2021 *E3S Web Conf.* **306.**
[2] Iskandar, S. and T. Sartika 2016. *Proceeding 16th AAAP Congr.*
[3] Ripon, M. M. R. *et al.* 2019 *J Adv Vet Anim Res.* **6**: 253–259.
[4] Tavangar, P., S. Gharahveysi, V. Rezaeipour, and M. Irani 2021 *Trop. Anim. Health Prod..* **53** (3).
[5] El-Hack, M. E. A. *et al.* 2020. *Animals.* **10** (3): 2020.
[6] Zidan, D. E., Khaled A. Kahilo., A. H. El-Far, and K. M. Sadek 2016 *Glob. Vet.* **16** (6): 530–538, 2016.
[7] Aikpitanyi, K. U. and N. O. Egweh 2020 *Nigerian J. Anim. Sci.* **22** (2): 114–125.
[8] Etim, N. N. 2014 *Agric. Sci.* **2** (1): 37–47.
[9] Pantaya, D., R. Y. Pratama1, T. A. Marjiatin, N. Ningsih, and G Syaikhullah 2020 *IOP Conf. Ser.: Earth Environ. Sci.* 672 012041
[10] I. P. Ogbuewu, C. A. Mbajiorgu, and I. C. Okoli 2019 *Comp. Clin. Path.*, **28** (3) : 655–660.
[11] Y. Wang, P. Saelao, K. Chanthavixay, R. Gallardo, D. Bunn, S. J. Lamont, J. M. Dekkers, T. Kelly, and H. Zhou, 2018 *Poult Sci* **97**:770–780.
[12] M.H. Tamzil, 2014 *Indonesian Bull. Anim. Vet. Sci..* **24** (2) :57– 66.
[13] N. Saleh, T. Allam, A. A. El-latif, and E. Ghazy,2014 *Glob. Vet.*, **12** (6) : 36–744, 2014.

IOP Conf. Series: Earth and Environmental Science **980** (2022) 012028 doi:10.1088/1755-1315/980/1/012028

The effect of instanization methods on rehydrated instant bose characteristics

L E Yulianti[1], H Sholechah[2], E Widowati[2], W Setiaboma[1], R Ekafitri[1] and N Afifah[1]

[1] Research Center for Appropriate Technology Indonesian Institute of Sciences, Jl. KS. Tubun No. 5 Subang West Java, Indonesia, 41213

[2] Department of Food and Science Technology, Faculty of Agricultural, University of Sebelas Maret, Jl. Ir. Sutami 36, Kentingan Jebres, Surakarta, Central Java, Indonesia, 57126

*email: listaeka0507@gmail.com

Abstract. Bose is a typical food from East Nusa Tenggara which is made from a mixture of corn and beans. The process of making bose requires a relatively long time. Instanization is a technology that can be applied to shorten the process of preparing bose corn before can be consumed. This study aims to determine the effect of the cooking methods in the instanization process on the physical and chemical properties of rehydrated instant bose corn. The factors tested were cooking methods: mixed and boiled-steamed; mixed and boiled-autoclaved; mixed and steamed-autoclaved; unmixed and boiled-steamed; unmixed and boiled-autoclaved; unmixed and steamed-autoclaved. The mixed and boiled-autoclaved method produced the best physical properties of instant bose corn compared to other treatments. This instant bose corn had 6.34 minutes of rehydration time, 4.45 of rehydration ratio, 190.01% of volume expansion, 261.65 N of hardness and contains of 3.41% of protein, 0.56 mg/100 g of iron and 0.42% of fat.

1. Introduction

Besides rice, several regions in Indonesia consumed various staple food, such as corn, cassava and sago. East Nusa Tenggara is one of the region with the highest corn productivity [1]. Corn is one of the main carbohydrate sources which is consumed by East Nusa Tenggara community. Generally, corn processed into corn rice which comes from dry corn that has been shelled and coarsely grinded to become grits [2].

Bose is one of processed food from corn rice which is a typical food of East Nusa Tenggara. Bose is made from a mixture of corn grits with beans and takes more than 4 hours to prepare [3]. Making bose is classified as impractical because it takes a long time. This is incompatible with the current lifestyle of society which is require the convenience and fast-preparation foods. Therefore, proper processing technology is required to shorten the preparing of Bose and make it into an instant product.

Kumalasari *et al.* [4] conducted study on making instant corn rice in which the stages included were washing, soaking, drying, steaming, cooling, freezing, thawing and drying. Similar research is conducted by Prapluettrakul *et al.* [5] in the diamond rice production. The stages of this study consists of soaking in Na-citrate solution, cooking, pre-treatment before drying, drying and rehydrating. Lailiyati *et al.* [6] also conducts a study on making instant corn rice with the addition of tempeh flour.

The cooking process in making instant rice or instant corn can be done by several methods, such as boiling [6], boiling and steaming [4,7] and pressure cooking [8,9]. However, to the best of our knowledge, study about the preparation methods of instant bose and its effect on the characteristics of instant bose is still limited. Therefore, this study aims to determine the effect of the cooking methods in the instanization process on the physical and chemical properties of rehydrated instant bose corn.

2. Materials and Methods

2.1. Materials

The main ingredients used in this study were corn grits, rice beans, cowpeas and sodium citrate. Corn grits, rice beans and cowpeas were obtained from East Nusa Tenggara, Indonesia.

2.2. Methods

2.2.1. Instant bose preparation

A method to prepare instant bose was adopted from method to prepare instant corn rice as described by Husain *et al.* [7] and Kumalasari *et al.* [4] with modifications. As much as 60% corn grits, 20% cowpeas and 20% rice beans were soaked in 1% Na-citrate solution for 2 hours. Started from the soaking stage, the materials (corn grits and beans) were divided at 2 methods, namely mixed and non-mixed. Then, the soaked materials was washed and drained. The materials was cooked at 3 methods: boiled-steamed; boiled-autoclaved; steamed-autoclaved. Boiling process used the water at 1:10 ratio until the water was shrinked. Steaming and autoclaving carried out for 30 minutes, for each process. After the cooking process, the sample was freezed for 16 hours, thawing and drying later. The treatment in this study was the method of material mixing (mixed and non-mixed) and the cooking method (boiling-steaming; boiling-autoclaving; steaming-autoclaving).

2.2.2. Sample analysis

The physical properties that are analyzed were rehydration time [10], rehydration ratio and expansion volume [4]. Hardness was tested using Texture Profile Analyser (Stable Micro Systems, TA.XT plus) and the method refers to Prasert and Suwannaporn [11] with modifications. As much as 10 grams of rehydrated *Bose* samples were molded in the tube (1 cm of height and 1.5 cm of diameter). The sample was pressed with P 36 probe at 2.00 mm/s of pretest speed, 1.7 mm/s of test speed, 1.00 mm/s of post test speed, 30% of strain, and 5.0 gr of trigger force. Chemical analysis was carried out included protein, fat and fe content. The analysis refers to SNI 01-2891-1992 [12] concerning the analytical methods for food and beverage.

This study used a completely randomized design to evaluate the effect of mixing methods and cooking methods on the Bose physicochemical characteristics. Statistical analysis by using Analysis of Variance (ANOVA) method was performed to determine the mean difference between samples (IBM SPSS Statistics 25, 2018). A post hoc by using a Duncan Multiple Range Test (DMRT) method was conducted to confirm the significant differences between samples at a confidence level of 95% ($p \leq 0.05$).

3. Result and Discussion

3.1. Physical characteristics

The physical characteristics of instant Bose consisting of rehydration time, rehydration ratio, expansion volume and hardness were shown in the Table 1. The rehydration time is the time required for the product to re-absorb the water, so the homogenous texture is obtained [13]. It is the main parameter of a instant product. Instant food is a food product which can be served in a relatively short time. Based on Table 1, mixing and cooking method significantly affected the rehydration time. The shortest rehydration time was produced by mixed ingredients and boiling-autoclaving method. This method produces instant bose corn product with a rehydration time of 6.34 minutes. Treatment combination of mixed ingredients and boiling-autoclaving method make the homogeneous and perfectly gelatinized material. On the mixed ingredients treatment, all the ingredients used were blended into one in the same container. This makes the material absorb the water more evenly, so that the gelatinization of the material is more homogeneous [14].

In boiled-autoclaved treatment, the materials gelatinized more quickly and homogeneously than the other treatments. This happens because the heat conducting medium in the boiling process is water in direct contact with the material. Sipayung *et al.* [15] revealed that gelatinization in the steaming process is slower than boiling process which uses water as the heat conductor. Rehydration time is closely related to the porosity of a material. Porosity determines the rehydration time, the more porous the material, the shorter the rehydration time [6]. The porous dry materials will be able to absorb the water rapidly due to the opening of pores in the material, so that the porosity is an important role in the instanization [7].

Table 1. Physical characteristics of rehydrated instant bose corn

Mixing method	Cooking method	Rehydration time (min)	Rehydration ratio (g/g)	Expansion volume (%)	Hardness (N)
Mix	Boil-steam	7.38 ± 0.25^b	4.36 ± 0.09^e	185.95 ± 0.09^d	337.72 ± 1.83^b
	Boil-autoclave	6.34 ± 0.14^a	4.45 ± 0.07^f	190.01 ± 0.02^e	261.65 ± 3.32^a
	Steam-autoclave	29.22 ± 0.11^e	2.49 ± 0.18^b	115.77 ± 0.01^a	495.01 ± 9.91^d
Non-mix	Boil-steam	14.70 ± 0.42^d	3.49 ± 0.12^c	150.87 ± 0.01^b	389.07 ± 7.38^c
	Boil-autoclave	12.38 ± 0.12^c	3.60 ± 0.21^d	154.83 ± 0.06^c	477.05 ± 0.66^d
	Steam-autoclave	29.32 ± 0.31^e	2.19 ± 0.06^a	115.06 ± 0.09^a	947.91 ± 3.36^e

Means with different letters in the same column differ significantly at $p \leq 0.05$

The rehydration ratio is a value that shows the ratio of the sample weight after and before rehydration. This value indicates the water absorption ability of the dried product. A high rehydration ratio is expected in instant products [4]. Based on Table 1, the rehydration ratio significantly be affected by mixing and cooking method. Rehydration ratio ranged from 2.19 to 4.45, of which the highest value is produced by mixed ingredients and boiling-autoclaving method. The higher rehydration ratio value, the more water is absorbed by the material [16].

The high value on sample with mixed ingredients and boiling-autoclaving treatment is probably due to the porosity. On materials with high porosity, the pores are open and made it easier for water absorption. A high porosity is formed due to the complete gelatinization process. Santosa et al. at Kumalasari *et al.* [4] stated that gelatinized starch can increased the water absorption due to the breaking of hydrogen bonds between starch molecules, so the water will more easily to enter the starch molecules in the material. According to Lailiyati *et al.* [6], a good instant corn rice had a high rehydration ratio, so it has a high water absorption and a shorter cooking time.

The expansion volume is a value that indicates a change in the shape of the material particles caused by the water absorption process [17]. Based on Table 1, mixing and cooking method significantly affected the expansion volume. The highest expansion volume is 190.01%, was produced by mixed ingredients and boiling-autoclaving method. This value was higher than the volume expansion of instant corn rice from the study of Kumalasari *et al.* [4], which ranged from 44% to 119%.

According to Azizah *et al.* [18], the expansion volume of a product is related to the gelatinization process of the materials. In the boiling-autoclaving method, the materials were maximally gelatinized. This occured because the material that has been gelatinized in the boiling process was continued into the autoclaving process. In pressurized cooking with an autoclave, the gelatinization occurs while the texture of the material was maintained [19]. Sugiyono *et al.* [20] stated that intramolecular hydrogen bonds in gelatinized starch were damaged so that many hydroxyl groups are released. The more hydroxyl groups were released, the greater the bond with the water and it can increase the material expansion.

Hardness is the value of the maximum compressive force required for the first bite or pressure [21]. The higher value of hardness indicates that the product is getting harder [22]. Based on Table 1, instant bose corn samples had the hardness value ranged from 261.65 N to 947.91 N. The lowest hardness value was produced by mixed ingredients and boiling-autoclaving method, whereas the highest hardness value was produced by non mix ingredients and steaming-autoclaving method. The lowest hardness value resulted from this study is still lower than the hardness value of instant rice from the Prapluettakul *et al.* [5] research which ranges from 275.93-454.62 N and Chen *et al.* [23] research which ranges from 983.27-1159.54 N.

According to Cameron and Ya [24] the hardness level differences in a rehydrated product may caused by the degree of starch swelling. The gelatinization process begins with swelling of the starch granules caused by the water absorption. In the boiling method, the water is more easy to absorbed because of direct contact with the material [25]. The boiled material then continue with pressure cooked made the maximum gelatinized material. Prasert and Suwannaporn[11] stated that the high pressure can caused the maximum gelatinization and can produces a softer rehydrated rice.

The hardness level had a positive correlation with the material density [11]. The porous material allows the water and air to get in and activates the heat process during cooking, it can resulted the soft texture after rehydration. According to Rewthong *et al.* [26] instant rice grains should have small pores that spread evenly throughout the rice center. In the rehydration process, the water will get into the grain and be trapped in the middle of instant rice. The water press the cell structure towards the surface, thus forming a large cavity in the dry center. These large cavities can weaken the instant rice structure after being rehydrated and can caused the instant rice to become mushy [10].

3.2. *Chemical characteristics*

Table 2 below contains the chemical characteristics of instant Bose consisting of protein, fat and iron content. The result showed that mixing and cooking method were significantly affected to the protein, fat and iron content of rehydrated instant bose corn. Protein content of rehydrated instant bose corn ranged from 3.41% to 6.49%. The lowest protein content produced by the mixed ingredients and boiling-autoclaving method. The low protein content that produced by boiling-autoclaving method is thought to be due of cooking method combination.

Table 2. Chemical characteristics of rehydrated instant bose corn

Mixing method	Cooking method	Protein (%)	Fat (%)	Iron (mg/100g)
Mix	Boil-steam	3.48 ± 0.02^{a}	0.32 ± 0.11^{ab}	0.57 ± 0.06^{a}
	Boil-autoclave	3.41 ± 0.09^{a}	0.42 ± 0.01^{bc}	0.56 ± 0.03^{a}
	Steam-autoclave	6.49 ± 0.78^{b}	0.63 ± 0.02^{d}	0.87 ± 0.07^{b}
Non-mix	Boil-steam	4.08 ± 0.56^{a}	0.35 ± 0.16^{abc}	0.66 ± 0.02^{a}
	Boil-autoclave	4.07 ± 1.05^{a}	0.19 ± 0.06^{a}	0.64 ± 0.11^{a}
	Steam-autoclave	5.96 ± 0.54^{b}	0.47 ± 0.02^{bcd}	0.98 ± 0.09^{b}

Means with different letters in the same column differ significantly at $p \leq 0.05$

According to Sundari *et al.* [27] the cooking process involves the use of heat, the longer cooking time can lead the protein denaturation to a high levels. The use of heat in all cooking techniques can affect the nutritional value of foodstuffs. Furthermore, the use of water in the boiling method can accelerate the protein denaturation process because the heat transfer process occurs directly from the water to the material. This is evidenced by the protein content in the boiled-steamed and steamed-autoclaved method is higher than the protein content in the boiled-autoclaved method.

Based on Table 2, the fat content of rehydrated instant bose corn ranged from 0.19% to 0.63%. This result was lower than the instant corn rice research by Azizah *et al.* [18] which had 0.79% of fat content. The low fat content in instant bose corn is due to the combination of cooking methods, whereas each sample got two times of cooking. Food processing such as in the cooking process can caused a damage to the fat content of a food materials. The decrease in fat content after the cooking process is due to the fat characteristics that is not heat resistant. During the cooking process the fat would melt and even evaporate (volatile) into other components [27].

According to Nguju *et al.* [28] the heating process can reduce the fat content by melting the fat. The fat components can be break into volatile products such as aldehyds, ketons, alcohols, acids and hydrocarbons. High temperature and use of water during cooking can occurs the fat come out along with the water. Fats can be hydrolyzed to produced glycerol which is soluble in water and fatty acids.

Iron (Fe) is an essential microelement for the body, needed in the formation of hemoglobin molecules [29]. Based on Table 2, the iron content of rehydrated instant bose corn ranged from 0.56 mg/100 g to 0.98 mg/100 g. This result was lower than the mocaf rice research by Loebis *et al.* [30] which had 2.83 mg/100 g of iron content. Cooking process which using heat can reduce the nutritional content of food material. Nutrients can be removed by the water used during the cooking process. The heating process can caused a decrease in minerals ranging from 5-40%, one of which is iron [27,31].

Furthermore, the iron content of a product is affected by the raw materials used. Loebis *et al.* [30] stated that the more proportion of ingredients that had a high iron content, the iron content in the product will increase. In instant bose corn, the proportion of corn is higher than beans. This may caused the low content of iron in instant bose corn.

4. Conclusion

In summary, cooking method largely affected the physical and chemical characteristics of rehydrated instant bose corn. Mixed ingredients and boiling-autoclaving method produced the instant bose corn with the best physical characteristics but the development is needed to improve its chemical characteristics. This instant bose corn had 6.34 minutes of rehydration time, 4.45 of rehydration ratio, 190.01% of volume expansion, 261.65 N of hardness and contains of 3.41% of protein, 0.56 mg/100 g of iron and 0.42% of fat.

References

[1] Badan Pusat Statistik 2016 Produksi Jagung Menurut Provinsi (1993-2015) https://www.bps.go.id/dynamictable/2015/09/09/868/produksi-jagung-menurut-provinsi-ton-1993-2015.html.

[2] Hamaisa A 2016 Prospek teknologi pengolahan beras jagung instan di Provinsi Nusa Tenggara Timur Seminar Nasional Agroinovasi Spesifik Lokasi untuk Ketahanan Pangan pada Era Masyarakat Ekonomi ASEAN

[3] Gasong LS 2019 Pengembangan produk jagung bose instan diperkaya zat besi untuk penanganan anemia remaja putri di Kupang [disertasi] Sekolah Pascasarjana Institut Pertanian Bogor

[4] Kumalasari R, Setyoningrum F and Ekafitri R 2015 Karakteristik fisik dan sifat fungsional beras jagung instan akibat penambahan jenis serat dan lama pembekuan *Jurnal Pangan* **24(1)** 37-48

[5] Prapluettrakul B, Patcharee T, Sukamol P and Tasanee L 2012 Development of instant rice for young children *Silpakorn U Science & Tech J* **6(1)** 49-58

[6] Lailiyati SN Rahmawanti D and Andriani MAM 2014 Formulasi dan kajian karakteristik nasi jagung (*Zea mays* L) instan dengan penambahan tepung tempe *Jurnal Teknosains Pangan* **3(1)** 155-163

[7] Husain H, Muchtadi TR, Sugiyono and Haryanto B 2006 Pengaruh metode pembekuan dan pengeringan terhadap karakteristik grits jagung instan *Jurnal Teknologi dan Industri Pangan* **17** 189-196

[8] Widowati S, Rahmawati N and Wiwit A 2010 Proses Pembuatan dan Karakterisasi Nasi Sorgum Instan Prosiding Pekan Serealia Nasional

[9] Ali MA, Hasan SMK, Mahomud MS and Sayed MA 2012 Processing and storage of instant cooked rice *Bangladesh Research Publications Journal* **7(3)** 300-305

[10] Bui LTT, Coad RA and Stanley RA 2018 Properties of rehydrated freeze dried rice as a function of processing treatments *LWT - Food Science and Technology* **91** 143-150
[11] Prasert W and P. Suwannaporn 2009 Optimization of instant jasmine rice process and its physicochemical properties *Journal of Food Engineering* **95** 54-61
[12] Badan Standarisasi Nasional (BSN) 1992 SNI 01-2891-1992 *Cara Uji Makanan dan Minuman* Jakarta
[13] Sasmitaloka KS, S Widowati and E. Sukasih 2020 Karakterisasi sifat fisikokimia, sensori, dan fungsional nasi instan dari beras amilosa rendah *Jurnal Penelitian Pascapanen Pertanian* **17(1)** 1-14
[14] Priyati A, SH Abdullah and GMD Putra 2016 Pengaruh kecepatan putar pengadukan adonan terhadap sifat fisik roti *Jurnal Ilmiah Rekayasa Pertanian dan Biosistem* **4(1)** 217-221
[15] Sipayung MY, Suparmi and Dahlia 2015 Pengaruh suhu pengukusan terhadap sifat fisika Kimia tepung ikan rucah *Jurnal Online Mahasiswa Fakultas Perikanan dan Ilmu Kelautan* **2(1)** 1-13
[16] Asgar A and D. Musaddad 2006 Optimalisasi cara, suhu, dan lama blansing sebelum pengeringan pada wortel *J. Hort.* **16(3)** 1-8
[17] Muramatsu Y, A Tagawa, E Sakaguchi and T Kasai 2006 Water absorption characteristic and volume changes of milled and rown rice during soaking *Cereal Chemistry Journal* **83(6)** 624-631
[18] Azizah YN, Dian RA and Dimas RAM 2014 Formulasi dan kajian karakteristik nasi jagung (*Zea mays* L.) instan yang disubtitusi tepung kacang hijau (*Phaseolus radiatus*) *Jurnal Teknosains Pangan* **3(1)** 84-95
[19] Sukasih E, Kirana SS and Sri W 2020 Karakteristik fisikokimia dan organoleptik kacang hijau instan dengan teknologi pembekuan *Jurnal Penelitian Pascapanen Pertanian* **17(1)** 37-47
[20] Sugiyono, Soewarno T, Purwiyatno H and Agus S 2004 Kajian optimasi teknologi pengolahan beras jagung instan *Jurnal Teknologi dan Industri Pangan Perhimpunan Ahli Teknologi Pangan Indonesia* **15** 119-128
[21] Indiarto R, Bambang N and Edy S 2012 Kajian karakteristik tekstur (*Texture Profil Analysis*) dan organoleptik daging ayam asap berbasis teknologi asap cair tempurung kelapa *Jurnal Teknologi Hasil Pertanian* **V(2)** 106-116
[22] Haliza W, Sari IK and Sri Y 2012 Penggunaan mixture response surface methodology pada optimasi formula brownies berbasis tepung talas banten (*Xanthosoma Undipes* K. Koch) sebagai alternatif pangan sumber serat *J Pascapanen* **9(2)** 96-106
[23] Chen F, Chunhua Y, Linlin L, Tianyi L, Yang W, Lichao W, Yanguo S and Osvaldo HC 2017 Differences, correlation of compositions, taste and texture characteristics of rice from Heilongjiang China *J Rice Research* **5(1)** 1-5
[24] Cameron DK and Ya JW 2005 A better understanding of factors that affect the hardness and stickiness of long-grain rice *Cereal Chemistry* **82(2)**113-119
[25] Radiati A and Sumarto 2016 Analisis sifat fisik, sifat organoleptik, dan kandungan gizi pada produk tempe dari kacang non-kedelai *Jurnal Aplikasi Teknologi Pangan* **5(1)** 16-22
[26] Rewthong O, Somchart S, Chaiyong T, Patcharee T and Somkiat P 2011 Effects of cooking, drying and pretreatment methods on texture and starch digestibility of instant rice *Journal of Food Engineering* **103** 258-264
[27] Sundari D, Almasyhuri and Astuti L 2015 Pengaruh proses pemasakan terhadap komposisi zat gizi bahan pangan sumber protein *Media Litbangkes* **25(4)** 235-242
[28] Nguju AL, Pieter RK and Bastari S 2018 Pengaruh cara memasak yang berbeda terhadap kadar protein, lemak, kolesterol dan rasa daging sapi Bali.*Jurnal Nukleus Peternakan* **5(1)**17-23
[29] Sundari E and Nuryanto 2016 Hubungan asupan protein, seng, zat besi, dan riwayat penyakit infeksi dengan z-score TB/U pada balita *Journal of Nutrition College* **5(4)** 520-529
[30] Loebis EH, Lukman J and Irma S 2017 Karakteristik mutu dan nilai gizi nasi mocaf dari beras analog *Biopropal Industri* **8(1)** 33-46

[31] Pereira, Elenilda J, Lucia MJC, Gisela MD, Flávio SNC, José LVC., Daniela SV, Sidinea CF and Maurisrael MR 2014 Effects of cooking methods on the iron and zinc contents in cowpea (*Vigna unguiculata)* to combat nutritional deficiencies in Brazil *Food and Nutrition Research* **58** 1-8

Raspberry Pi based crop and weed classification using Machine Learning Algorithm

K L Joshitha [1*], **S Nishanthini** [1], **B J Harshini** [1], **T Sadhana** [1]

[1] Department of Electronics and Communication Engineering, Sri Sai Ram Engineering College Chennai, Tamil Nadu, India

* email: lakshmi.ece@sairam.edu.in

Abstract. Farming with fun has uplifted the agritourism in our country. The technological development and support have made farming a passion nowadays. Farming is not only important to cater the needs of the growing population but also to develop agritourism. The major challenge in crop production like irrigation, de-weeding, etc can be handled with good improvement in quality crops through the implementation of automated systems in monitoring and controlling application. With more Techies joining into the farming wishing to lead a stress-free life, automation. The work presented aims at identifying the weeds from the desired crop using the learning algorithm. The weed location identified can be used as the inputs to the de-weeding Robots. Besides, the work can equally be applicable in intercropping process, irrigation management, quality assurance in crops, etc. The input images of the crop to be classified are taken using the camera interfaced with Raspberry pi. Classification is done using the supervised learning algorithm with a set of target classes. The model is trained to recognize the target class. The accuracy of the classification varies from 75% to 100% as the proportion of the training and test data set varies from 60-40% to 80-20%.

1. Introduction

Any undesired crop that grows in the surrounding that sucks the nutrition of the desired crop is called the weed. Manually, it takes a long time to separate crop and weed from the harvesting fields. In order to reduce the time consumption and man power, this technique can be implemented in the agricultural sector. The careful use of the available resources retaining some for the future generation and the proper disposal of the processed waste and pollutants, are important targets to achieve the sustainable development goals (SDG). There are quite a large amount of population who consume too little and barely are able to meet their basic needs .Reducing the food wastage in the retailer and consumer levels is equivalent to doubling up the food production globally. Besides it is mandatory for maintaining proper supply chain to distribute the quality crop produced within the stipulated timing. The economy and health of a country mainly depends on the storage and supply chain of food at all levels of the population. The source of the income in India is basically the farming that strongly decides the economy of the country. The issues and challenges in the agro business is of many folds staring with the unpredictable monsoon, power distribution for the electrification process, unused technology for automation, over usage of the pesticides, manual labors and so on. One of the serious issues in agribusiness is the control of weeds growing among the plantation crops. Weed basically is the 'wrong growth in the wrong spot which bothers many farmers.

In some cases these weeds are deliberately grown in farms and are called as the useful weeds. Weed may be well tangled with any plant that develops or replicates forcefully. All the essential nourishment that is required for desired plant Weeds also require all essential nutrients that are required by plantation crops. These weeds if not removed at the right time may compete with desired plantation for nutrients, space, water, soil, and restrict their growth. Weeds besides acting as the host to the plants may infect and degrade the quality of crops. They grow wildly and cause the root damage of the pumps used for watering and the nearby road surfaces etc. Hence the weeds are to be removed at the earlier stage of identification either manually or using herbicides sprayed all over the field in a controlled manner. This technique causes a compromise in the quality of the crop as both plants and weeds are exposed to the sprayed herbicides. This definitely has an adverse effect on the environment and causes health problems in human. The smart weed control system can be designed to identify and remove the crop. The system must be capable of locating weeds in the field and returning the position using local or global positioning system. The automated weed removal Bot can take this input to remove the same or to do the selective spraying of herbicide. The serious health issues posed in humans are reduced due to the controlled use of the pesticides.

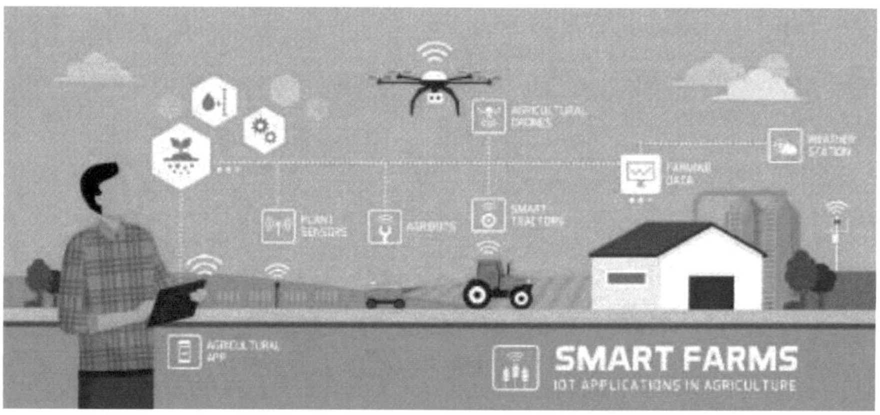

Figure 1. Precision Agriculture (courtesy Wikipedia)

Precision agriculture as in figure 1 began in the mid of the 19th century, as shown in the above picture and since then technologies used in agriculture has evolved to a great extent. The following section deals with related work

2. Methods

2.1. Related Work

The existing technologies consist of manual removal of weeds which can very hectic to the farmers. Author in [1] has followed the image processing technique to locate the weed with specific algorithm of dilation and erosion. The herbicide is sprayed through a nozzle only on the identified weed. The work of [2] covers finding the essentials like the ground water quality, the temperature humidity level, nitrite content and so on with wireless sensor network set up. Base station communication is through wireless mode and the Raspberry pi is used for the implementation. The goal of work in [3] is to have the high productivity with less wastage using precision agriculture. The work flow includes the training with the images of the plants using deep learning algorithm. However it requires the high speed GPU for processing. Author in [4] uses a specific architecture like the U- Net, Dens net, mobile Net for the implementation of the automated de-weeding process. The tensor flow is used along with the deep Neural Network in the work.

Authors of [5, 6] strive to use the image processing based Raspberry pi arrangement for the implementation of de weeding process. In addition the plant deficiency of protein is also monitored. The work of [7] is to have an intelligent sprayer for herbicide spraying. The review of the different techniques like the neural network, SVM and Discriminant analysis are compared in their performance in the work of [8]. Use of 5 different features like entropy, local homogeneity etc. are used to identify the weed using the wavelet transform in work [9]. The work also has a design of manipulator for the herbicide spraying process. The parameters like the misclassification rate is evaluated in the process of weed classification in the work of [10].

The author uses the image processing for the operation in Mat Lab. Statistical approach has been used to get a best solution of the classification of weed [11] in oil palm plant field in work using image processing techniques. The work in [12] is focused on de weeding using Delta Robot. The bot is designed to take the input from the image captured and the Bot is moved through the steps of the stepper motor calculate using the inverse kinematics. The same has been tried with Dilation and Erosion algorithm in [13] and using the ROI and quadratic polynomial in [14]. In general every work has a image processing unit, weed identification unit and the weed removal unit or a sprayer unit. The smooth device control can be done if 10 frames per second are used for the segmentation of the process. Appropriate width and length of the camera view at the ground level is required for proper image acquisition in a mobile condition.

The proposed work is done for the system for Weeds and Crops Identification. This can also be used in the intercropping process where two or more crops are grown in specific pattern. The two crops may be classified and accordingly the required sprayer can be activated delivering the required amount of the herbicides. The existing machines and accessories can be well modified with the de weeder which would be cost effective than standalone solutions. The procedure of de-weeding can be done in multiple rows to conserve time. Weed removal should be at least 90 percent done for the free growth of the desired crop. This process becomes too difficult if the scenario involves the intercropping method.

2.2. Proposed Work

The presented work is elucidated as the work flow diagram in figure 2. The hardware consists of a vision system connected to the Raspberry pi for the Image acquisition. The image captured by camera is sent to raspberry pi which performs Image pre-processing and Image segmentation. A large set of such data is collected and is located for the training and testing process. The proposed model uses the binary classification algorithm to segregate the crop from the weed on comparing their features like color size etc. Combinations of training and test data set is varied in the classification algorithm used to obtain the accuracy. The proposed model has a good percentage of accuracy at 80-20 combination of training and test data for the classification.

2.3. Binary Classification Algorithm

It is a type of supervised learning algorithm where each instance has a label and predefined. New instances can easily be classified based on the trained model. The more the number of data for training better is the convergence time and the accuracy of the classification. When there are only two categories the problem is known as statistical binary classification. Figure 3 shows the connection diagram of the presented model. The binary classification can be done based on the different features like the color, shape etc.

The data set comprises of set of images as shown in figure.4. Initial preprocessing of the images are done which includes the Hog transform, segmentation. The segmentation process includes the watershed algorithm in this work that has the flooding process applied to the images. The segmented images have the marked labels that can be saved as the .csv file. This file can be accessed by the classifier for further processing. This is followed by the process of training and testing done with various combinations. The percentage error is tabulated for the various combinations of training and test data set.

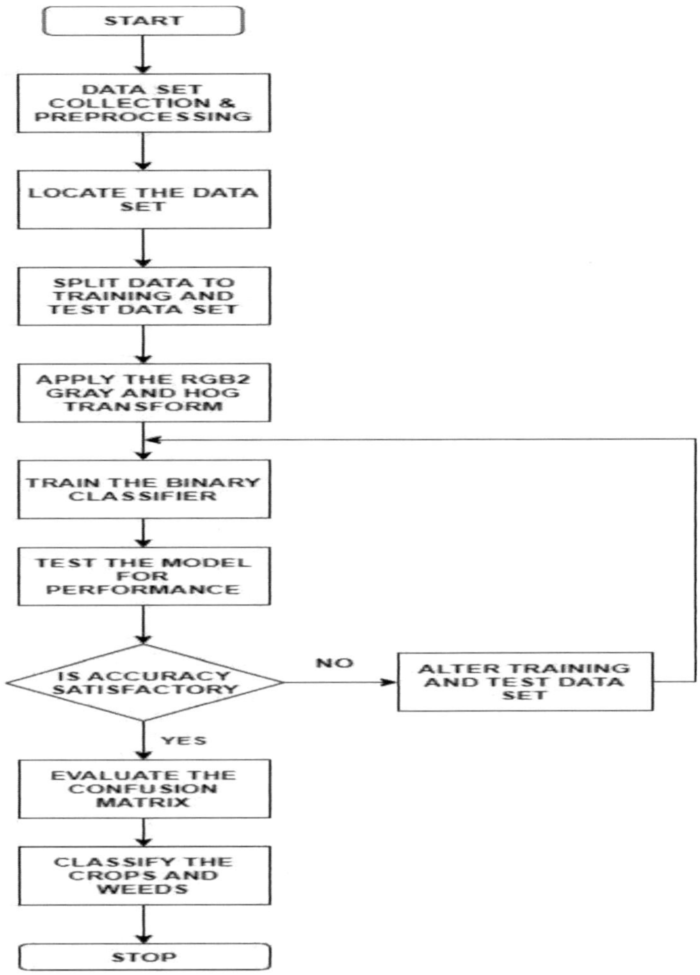

Figure 2. Work Flow

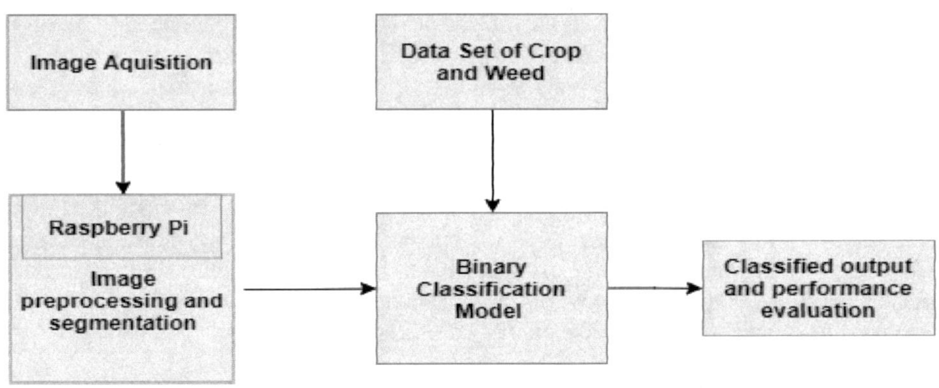

Figure 3. Block Diagram of proposed system

Figure 4. Crop And Weed Image

2.4. Hardware and Software Requirement
The Techno stack of the work includes the following hardware and software items.

2.4.1. Raspberry Pi
Raspberry Pi works with the Raspbian OS which has the Arm 7 Quad core processor that works at 900 MHz. It has two USB ports along with display and camera options.

2.4.2. Camera.
The camera with the resolution of 2048X1056 at 30/15 fps is used here to acquire the input images. It can easily be interfaced with the raspberry Pi to input the images through the port. The work also needs the display device like a monitor to view the output.

2.4.3. Python anaconda
The python program is used here with Anaconda Jupyter notebook. Jupyter is the open source interactive web tool to view the computed output. The open CV can also be used for providing the common infrastructure for computer vision application.

3. Results and Discussions

The efficiency of the presented work is evaluated using the accuracy of the classification scheme. The five different data set is taken each with about 60 to 65 images. Accuracy is calculated as the proportion of the weeds identified correctly to that of the overall crop and weeds data. Table 1 and figure 5 shows the simulated accuracy for a different combination of data sets.

Table 1. Accuracy of classification

Sl No	Training : Test data in %	Accuracy in %
1	70:30	85.714
2	80:20	100
3	55:45	75
4	60:40	75

Table 2. Success Rate

Description	Total Trials Done	No. of Success Trials	No. of Failure Trials
Images Used For Testing The Model.	7	6	1

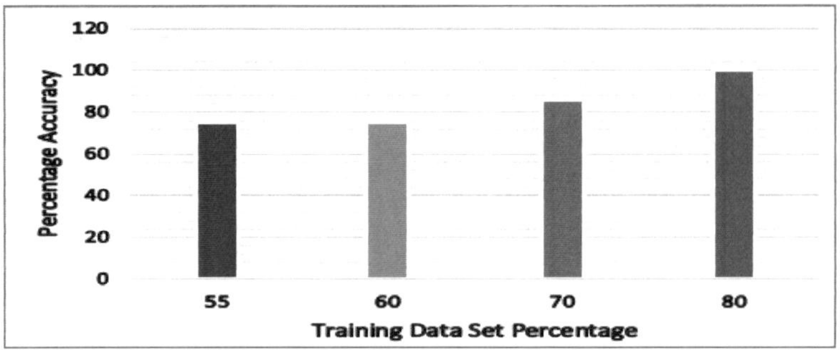

Figure 5. Accuracy plot

The proposed work is based on image classification of plants as crops and weeds. The simulation as an example takes in 33 samples form the dataset out of which 29 images are classified as crops and 4 images are classified as weeds. The table shows the observation and result of the simulation. The simulation results shows the accuracy and reliability of the model developed. Table.2 shows the success rate of the simulation of 7 runs of classification. The figure 6 shows the simulated value of the percentage accuracy calculation which is variable with the training and test data set values

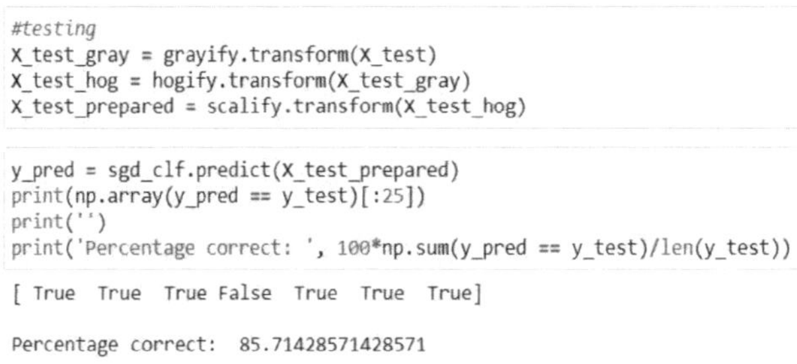

Figure 6. Percentage Accuracy

The confusion matrix of the same can also be calculated which is a matrix that gives us the value of True, Negatives and False Positives and negatives. These parameters can be further used to evaluate the Accuracy, Precision, Recall, Specificity and F1 score. The confusion matrix was calculated for different combinations of input and are tabulated in Table.3 and Table 4. Accuracy gives the correct classification. Precision refers to the number of positive identification that are actually correct on the other hand recall denotes the proportion of the positive identifications identified correctly as shown in Table 4. The harmonic mean of the precision and Recall is F1 score. Specificity denotes the portion of negatives correctly identified

Table 3. Confusion matrix and metric evaluation 1

Train: Test (%)	Accuracy (%)	Precision (%)
55:45	75	75
60:40	75	100
70:30	85.71	80
80:20	100	100

Table 4. Confusion matrix and metric evaluation 2

Recall (%)	Specificity (%)	F1-Score (%)
100	0	85.71
75	0	85.71
100	66	88.88
100	100	100

4. Conclusion and Future Work

The proposed model was used to classify the weeds from the crop. The same can also be applied to intercrop farming for the identification of the different crops in a specific pattern. The model can be used for the spraying of the herbicides in the intercropping, irrigation management of different sorts of crops, etc. The same can also serve as a valid input to the design of Robots for the weed removal process. More accurate is the classification more precise would be the weed positioning and movement of robot. The design of a delta Robot for weed removal is being designed as a continuation of this work. The IoT system can be used along with this sort of classifier to monitor the ambiance and for control applications. In industries where human intervention is not possible the flawed items manufactured can be easily and swiftly removed and a perfect batch processing can also be done.

References

[1] Aravind R, Daman M and Kariyappa B S., "Design and Development of Automatic Weed Detection and smart Herbicide sprayer Robot," IEEE recent Advanes in Intelligent ComputationalSystems (RAICS),Trivandrum, Dec,2015,pp.257-261.

[2] R.Kamath, M.Balachandra and S Prabhu,"Raspberry Pi as Visual Sensor Nodes in Precision Agriculture,A Study.1'in IEEE Access, vol.7,pp.45110-45122,2019,doi:10.1109/ACCESS.2019.2908846.

[3] S Umamaheswari,R. Arjun and D.Meganathan, "Weed Detection in Farm Crops using Parallel Image Processing," 2018 Conference on Information and Communication Technology(CICT),Jabalpur India,2018,pp. 1-4 doi:10.1109/INFOCOMTECH.2018.8722369.

[4] Chechlinski, L, Siemiatkowska, B Majewski, M. A system for weed and crop Identification – Reaching over 10 FPS on Raspberry Pi with the usage of MobileNets,DenseNet and Custom Modifications.Sensors 2019, 19, 3787.

[5] Ajinkya Paikekari, Vrushali Ghule, Rani Meshram, and V B Raskar, "Weed Detection using Image Processing," Int.Res. J.of Engg.& Tech.(IRJET),vol.3, no. 3.Mar.2016,pp.1220-1222.

[6] Amir H Kargar B, Ali M Shirzadifar, "Automatic Weed Detection system and Smart Herbicide Sprayer Robot for corn fields" Proc.2013 RSI/ISM Int.Conf. Robotics & Mechatronics. Tehran, Feb 2013, pp.468-473

[7] G Y Rajaa Vikhram, Rakshit Agarwal, Rohan Uprety and V N S Prasanth,"Automatic Weed Detection and Smart Herbicide Sprayer Robot," Int. J. Engg.& Tech., 2018,pp.115-118

[8] Riya Desai,Kruti Desai, Shaishavi Desai, Zinal Solanki, Densi Patel, and Vikas Patel,"Removal of weeds using Image Processing: A Technical Review," Int. J. Adv. Comp.Tech.(IJACT), vol.4,no.1,pp.27-31

[9] Amrutha A. Aware, Kavitha Joshi,"Crop and Weed Detection Based on Texture and Size Features and Automatic Spraying of Herbicides,"Int J Adv Res Comput Sci SoftwEng,6(1).

[10] Hlaing, S H.,&Khaing, A S (2014). Weed and crop segmentaton and classification using area thresholding. Internationa Journal of Research in Engineering and Technology,3(3),375-380.

[11] Kamarul Hawari Ghazali, Mohd.Marzuki Mustafa,Aini Hussain(2008) – Machine VisionSystem for Automatic weeding Strategy in oil Palm Plantation using Filtering Technique, American Eurasian Journal of Agriculture & Environmental Science,Vol.3,no.3

[12] Shiva R, Vimal G, Kaviyarasu M, (2020,December),Intelligent Farming using Delta Robot.In 2020 International Conference on Power, Energy, Control and Transmission systems(ICPECTS), pp1-5, IEEE.

[13] Muhammad H Siddiqi, Irshad Ahmad, Suziah Bt Sulaiman (Apr 2009) – Weed Recognition Based on Erosion and Dilation Segmentation Algorithm, International Conference on Education Technology and Computer Singapore.

[14] Asnor Juraiza Ishak, Siti Salasiah Mokri, mohd. Marzuki Mustafa and Aini hussain, (Dec 2007) Weed Detection using Quadratic Polynomial and ROI Techniques, in proc. Of 5th student Conference on Research and Development, Malaysia.

IOP Conf. Series: Earth and Environmental Science **980** (2022) 012030 doi:10.1088/1755-1315/980/1/012030

Effect of drying method on the chemical properties of local soy flour

A L Suryana [1], N M Rosiana [1*] and Z Olivia [1]

[1] Clinical Nutrition Study Program, Health Department, Politeknik Negeri Jember, Jalan Mastrip Kotak Pos 164 Jember 68101, Indonesia

[*] email: nita.maria.r@polije.ac.id

Abstract. Soybean is a good source of protein because the nutrition value of soybean protein is equivalent to animal protein. Soy foods have been generally proven to prevent chronic disease, such as antiobesity, antihypertensive, immunity regulation, cholesterol-lowering, lipid-lowering, anticarcinogenic and antioxidant. Moreover, soy food also decreases the prevalence of stunting in toddlers. Soybean in the form of flour can make it easier to process into food, but soy has a bad smell that they don't like. This aroma can be reduced by heating treatment such as the drying method. So this study aimed to determine the effect of the drying method on the chemical content of soybean flour. Four methods to dry the soybean are sun drying, oven drying at 50o C, oven drying at 70o C hours, and oven drying at 100o C. After drying, the soybeans are roasted for 2 minutes with medium heat. The result shows the drying method affects the moisture, protein, fat, ash, and carbohydrate of soybean flour. Oven drying at 50oC for 3 hours is recommended to use as the best method to produce soy flour due to its nutritional value and has better aroma, sweet taste, light color, and good nutritional value.

1. Introduction

The soybean, soya bean or more famous with soybean, is a species of legume from East Asia. Traditional food uses of soybeans are soy milk, tofu, tempeh, soy sauce, natto, miso, oncom, etc. Soybean is a source of nutrition due to its amazing chemical properties. Between cereal and other legumes, soybean has the highest protein content. Raw dried soybean contains 40,4% protein, higher than dried mungbean (22,9%), kidney bean (22,1%), beef (18,8%), sesame seed (19,3%), and corn (9,8%) [1]. Soybean protein can be separated from other components and is called isolate soybean protein (ISP). Soybean protein is used as an alternative protein to increase the nutritional composition of various food such as beverage powders, cheeses, salad dressings, infant formulas, bars, pasta, meat analogs, bread, nondairy creamers, frozen desserts, and breakfast cereals [2]. In food, soybean proteins function as an emulsifying agent, foaming agent, gelling agent replacement of eggs, hydration, oil absorption, solubility, and film properties [3–7]. Soybean protein does not only affect the sensory properties of the product but also affects human health such as lower chronic diseases such as cardiovascular disease, obesity, type II diabetes, immune disorders, and cancers [8].

Other important components discovered in soybeans are polysaccharides, oligosaccharides, phospholipids, minerals, vitamins, and isoflavones. Soybean polysaccharides have antioxidant activity and are composed of mannose, galactose, arabinose, rhamnose, glucose, and galacturonic acid. Soybean polysaccharides also have considerable protective effects against CCl4-induced liver damage and prevent obesity and metabolic syndrome in mice [9,10]. The main active ingredient of phospholipids in soybeans is 1,2-di-linoleoyl-phosphatidylcholine (DLPC). Soybean phospholipids influenced membrane-dependent cellular functions and showed antioxidant, antiinflammatory, anti-fibrotic, apoptosis-modulating, regenerative, membrane-repairing and -protective, cell-signaling and receptor-influencing, as well as lipid-regulating effects in intoxication models with chemicals or drugs [11]. The vitamin and mineral content in soybeans includes calcium, phosphor, iron, sodium, potassium, zink, thiamin, riboflavin, and niacin [1]. Soybean flour has 250,33 mg/100g phenol, 476,67 mg/100g flavonoids and antioxidant activity as IC_{50} is 18,60 [12].

Various nutrients can be obtained from soybeans that are processed into functional foods. Soy flour can be processed into snack bars containing up to 12, 55% fiber [13]. Soybeans are turned into flour to make it easier to use them as food ingredients such as nuggets, brownies, food bars, cookies [14–18]. In the process of making flour using a drying process which greatly affects the functional properties of the product. Comparably drying methods such as microwave, vacuum and freeze drying, sun drying and oven drying methods are the most preferenced food processing technology in developing societies. The drying process uses low energy so it is cheap and suitable for the agricultural environment with limited resources like in Indonesia. The food drying process purposes to dehydrate the fresh produce, so it can be stored at minimum moisture content which extends the shelf life of the products [19]. Soybean flour in the market generally has a beany flavor, so it needed a flour processing process that can reduce this aroma. So, the purpose of this study was to determine the effect of the drying process on the proximate content of soybean flour.

2. Materials and Methods

2.1. Materials
Local soybeans (*Glycine max*) are purchased from Jember local market with the characteristics of whole seeds, uniform size, yellowish color, not broken. Soybeans flour-making equipment includes an oven, blender, and 80 mesh sieve. 600g soybeans, wash thoroughly with water flow. Then drain. Furthermore, there are 4 treatments for drying, that are open sun drying, oven drying at 50°C, 70°C, 100°C. The ambient conditions for open sun drying were a dry season, without rain, and an average temperature of around 30-33°C. After drying, dry soybeans roasted at low heat for 2 minutes. Grind soybeans with a blender until they become fine flour and then sieve with an 80 mesh sieve. Soybean flour was stored in a dry and airtight place.

2.2. Analysis
The samples were analyzed for proximate analysis based on the AOAC standard [20]. Moisture determination used the oven drying method. Protein content analysis used the Kjeldahl method. Fat analysis used the soxhlet method. Ashing determination used dry method with muffle at 600°C. Carbohydrates by different. This analysis was done in duplicate. Organoleptic observations were carried out by 3 trained panelists with aroma and taste parameters using descriptive methods.

2.3. Statistics analysis
All experiments used a completely random design. The results were shown as mean values with standard deviations. Statistical analysis used ANOVA (α = 0,05) and Duncan's post hoc test.

3. Result and Discussion
Drying process using sun drying and oven. The temperature of the oven for drying is 50°C, 70°C, and 100°C. In the traditional method, soybeans are directly exposed to the sun by spreading the beans on a mat. Oven drying is the easiest method to dry a product because of almost doesn't need particular equipment. This method is faster than sun drying. But oven drying has a limited capacity. Home electric oven usually has limited capacity, between 9-60 L. A home electric oven can be used as a drying method for small-scale proses or the home industry.

The drying process is very time-dependent. Sun-drying method for 12 hours, oven drying at 50°C for 3 hours, oven drying at 70°C for 2,5 hours, and oven drying at 100°C for 1 hour. When the drying temperature is increased, the drying time is shorter. The results of the proximate analysis of soybean flour can be seen in Table 1.

Table 1. Proximate characteristics of soybean flour

Parameter	**Sun-drying**	**Oven drying**		
		50°C	**70°C**	**100°C**
Moisture (%)	9,15 ± 0,08^{d}	8,87 ± 0,08^{c}	8,68 ± 0,03^{b}	8,52 ± 0,03^{a}
Protein content (%)	37,56 ± 0,05^{a}	37,76 ± 0,03^{b}	37,89 ± 0,09^{c}	38,09 ± 0,06^{d}
Fat content (%)	19,21 ± 0,04^{a}	19,37 ± 0,03^{b}	19,45 ±0,03^{c}	19,59 ± 0,05^{d}
Ash content (%)	3,76 ± 0,76^{a}	3,91 ± 0,03^{b}	3,97 ± 0,02^{bc}	4,05 ± 0,02^{c}
Carbohydrate (%)	30,32 ± 0,07^{c}	30,08 ± 0,49^{b}	30,00 ± 0,08^{b}	29,74 ± 0,15^{a}

The same lowercase notation in the same line showed no significant difference at the 95% (Duncan test, $p<0.05$)

Based on Table 1 the method of drying has a significant effect on the moisture, protein, fat, ash, and carbohydrate of soybean flour. The moisture content of the flour is between 8,52-9,15%. The sun-drying method has the highest moisture content and the oven drying at 100°C has the lowest moisture content. The moisture content of the dried product varies depending on the type of commodity and the drying method. The sun-drying method gives 9,5-12,5% moisture content for germinated brown rice flour and 10,00% for waxy maize starch has 10,00% moisture content with the same method [21,22]. Meanwhile, the oven drying method provides a moisture content of coconut flour between 3,42-5,70% at 70°C for 4-5 hours [23]. First, the drying process removes free moisture on the surface of the food product. Then it removes bound moisture in the inside or matrix of the food product until the lowest moisture is reached. It involves heat and mass transfer operations together [24].

Soybean flour had 37,56-38,09% of protein, 2,76% of ash and 29,74-30,32% of carbohydrate. Proximate content on the sun drying method was lower than oven drying. Drying methods, temperatures, and time had a significant effect on the nutritional components of products. Usually, protein and fat decreased as increased drying temperature whereas ash and carbohydrate increased as increased drying temperature [24]. Proteins and fats have been hydrolyzed by heat. But based on Table 1, protein and fat increased significantly with drying temperature. This showed that the drying temperature did not hydrolyze the protein and fat. Decreased water content caused an increase in protein and fat content in soybean flour.

Table 2. Organoleptic characteristics of soybean flour

Parameter	**Organoleptic characteristics**
Sun-drying	Beany flavors strong, not sweet taste, yellowish color
Oven drying at 50°C	Beany flavors not too strong, sweet taste and aroma, yellowish color
Oven drying at 60°C	Beany flavors not too strong, not sweet taste, sweet aroma yellowish color
Oven drying at 70°C	Beany flavors were little smell, the smell was gone, darker yellow color

Generally, soybean flour had beany flavors that decrease panelist preference when it was made into a product (Table 2). After drying, soybean roasted for a moment to decrease the beany flavors. The result showed that the soybean flour that was roasted after the drying process has low beany flavors and sweet flavors and sweet taste appears. Oven drying at 50° C for 3 hours was recommended to use as the best method to produce soybean flour due to better its aroma, sweet taste, light color, and good nutritional value. This result was suitable with soymilk that the bean has been roasted before. Roasting soybeans improved sensory characteristics by decreasing the beany flavors, increasing sweet taste, and roasted flavor [25].

4. Conclusion
In developing societies, using the sun to dry soybeans is easy to use. However, there are obstacles such as inconsistent weather. Oven drying is an alternative even though its capacity is limited and more expensive. The research result shows that this method of drying has a significant effect on the moisture, protein, fat, ash, and carbohydrate of soybean flour. The roasted process after drying decreases beany flavor and increases sweet taste and light color.

5. Acknowledgment
Special thanks to DIPA Politeknik Negeri Jember for funding this research.

References
[1] Kementerian Kesehatan RI 2021 Data Komposisi Pangan Indonesia
[2] Thrane M, Paulsen P V., Orcutt M W and Krieger T M 2017 *Soy Protein: Impacts, Production, and Applications* (Elsevier Inc.)
[3] Nishinari K, Fang Y, Guo S and Phillips G O 2014 Soy proteins: A review on composition, aggregation and emulsification *Food Hydrocoll.* **39** 301–18
[4] Matsumiya K and Murray B S 2016 Soybean protein isolate gel particles as foaming and emulsifying agents *Food Hydrocoll.* **60** 206–15
[5] Sun D, Fan W, Xu Y and Wang L 2016 Functional Properties of Soy Protein Isolate and Its Application in Foods *Biotechnol. Med. Sci.* 630–6
[6] Lin M, Tay S H, Yang H, Yang B and Li H 2017 Replacement of eggs with soybean protein isolates and polysaccharides to prepare yellow cakes suitable for vegetarians *Food Chem.* **229** 663–73
[7] Zhang L, Chen F, Lai S, Wang H and Yang H 2018 Impact of soybean protein isolate-chitosan edible coating on the softening of apricot fruit during storage *Lwt* **96** 604–11
[8] Chatterjee C, Gleddie S and Xiao C W 2018 Soybean bioactive peptides and their functional properties *Nutrients* **10** 8–11
[9] Han L, Song H, Fu L, Li J, Yang L and Liu H 2021 Effect of extraction method on the chemical profiles and bioactivities of soybean hull polysaccharides *Food Sci. Nutr.* 1–11
[10] Lu Y, Zhao A, Wu Y, Zhao Y and Yang X 2019 Soybean soluble polysaccharides enhance bioavailability of genistein and its prevention against obesity and metabolic syndrome of mice with chronic high fat consumption *Food Funct.* **10** 4153–65
[11] Gundermann K J, Kuenker A, Kuntz E and Droździk M 2011 Activity of essential phospholipids (EPL) from soybean in liver diseases *Pharmacol. Reports* **63** 643–59
[12] Rosiana N M, Suryana A L and Olivia Z 2021 Polyphenol content and antioxidant activity of beverage from dragron fruit peel powder and soy powder *IOP Conf. Ser. Earth Environ. Sci.* **672**
[13] Cahyani W and Rosiana N M 2020 Kajian Pembuatan Snack BarTepung Gembili (Dioscorea Esculenta) Dan Tepung Kedelai (Glycine Max) Sebagai Makanan SelinganTinggi Serat *J. Kesehat.* **8** 1–9
[14] Simbolon M V T, Pato U and Restuhadi F 2016 Kajian Pembuatan Nugget dari Jantung Pisang dan Tepung Kedelai dengan Penambahan Ikan Gabus (Opiochepalus striatus) *J. Online Mhs. Fak. Pertan. Univ. Riau* **3** 1–15
[15] Eni W, Karimuna L and Isamu K T 2017 Pengaruh Formulasi Tepung Kedelai dan Tepung Tapioka Terhadap Karakteristik Organoleptik dan Nilai Gizi Nugget Ikan Kakap Putih (Lates carcarifer, Bloch) *J. Sains dan Teknol. Pangan* **2** 615–30

[16] Fibriafi R and Ismawati R 2018 Pengaruh Substitusi Tepung Kedelai, Tepung Bekatul Dan Tepung Rumput Laut (Gracilaria Sp) Terhadap Daya Terima, Zat Besi Dan Vitamin B12 Brownies *Media Gizi Indones.* **13** 12

[17] Jariyah, B.S E K and Pertiwi Y A 2017 Evaluation of Physicochemical Properties of Food bar from Composite Flour (Pedada, Taro and Soybeans) as an Emergency Food Alternative *J. Rekapangan* **11** 70-75.

[18] Jaya I K S 2016 Pengaruh Penambahan Tepung Kedelai Terhadap Cita Rasa Dan Kadar Air Cookies Ubi Jalar Ungu *J. Gizi Prima* **1** 37–9

[19] Mohammed S, Edna M and Siraj K 2020 The effect of traditional and improved solar drying methods on the sensory quality and nutritional composition of fruits: A case of mangoes and pineapples *Heliyon* **6** e04163

[20] AOAC 1995 *Official Method of Analysis of The Association Analitycal Chemist* (Whasington DC)

[21] Teixeira C S, Neves G A da R, Caliari M and Soares Júnior M S 2019 Waxy maize starch modified by sun-drying after spontaneous or backslopping fermentation *Int. J. Biol. Macromol.* **135** 553–9

[22] Cáceres P J, Peñas E, Martinez-Villaluenga C, Amigo L and Frias J 2017 Enhancement of biologically active compounds in germinated brown rice and the effect of sun-drying *J. Cereal Sci.* **73** 1–9

[23] Septiani S and Rousmaliana R 2019 Identifikasi Tepung Ampas Kelapa Terhadap Kadar Proksimat Menggunakan Metode Pengeringan Oven *J. Ilm. Kesehat.* **1** 18–31

[24] Babu A K, Kumaresan G, Raj V A A and Velraj R 2018 Review of leaf drying: Mechanism and influencing parameters, drying methods, nutrient preservation, and mathematical models *Renew. Sustain. Energy Rev.* **90** 536–56

[25] Navicha W, Hua Y, Masamba K G, Kong X and Zhang C 2018 Effect of soybean roasting on soymilk sensory properties *Br. Food J.* **120** 2832–42

Physical quality of coated and uncoated breadcrumbs of *Moringa oleifera* leaf-fish nugget with various fillers

D P Putri[1*], W Setiaboma[1], D Kristanti[1]

[1] Research Center for Appropriate Technology, BRIN
KS Tubun Street No. 5 Subang 41213

E-mail: devry26@gmail.com

Abstract. Nugget is a restructured meat product made by adding various spices and flour as filling and coating materials. Flour as a filler plays an important role in the physical quality of nuggets. The nuggets are coated with breadcrumbs that have a function to improve appearance and increase acceptability of product. This study aims to determine effect of various fillers on physical properties of coated and uncoated breadcrumbs of Moringa leaf fish nuggets. The factors were different ratio of fillers (comparison of corn starch and sago flour) and breadcrumbs coating (coated and uncoated). Nuggets with 100% wheat flour filler were used as a control. The results showed that the use of sago and corn starch flour as a filler increased the ability of WHC, hardness, gumminess, and chewiness but reduced the cooking loss of nuggets. The used of 100% sago flour was the highest of WHC (139.36%). The coating breadcrumbs had lower WHC (111.87%) ability. The cooking loss value ranged from 8.65–17.66%. The cooking loss was more visible in the samples without breadcrumb coating. The different ratio of filler significantly affected increasing the hardness, the springiness, and the cohesiveness while coated of breadcrumb reduced the hardness.

1. Introduction

Nugget is fried food meat product that is popular frozen food in Indonesia. The nuggets are restructured meat commonly made from chicken meat, but now days the nugget is made by white, red or mix of both them[1]. Now days, consumers concern to healthy food however convenient food is still trended due to busy live style. *Moringa oleifera* leaf contains functional properties that can be applied in food product[2]. The development of nugget with addition moringa leaf have been investigated such as chicken nugget [3], nugget goat meat[4], and giant catfish [5].

Batter of nugget was combination of meat, flour or starch, and seasoning forms. The batter formulation affects to quality of the nuggets. Wheat flour is original flour that used as filler in the batter, but Indonesia must import it. The study of filler no wheat flour or mix flour have been done. Corn starch is local flour that potential applied in restructure meat product. Application of corn starch in restructure product have investigated i.e.corn starch of meat patties [6], corn starch as fat replacer in buffalo sausage[7], and various different corn in nugget[8]. Corn starch could act water binding among meat particles even though added corn starch impact on performance texture, appearance and sensory[6,7]. Another local flour that potential as filler is sago flour. Sago flour have been used in restructured product such as sausage[9] and chicken nugget[10].

Generally, nuggets are coated with breadcrumbs/tempura that contained protein and starch which are important components as coating materials. Interaction between both can affect the level of adhesion of nugget system. The coating commonly used in deep frying product due to increasing appearance and texture, moreover it behavior as barrier resis to loss of moisture[11,12].

The different filler effect on adheres of the batter to breadcrumbs. However, preparation of nugget addition Moringa with different filler (corn starch and sago flour) moreover effect of coated and uncoated breadcrumbs has not been performed. This study aims to determine effect of various fillers on physical properties of coated and uncoated breadcrumbs of Moringa leaf-fish nuggets.

2. Materials and Methods

2.1. Materials

Fresh of giant catfish (*Arius thalassinus*) fillet were purchased from a local farmer from Subang, West Java, Indonesia. The meat pieces were washed and then mashed with meat grinder. The fish was stored in a freezer at -18◦C until used. Fresh *Moringa oleifera* leaves were collected from Dawuan District, Subang, West Java Province, Indonesia. Corn starch, sago flour, wheat flour, garlic, onion, shallot, salt, sugar, pepper powder, nutmeg powder, eggs, and breadcrumbs were obtained from market in Subang, West Java, Indonesia.

2.2. Sample preparation

Sample formula is presented in Table 1. Garlic, onion, shallot, ground pepper, nutmeg powder, and salt as a seasoning were pounded using a cobek. Fillet of giant catfish, spices and *M. oleifera* leaves were added to the chopper and mixed. Pour the flour and mixed well. The batter placed on the bowl added the scallion and mixed well. The batter was placed on an aluminum pan and steamed for 30 minutes. The mixture was cooled at room temperature then cut to size 1 x 4 cm. The next process was coated (BC) and uncoated (NBC) nugget. The nugget coated using flour, egg whites, and breadcrumbs. The nuggets were pre frying for 15 second and then kept in the freezer. The factors were different ratio of fillers (comparison of corn starch and sago flour) and breadcrumbs coating (coated and uncoated). Nuggets with 100% wheat flour filler were used as a control.

Table 1. Formulation of fish nugget with various filler and breadcrumbs.

Ingredients	Formula (g)					
	1	2	3	4	5	6
Corn starch	-	11.25	22.5	33.75	45	-
Sago flour	45	33.75	22.5	11.25	-	-
Wheat Flour	-	-	-	-	-	45
Fish	100	100	100	100	100	100
Moringa leaves	20	20	20	20	20	20
Leek	15	15	15	15	15	15
Seasoning	44.025	44.025	44025	44.025	44.025	44.025
Ice cube	10	10	10	10	10	10

2.3. Water Holding Capacity (WHC)[13]

The WHC was measured based on the Excess-Water method with modification. Five grams of sample were weighed into centrifugation tubes and there after centrifuged at 5° C at low speed (1000g for 15 min). The WHC was determined as liquid loss and expressed as percentage of weight of liquid release. %WHC = (before centrifuge weight - after centrifuge weight)/(before centrifuge weight) x 100.

2.4. Cooking yield [14]

Fish nuggets (initial weight) are fried at 180±2°C for 5 minutes. The sample was lowered for 1 hour, then weighed (the weight of the frying pan), and the percentage of cooking yield was calculated from the weight with formula (1).

$$\text{Cooking yield (\%)} = (\text{frying weight} / \text{initial weight}) \times 100 \quad (1)$$

2.5. Texture Profile [15]
The textural properties of nuggets were evaluated using a texture analyzer model TA-XT2 (Stable Microsystems Ltd. Surrey, England, UK). Texture profile analysis was performed using central cores of five pieces of each sample (1.5 cm x 1.5 cm x 1.5 cm), which were compressed twice to 80% of the original height [6] by a compression probe (P 75). A crosshead speed of 2 mm/s was used. The following parameters were determined: hardness (N) = maximum force required to compress the sample (H); springiness (mm) = ability of sample to recover its original form after a deforming force was removed (S); cohesiveness = extent to which sample could be deformed prior to rupture (A2/A1, A1 being the total energy required for first compression and A2 the total energy required for the second compression); adhesiveness (Ns) = work necessary to pull the compressing plunger away from sample; gumminess (N/mm2) = force necessary to disintegrate a semisolid sample for swallowing (H x cohesiveness); chewiness (N/mm) = work to masticate the sample for swallowing (S x gumminess). 5-bladed Kramer Shear Cell was used to compress the nugget to 35% of its original height. The test speed was 3.0 mm/s. The average of the ten samples is reported as toughness.

2.6. Statistical analysis
The data analysis in this study are statistically tested using analysis of variance ANOVA using SPSS. If there is a difference, then it uses the *Duncan Multiple Range Test* the level of accuracy of 5%.

3. Result and Discussion

3.1. Water holding capacity and cooking loss nugget coated and uncoated with breadcrumb
The WHC analysis showed the ability of the meat protein in the nuggets to bind the added water during the influence of external forces. The ability of WHC of nuggets was influenced by the use of flour as filler. The results of the analysis of WHC and cooking loss of nuggets with coated and uncoated treatments were in Table 2.

Table 2. WHC and cooking loss of nuggets with various fillers and breadcrumbs.

		F1	F2	F3	F 4	F5	F6
WHC (%)	BC	111.87 ± 2.94^{cA}	120.24 ± 3.59^{dA}	120.25 ± 2.98^{dB}	109.80 ± 0.44^{cA}	101.22 ± 6.12^{bA}	81.69 ± 1.48^{aA}
	NBC	139.36 ± 4.47^{dB}	136.33 ± 1.53^{dB}	108.72 ± 3.07^{bA}	125.14 ± 1.94^{cB}	110.63 ± 3.67^{bB}	103.59 ± 2.19^{aB}
Cooking loss (%)	BC	12.93 ± 0.23^{bB}	11.21 ± 0.55^{aA}	11.12 ± 0.70^{aA}	12.48 ± 0.70^{bB}	12.71 ± 0.58^{bB}	12.95 ± 0.77^{bA}
	NBC	10.91 ± 0.42^{bA}	11.05 ± 0.41^{bA}	17.12 ± 0.42^{cB}	8.65 ± 0.42^{aA}	10.22 ± 0.58^{bA}	17.66 ± 0.81^{cB}

Values are expressed as mean ± standard deviation. Means in the same column (uppercase letters) and row (lowercase letters) with different letters were significantly different at $p < 0.05$. F1 = moringa leaf fish nugget with 100% of sago flour as a filler; F2 = moringa leaf fish nugget with 75% of sago flour and 25% of corn starch as a filler; F3 = moringa leaf fish nugget with 50% of sago flour and 50% of corn starch as a filler; F4 = moringa leaf fish nugget with 25% of sago flour and 75% of corn starch as a filler; F5 = moringa leaf fish nugget with 100% of corn starch as a filler; and F6 = moringa leaf fish nugget with 100% of wheat flour as a filler (control).

The results of the analysis showed that the control nuggets had the smallest WHC, which was 81.69% for coated nuggets and 103.59% for uncoated nuggets. The use of sago flour and corn starch produced nuggets with higher WHC than control, namely 101.22%-120.25% on coated nuggets and 108.72-139.36% on uncoated nuggets. The results consistent with previous study which showed that there was no difference in cooking loss and WHC on chicken nuggets made with 4 different types of fillers, namely sago flour, corn starch, glutinous rice flour and tapioca [16]. Nugget made from sago flour filler has the highest WHC value, which is 139.36% in uncoated breadcrumb nuggets.

In this study, it was suspected that the ability of WHC nuggets was influenced by the type of filler used. Sago flour has starch containing 27% amylose and 73% amylopectin. Corn starch contained 74-76% amylopectin and 24-26% amylose. Sago flour has higher amylose content than corn starch. This causes the interaction between protein from fish and sago starch to produce a matrix that can trap more water, resulting in the highest WHC value [17]. This was supported by [18] which states that corn can be soluble in water but less able to hold water. The water absorption capacity determines the amount of water available for the process gelatinization of starch during cooking. If the amount of water was less than the gel formation does not reach the optimum condition. The water absorption capacity also affects the ease of homogenizing the flour mixture when mixed with water. Flour with high water absorption tends to be homogenized faster. The high amylose content in sago makes the starch not too wet, not too sticky, and easy to absorb water [19]. The high water binding capacity of products containing tapioca flour and sago flour was due to the size of the granules and the very high amylose content because it contains a lot of hydroxyl groups in starch molecules so that it allows more bound water. Fillers can increase water binding capacity because of their ability to hold water during processing and cooking [16].

Breadcrumb dough adhesive provides an excellent barrier to mass transfer from the food matrix. They increase the level of adhesion and yield value, and reduce the frying loss value [6]. However, in this study the coating treatment with breadcrumb didn't showed a trend towards cooking loss This is presumably due to the less than optimal breadcrumb attachment. So that breadcrumb as a barrier is not optimal. On the other hand, the double frying process causes a decrease in yield and an increase in frying loss due to the decomposition of coating materials and meat. So that there was increasing mass transfer from the nugget to the outside during frying.

3.2. Texture Profile

The results of texture profile analysis of nuggets with coated and uncoated treatments are in Table 3. The results showed that the hardness, gumminess, and chewiness values of control nuggets were 1417.36 (N/cm^2) 1043.76 (N/cm^2) and 795.79 (Ncm), respectively. This value was significantly smallest than nuggets with corn flour and sago flour fillers.

Table 3. Texture profile of nuggets with various fillers and breadcrumbs.

		F1	F2	F3	F4	F5	F6
Hardness (N/cm^2)	BC	2680.51 ± 29.96^{cA}	2232.51 ± 80.21^{bA}	2304.23 ± 66.50^{bA}	2249.50 ± 42.36^{bA}	2317.88 ± 138.68^{bA}	1417.36 ± 39.93^{aA}
	NBC	2917.22 ± 29.32^{dB}	3110.81 ± 8.92^{eB}	2430.48 ± 15.16^{bB}	2533.66 ± 85.52^{cB}	2383.32 ± 60.22^{bA}	1949.41 ± 58.71^{aB}
Springiness (cm)	BC	0.77 ±0.05 ^{aA}	0.81 ± 0.07^{aA}	0.81 ± 0.03^{aA}	0.76 ± 0.04^{aA}	0.78 ± 0.01^{aA}	0.76 ± 0.01^{aA}
	NBC	0.93 ± 0.05 ^{aB}	0.95 ± 0.01^{aB}	0.88 ± 0.04^{aB}	0.87 ± 0.06^{aB}	0.88 ± 0.05^{aB}	0.92 ± 0.02^{aB}
Cohesiveness (ratio)	BC	0.74 ±0.01^{aA}	0.73 ± 0.03^{aA}	0.72 ± 0.02^{aA}	0.72 ± 0.02^{aA}	0.72 ± 0.02^{aA}	0.74 ± 0.01^{aA}
	NBC	0.77 ± 0.04^{aA}	0.76 ± 0.05^{aA}	0.76 ± 0.04^{aA}	0.73 ± 0.03^{aA}	0.74 ± 0.02^{aA}	0.72 ± 0.03^{aA}
Gumminess (N/cm^2)	BC	1973.68 ± 30.14^{cA}	1638.05 ± 68.31^{bA}	1670.44 ± 94.31^{bA}	1629.87 ± 47.97^{bA}	1665.31 ± 119.62^{bA}	1043.76 ± 38.11^{aA}
	NBC	2250.33 ± 111.12^{cB}	2377.29 ± 163.89^{cB}	1839.02 ± 91.18^{bB}	1860.57 ± 141.92^{bB}	1759.07 ± 78.13^{bB}	1397.15 ± 84.54^{aB}
Chewiness (Ncm)	BC	1527.05 ± 101.32^{cA}	1333.95 ± 157.37^{bA}	1347.44 ± 92.03^{bA}	1242.90 ± 81.90^{bA}	1302.19 ± 114.55^{bA}	795.79 ± 38.18^{aA}
	NBC	2080.42 ± 50.07^{cB}	2247.68 ± 177.04^{cB}	1621.75 ± 111.40^{bB}	1611.76 ± 164.99^{bB}	1547.23 ± 54.65^{bB}	1282.99 ± 77.46^{aB}

Values are expressed as mean ± standard deviation. Means in the same column (uppercase letters) and row (lowercase letters) with different letters were significantly different at $p < 0.05$. F1 = moringa leaf fish nugget with 100% of sago flour as a filler; F2 = moringa leaf fish nugget with 75% of sago flour and 25% of corn starch as a filler; F3 = moringa leaf fish nugget with 50% of sago flour and 50% of corn starch as a filler; F4 = moringa leaf fish nugget with 25% of sago flour and 75% of corn starch as a filler; F5 = moringa leaf fish nugget with 100% of corn starch as a filler; and F6 = moringa leaf fish nugget with 100% of wheat flour as a filler (control)

Treatment of uncoated breadcrumb on nuggets significantly increased the value of the texture profile which included hardness, springiness, cohesiveness, gumminess, and chewiness. Comparison of filler types significantly affected the texture of nugget. Nugget made with 100% sago flour filler had the hardest texture. The use of a combination of filler between corn starch and sago flour can reduce the hardness of the nugget, even though the resulting nugget texture is not as soft as that made from 100% wheat. The hardness is force that needed to deformation of sample. The gumminess is the density that persists during chewing. The chewiness is energy that needed to chew the sample until ready to swallow. The higher of hardness, gumminess, and chewiness, the harder to bite and chew. The hardness, gumminess, and chewiness of F6 nugget were lower than other, while F1 nugget higher. The hardness, gumminess, and chewiness of Moringa fish nuggets were not significantly different. The hardness, gumminess, and chewiness of F1, F2, F3, F4, and F5 nuggets were higher than F6 nugget. The result indicated that sago and cornstarch caused the harder texture of nugget than wheat flour.

The results were consistent with the previous study results. [20] Devadason et al., 2010 reported that score of gumminess, and chewiness of buffalo meat nuggets with corn starch as filler higher than wheat flour. The use of sago as a filler caused an increase in hardness and elasticity of duck sausages [9]. The non wheat flour as a filler lead an increase in hardness, gumminess, and chewiness of mushroom chicken nuggets due to the lower of amylose content [12]. The amylose content of corn starch and sago are 24-26% and 27%, respectively. The amylose content of corn starch and sago are lower than wheat flour, 38.6% [21].

Springiness is ability of sample to recover its original form after deforming force removed. Cohesiveness is ratio of energy for second and first compression. The springiness and cohesiveness of moringa fish nuggets were not significantly different. The results were similar with the previous study. The use of cornstarch in buffalo meat nuggets and sago in duck sausages did not affect the springiness and cohesiveness [20] [9].

The breadcrumbs coating affected the texture profile of moringa leaf fish nuggets. The hardness, springiness, cohesiveness, gumminess, and chewiness of coated moringa leaf fish nuggets were lower than uncoated. [12] Kristanti and Setiaboma (2021) reported similar result, the texture properties of breadcrumbs coated mushroom chicken nuggets were significantly lower than non-coated. The breadcrumbs have been shown to improve the texture profile of nuggets.

4. Conclusion

The use of sago flour and cornstarch as a filler increased the ability of WHC, hardness, gumminess, and chewiness but reduced the cooking loss of nuggets. The coated of breadcrumbs reduced the profile texture of Moringa leaf-fish nugget. Furthermore, it needs to sensory evaluation on Moringa leaf-fish nugget to determine the effect of the filler and physical properties changes on customer acceptance.

Reference

[1] Lukman I, Huda N and Ismail N 2009 Physicochemical and Sensory Properties of Commercial Chicken Nuggets *J. Food Ag-Ind* **2** 171–80

[2] Devisetti R, Sreerama Y N and Bhattacharya S 2016 Processing effects on bioactive components and functional properties of moringa leaves: development of a snack and quality evaluation *J. Food Sci. Technol.* **53** 649–57

[3] Madane P, Das A K, Pateiro M, Nanda P K, Bandyopadhyay S, Jagtap P, Barba F J,

Shewalkar A, Maity B and Lorenzo J M 2019 Drumstick (Moringa oleifera) flower as an antioxidant dietary fibre in chicken meat nuggets *Foods* **8** 1–19

[4] Verma A K, Rajkumar V, Kumar M S and Jayant S K 2019 Antioxidative effect of drumstick (Moringa oleifera L.) flower on the quality and stability of goat meat nuggets *Nutr. Food Sci.* **50** 84–95

[5] Solichah E, Iwansyah A C, Pramesti D and Desnilasari D 2021 Evaluation of physicochemical , nutritional , and organoleptic properties of nuggets based on moringa (Moringa oleifera) leaves and giant catfish (Arius thalassinus) *Food Sci. Technol.* **A head of** 1–6

[6] Kilincceker O 2018 Effects of different starches on some of the frying and storage properties of meat patties *Adv. Food Sci.* **40** 35–41

[7] Jairath G, Sharma D P, Dabur R S, Singh P K and Bishnoi S 2018 Standardization of corn starch as a fat replacer in buffalo calf meat sausages and its effect on the quality attributes *Indian J. Anim. Res.* **52** 1521–5

[8] Ma'ruf W, Rosyidi D, Radiati L E and Purwadi P 2019 Physical and Organoleptic Properties of Chicken Nugget from Domestic Chicken (Gallus domesticus) Meat with Different Corn Flours as Filler *Res. J. Life Sci.* **6** 162–71

[9] Muthia D, Nurul H and Noryati I 2010 The effects of tapioca, wheat, sago and potato flours on the physicochemical and sensory properties of duck sausage *Int. Food Res. J.* **17** 877–84

[10] Nurlela S, Hastuti H and Suparman S 2018 The Quality of Nugget of Broiler Chicken Meat with Addition of Sago Flour (Metroxylon Sp.) *Chalaza J. Anim. Husb.* **3** 67–72

[11] Sarteshnizi A and Khaneghah M 2015 Mini Review A review on application of hydrocolloids in meat and poultry products *Int. Food Reserach J.* **22** 872–87

[12] Kristanti D and Setiaboma W 2021 The colour and texture properties of mushroom chicken nugget with various flour as a filler *IOP Conf. Ser. Earth Environ. Sci.* **672**

[13] Verbeken D, Neirinck N, Van Der Meeren P and Dewettinck K 2005 Influence of κ-carrageenan on the thermal gelation of salt-soluble meat proteins *Meat Sci.* **70** 161–6

[14] Kim H Y, Kim K J, Lee J W, Kim G W, Choe J H, Kim H W, Yoon Y and Kim C J 2015 Quality evaluation of chicken nugget formulated with various contents of chicken skin and wheat fiber mixture *Korean J. Food Sci. Anim. Resour.* **35** 19–26

[15] Mittala G S, Nadulskib R, Barbut' S and Negi S C 1992 *Textural profile analysis test conditions for meat products* vol 25

[16] Komansilan S 2015 Pengaruh Penggunaan Beberapa Jenis Filler Terhadap Sifat Fisik Chicken Nugget Ayam Petelur Afkir *Zootec* **35** 106

[17] Utami E Y, Rosyidi D and Widyastuti E S 2015 Pengaruh substitusi daging ayam broiler dengan jamur salju (Tremella fuciformis) pada kualitas nugget ayam *J. Ilmu dan Teknol. Has. Ternak* **10** 63–75

[18] Aini N, Wijonarko G and Sustriawan B 2016 SIFAT FISIK, KIMIA, DAN FUNGSIONAL TEPUNG JAGUNG YANG DIPROSES MELALUI FERMENTASI (Physical, Chemical, and Functional Properties of Corn Flour Processed by Fermentation) *J. Agritech* **36** 160

[19] Patricia Caesy C, Kathleen Sitania C, Gunawan S and Aparamarta H W 2018 Pengolahan Tepung Sagu dengan Fermentasi Aerobik menggunakan Rhizopus sp. *J. Tek. ITS* **7** 7–9

[20] Devadason I P, Anjaneyulu A S R and Babji Y 2010 Effect of different binders on the physico-chemical, textural, histological, and sensory qualities of retort pouched buffalo meat nuggets *J. Food Sci.* **75** 31–5
[21] Blazek J and Copeland L 2008 Pasting and swelling properties of wheat flour and starch in relation to amylose content *Carbohydr. Polym.* **71** 380–7

IOP Conf. Series: Earth and Environmental Science **980** (2022) 012032 doi:10.1088/1755-1315/980/1/012032

Fatty acids and amino acids profile of organic black rice (*Oryza sativa* L.) milk

A Romulo [1*], and N F Sadek [1]

[1] Food Technology Department, Faculty of Engineering, Bina Nusantara University, Jakarta, Indonesia, 11480

[*] email: andreas.romulo@binus.ac.id

Abstract. The popularity of plant-based milk has been increasing over the last few years to substitute animal milk. Cereal such as black rice (*Oryza sativa* L.) is a plant material that can be used to produce rice milk. Black rice has been reported to have high vitamin and mineral content and high fiber. Previous research also has shown the functionality of black rice, such as antioxidant, antihypertensive, and antihyperlipidemic. For this reason, black rice has the potency to be further processed into functional food such as rice milk. However, there is still a lack of basic information about the nutritional profile of black rice milk. Therefore, this research aimed to examine the fatty acids and amino acids profile of black rice milk. Fatty acid analysis was carried out using GC-FID. Amino acid content was analyzed using UPLC. The fatty acid profile analysis revealed that polyunsaturated fatty acid was the most abundant (0.1062%) in black rice milk, followed by saturated fatty acid (0.062%). The highest amino acid found in black rice milk was glutamic acid (0.0045 g/100 mL), aspartic acid (0.00269 g/100 mL), and arginine (0.0228 g/100 mL)

1. Introduction

The need to obtain nutritional benefits and, at the same time, combined with the sensorial acceptability are the main driving force of shifting in food choices [1]. These days, many people are more aware of their health and therefore influences their preferences to choose healthier food product. For example, an animal-based diet has been running for centuries and has negatively impacted health and the environment [2]. For a long time, milk has been described as a complete food with high nutritional contents and bioactive properties [3]. However, some people avoid consuming milk due to medical reasons such as lactose intolerance and allergy. Over the last few years, the emergence of new lifestyles, such as vegetarian and vegan, environmental issues, and antibiotic residues in milk have pushed researchers and food industries to develop alternative milk made from plant-based [4,5]. In several countries, the limited access to animal milk in some areas, high prices, and the contamination of pathogens are other reasons for realizing new milk analogs [6].

A liquid obtained from the extraction of a plant that resembles cow milk is called plant-based milk. It is produced by processing soluble material from grounded plant material with water to make a colloidal suspension or emulsion [5]. Although there are various preparation techniques to produce plant-based milk, the general processing method is the same such as soaking, grinding, filtration, mixing, homogenization, heat treatment, and aseptic packing [7]. The earliest use of plant-based milk substitute for cow milk was recorded in China through soybean milk [5]. Nowadays, several plant materials are used to make plant milk alternatives, such as legumes, cereals, seeds, nuts, and pseudocereals [8]. The popularity of soybean is due to the nutrient content like protein that is comparable with cow milk and contains bioactive compounds such as isoflavones [8]. However, soybean also could cause allergy, and beany flavor limited the consumption of this product. Therefore, the use of cereal plants such as rice (*Oryza sativa* L.) is gradually increasing. Consumers perceive rice as a non-allergenic material and have the potency to be developed as rice milk [1].

Rice is a cereal grain widely cultivated in Indonesia and consumed by the majority of Indonesian as a staple food [9]. There are several types of rice in Indonesia, such as white rice, red rice, brown rice, and black rice. Black rice has been a significant interest in nutritional quality, bioactive compounds, and pigment [10]. It has been reported to contain higher macronutrients and micronutrients than white rice [11]. Previous studies also revealed that black rice exerted health benefits such as anti-inflammatory, antioxidant, and anti-cancer [12]. Based on that fact, there is an attempt to innovate a new black rice-based functional food. To successfully create the new food product, it is essential to know the behavior and characteristic of black rice milk. The data on the amino acids and fatty acids composition of black rice milk is lacking. Therefore, this study aimed to analyze the amino acids and fatty acids profile on black rice milk.

2. Materials and methods

2.1. Materials

Black rice var. Padi Hitam Mutiara (Mama Kamu, Karanganyar) was purchased from online marketplace. Chemicals used for fatty acid analysis are GC-MS grade, and for amino acid, analysis is HPLC grade.

2.2. Black rice milk preparation

The production of black rice milk was following the method mentioned in a previous study [16]. The first stepof production was cleaning the black rice with water thoroughly and then soaking it for 180 minutes. Afterthesoaking process, size reduction occurred by mixing the rice with water (ratio 3:10), then blending usingacommercial blender for 120 seconds until the appearance turned into a slurry. Then, the slurry obtained wasfiltered and the liquid obtained was undergone heat treatment at 100 °C for 10 minutes. The cooked liquid rice slurry was then put aseptically into a packaging, cooled, and stored at 4 °C. The sample wasthen ready to use for analysis.

2.3. Determination of fatty acids profile of black rice milk

Analysis of fatty acid profile has followed the method in 18-6-1/MU/SMM-SIG. Fatty acids profile of black rice milk was analyzed using gas chromatography (GC) equipped with DB FastFAME column and detector used was flame ionized detector (FID). The carrier gas used in this analysis was helium. The injection of the sample occurred at 240 °C. The oven temperature was programmed from 50 and gradually increased up to 230 °C. The amount of sample injected was 1 μL in split mode. The interpretation of data was based on a comparison of retention time with fatty acid standards (C4–C24). The fatty acid content was calculated following the equation:

$$Fatty\ Acid\ Content\ (\%) = \frac{Area\ of\ Fatty\ Acid}{Total\ Area\ of\ Fatty\ Acids}\ x100\%$$

2.4. Determination of amino acids profile of black rice milk

The determination of amino acids composition in black rice milk was conducted following the method in a previous study [12]. Ultra-performance liquid chromatography (UPLC) coupled with a C18 column was used to separate the amino acids. The results of separation were analyzed using a photodiode array detector (PDA). The sample was first hydrolyzed using 6 M HCl containing 0.5% (w/v) at 110 °C for 24 h. After hydrolyzed, the sample was dried at room temperature for 20 minutes using the mixture of triethylamine: water: ethanol (1:2:2, v/v). For the injection to UPLC, the sample was prepared by dissolving the sample with buffer solution. The mobile phase was a mixture of AccQ. Tag UltraTM and aqua bikes. The column temperature was programmed at 49 °C, and a peak of the compounds was detected at 254 nm wavelength. The comparison of the peak with amino acid standards was conducted to quantifythe amino acid.

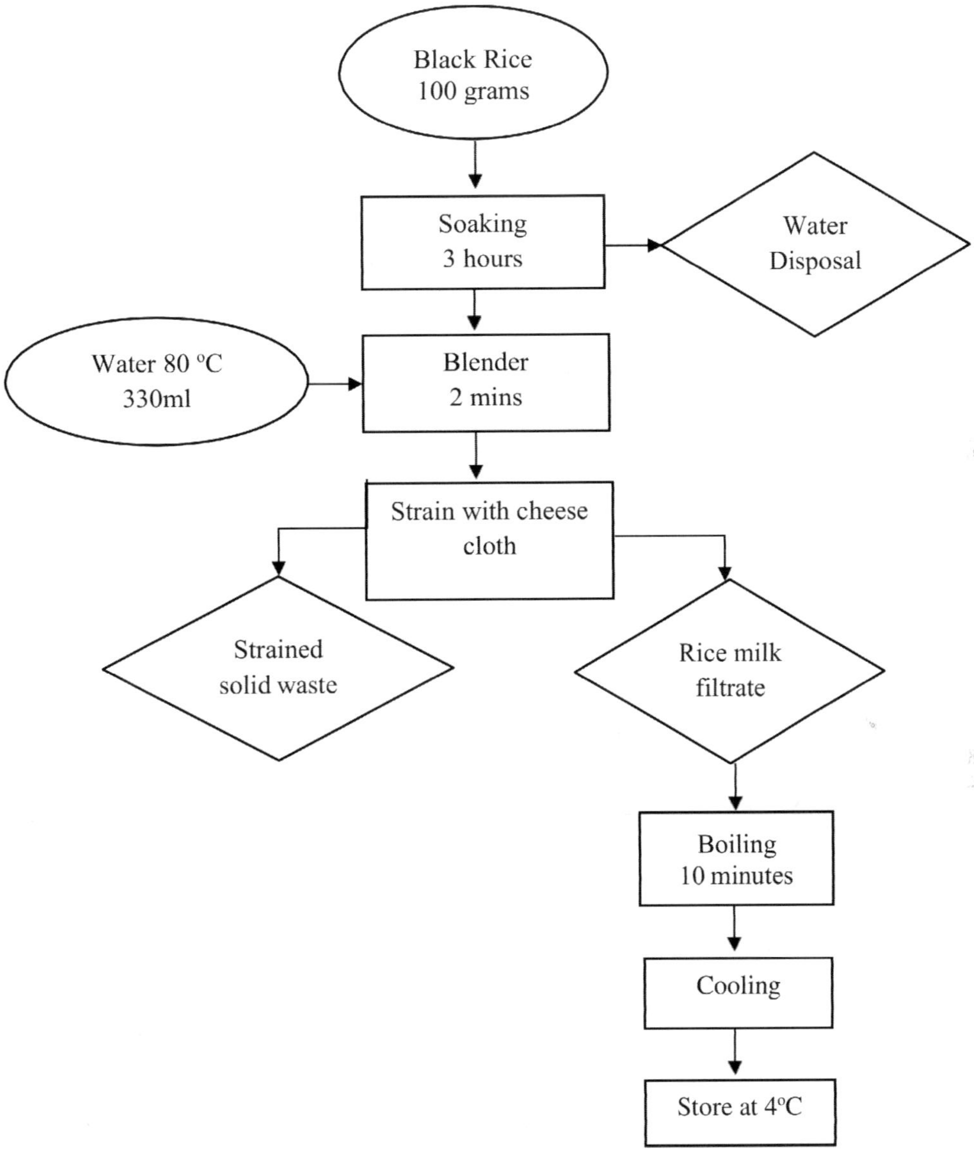

Figure 1. Flowchart of black rice milk production

3. Results and discussions

The fatty acid contents of black rice milk are expressed as a percentage of g total fatty acids per 100 mL of sample (Table 1). The result showed that polyunsaturated fatty acids (PUFAs) were the most abundant fatty acids in black rice milk with a total of 0.1062%, followed by monounsaturated fatty acids (0.00951%), and saturated fatty acids (0.062%). Linoleic acid (0.0982%) and docosadienoic acid

(0.008%) were thecompounds of PUFAs observed. In a previous study, saturated fatty acids such as stearic acid, palmiticacid, and myristic acid, were the dominant fatty acid constituents and responsible for 50% of cow milk'stotal fatty acid content [13]. Compared to cow milk, the fatty acid content in black rice milk is higher interms of unsaturated fatty acid. It is recommended to reduce the consumption of saturated fatty acid-containing products like milk, as excessive intake of saturated fatty acid has been proven to increase the risk of heart diseases [14].

Table 1. Fatty acid profile of black rice milk

Saturated Fatty Acid	**% (w/v)**	**Monounsaturated Fatty Acid**	**% (w/v)**	**Polyunsaturated Fatty Acid**	**% (w/v)**
Butyric acid	-*	Myristoleic acid	-	Linoleic acid	0.0982
Caproic acid	-	Pentadecenoic acid	-	Linolenic acid	-
Capric acid	-	Palmitoleic acid	-	Eicosadienoic acid	-
Undecylic acid	-	Heptadecenoic acid	-	Eicosatrienoic acid	-
Pelargonic acid	-	Oleic acid	0.00951	Arachidonic acid	-
Enanthic acid	-	Eicocyanic acid	-	Eicosapentaenoic acid	-
Caprylic acid	-	Erucic acid	-	Docosadienoic acid	0.008
Lauric acid	-	Capric acid		Docosahexaenoic acid	-
Tridecylic acid	-	Undecylic acid			
Myristic acid	-				
Pentadecylic acid	-				
Palmitic acid	0.0538				
Margaric acid	-				
Stearic acid	0.0062				
Nonadecylic acid	-				
Arachidic acid	0.0022				
Heneicosylic acid	n.d.				
Total	0.0622	Total	0.00951	Total	0.1062

Note: - = not detected

The amino acid profiles of black rice milk are reported in Table 2. Amino acid content is expressed as g of amino acid content per 100 mL of black rice milk. The most abundant amino acids in black rice milk are glutamic acid (0.00455 g/100 mL), aspartic acid (0.0269 g/100 mL), serine (0.0202 g/100 mL), and leucine (0.0201 g/100 mL). The number of amino acids in black rice was lower compared to animal milk [15]. It is expected that the low protein content in black rice milk contributed to the low number of amino acids. A previous study reported the low protein content of rice milk [16], complementing the findings in this study. Generally, plant protein is low compared to cow milk and affects the low nutritional quality such as amino acid [17].

Table 2. Amino acid profile of black rice milk (g/100 mL)

Amino acid	Black Rice Milk	Cow Milk [15]
Lysine	0.0142	0.261
Threonine	0.0122	0.149
Valine	0.0138	0.220
Isoleusine	0.0090	0.199
Leusine	0.0201	0.322
Tyrosine	0.0119	0.159
Phenylalanine	0.0144	0.159
Histidine	0.0085	0.089
Arginine	0.0228	0.119
Aspartic acid	0.0269	0.250
Serine	0.0202	0.179
Glutamic acid	0.0455	0.689
Proline	0.0120	0.319
Glycine	0.0171	0.070
Alanine	0.0187	0.113

n.d. = not determined

4. Conclusion
Analysis of the fatty acid profile of black rice milk showed that the major fatty acid was polyunsaturated fatty acids (linoleic acid and docosadienoic acid). Analysis of the amino acid profile of black rice milk showed that glutamic acid was the highest, and tyrosine was the lowest amino acid content.

5. Acknowledgment
The authors are grateful to Bina Nusantara University for the support through the 2021 Internal Research Grant 017/VR.RTT/III/2021

References
[1] Lee Y, Dias-Morse P N and Meullenet J-F 2019 Effect of rice variety and milling fraction on the starch gelatinization and rheological properties of rice milk *Food Sci. Technol.* **39** 1047–51
[2] Poore J and Nemecek T 2018 Reducing food's environmental impacts through producers and consumers *Science* **360** 987–92
[3] Paul A A, Kumar S, Kumar V and Sharma R 2020 Milk analog: plant based alternatives to conventional milk, production, potential and health concerns *Crit. Rev. Food Sci. Nutr.* **60** 3005–23
[4] Valencia-Flores D C, Hernández-Herrero M, Guamis B and Ferragut V 2013 Comparing the effects of ultra-high-pressure homogenization and conventional thermal treatments on the microbiological, physical, and chemical quality of almond beverages *J. Food Sci.* **78** 199-205
[5] Sethi S, Tyagi S K and Anurag R K 2016 Plant-based milk alternatives an emerging segment of functional beverages: a review *J Food Sci. Technol.* **53** 3408–23
[6] Vanga S K and Raghavan V 2018 How well do plant based alternatives fare nutritionally compared to cow's milk? *J Food Sci. Technol.* **55** 10–20

[7] Aydar E F, Tutuncu S and Ozcelik B 2020 Plant-based milk substitutes: Bioactive compounds, conventional and novel processes, bioavailability studies, and health effects *J. Funct. Foods* **70** 103975

[8] Mäkinen O E, Uniacke-Lowe T, O'Mahony J A and Arendt E K 2015 Physicochemical and acid gelation properties of commercial UHT-treated plant-based milk substitutes and lactose free bovine milk *Food Chem.* **168** 630–8

[9] Bantacut T 2014 Indonesian staple food adaptations for sustainability in continuously changing climates *J. Environ. Earth Sci.* **4** 202–15

[10] Pratiwi R and Purwestri Y A 2017 Black rice as a functional food in Indonesia *Funct. Food Health Dis.* **7** 182–94

[11] Apridamayanti P, Pratiwi R, Purwestri Y A, Tunjung W A S and Rumiyati R 2017 Anthocyanin, nutrient contents, and antioxidant activity of black rice bran of *Oryza sativa* L. 'Cempo Ireng' from Sleman, Yogyakarta, Indonesia *Indonesian J. Biotechnol.* **22** 49–54

[12] Alen Y A Y, Suarmin O S O, Suci L N S L N and Rivai H R H 2017 Analysis of fatty acids levels of freeze-dried termite queen *Macrotermes gilvus* Hagen using gas chromatography-mass spectrometry *Int. J. Indigenous Herbs Drugs* **2** 1–4

[13] Markiewicz-Kęszycka M, Czyżak-Runowska G, Lipińska P and Wójtowski J 2013 Fatty acid profile of milk - a review *J. Veter. Res.* **57** 135–9

[14] German J B, Gibson R A, Krauss R M, Nestel P, Lamarche B, van Staveren W A, Steijns J M, de Groot L C P G M, Lock A L and Destaillats F 2009 A reappraisal of the impact of dairy foods and milk fat on cardiovascular disease risk *Eur. J. Nutr.* **48** 191–203

[15] Park Y 2007 Impact of goat milk and milk products on human nutrition *CAB Rev. Perspect.* **81** 1–19

[16] Wijaya C and Romulo A 2021 Proximate analysis and antioxidant activity of red rice (*Oryza sativa* L.) Milk *J. Phys.: Conf. Ser.* **2049** 012012

[17] Mäkinen O E, Wanhalinna V, Zannini E and Arendt E K 2016 Foods for special dietary needs: non-dairy plant-based milk substitutes and fermented dairy-type products *Crit. Rev. Food Sci. Nutr.* **56** 339–49

Mini-Review: The potential of raffinose as a prebiotic

A A Anggraeni

Culinary Science Study Program, Faculty of Engineering, Universitas Negeri Yogyakarta, Indonesia

Email: andian_ari@uny.ac.id

Abstract. Prebiotics are dietary fiber components that cannot be digested by the human gastrointestinal tract but can be selectively fermented by bacteria in the gastrointestinal tract. Therefore, prebiotics provides health effects for humans. The specificity of prebiotics is determined by the bacteria that are specifically capable of fermenting the prebiotic substrate. The characteristics of prebiotic substrate need to be investigated in-vitro and in-vivo to determine the function and effectiveness of the substrate as a prebiotic. The prebiotic production process must be able to keep the prebiotic components stable. Raffinose is an oligosaccharide that has potential as a prebiotic. This article will discuss raffinose chemical structure, degrading enzymes, and health benefits as a prebiotic. Raffinose consists of 3 monomers, namely α-D-galactose, α-D-glucose, and α-D-fructose. Substrates containing raffinose-family oligosaccharides (RFOs) are the source of raffinose. Those substrates can be degraded to raffinose by the α-1,6-galactosidase. Raffinose can increase the growth of lactic acid bacteria, suppress the growth of pathogenic bacteria, increase short-chain fatty acids (SCFA), reduce constipation, inhibit the formation of putrefactive compounds from protein, and reduce the risk of cardiovascular diseases.

1. Introduction

Food processing, whether in the food processing industry or at home, often takes a long time. Thus, it reduces the content of functional compounds. This phenomenon then causes unhealthy eating habits. This unhealthy eating habit increases the diseases related to diet. However, lately, consumer awareness is increasing on the importance of healthy food. Several studies have shown a relationship between the health of the microbiota and several diseases. One component of functional food that maintains the health of the gut microbiota is prebiotics.

Prebiotics cannot be digested in the upper human gastrointestinal tract. However, according to ISAPP (International Scientific Association for Probiotics and Prebiotics), prebiotics act as substrates that can be used selectively by microorganisms in the colon and benefit the host of these microorganisms. The host is a human or an animal [1].

Lactobacillus and *Bifidobacterium* can selectively use prebiotics. Based on the ISAPP definition, the term selective is not limited to these two types of bacteria but can be extended to several other bacteria. However, the types of microbes whose growth is increased due to prebiotics must be limited. The latest technological development in the form of 16S rDNA sequencing has proven the existence of other microbes whose growth is increased due to the effect of prebiotics, namely *Pediococcus, Lactococcus, Streptococcus* [2], *Anaerostipes* [3], and *Bacteroidetes* [4].

Some prebiotic components are indigestible oligosaccharides such as fructooligosaccharides and galactooligosaccharides. Those are found in breast milk or cow's milk. Inulin is a commercial prebiotic extracted from the tubers of Jerusalem artichoke, chicory, dahlia, and yacon. Other prebiotic components that have been widely studied include soluble fiber and dietary fiber. Polyphenols also have criteria as prebiotics, but further studies are still needed to support this hypothesis. Substrates that affect the composition of the microbiota through mechanisms that do not involve selective use of the substrate by microorganisms cannot be called prebiotics. These substrates include antibiotics, minerals, vitamins, and bacteriophages [1].

The development of new prebiotic sources continues today, one of which is raffinose [5]. Reports show that raffinose can increase the composition of *Lactobacillus* and *Bifidobacterium* [6,7], decrease the composition of *Clostridium perfringens* [7] and ammonia-producing bacteria [6], and increase the production of SCFA [6]. Raffinose has health benefits for modulating the gut microbiota composition [6], treating constipation [7], lowering blood cholesterol levels [8], reducing the formation of putrefactive compounds from protein [9], and others.

Initially, raffinose was considered an anti-nutritional compound because it caused flatulence due to excessive gas production. The production of CO_2 and H_2 gases is carried out by the gut microbiota, which indicates a fermentation process by the gut microbiota. This article will discuss the latest research related to the chemical structure, degrading enzyme, and benefits of raffinose as a prebiotic.

2. Chemical structure of raffinose

Raffinose is an oligosaccharide found in legumes, peas, lentils, soybeans [10], seeds, roots[11], and leaves, such as the leaves of the herbal plant *Eupatorium* [12]. The raffinose content in lupine nuts depends on the type of cultivar. Some cultivars are high sources of raffinose. The raffinose content in lupine nuts can be increased by dehulling, which is a process to remove the outer shell of lupine nuts [13].

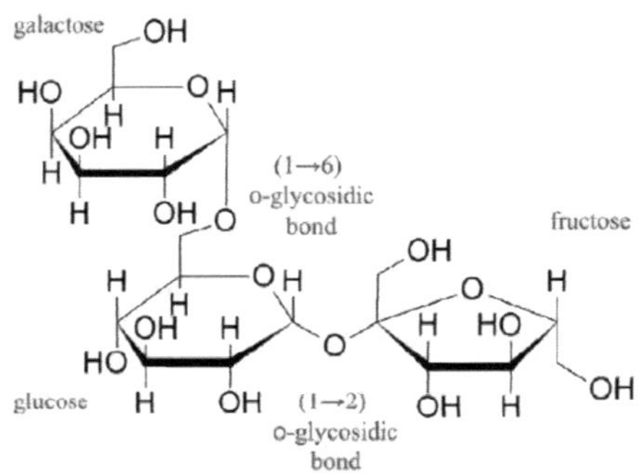

Figure 1. Chemical structure of raffinose [11].

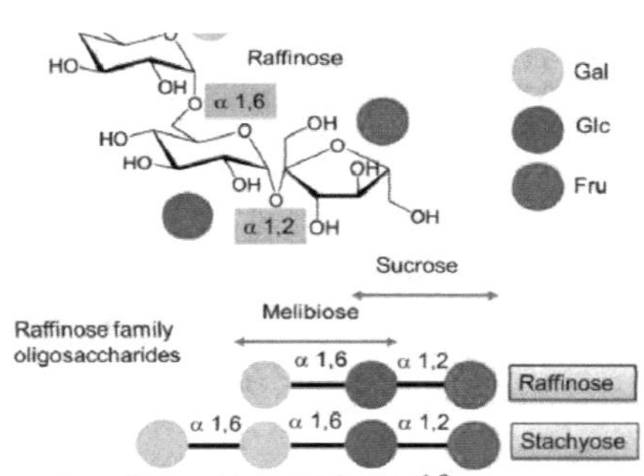

Figure 2. Structure of RFOs [10].

D-(+)-raffinose is a pentahydrate crystalline oligosaccharide. It consists of three monosaccharide units, namely: α-D-galactose, α-D-glucose, and β-D-fructose with 1→6 and 1→2 glycosidic bonds

(Figure 1). The full name of raffinose is O-α-D-galactopyranosyl-[1-6]-α-D-glucopyranosyl-[1-2]-β-D-fructofuranoside. The five water molecules in raffinose hydrate are linked in a chain of sugar molecules gula [11]. Raffinose contains a disaccharide, which is either melibiose or sucrose (Figure 2). Raffinose is the smallest compound in the raffinose-family oligosaccharides (RFOs). While members of RFOs are raffinose, stachyose, verbascose, and others. RFOs consist of several α-D-galactose bound to α-1,6 galactoside, α-D-glucose, and β-D-fructose (Figure 2).

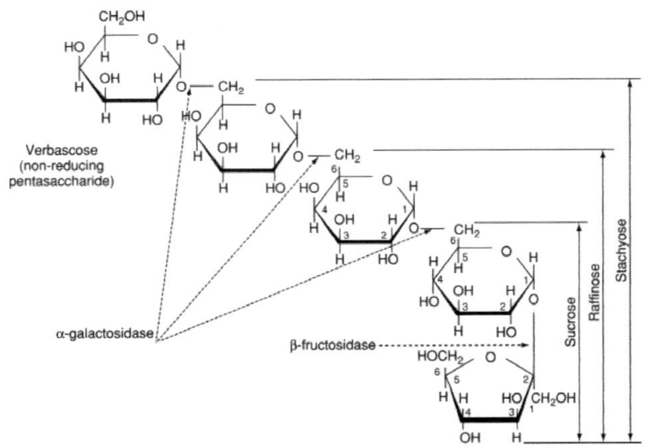

Figure 3. RFOs' degrading enzymes (http://bit.ly/38hY1Br).

3. Raffinose degrading enzyme

The degradation of raffinose (and RFOs in general) requires the enzyme α-1,6-galactosidase (EC 3.2.1.22) to hydrolyze the α-1,6-galactosidic linkage (Figure 3). The human digestive system cannot produce the α-1,6-galactosidase enzyme so that humans can not digest raffinose. However, gut microbiota can metabolize raffinose [14]. The metabolism of raffinose by these microbes produces CO_2 and H_2 gases, causing flatulence in the stomach [6]. Thus, raffinose was formerly referred to as an anti-nutritional compound. However, recent research shows that raffinose has the potential as a prebiotic by specifically stimulating the growth of microbes in the gastrointestinal tract of humans and animals.

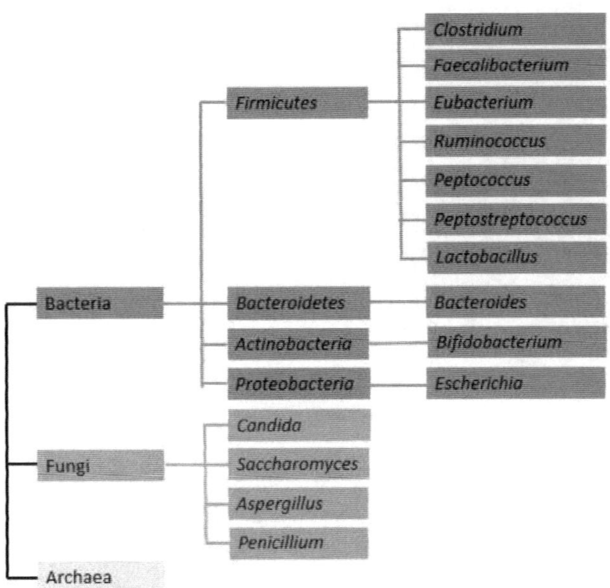

Figure 4. Microorganisms in the human gut [14].

4. Benefits of raffinose as a prebiotic

4.1. Modulate the composition of the gut microbiota

There are 300-500 species of bacteria in the human gut. The most dominant bacterial phyla are Bacteroidetes and Firmicutes, constituting>90% of the microbiota population. Bacteroidetes include *Bacteroides* and *Prevotella*, while Firmicutes contain *Clostridium*, *Eubacterium*, and *Ruminococcus*. The phyla Actinobacteria, Proteobacteria, Fusobacteria, Spirochaetes, Verrucomicrobia, and Lentisphaerae are also present in the human gut, although in lower composition (Figure 4). Every human being has a different composition of gut bacteria [14].

An in vitro study of raffinose fermentation using human feces from 2 healthy adult donors was carried out. Fermentation was carried out for 48 hours under anaerobic conditions. It showed changes in the composition of probiotics, namely *Lactobacillus* and *Bifidobacterium*. There was an increase of *Lactobacillus* in donor 1 (Figure 5A and 5B) and *Bifidobacterium* in donor 2 (Figure 5C and 5D) [6]. In addition, Proteobacteria pathogenic bacteria decreased during the fermentation process. Proteobacteria are the cause of various gastrointestinal tract diseases, for example, *Escherichia coli* [6].

Besides, ammonia gas produced during the raffinose fermentation decreased significantly compared to the control (Figure 5). It indicated that the population of ammonia-producing bacteria, namely *Escherichia coli* and *Clostridium sp*, decreased after consuming raffinose [6]. Therefore, it could be concluded that raffinose could change the microbiota composition in the human gut.

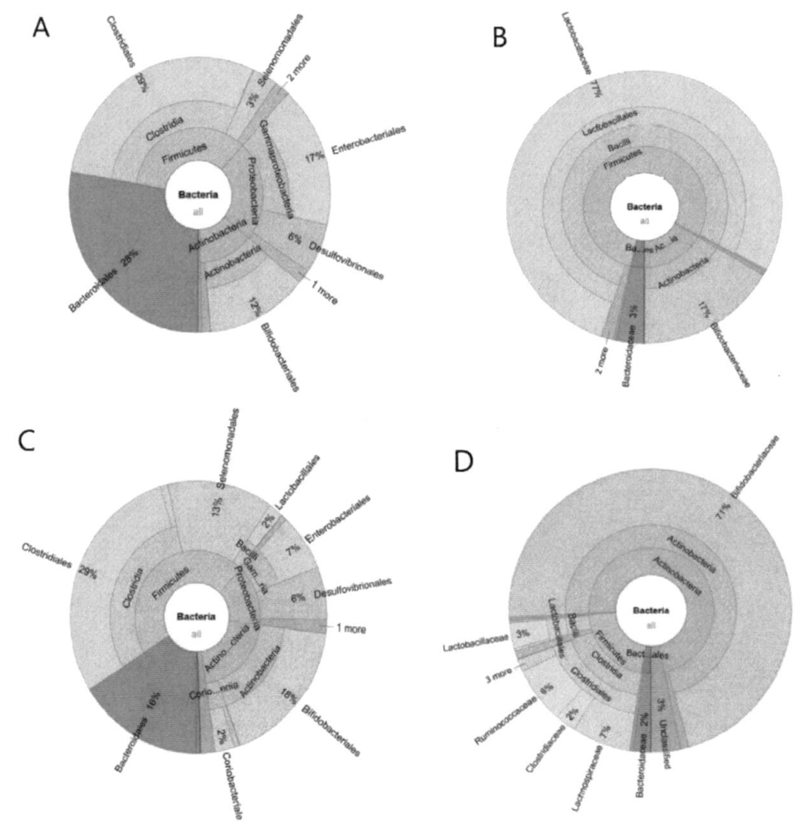

Figure 5. The gut microbiota composition after 48 hours of in vitro fermentation from healthy adult donor 1: A, in control media without prebiotics and B, in media with raffinose. The gut microbia composition after 48 hours of in vitro fermentation from healthy adult donor 2: C, in control media without prebiotics and D, in media with raffinose (D) [6].

The composition of microbes whose growth is influenced by raffinose also has differences in each person. An in vivo study reported the effect of raffinose on the composition of the pool fecal microbiota in normal-weight children, over-weight children, and healthy adults (Figure 6). It showed that raffinose consumption increased the Firmicutes bacteria composition and decreased the Actinobacteria in over-weight children. Enriched growth of Firmicutes in over-weight children was related to the growth of *Streptococcus* and *Solobacterium*. The raffinose fermentation increased the Actinobacteria in normal-weight children. The expanded population of *Bifidobacterium*, *Collinsella,* and *Senegalimassilia* could be responsible for the high Actinobacteria growth. Raffinose also stimulated the growth of Actinobacteria and Proteobacteria in healthy adults [10].

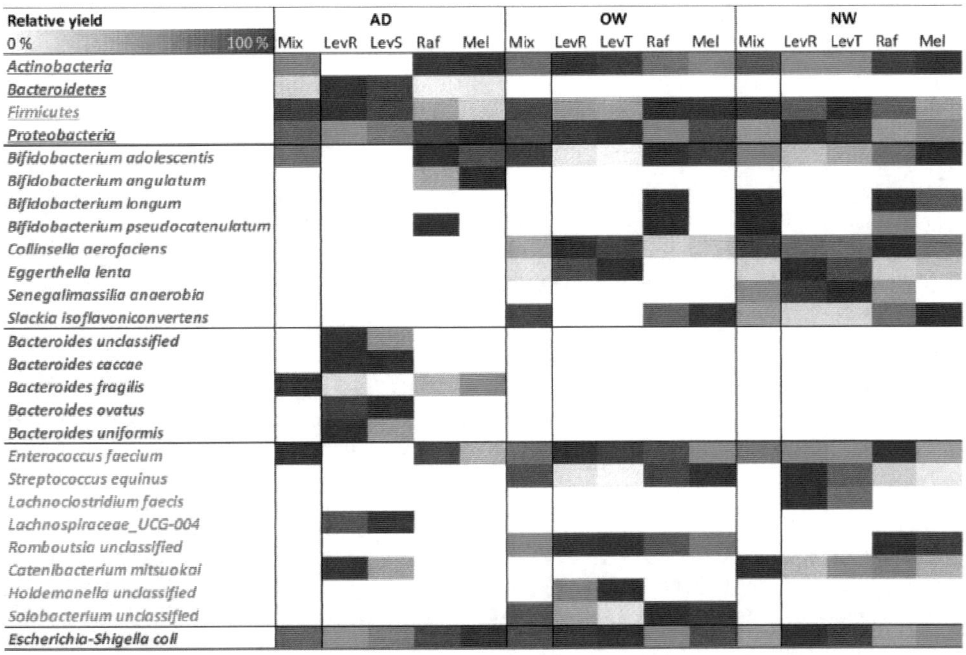

Figure 6. Gut microbiota population in the fermentation of levans (LevR, LevS and LevT), raffinose (Raf), melibiose (Mel), and raffinose-mix (Mix) from the fecal pools of healthy adults (AD), over-weight children (OW), and normal-weight children (NW). The darkest the blue color, the more population of the microbiota [10].

4.2. Increase SCFA production

Prebiotic fermentation by gut microbiota will produce SCFA. Production of lactic acid and other SCFA increased during the in vitro fermentation of raffinose for 48 hours under anaerobic conditions, using normal adult human feces. The maximum production of lactic acid and other SCFA was reached when the fermentation lasted between 36-48 hours. Most of the raffinose was consumed rapidly during the 24-hour fermentation duration. Figure 7 showed that acetic acid was the most abundant SCFA produced in the raffinose fermentation process. Acetic acid was reported to have an essential role in the prevention of colorectal cancer. Lactic acid was the second dominant metabolite produced in raffinose fermentation. However, butyric acid was not produced in the raffinose fermentation process [6].

Besides, normal-weight children and over-weight children who consumed raffinose also showed differences in the composition of SCFA. Metabolites in normal-weight children were primarily acetic acid and butyric acid. In contrast, the dominant metabolites in over-weight children were lactic acid and acetic acid. The difference in SCFA produced by normal-weight children and over-weight children indicated a difference in the microbiota composition in their gastrointestinal tract [10].

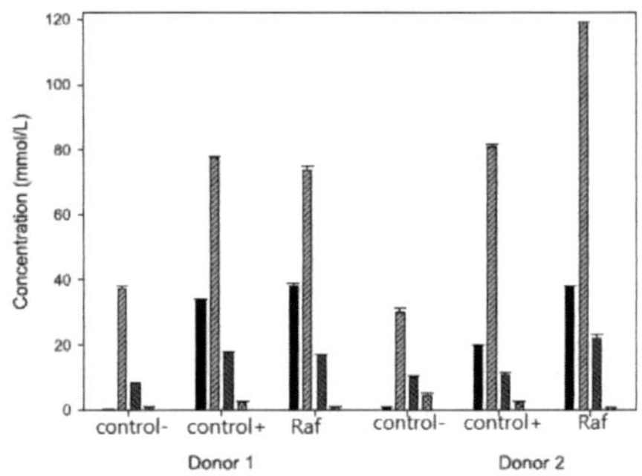

Figure 7. SCFA production during fermentation on raffinose for 48 hours, using feces from two donors. A negative control did not contain any prebiotics, while a positive control contained 10 g/L lactulose. Raf sample contained 10 g/L raffinose. Analyzed SCFA included lactic acid (■), acetic acid (▨), propionic acid (▩), and butyric acid (▦) [6].

Lactic acid is an SCFA precursor that is reported to have bioactive components for human health, namely regulating the critical functions of macrophages and dendritic cells in the immune system and regulating inflammatory activity in epithelial cells. Lactic acid and lactic acid bacteria can affect parts outside the intestine, such as the vagina. It also defends the system against pathogenic microbes in the vagina, such as bacterial vaginosis, aerobic vaginitis bacteria, viruses, fungi, and protozoa [15].

4.3. Reduce constipation

The characteristics of constipation are the frequency of defecation less than 2-3 times a week, small, dry or hard stools, and the process of defecation is difficult. The microbiota composition in the intestines of constipated patients shows a decrease in the population of obligate bacteria and an increase in pathogenic bacteria. Therefore, one way to reduce constipation is to regulate the composition of the gut microbiota.

A total of 103 patients with constipation were treated to consume 5 g of DSG (Deshipu stachyose granules) for 30 days. DSG contained RFOs such as stachyose (55.3%), raffinose (25.8%), verbascose (9.7%), and sucrose (6.9%). Consumption of DSG rich in RFOs could improve bowel function in constipated patients by increasing bowel movement frequency, softer stools, and easier defecation [7].

Cecum weight in RD rats fed a high-fat, high-calorie diet decreased compared to RD rats fed a regular diet. However, raffinose supplementation on a high-calorie, high-fat diet increased cecum weight in RD rats [16].

4.4. Inhibit the putrefactive compounds from protein

Intestinal microbes can be beneficial for humans as well as harmful effects. Gut microbes can convert protein residues in the intestine into putrefactive compounds, such as ammonia, amines, phenols, and indole. These putrefactive compounds are thought to contribute to tumor formation. Wistar rats fed with a soy protein diet experienced an increase in indole in the intestine. Tryptophan-rich soy protein can be converted to indole by the gut microbiota. In vitro study was performed using media inoculated with human feces using soy protein substrate. This study reported the effect of raffinose on the formation of putrefactive compounds. Gut microbiota had grown on media containing raffinose. It resulted in lower concentrations of ammonia, indole, and phenol. It might be caused by the high lactic acid production leading to a pH decrease [9]. Raffinose could prevent the formation of putrefactive compounds resulting from protein metabolism by the gut microbiota.

4.5. Reduce the risk of cardiovascular disease

Cardiovascular disease is a threat to human health. This disease is caused by dietary habits, especially the consumption of saturated fatty acids and animal cholesterol. However, recent data suggest that L-carnitine had a positive correlation with the accumulation of endothelial dysfunction, which was an early phase of cardiovascular disease. L-carnitine was abundant in red meat and contained a trimethylamine structure. The gut microbiota then metabolized trimethylamine to trimethylamine-N-oxide (TMAO), which caused cardiovascular disease. Consumption of L-carnitine in high concentrations also caused metabolic disturbances in cholesterol, sterol, and lipid levels in the blood and reduces nitric oxide (NO) synthesis. NO was a major relaxing factor derived from vascular endothelial cells and essential in maintaining vascular homeostasis and endothelial function [8].

Mice were allowed to consume 3% L-carnitine for three weeks. Then the mice consumed RFOs as much as 200, 400, and 800 mg/kg bw. The results of this study showed that RFOs at concentrations of 400 and 800 mg/kg bw were significantly ($p<0.01$) able to reduce cholesterol, triglyceride, and low-density lipoprotein concentrations, as well as increase high-density lipoproteins. In addition, these concentrations of RFOs were also able to prevent a decrease in NO and an increase in C-reactive protein markers, which was a marker of cardiovascular disease [8].

High blood cholesterol levels are also one of the causes of cardiovascular disease. There are two types of cholesterol sources in the human body: 1) derived from biosynthesis in the liver, and 2) derived from the diet, especially from animal fat. Currently, drugs to lower cholesterol are available, such as statins. However, there are some reports of side effects from using these drugs. Prebiotics and probiotics are safe alternatives to lower blood cholesterol [17].

The mechanism of reducing blood cholesterol levels caused by prebiotics and probiotics is still being debated. This cholesterol-lowering mechanism requires the synergistic performance of prebiotics and probiotics. The proposed mechanism was as follows [17]:

4.5.1. Bile salt deconjugation by BSH. The bile salt deconjugation process is carried out by probiotic bacteria's bile salt hydrolase (BSH) enzyme. The deconjugation process will reduce cholesterol. Generally, conjugated bile salts are circulated through the enterohepatic tract. In contrast, deconjugated bile salts are more soluble in water and can be excreted in the feces. Bile salts that are excreted in the feces must be replaced by new bile salts synthesized from blood cholesterol. Therefore, the more bile salts that come out through the feces, the more cholesterol is taken from the blood, lowering cholesterol levels in the blood.

4.5.2. Coprecipitation of cholesterol with deconjugated bile. Prebiotic fermentation by probiotic bacteria produces SCFA, which results in a decrease in pH. Cholesterol will coprecipitate with deconjugated bile salts at a pH lower than 5.5.

4.5.3. Cholesterol use by cellular membranes. Cholesterol is absorbed by the cellular membrane of probiotic bacteria and used for the growth of these bacteria. *Lactobacillus* carries out this mechanism. This mechanism causes a decrease in cholesterol absorption by the blood in the intestines, thereby reducing cholesterol levels in the blood.

4.5.4. Cholesterol assimilation by probiotics. Probiotic bacteria such as lactic acid bacteria carry out cholesterol assimilation. Cholesterol is bound to the cellular membrane of probiotics and is used in the phospholipid layer of the bacterial membrane. This assimilation process occurs under anaerobic conditions.

5. Conclusion

Prebiotics benefit from maintaining the composition of the gut microbiota by becoming substrates for the fermentation of probiotic bacteria. The fermentation can produce metabolites such as SCFA and offer health benefits to both humans and animals. Prebiotics are functional food components that can prevent cardiovascular disease and other degenerative diseases. The search for new sources of prebiotics continues. Raffinose is one of the novel prebiotics. The potential of raffinose as a prebiotic has been studied recently. Raffinose has many benefits for human health. Furthermore, it is necessary to conduct further research on the use of raffinose in various food products to maintain the benefits of prebiotics and maintain the sensory properties of these food products.

6. Acknowledgment

The author would like to express gratitude to the reviewers for their helpful suggestions to improve this article.

7. References

[1] Gibson G R, Hutkins R, Sanders M E, Prescott S L, Reimer R A, Salminen S J, Scott K, Stanton C, Swanson K S, Cani P D, Verbeke K and Reid G 2017 Expert consensus document: The International Scientific Association for Probiotics and Prebiotics (ISAPP) consensus statement on the definition and scope of prebiotics *Nat. Rev. Gastroenterol. Hepatol.* **14** 491–502

[2] Zartl B, Silberbauer K, Loeppert R, Viernstein H, Praznik W and Mueller M 2018 Fermentation of non-digestible raffinose family oligosaccharides and galactomannans by probiotics *Food Funct.* **9** 1638–46

[3] Vandeputte D, Falony G, Vieira-Silva S, Wang J, Sailer M, Theis S, Verbeke K and Raes J 2017 Prebiotic inulin-type fructans induce specific changes in the human gut microbiota *Gut* **66** 1968–74

[4] Lan J, Wang K, Chen G, Cao G and Yang C 2020 Effects of inulin and isomalto-oligosaccharide on diphenoxylate-induced constipation, gastrointestinal motility-related hormones, short-chain fatty acids, and the intestinal flora in rats *Food Funct.* **11** 9216–25

[5] Celebioglu H U, Ejby M, Majumder A, Købler C, Goh Y J, Thorsen K, Schmidt B, O'Flaherty S, Abou Hachem M, Lahtinen S J, Jacobsen S, Klaenhammer T R, Brix S, Mølhave K and Svensson B 2016 Differential proteome and cellular adhesion analyses of the probiotic bacterium Lactobacillus acidophilus NCFM grown on raffinose - an emerging prebiotic *Proteomics* **16** 1361–75

[6] Amorim C, Silvério S C, Cardoso B B, Alves J I, Pereira M A and Rodrigues L R 2020 In vitro fermentation of raffinose to unravel its potential as prebiotic ingredient *LWT - Food Sci. Technol.* **126** 109322

[7] Li T, Lu X and Yang X 2017 Evaluation of clinical safety and beneficial effects of stachyose-enriched α-galacto-oligosaccharides on gut microbiota and bowel function in humans *Food Funct.* **8** 262–9

[8] Li W, Zhang R, Guo J, Shao H and Yang X 2016 Protective effect of R. glutinosa oligosaccharides against high l-carnitine diet-induced endothelial dysfunction and hepatic injury in mice *Int. J. Biol. Macromol.* **85** 285–93

[9] Nakata T, Kyoui D, Takahashi H, Kimura B and Kuda T 2017 Inhibitory effects of soybean oligosaccharides and water-soluble soybean fibre on formation of putrefactive compounds from soy protein by gut microbiota *Int. J. Biol. Macromol.* **97** 173–80

[10] Adamberg K, Adamberg S, Ernits K, Larionova A, Voor T, Jaagura M, Visnapuu T and Alamäe T 2018 Composition and metabolism of fecal microbiota from normal and overweight children are differentially affected by melibiose, raffinose and raffinose-derived fructans *Anaerobe* **52** 100–10

[11] Wiercigroch E, Szafraniec E, Czamara K, Pacia M Z, Majzner K, Kochan K, Kaczor A, Baranska M and Malek K 2017 Raman and infrared spectroscopy of carbohydrates: A review *Spectrochim. Acta - Part A Mol. Biomol. Spectrosc.* **185** 317–35

[12] Wang Y, Han W, Song L and Zhao X 2019 Compositional analysis and structural characterization of raffinose family oligosaccharides from Eupatorium *J. Food Compos. Anal.* **84** 103298

[13] Karnpanit W, Coorey R, Clements J, Nasar-Abbas S M, Khan M K and Jayasena V 2016 Effect of cultivar, cultivation year and dehulling on raffinose family oligosaccharides in Australian sweet lupin (Lupinus angustifolius L.) *Int. J. Food Sci. Technol.* **51** 1386–92

[14] Huang H, Krishnan H B, Pham Q, Yu L L and Wang T T Y 2016 Soy and Gut Microbiota: Interaction and Implication for Human Health *J. Agric. Food Chem.* **64** 8695–709

[15] Amabebe E and Anumba D O C 2018 The vaginal microenvironment: The physiologic role of Lactobacilli *Front. Med.* **5** 1–11

[16] Hira T, Yanagihara K, Koga T, Takahashi K, Nagura T, Uchino H and Hara H 2017 Impact of difructose anhydride III, raffinose, and fructooligosaccharides on energy intake, gut hormones, and cecal fermentation in rats fed a high-fat and high-sucrose diet *Biosci. Biotechnol. Biochem.* **81** 2186–94

[17] Shehata M G, El-Sahn M A, El Sohaimy S A and Youssef M M 2019 Role and Mechanisms Lowering Cholesterol by Dietary of Probiotics and Prebiotics: A Review *J. Appl. Sci.* **19** 737–46

IOP Conf. Series: Earth and Environmental Science **980** (2022) 012034 doi:10.1088/1755-1315/980/1/012034

Packaging development of dehydrated strawberry using quality function deployment for e-commerce

S Mentari[1], M Ainuri[1], M A F Falah[1*]

[1]Agroindustrial Technology Department, Universitas Gadjah Mada, Jl Flora No 1 Bulaksumur Yogyakarta 55281, Indonesia

*Email's corresponding author : affan_tip@ugm.ac.id

Abstract. Dehydrated strawberry is product development from strawberry fresh fruit to improve, increase the shelf-life and value of the product. During the Covid-19 pandemic, shopping lifestyle has changed gradually into online buying through e-commerce that was growth significantly, it must be followed by producer to adapt and adopt the situation especially for purchasing decisions, while customers tend to choose products based on the packaging appearance. This study aimed to determine attributes of the packaging and technical needs for producer selling preparation of dehydrated strawberry in e-commerce based on customer needs. The method in this study that has proven effective in fulfilling customer needs is Quality Function Deployment (QFD) in product planning phase. QFD analysis begins with filtering Voice of Customer (VoC) using open questionnaire then at the stage of customer importance level and product performance using closed questionnaire, then compiled into attributes of customer needs. Technical requirements were brainstormed using discussion with local sellers on e-commerce. Two attributes with the highest priority to be realized is product information on the packaging is easy to read and understand and product-related information fully listed on the outer product packaging and the highest priority for the technical needs is addition of secondary packaging.

1. Introduction

Strawberry production in Indonesia encountered with many post-harvest problems such as distribution chain length that makes the strawberries are damaged during the transportation process and the lack of processing to turn fresh strawberry fruit into a new product with higher economic value [1]. In postharvest handling of fruits, the application of osmotic dehydration (OD) is currently widely applied to extend the shelf life by which new products with higher economic value can be produced. Processing of fruits using the OD technique is then continued with other preservation techniques such as hot air-drying using tray dryer [2]. From the processing of strawberries with this technique, a new product is produced in the form of dehydrated strawberry which can be stored for up to 6 months at room temperature of 26°C [3]. Dehydrated strawberry can be consumed in various ways, either mixed into other foods such as cereals, granola, salads, or cakes, dairy products, sweets or made into snacks that are rich in antioxidants [4].

The COVID-19 pandemic that is currently sweeping the world has caused significant changes in human lifestyles. Along with the policy implementation to stay at home as a step to prevent the spread of the virus, as many as 31% of respondents experienced an increase in online shopping activities [5]. In product marketing through e-commerce, producer must be able to adapt and adopt this situation especially for purchasing decisions, customers tend to choose the products based on the packaging appearance [6].

In terms of packaging handling, current product packaging that has the best type of packaging that can maintain the product quality is 75micron thick nylon monolayer plastic with vacuum handling. The packaging is able to maintaining the quality of dehydrated strawberry in terms of quality parameters including physical, chemical, and microbiological [7]. Of course, the packaging needs to be developed for suitable to be marketed through e-commerce, that the packaging should not be only meets the requirements of the packaging function which can protect the product (functional), but also meets the actual customer needs of product packaging, especially in terms of visuals, so that customers are more interested in buying the products through e-commerce.

The right method that has proven to be effective to solve these problems is to use the Quality Function Deployment (QFD) method that uses the basis of customer needs, finds innovative responses to these customer needs, then improves the quality of the products developed in order to meet customer expectations, which are in accordance with the needs of the research. This study uses the QFD method at the product planning process stage using House of Quality (HoQ) matrix to filter Voice of Customer (VoC) into Attributes of Customer Needs (ACN) and translate it into technical needs for producer to develop dehydrated strawberry packaging for e-commerce.

The purpose of this study was to determine attributes of packaging and technical needs for selling preparation of dehydrated strawberry products in the e-commerce according to customer needs and product quality characterization.

2. Method

Development of dehydrated strawberry packaging for e-commerce in this study uses Quality Function Deployment (QFD) method because in accordance with the purpose of the study, the method uses the basis of customer needs, find innovative responses to these customer needs, then improves the quality of the product in order to meet customer requirement. Specifically, this study uses product planning stage of QFD, using HoQ matrix. The complete structure of HoQ can be seen in Figure 1.

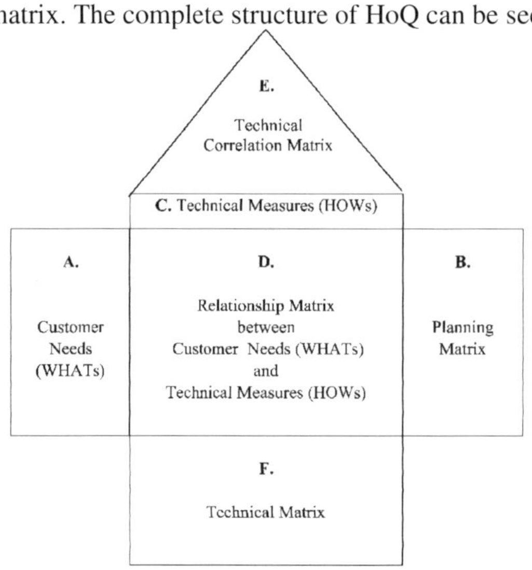

Figure 1. Brief structure of HoQ [8].

2.1. Types and sources of data

The data used in this study consisted of two data types, primary and secondary data. Primary data obtained directly from respondents i.e. people who have sold or bought or consumed packaged dehydrated strawberry product. The primary data needed in this study include respondent information, customer importance and satisfaction level, as well as customer needs and specifications of the dehydrated strawberry packaging in e-commerce obtained through the results of observations and surveys using a questionnaire. While the secondary data needed in this study are data that related to the research object, validity and reliability testing, product quality standards, packaging attributes, packaging, and product development methods obtained through electronic sources.

2.2. Data collection technique

The data of this study were obtained through field and literature studies. For field studies, data were collected through observation, discussion, and survey using questionnaire filled by respondents. At the stage of identifying customer needs, an open questionnaire was distributed to 10 respondents to filter VoC so that respondents could provide in-depth answers according to their experience and knowledge. Meanwhile, in the validity and reliability test, a closed questionnaire was used and an initial distribution of the main questionnaire was carried out to 30 respondents. Then in the advanced stage, analysis of the customer importance level and product performance was carried out using the main questionnaire through further distribution to 100 respondents. While literature studies were obtained to collect secondary data, such as journal, book, or electronic source.

2.3. Data analysis

Data analysis in this study uses HoQ matrix which can be seen in Figure 1. There are six main steps in data analysis using QFD product planning stage that are:

1) *Identification of customer needs.* Identification is done by observing and distributing an open questionnaire in order to obtain WHATs from the matrix. The use of an open questionnaire aims to obtain true unlimited information from respondents about the object and obtain the actual VoC. At this stage, after obtaining VoC which is then compiled into ACN, data analysis is carried out by validity and reliability test of the attributes that have been obtained before proceeding to the next stage. The importance of each attribute is measured based on the results of the ACN importance level questionnaire after the test shows the results of the questionnaire are valid and reliable.
2) *Planning matrix.* Each ACN that has been obtained is then measured the level of product performance through benchmarking stages between the product packaging developed and competing import product packaging (Doi Kham Brand, Thailand). Then, the calculation of improvement ratio and determination of future goals also sales point is carried out through discussion with the seller. To determine the priority level of each ACN, the scale of customer importance and its normalization (%) is calculated for each ACN. The greater the value generated, the higher the priority of the ACN to be realized.
3) *Framing technical needs.* The ACN that has been obtained is then translated into technical needs (HOWs) to make it easier for sellers to realize ACN in developing product packaging so it can meet customer needs. Identification of technical needs is carried out by researchers taking into account the results of literature studies and discussions with sellers related to repair direction of each technical needs.
4) *Relationship matrix.* Determination of the relationship between ACN and technical needs is carried out using the relationship matrix symbol (none, weak, moderate, strong). Every WHATs must be answered by HOWs, as of to increasing satisfaction with ACN, the parameters in technical needs must be increased. Which one is more likely to be improved can be seen based on the relationship.
5) *Technical needs correlation matrix.* Analysis of the relationship between technical needs is carried out to determine the relationship between each of the technical needs that can be either positive or negative. Technical requirements with a positive relationship indicate that these technical needs support each other, while a negative relationship indicates that these technical needs inhibit each other in the product development process.
6) *Technical needs matrix.* In this matrix, determination of the level of difficulty of technical needs is carried out by discussing with the seller and calculating the priority and its normalization (%) to

determine the priority level of each technical need. The greater the value generated, the higher the priority of the technical needs to be realized.

3. Result and Discussion

3.1. Dehydrated strawberry and online product packaging

Dehydrated strawberry is one of the products from the preservation of fresh strawberries in order to obtain a new product that is more durable. The product is obtained from the results of pre-treatment using an osmotic dehydration technique, then followed by convective thermal drying using a tray dryer [7].

The standard of dried fruit packaging is the product packaged in a tightly closed container, not affected or affected by the contents, safe during storage and transportation [9]. That packaging standard has not been detailed related to such as materials, material properties, material characteristics, labels, packaging forms, and others. Specifically, dried fruit requires packaging with materials that have a good barrier to moisture, O_2, SO_2, and other volatiles. This is because the surface of dried fruit is very sensitive to moisture and oxidation [10]. Meanwhile in general, related to packaging labels, processed food packaging labels must contain information on at least 8 main aspects of the product [11].

For online marketing, the product packaging needs to be given additional packaging so that the product remains safe and shocks that occur to the product can be minimized during the distribution process [12]. Packaging of products that marketed through e-commerce must be able to protect the product in terms of literal function, but packaging has marketing potential that is quite influential on product sales. Compared to offline sellers, online sellers have less contact power in providing satisfaction to consumers, as of online sellers must be able to provide a memorable experience for consumers and outperform competitors in making online product purchase decisions, especially through packaging [13].

3.2. Data collection records

In this study, two types of questionnaires were used at several stages in the QFD analysis. At the stage of identifying consumer needs, an open questionnaire was distributed to 10 respondents and homogeneous answers were obtained regarding customer needs for dehydrated strawberry packaging for e-commerce, which then became input for the WHATs section.

Then at the initial stage of distributing the main questionnaire, a closed questionnaire was used which was filled out by 30 respondents to then be tested for validity and reliability. Of the 10 attributes, all of them were declared valid and reliable, so that they could proceed to the next stage, the continued distribution of the main questionnaire to 100 respondents. In the follow-up distribution, a questionnaire was also distributed which functions to analyze customer importance level and product performance which was also filled out by the 100 respondents. The results of the analysis become input to the planning matrix.

3.3. Customer needs (WHATs)

Customer needs were identified based on the results of literature studies and survey observations using an open questionnaire to filter VoC as WHATs. Product packaging attributes are divided into two main attributes, namely verbal and visual attributes. Verbal attributes related to product information and to address the cognitive side of decision making. While visual attributes can be divided into several sub-attributes includes: (a) graphics, (b) color, (c) shape, (d) size, and (e) packaging materials [14].

ACN that has been compiled from the development of verbal and visual attributes obtained as many as 10 attributes includes: (a) packaging gives the impression of fresh and quality product, (b) simple packaging design, (c) the color of the packaging is not too flashy, (d) easy to use packaging, (e) packaging has unique shape, (f) packaging does not take up space when stored, (g) packaging materials not easily damaged, (h) packaging can protect the product, (i) product-related information fully listed on the outer packaging of the product, and (j) product information on the packaging easy to read and understand.

Based on the results of customer importance level measurement of those 10 attributes, it is found that attribute (d) has the highest level of importance and attribute (i) has the lowest level of importance in the development process of dehydrated strawberry packaging for e-commerce.

3.4. Planning matrix and WHATs priorities

The results of product performance level measurement indicate that all of the current packaging attributes have lower performance values than import , except for attribute (g). This shows that customers are satisfied with the current packaging materials which are not easily damaged compared to the import packaging. However, based on the overall results, current product packaging must be developed in order to be more competitive in the market.

The results of the improvement ratio and product performance level calculation as well as the value of future goals and sales points determination are then compiled in planning matrix. Meanwhile, in arranging the priority order of each WHATs attribute, it was found that the attribute with the largest normalized value of the customer importance level scale was attribute (j) and the lowest was attribute (c). Therefore, it can be concluded that the attribute of WHATs with the highest priority level to meet consumer needs in the packaging development process is attribute (j).

3.5. Technical needs (HOWs)

The arrangement of technical needs for the translation of 10 WHATs attributes resulted in 10 HOWs includes: (a) packaging visual design represents the product; (b) addition of window pack features; (c) application of simple, clean, and minimalist design themes; (d) addition of opening and closing features on the packaging; (e) adjustment of packaging form and product net weight; (f) addition of secondary packaging; (g) airtight and waterproof primary packaging; (h) packaging labels based on BPOM RI Regulations No. 31 2018; (i) font clearly legible; and (j) use of frequently used language.

3.6. Relationship matrix (WHATs and HOWs)

Analysis of the relationship between HOWs and WHATs in relationship matrix is carried out to find out how strong the relationship between technical needs and customer needs are. In addition, the analysis is carried out to assist in determining the priority order of technical needs to be made in the product packaging development process.

The results of the analysis show that there are 26 strong relationships, 12 moderate relationships, and 5 weak relationships between technical needs and customer needs. While the technical needs with the most relationship with customer needs is the addition of secondary packaging.

3.7. Correlation matrix (HOWs)

Analysis of the relationship between HOWs is carried out to determine the relationship between each of the technical needs that have been designed. The relationship between these technical needs can be either positive or negative. Technical requirements with a positive relationship indicate that these technical needs support each other, while a negative relationship indicates that these technical needs inhibit each other in the product development process carried out.

The results of the analysis show that there are 8 strong positive correlations and 7 moderate positive correlations. This shows that of all the technical requirements designed, there is no technical need that hinders the realization of other technical needs.

3.8. Technical needs matrix and HOWs priorities

The technical requirement matrix consists of determining the level of difficulty and calculating the priority of technical needs. Based on the results of discussions with the seller, the technical need with the highest level of difficulty is the addition of secondary packaging (quite easy) compared to other technical needs (easy and very easy).

While in arranging the priority order of HOWs, it is known that the technical needs with the highest percentage of normalization of priority are the technical needs of addition secondary packaging and the lowest is addition of opening & closing features on the packaging. Therefore, it can be concluded that HOWs with the highest priority to be carried out in the packaging development process is the technical need of addition of secondary packaging.

3.9. House of Quality (HoQ) matrix

All the results of six stages in product planning QFD analysis process that have been carried out are then inputted into the HoQ matrix according to the placement matrix. The HoQ matrix of dehydrated strawberry packaging development for e-commerce can be seen in Figure 2.

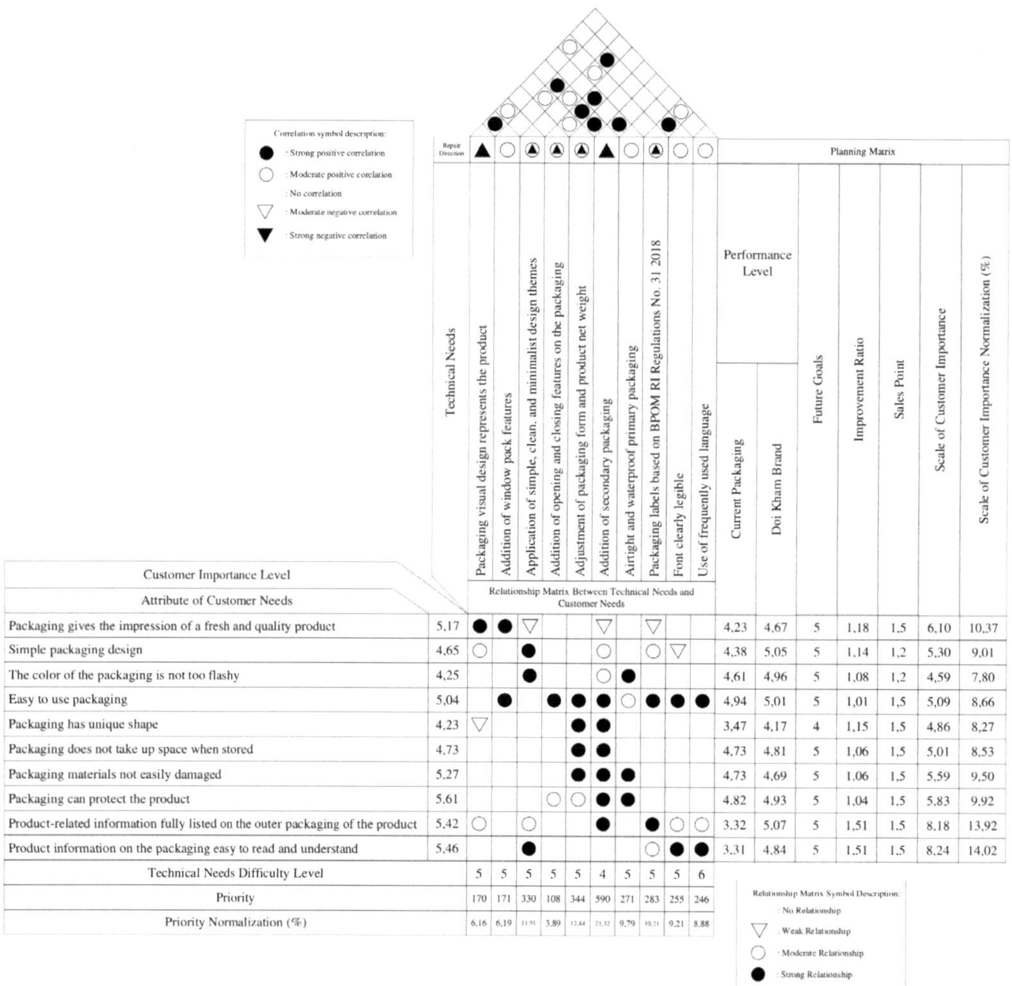

Figure 2. Dehydrated Strawberry Packaging for E-commerce Development House of Quality Matrix.

3.10. Result Analysis

Based on the HOQ matrix, it can be seen that ACN with the highest Importance Customer Level is attribute packaging can protect the product. This shows that customers want to receive the product intact with packaging characteristics that can protect the product. This attribute then translated into technical need addition of secondary packaging, which from the calculation results are technical needs with the highest priority level to be immediately realized in the packaging development process. This meets the standards of online sales product packaging [12], as of the product can be protected and safe until it reaches the customer. In addition, with the addition of secondary packaging, the development of visual and verbal attributes of the packaging can be focused on secondary packaging, as of product packaging can attract customers because of attractive packaging designs as product marketing tool [13].

However, after further calculations that have been carried out by considering other aspects of the calculation, ACN that has the highest priority to be realized immediately is attribute product information on the packaging easy to read and understand. By including complete product information on the packaging according to the packaging label requirements [11], product information must be easy to read and understand because product information has an important role in making product purchase decisions in e-commerce [15].

3.11. Packaging development concept

Based on the results of the QFD analysis related to the development of dehydrated strawberry packaging for e-commerce, from the technical requirements that have been prepared to meet consumer needs, a packaging concept can be made according to the results of the analysis. The visual design of the packaging can be seen in Figure 3 and the packaging used can be seen in Figure 4.

(A) (B)

Figure 3. Visual design of the outer packaging

(A): Front of packaging and (B): Back of packaging

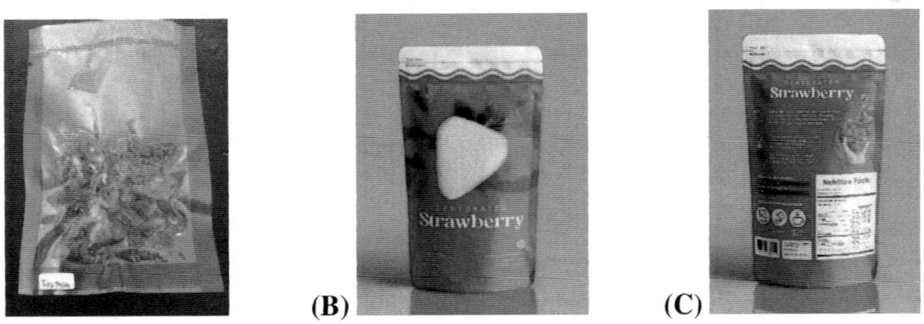

(A) (B) (C)

Figure 4. Dehydrated strawberry for e-commerce packaging concept

(A): Primary packaging [7], (B): Front of secondary packaging, and (C): Back of secondary packaging

4. Conclusions

1. Attributes of dehydrated strawberry packaging for e-commerce according to customer needs in order by priority are product information on the packaging is easy to read and understand, product-related information fully listed on the outer product packaging, packaging gives the impression of a fresh and quality product, packaging can protect the product, packaging is made

from materials that are not easily damaged, the packaging design is simple, the packaging is easy to use, the packaging does not take up too much space when stored, the packaging has a unique shape, and the color of the packaging is not too flashy.

2. The technical needs that must be done in developing dehydrated strawberry packaging for e-commerce in order of priority are the addition of secondary packaging; adjustment of packaging form and product net weight; application of simple, clean, and minimalist design themes; packaging labels based on BPOM RI Regulations No. 31 2018; airtight and waterproof primary packaging; the font is clearly legible; use of frequently used language; addition of window pack features; visual packaging design represents the product; addition of opening and closing features on the packaging.

Acknowledgments
Authors offers gratitude thanks to Universitas Gadjah Mada for their financial support with contract number: 3143/UN1.P.III/DIT-LIT/PT/2021, and also for Faculty of Agricultural Technology Universitas Gadjah Mada for the support of research during Pandemic Covid-19.

Reference

[1] Herawati E, Etikawati N, Setyaningsih R and Sugiyarto 2019 *Journal SEMAR.* **8** 6
[2] Guin é, R P F 2018 The drying of foods and its effect on the physical-chemical, sensorial, and nutritional properties *International Journal of Food Engineering*
[3] Andress E L and Harrison J A 2014 Drying: Packaging and storing dried foods *National Center for Home Food Preservation* (https://nchfp.uga.edu/how/dry/pack_store)
[4] Kowalska J, Kowalska H, Marzec A, Brzezin´ski T, Samborska K and Lenart, A. 2018 Dried Strawberries as a High Nutritional Value Fruit Snack *J. Food Science Biotechnology* 799
[5] Central Statistics Agency 2020 *Results of the social demographic survey on the impact of Covid-19* (Indonesia: Central Statistics Agency)
[6] Raheem A, Ahmad N, Vishnu P and Imamuddin K 2014 Role of Packaging and Labelling on Pakistani Consumers Purchase Decision *European Journal of Scientific Research* 464
[7] Putri R M 2021 Physical quality changes of dehydrated strawberry affected by different packaging in a tropical environment *IOP Conf. Series: Earth and Enviromental Science* **759**
[8] Chan L and Wu M 2002 *Quality Engineering* **15** 23
[9] National Standardization Agency 1995 *Dried Fruit: SNI 01-3710-1995* (Indonesia: National Standardization Agency)
[10] Gonzalo M, Berna A, and Antonio 2019 Dried-Fruit Storage: An Analysis of Package Headspace Atmosphere Changes *Journal Foods*
[11] Food and Drug Supervisory Agency 2018 *Regulation of the Food and Drug Supervisory Agency about Processed Food Labels* (Indonesia: Food and Drug Supervisory Agency)
[12] NDE Courier and Cargo 2020 How to Wrap Dry Food for Shipping *NDE Courier and Cargo* (https://ekspedisimurah.com/cara-membungkus-makanan-kering-untuk-dikirim/)
[13] Jet Commerce 2020 The Importance of Product Packaging & Memorable Shopping Experiences for Online Customers *Jet Commerce* (https://jetcommerce.co.id/update/pentingnya-product-packaging-pengalaman-berbelanja-yang-berkesan-bagi-online-customer/)
[14] Hassan S H, Leng L W and Peng W W 2012 *Journal of Agribusiness Marketing* **5** 14
[15] Raheem A R, Ahmad N, Vishnu P, and Imamuddin K 2014 Role of Packaging and Labelling on Pakistani Consumers Purchase Decision *European Journal of Scientific Research* 464

Consumer acceptance of quality characterization of dehydrated strawberry product

S S Hasna [1], A D Nugrahini [1], M A F Falah [1*]

[1] Department of Agroindustrial Technology, Faculty of Agricultural Technology, Universitas Gadjah Mada, Jl. Flora No. 1 Bulaksumur Yogyakarta Indonesia 55281

*email: affan_tip@ugm.ac.id

Abstract. Strawberry is a horticultural fruit that has high nutritious content but a short shelf-life. Strawberries can be dried to extend their shelf-life. However, the drying process can cause quality change that can affect consumer acceptance. This research aimed to identify the quality of dehydrated strawberries based on consumer acceptance. Samples were produced using osmosis dehydration pre-treatment and dried by tray dryer. Imported and local products were used as a comparison. An untrained panelist was used to make a sensory analysis on five attributes on color, aroma, flavor, texture, and overall. Then, sample with highest acceptance value for each attribute were tested for the quality that consists of color, total sugar content, texture, and water content test. Based on the sensory analysis, sample with highest acceptance value for each attribute is as follows: local sample for color attribute, imported sample for aroma attribute, self- produced a sample for flavor, texture, and overall attributes. Based on the quality test, the dehydrated strawberry had discoloration due to maillard reaction. The sugar content, texture, and water content of the sample is in accordance with previous research. Sample with the highest acceptable value for overall is self-produced sample and can compete with local and imported sample.

1. Introduction

Strawberry (Fragaria x ananassa) is one of the horticultural fruits from Europe and is in great demand by Indonesian consumers because of its attractive appearance and nutritional content that is beneficial to health. Based on data from Indonesia Statistics Bureau, strawberry consumption in Indonesia in 2015 reached 63.74 ton and continued to increase until in 2018 strawberry consumption in Indonesia reached 81.28 ton [1]. Strawberry are good for heart health, reduce the risk of cancer, arthritis, and improve memory function [2]. In Indonesia, strawberry are cultivated commercially and have promising business opportunities. This fruit can be consumed directly or used as a food additive and other products such as yogurt, jam, and fruit salad. Strawberry have a short shelf life of 3-4 days after harvesting, this happens because their water content reaches 90% [3]. Therefore, post-harvest handling is needed to extend the shelf life of strawberry. Some post-harvest handling that can be done is processing into strawberry jam, frozen strawberry, dehydrated strawberry, and cosmetic [4].

Drying is an alternative process that can extend strawberry's shelf life by reducing or eliminating the water content contained in the fruit. In addition to extending the shelf life of strawberry, drying can also cause shrinkage of the dried material so that it can reduce the cost of storage and transportation of materials so that dehydrated strawberry can be sold oversea because they aren't easily damaged during storage and transportation [5]. Dehydrated strawberry have many benefits, both consumed directly and as additional ingredients. Similar to fresh strawberry, dehydrated strawberry are useful as antioxidants, prevent radicals, and contain soluble fiber [6]. Other than that, the drying process can cause the sugar content in the product to increase and it is easier to digest [7]. Dehydrated strawberry can be used as an ingredient in cereal mixes as breakfast, topping on ice cream, cakes, yogurt, and as a dietery snack.

In addition to having several advantages, the drying process can causes several adverse effects as changes in the shape, physical and chemical properties of the material, as well as a decrease in the quality of the dried material [8]. The drying process on strawberry can cause changes in the quality of the fruit such as oxidation which can changes the color of the product to become darker or browned. The drying process can also cause changes in the flavor, aroma, and texture of dehydrated strawberry. This quality changes can affect the consumer's acceptance to the dehydrated strawberry product. The decrease in product quality during the drying process can reduce the added value and acceptance of dried fruit products [9]. Therefore, sensory analysis was used to determined consumer acceptance of dehydrated strawberry and quality tests were carried out including color tests, aroma tests, texture tests, total sugar content tests, and water content tests to determined the quality characteristics of dehydrated strawberry. The aim of this research was to identify the quality of dehydrated strawberry product based on consumer acceptance using sensory analysis and quality analysis of dehydrated strawberry product.

2. Materials and Methods

2.1. Sample Preparation

The dehydrated strawberry samples used in this study consisted of 3 samples, there are self-produced sample, local sample, and imported sample that used as a comparison. In production of dehydrated strawberry samples, california variety strawberry were used with an average weight of 6.29 grams and a weight range of 4.18 grams to 12.69 grams. Strawberry are obtained from a commercial strawberry garden located in Magelang, Central Java and then brought and processed at the Faculty of Agricultural Technology. The dehydrated strawberry samples were made in two stages, namely the osmotic dehydration pre-treatment stage and the tray dryer drying stage. The process of samples production was based on previous research about production factor of dehydrated strawberry that produced by osmotic dehydration and dried by convection dryer. According to this research, the best combination to produce dehydrated strawberry is by osmotic dehydration with strawberry concentrate for 60 minutes and dried by convective dryer with 70°C [9]. Other than that, another previous research stated combination of 60 minutes osmotic dehydration and 9 hours convection drying is good combination to produce dehydrated strawberry [11]

The osmotic dehydration stage is carried out by immersing the strawberry in a 50°Brix strawberry concentrate solution as osmotic agent for 90 minutes. This solution was obtained by blending 60 grams of strawberry with 230 grams of sugar and 200 ml of aquadest using a blender (HR2116/01, Philips, The Netherlands) for 5 minutes. Before being used for osmotic dehydration, the osmotic agent was filtered for 2 repetitions. Then the osmotic agent is used to immerse the strawberry with the Magnetic Heated Strirer (IKA C-MAG HS7, IKA, China). This process lasts for 90 minutes at a temperature of 35°C. After being processed by osmosis dehydration pre-treatment, strawberry will experience a change in color and texture to become softer. Then the strawberry are arranged on a tray and dried using a tray dyer (MKS-FDH6, Maksindo, Indonesia). This tool uses the principle of convection drying, which is direct contact between hot air and the material. This dryer is designed with a parallel flow tray and works by circulating the hot air parallel to the surface of the tray inside [12]. Drying was carried out for 12 hours at a temperature of 70°C. After drying with a tray dryer, the dehydrated strawberry were stored in a desiccator for 36 – 48 hours to reduce the water content that was still contained in the strawberry.

The samples used in this study were given a random code for each sample. The first sample is sample 172 which is a local dehydrated strawberry that obtained from the Aremafood (Malang, Indonesia). This sample has sheets form strawberry. The second sample is 758 which is a self-produced sample of strawberry in whole form. The last sample is 637 which is an imported dehydrated strawberry that obtained from Enjoy All The Bear (China) and has a whole form strawberry.

(a) (b) (c)

Figure 1. Samples of dehydrated strawberry that consist (a) local sample (Aremafood, Malang, Indonesia); (b) self-produced sample produced by osmotic dehydration and convection drying' and (c) imported sample (Enjoy All The Bear, China).

2.2. Sensory Analysis

Data was collected by a sensory analysis to determine consumer acceptance of the dehydrated strawberry samples. The type of test carried out is the affective test with the hedonic test. A hedonic scale of 1-7 is used in assessing the level of consumer preference for the sample for each attribute. Sensory analysis was conducted at the Faculty of Agricultural Technology, Universitas Gadjah Mada. The panelists used were untrained panelists as many as 45 students of the Faculty of Agricultural Technology who had met the selection requirements according to SNI 01-2346-2006. In this analysis, panelists were asked to rate each sample based on their preferences for five attributes: color, aroma, flavor, texture, and overall.

2.3. Statistical Test

The data of sensory analysis were analyzed statistically using one way analysis of variance (ANOVA) or Kruskal-Wallis test to determine whether there was a significant difference in the acceptance of each sample. But first, a normality test was carried out on the data for each attribute. Tests are carried out to determine whether the data distribution is normal or not. In this study, the normality test using the Shapiro-Wilk method. The Shapiro-Wilk method is used to determine the data distribution with small data sizes. The Shapiro-Wilk test is recommended to be used for testing when the data size is less than 50 ($N<50$). If the data is normally distributed, then the test is continued with one way ANOVA but if the data is not normally distributed, then the test was continued with Kruskal-Wallis and the Post Hoc Pairwise Comparison test. Statistical data testing was carried out with SPSS ver. 24.

2.4. Dehydrated Strawberry Quality Analysis

The quality analysis consists of color, aroma, flavor, texture, and water content test on selected sample for each attribute. The quality analysis was carried out on selected samples for each attributes based on acceptance value. Color testing was carried out using the chromameter (NH310, 3nh, China) to obtain L*a*b values, ΔE value, and Hue (h°) value from selected samples. Testing is done by shooting at the edge of the sample because consumers judge the color of the product from the skin color. The L* notation states the brightness parameter (light) which has a value of 0 (black) to 100 (white). The notation a* denotes a chromatic mixture of red and green, the value of +a (positive) from 0 to +100 for red and -a (negative) from 0 to -80 for green. The b* notation denotes mixed chromatic blue-yellow, +b (positive) values from 0 to +70 for yellow and -b (negative) values from 0 to -70 for blue [13]. Value of ΔE shows the value of the difference in the appearance of the sample from the L*a*b value. The higher ΔE velue obtained, the greater difference value of the sample. Meanwhile, hue value (h°) is the angle that shows the saturated color in the color space. Hue value (h°) has a range values from 0° (red), 90° (yellow), to 180° (green). Value of ΔE and hue value (h°) are calculated using formulas:

$$\Delta E = \sqrt{(\Delta a)^2 + (\Delta b)^2} \quad (1)$$

$$h^o = tan^{-1} (b *)/ (a *) \quad (2)$$

Aroma testing is not carried out in a laboratory because of the limitations of the instruments. Aroma testing was carried out by literature related to the content of volatile compounds that cause the aroma in strawberry as well as the concentration and odor description of each compound. Meanwhile, the flavor test was carried out by testing the total sugar content contained in the selected sample. The tests were carried out using the spectrophotometer (GENESYS 30, Thermo Fisher Scientific, USA). The results of the total sugar content are expressed in percent (%). Texture testing was carried out with a fruit hardness tester (FHT200, Extech, Taiwan). The texture of the sample is expressed in the hardness value in Newton (N). Testing is done by inserting a needle at the center of sample and using its peak value. The last test is water content test. This test was carried out using the thermogravimetric method with universal oven (UN55, Memmert, Germany). The value of water content is expressed in percent (%).

3. Results and Discussion

3.1. Sensory Analysis

The panelist used in this study were untrained panelists of 45 students from Faculty of Agricultural Technology, Universitas Gadjah Mada that recently domicilied in Yogyakarta. The panelist are limited due to the covid pandemic that happened recently. The profile of the panelists used in this research is shown in Figure 2.

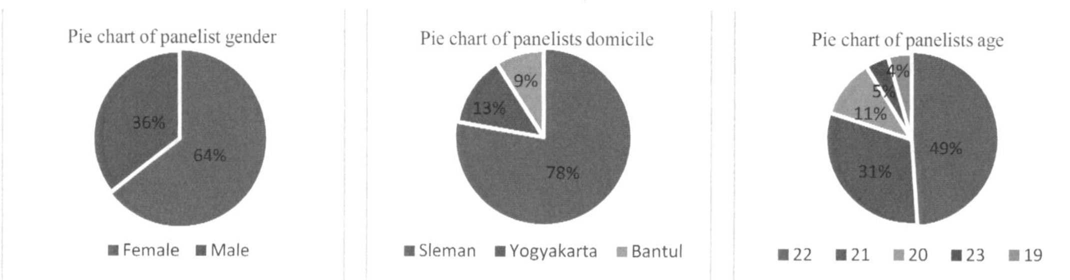

Figure 2. Profile of the used untrained panelists consist of gender, domicile, and age of the panelists.

Table 1. Sensory analysis results

Attribute	Sample		
	Local sample (172)	**Self-produced sample (758)**	**Imported sample (637)**
Color	5.27±1.37[a]	4,42±1,51[a]	5.02±1,67[a]
Aroma	4.82±1.66	4,60±1,48	6.02±1,02
Flavor	3.04±1.43	5,80±1,28[a]	5.27±1,50[a]
Texture	3.18±1.92	5,44±1,29[a]	4.80±1,69[a]
Overall	3.04±1.55	5,60±1,06[a]	5.22±1,31[a]

[a] The same letter in the same row shows no significant difference at the 5% significance level using Kruskal-Wallis and Post Hoc Pairwise Comparisons test.

Based on the results of the sensory analysis, the sample with the highest acceptance value for all attributes is sample 758. This sample has a dark red appearance that shows the characteristics of dried fruit, with a strawberry, fruity, and sweet aroma. The flavor possessed by this sample is a sweet strawberry flavor and does not have an aftertaste, with a soft and not sticky texture. Overall, this self-produced sample has characteristics like fresh strawberry but still has characteristics of dried fruit.

Sample with the highest acceptance value for color attribute is the local sample (sample 172) with an average of the acceptance value of 5.27±1.37. This sample is bright red in the middle of the fruit and slightly dark red at the edges of the fruit. Panelists assess the color of the sample tends to be close to the color of fresh strawberry. This is in line with the research results about color preference of the strawberry which states that consumers prefer strawberry with red color and higher chroma value [14]. Sample with the highest acceptance value for aroma attribute is the imported sample (sample 637) with an average of the acceptance value of 6.02±1.02. The panelists assessed that this sample had a stronger aroma than the other samples, the aroma was smelled in this sampleis like fresh strawberry aroma. This is in line with the results of research related to volatile compounds in strawberry which state that several ester compounds contained in strawberry form fruity, sweet, floral aroma, and others [15]. Sample with the highest acceptance value for flavor attribute is self-produced sample (sample 758) that obtained an average of the acceptance value of 5.80±1.28. Panelists assessed that this sample had a sweet strawberry flavor and a slightly sour flavor. In addition, there is no aftertaste in the assessment this sample. This is in line with the results of research related to consumer preferences for strawberry which states that the attributes of sweetness, caramel, berry, fruity, and floral are attributes that contribute greatly to consumer acceptance of strawberry. Meanwhile, the sour, astringent, citrus flavor attributes can cause a decrease in consumer acceptance of strawberry [16].

Sample with the highest acceptance value for texture attribute is self-produced sampe (sample 758). This sample obtained an average of the acceptance value of 5.44±1.29. Panelists considered that this sample had a dry texture on the outside of the sample and a soft and non-sticky texture on the inside. According to previous research, the water content in dried fruit affects the hardness of the product. The higher the water content, the greater the force required to cut the product. This means that products with lower water content have a softer texture than products with higher water content [17]. Sample with highest average of acceptance value for overall attribute is self-produced sample (sample 758) with acceptance value of 5.60±1.06. This sample is a dehydrated strawberry sample in the whole form and made by a combination of osmotic dehydration (OD) for 90 min hours at a temperature of 35°C and drying using a tray dryer for 12 hours. Panelists assessed that this sample had characteristics such as color, aroma, flavor, texture according to consumer preferences for dried fruit but still had the typical characteristics of fresh strawberry. This sample has a dark red color with an even distribution of color, the aroma of strawberry and the non pungent aroma of sugar, sweet flavor, slightly sour, no aftertaste, and a dry texture on the outside of the sample and a soft and non-sticky texture on the inside.

3.2. *Statistical Analysis*

The normality test was used to determine the distribution of the sensory analysis data for dehydrated strawberry samples. If the test results obtained data are normally distributed, then proceed with the one way ANOVA test. However, if the result is that the data is not normally distributed, then it is continued with the Kruskal-Wallis test as an alternative to the one way ANOVA test. From the results of the normality test, if the significance value is > 0.05 (sig. > 0.05), then the data is said to be normally distributed. However, if the significance value is < 0.05 (sig. < 0.05), then the data is said to be not normally distributed [18]. Based on the results of the normality test, the significance value for the five samples and the five attributes was less than 0.05 (sig. <0.05). This means that the data from the sensory analysis results are not normally distributed and do not meet the assumption of normality. Therefore, the test was continued with the Kruskal-Wallis test and the Post Hoc Pairwise Comparison test to determine whether there was a difference in consumer acceptance of the five samples.

Based on the results of the Kruskal-Wallis test on consumer acceptance data of dehydrated strawberry for the five attributes, the significance value of the data for the five attributes was smaller than 0.05 (sig < 0.05). This means that there are significant differences in consumer acceptance of the samples presented for the five attributes. First, the selected sample for the color attribute is local sample. There is no significant difference between this sample with self-produced and imported sample. Panelists like samples that have a bright and even red color. According to previous research, dried fruit by osmosis dehydration process as pre-treatment had higher L* (lightness), a* (redness) and b* (yellowness) values than samples without osmosis dehydration process [19]. Second, the selected sample for the aroma attribute is imported sample. Based on the results of statistical tests, there is a significant difference in consumer acceptance between sample 637 and the other samples. Sample 637 had a strong strawberry aroma while the other samples did not have a strong aroma. This is in line with the results of research related to consumer preferences for strawberry which stated that consumers tend to like strawberry with a strong aroma and fruity aroma. The strong aroma of strawberry tends to be liked by consumers because it is considered as the main indicator when consumers make purchases of dried fruit [20]. Third, the sample selected for the flavor attribute was self-produced sample. There was no significant difference between this sample with imported sample. Panelists assessed that this samples tasted sweet and sour and there was no aftertaste. According to previous research, during the process of osmotic dehydration with sucrose as the osmotic agent, sugar absorption can occur by the sample when there is a rapid and significant release of water content. The absorption of sugar during this osmotic dehydration process can cause the resulting dehydrated strawberry product to have a sweet flavor [21]. Fourth, the selected sample for the texture attribute is self-produced sample. This sample has a dry texture on the outside and soft on the inside of the sample. There is a significant difference in consumer acceptance between this sample and local sample. This sample has whole form of strawberry while local sample has form of sheets. According to previous research,when drying Lichi fruit with a tray at a temperature of 70°C, it was found that the texture of dried fruit can be affected by the total sugar content contained in the product. In addition, the texture of the dry sample is also influenced by the content of glucose and pectin. The increase in temperature used in the drying process also greatly affects the increase in dry fruit hardness [22].

Finally, the selected sample for the overall attribute is self-producd sample. This sample had an even dark red color, a mild strawberry and sugar aroma, a sweet, slightly sour taste, and no aftertaste, as well as a dry texture on the outside of the sample and a soft texture on the inside of the sample. Previous research explains that consumers tend to like dried fruit that has a strong fruit aroma, sweetness or sweetness, and a sour taste that is not too high [23]. Based on the results of statistical tests, there is no significant difference in the acceptance of this sample and imported sample which can be concluded that sample with combination process of 90 minutes osmotic dehydration and dried for 12 hours with convection dryer is can compete with local and imported sample.

3.3. Dehydrated Strawberry Quality Analysis

Table 2. Color analysis result in L* (lighness), a* (redness), b* (yellowness), h° (hue), and ΔE values

Sample	Color value				
	L*	**a***	**b***	**ΔE**	**h°**
Dehydrated strawberry	23.33	10.87	3.42	0.28	17.47
	23.22	10.68	3.25	0.14	16.93
	23.27	10.64	3.38	0.24	17.62
Average value	23.27 ± 0.045	10.73 ± 0.10	3.35 ± 0.73	0.22 ± 0.06	17.34 ± 0.30

Based on the Table 2, the average value of L* is 23.27 ± 0.045; value of a* is 10.73 ± 0.10; value of b* is3.35 ± 0.73; value of ΔE is 0.22 ± 0.06; and value of h° is 17.34 ± 0.30. According to Shaw (1991),

fresh strawberry have an L*a*b value of 28 ± 1.00; 30 ± 1.00; and 22 ± 1.00 [13]. This means that there is a color difference between fresh strawberry and dehydrated strawberry. This color change can occur due to the Maillard reaction which causes browning of the strawberry color. Maillard reaction is a chemical reaction that occurs between amino acids and reducing sugars during the thermal processing [24]. Value 0.22 ± 0.06 of ΔE shows that there's no great difference in the result of dehydrated strawberry color analysis. The average of hue value (h°) of the sample is 17.34 ± 0.30 that means color of the sample has red color. Previous research about effect of osmotic dehydration on the color of freeze dried strawberry explain that hue value of strawberry can decreased because of increasing sucrose concentration in the osmotic agent and increasing immersion time. The decrease of the hue value is a positive and desirable result in strawbery because this can produced more reddish strawberry [25].

Table 3. Dominant esters volatile compounds, concentrate, and odor description in strawberry based on previous research about volatile ester composition in strawberry fruit [15]

Volatile compound	Concentrate (mg kg^{-1})	Odor description
Ethyl hexanoate	48.98	Fruity, pineapple, banana, gree apple, winy
Methyl hexanoate	19.98	Fruity, ether-like, pineapple
Methyl butanoate	16.62	Fruity, apple, banana, pineapple, ester-like, green
Ethyl acetate	13.81	Fruity, brandy, pineapple-like
Ethyl butanoate	13.67	Fruity, sweet, pineapple
Ethyl octanoate	11.43	Herbaceous, green, orange, parsley, melon
Methyl octanoate	9.10	Winy, fruity, orange
Octyl acetate	7.66	Neroli, jasmine, fruity
Hexyl acetate	7.23	Apple, cherry, pear, floral, sweetish
Hexyl butanoate	6.41	Fruity, apricot

Based on the results of sensory analysis, it was found that the aroma of dehydrated strawberry samples that the panelists liked the most was imported samples which had a strong aroma like fresh strawberry. According previous research, esters is dominant component that give rise to the arma of strawberry. Ethyl hexanoate is a volatile compound with the highest concentration contained in strawberry. This compound produces fruity, pineapple, banana, gree apple, and winy aroma [15]. This is in line with the previous research that obtained an odor description for the ethyl hexanoate compound, namely the aroma of apple peel fruit or the aroma of apples. In addition to these compounds, methyl hexanoate compounds also have high concentrations in strawberry. Methyl hexanoate compounds give rise to a fruity, ether-like, and pineapple aroma [26].

Table 4. Total sugar content, hardness value, and water content of dehydrated strawberry sample

Sample	Total sugar content (%)	Hardness value (N)	Water content (%)
Dehydrated strawberry	57.4897	6.60	25.9268
	57.6147	7.00	25.8824
	57.8646	7.30	25.1123
Average value	57.66 ± 0.16	6.97 ± 0.29	25.64 ± 0.37

Based on the results of the total sugar content test, it is known that the self-produced sample has an average total sugar content of 57.66 ± 0.16%. This is in line with the results of research related to dried fruit, it is known that dehydrated strawberry treated with the freeze-dried method have a total sugar content of 44.5±2.4% to 65.4±4.1% while dehydrated strawberry processed by other methods which is not known to have a total sugar content of 62.7±3.6% to 91.2±3.4% [27]. During the drying process there can be a decrease in sugar levels (glucose and fructose) which affects non-enzymatic reactions that cause discoloration of the material. The faster rate of decrease in sugar levels during this drying process, the faster the rate of discoloration to brown [28]

Based on the tests, the value of hardness or texture of dehydrated strawberry products is obtained in the range of 6.60 N - 7.30 N with an average hardness or texture value of 6.97 ± 0.29 N. The greater the value of texture (N) indicates that the sample has a harder texture. According to previous research, the water content contained in dried fruit affects the fragility of the product. The higher the moisture content, the greater the force required to cut the product. This means that products with lower water content have a softer texture than products with higher water content [17].

Based on the results of the water content test, it was obtained that the average water content is 25.64 ± 0.37%. Based on the results of research about dehydrated strawberry, the best combination of dehydrated strawberry products has water content of 35.76 which means that the outside of the product has dried but the inside of the product is still watery and has a soft texture [10]. Meanwhile, according to SNI 01-3701-1995 regarding dried fruit, the maximum water content contained in dried fruit is 31%. The results of testing the water content in the 758 sample are 25.641 ± 0.374%. This means that the water content of the sample is in accordance with the standard of dried fruit according to SNI (Indonesian National Standard).

4. Conclusions

Based on the results of the sensory analysis, the dehydrated strawberry sample with the highest overall acceptance value is self-produced sample that produced with combination of 90 min osmotic dehydration at 35°C and 12 hours convection drying using tray dyer at 70°C. In addition, the self-produced sample has no significant difference with local and imported sample which means this sample can compete with local and imported dehydrated sample.

5. Acknowledgments

Authors offer gratitude thanks to Universitas Gadjah Mada for their financial support with contract number : 3143/UN1.P.III/DIT-LIT/PT/2021, and also for Faculty of Agricultural Technology Universitas Gadjah Mada for the support of research during Pandemic Covid-19.

References

[1] Badan Pusat Statistik 2018 Konsumsi Buah Hortikultura Stroberi
[2] Kurnia A 2005 Petunjuk Praktis Budidaya Stroberi (Jakarta – Agro Media Pustaka)
[3] Utari R R D, Soedibyo D W, and Purbasari D 2018 J. Agrotek. 12
[4] Herawati E, Etikawati N, Setyaningsih R, and Sugiyarto 2019 J. SEMAR 8
[5] Waziroh E, Ali D Y, and Istianah N 2017 Proses Termal pada Pengolahan Pangan (Malang – Universitas Brawijaya Press)
[6] Muzaffar H, Rouf A, Kanojia V, Muzaffar Z, and Noor F 2018 Int. J. Curr. Microbiol. App. Sci.
[7] Salli G J and Joni F 2015 J. Kajian Tek. 11
[8] Martunis 2012 J. Tek. dan Industri Pertanian Indonesia 4
[9] Link J V, Tribuzi G, and Laurindo J B 2017 LWT – Food Sci. and Tech. 84
[10] Ardhani I C, Falah M A F, Widodo K H 2020 IOP Conf. Series: Earth and Envi. Sci. 653
[11] Putri R M, Aziz I W F, Falah M A F 2020 IOP Conf. Series: Earth and Envi. Sci. 759
[12] Tobing I F, Mustaqimah, and Agustina R 2019 J. Ilmiah Mahasiswa Pertanian 4
[13] Octavia L and Choo W S 2017 LWT – Food Sci. and Tech. 86
[14] Lee S M, Lee K T, Lee S H, and Song J K 2013 J. Food Engineering 119
[15] Dong J, Zhang Y, Tang X, Jin W, and Han Z 2013 Sci. Horti. 150
[16] Oliver P, Cicerale S, Pang E, and Keast R 2018 Journal of Food Science 83
[17] Olesiuk A K, Nowacka M, Wesoly M, and Ciosek P 2013 Sensor Actuators B 187
[18] Pramesti G 2009 Sukses Mengolah Data di Bidang Bisnis dan Industri (Jakarta – PT Elex Media Komputindo)
[19] Lamadrid M C, Lech K, Michalska A, Wasilewska M, Figiel A, Wojdylo A, and Barrachina A A C 2017 J Food Chemistry 232
[20] Bhat R, Geppert J, Funken E, and Stamminger R 2015 Int. J. Fruit Sci. 15
[21] El-Aouar A A, Azoubel P M, Barbosa A L, and Murr F E X 2006 J. Food Engineering 75
[22] Cao X, Islam M N, Xu W, Chen J, Chitrakar B, Jia X, Liu X, and Zhong S 2020 J. Food 9
[23] Alphonce R, Temu A, and Almli V L 2015 British Food J. 117
[24] Zhou Z and Langrish T 2021 J. Food Engineering 305
[25] Noguera J G, Oliveira F I P, Weller C L, Rodrigues S, and Fernandes F A N 2014 J Food Sci. Technol 51

[26] Niu Y, Wang P, Xiao Z, Zhu J, Sun X, and Wang R 2019 J. Food Chemistry 275
[27] Perez R M, Santos J G, Soria A C, Villamiel M, and Montilla A 2014 J. Food Chemistry 150
[28] Sansano M, Borras M J, Escriche I, Andres A, and Heredia A 2015 J. Food Science 80

Organoleptic Characteristics of Banana Flakes Based on Maturity Level of Banana (*Musa Paradisiaca*) CV. Ambon

D N Surahman[1], R Ekafitri[1], T Rahman[1], W Cahyadi[2], D A Setyadi[2], H M Astro[1], A Indriati[1], Y Andriana[3], C E W Anggara[1]

[1]Research Center for Appropriate Technology, National Research and Innovation Agency, Subang Indonesia
[2]Program Study of Food Technology, Pasundan University, Bandung Indonesia
[3]Research Unit for Natural Product Technology, Yogyakarta Indonesia

Email: diki.lucky@gmail.com

Abstract. Bananas are rich in minerals such as potassium, magnesium, iron, phosphorus, and calcium, contain vitamins A, B6 and C and contain serotonin which is active as a neurotransmitter for intelligence. Bananas can be processed into semi-finished products or finished products. One of the semi-finished products from bananas is Banana Flour. Banana flour is an alternative to banana preservation because bananas are perishable fruits. Banana flour can be made from unripe bananas and ripe bananas. The types of breakfast cereals that are most consumed or preferred by consumers are products in the form of breakfast drinks, extruded products and flakes. Flakes belong to the cereal milk food group. The purpose of this study was to determine the organoleptic characteristics of banana flakes made from banana flour with different ripeness levels of bananas. The research methodology included organoleptic test with the hedonic test method on a scale of 1-7, with a total of 30 panelists. The test parameters carried out include color, taste, aroma, hardness, crunchiness and overall acceptance. The results showed that the sample a1b3 (ripe banana flour, 25 minutes of roasting) was the most preferred by the panelists, with test values for color 5.40, taste 5.90, aroma, 5.63, hardness 5.50, crispness 5.67, overall admission of 6.07.

1. Introduction

Bananas are one of the most widely consumed fruit types in Indonesia, both in fresh and processed forms. The development of banana-based products continues to be carried out considering that bananas are a perishable fruit. One of them by making it into banana flour. Banana flour can be made from ripe or unripe bananas. The characteristics of the ripeness level of bananas that are processed into banana flour are also different, especially when viewed from the organoleptic characteristic. The most significant distinctive organoleptic parameters especially taste and smell, so that it can be used in the processing of various types of food that use flour (rice flour, wheat) in it [1]

Banana flour is produced from the process of grinding dried banana slices. Making bananas into banana flour will facilitate marketing, preserve bananas and expand the use of banana flour. The bananas that will be made into flour are bananas with a maturity level of ripe, whose skin is still green and the flesh is still hard [2]. Banana flour can be made from young bananas and unripe bananas. The principle of manufacture is drying in the sun or using a dryer, then milled and then filtered using a 100 mesh filter. Ripe banana flour has a high water content of 9.08%, carbohydrate content of 82.81% and protein content of 4.84% [3]. However, ripe banana flour has poor physical properties such as low yield, hygroscopic properties, and a more brown color than unripe banana flour due to enzymatic browning during the milling process.

Unripe banana flour has advantages such as the color produced is good, namely white and gray and has a carbohydrate content of 70.7% - 78.99% with a water content of 6.26% [4]. One type of banana that is widely available in Indonesia is the Ambon banana. Ambon bananas have a strong taste characteristic, so Ambon bananas are often consumed both fresh and processed. One of the processed

Ambon banana products developed by the Center for Appropriate Technology Research - BRIN is Banana Flakes. According to Nurjanah [5], the types of breakfast cereals most consumed or preferred by consumers are products in the form of breakfast drinks, extruded products and flakes. All of these products are instant products where the preparation time is less than 3 minutes. Flakes are included in the cereal milk food group. According to Indonesian standard number 01-4270-1996, the definition of cereal milk is instant powder made from powdered milk and cereal with the addition of other foods and or without permitted food additives.

2. Methodology

2.1. Materials
The materials used in this study including ripe banana flour, unripe banana flour, sugar (®Gulaku), chicken eggs, baking powder (® koepoe – koepoe) and skim milk were purchased commercially. Aquadest is used for WAI and WSI analysis.

2.2. Research stages
In this study, the research method using a Randomized Block Design (RAK) with a factorial pattern (3 x 3) and 3 repetitions. The variables used are the maturity level of banana flour (A) with 3 levels and baking time (B) with 3 levels. Determination of the selected product is done by chemically analysing the sample including water content, Water Absorption Index (WAI) and Water-Soluble Index (WSI) of each group. The results of chemical analysis obtained for each treatment is carried out using a scoring test with a score range of 1 - 6. After getting the selected product in each group, then sensory analysis is carried out.

2.2.1 Analysis
The analytical response used is the sensory response to the tested product consisting of 6 attributes including color, taste, aroma, hardness, crispness, and overall acceptance. The method used in the test is hedonic test using 30 panellists, with hedonic test criteria can be seen in (Table 1).

Table 1. Criteria for the hedonic scale

Hedonic Scale	Numeric scale
Really like	7
Like	6
Kinda like	5
Neutral	4
Kinda don't like it	3
Do not like	2
Very dislike	1

2.2.2 The process of making banana flour and banana flakes
The following is a flow chart of the process of making ripe and unripe banana flour and banana flakes, as you can see in Figure 1,2 and 3.

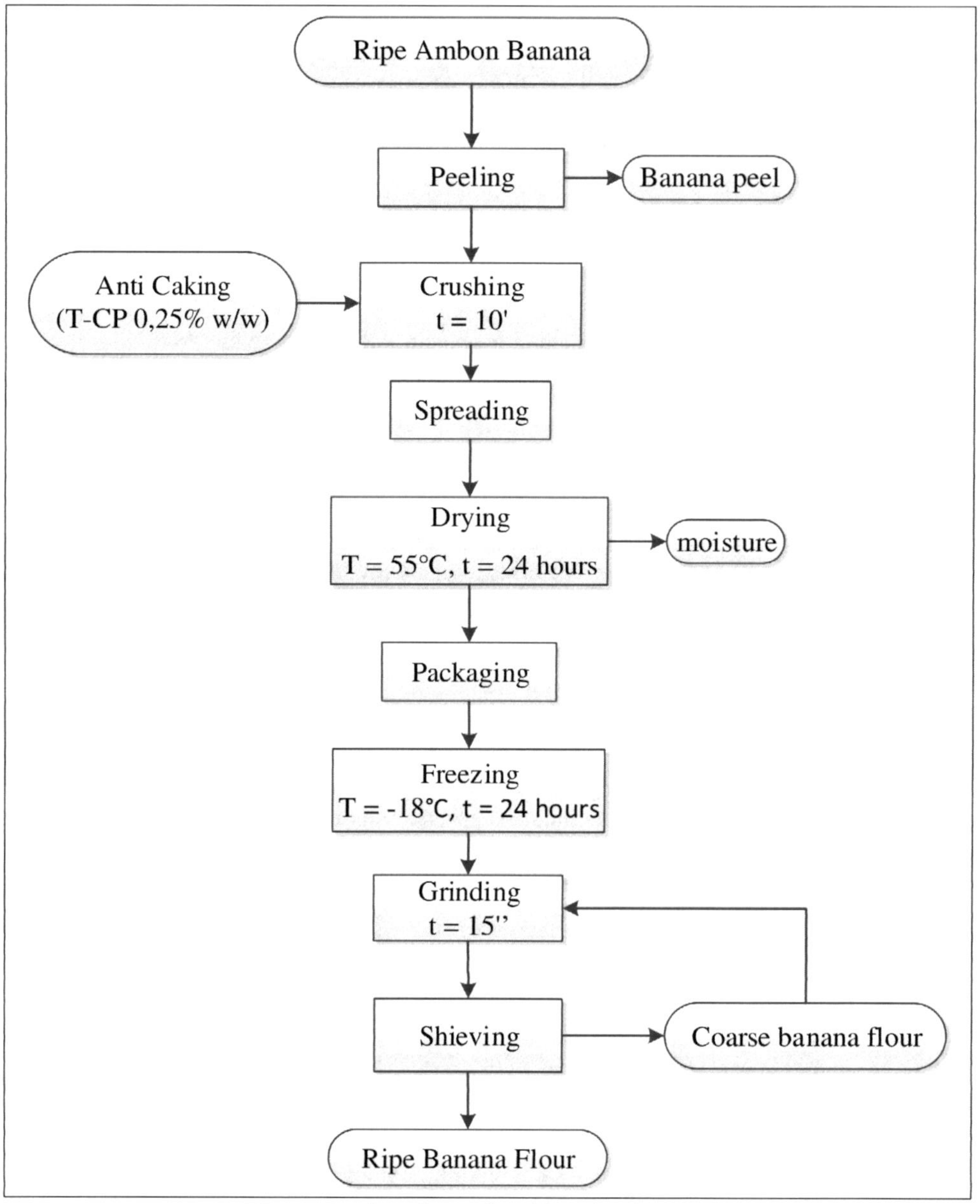

Figure 1. Flowchart of making ripe banana flour

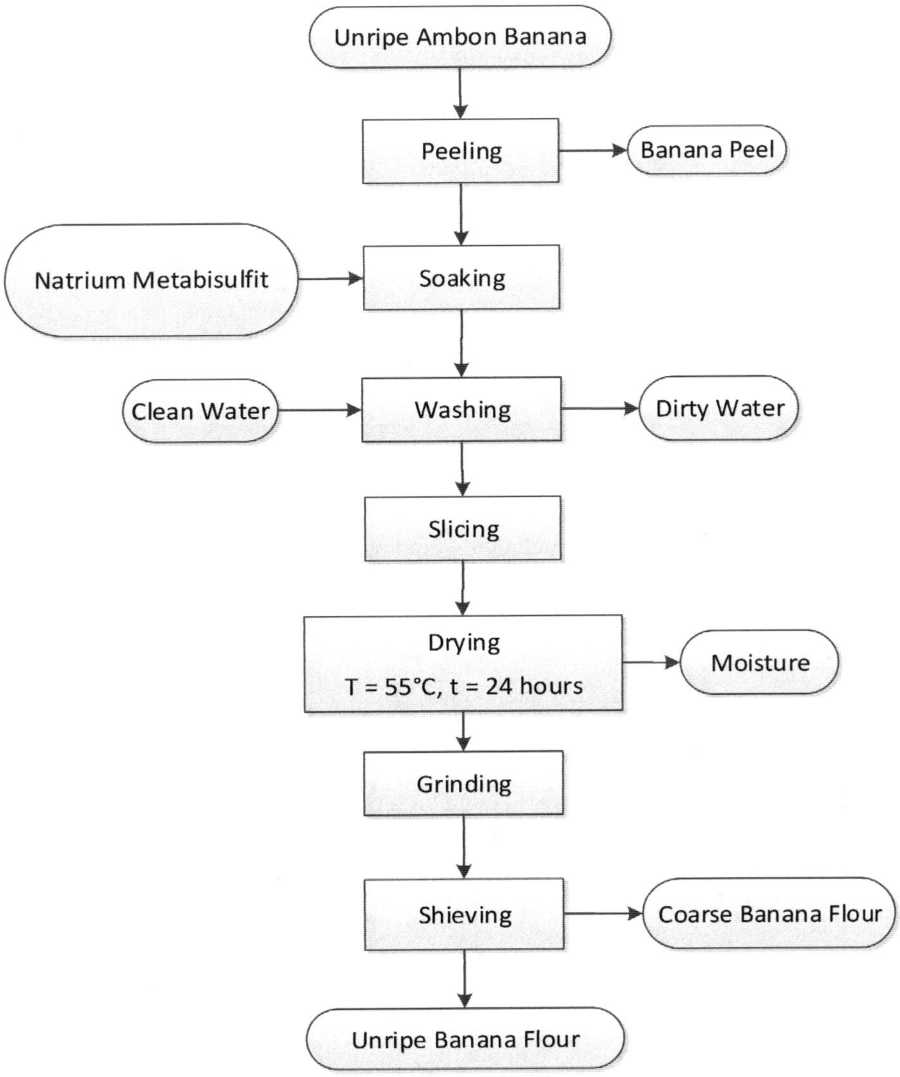

Figure 2. Flowchart of making unripe banana flour

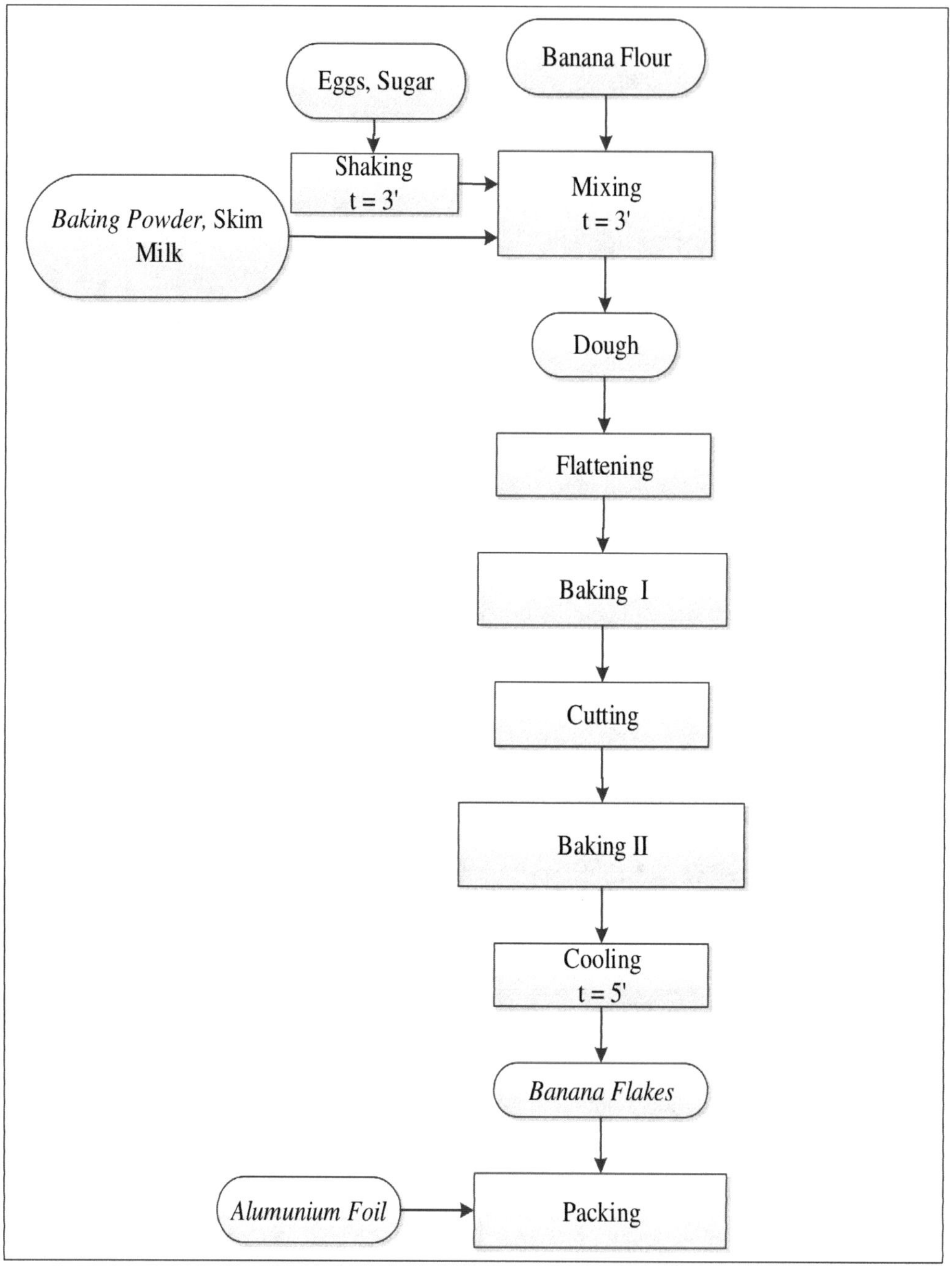

Figure 3. Flowchart of making banana flakes

3. Result and discussion

The results of the analysis of water content, WAI (Water Absorption Index) and WSI (Water-Soluble Index) of all samples, then a scoring test was carried out. From the results of the scoring test, the best 3 samples from each group were A1B3, A2B2, A3B2. The results of the scoring test can be seen in Table 2.

Table 2. The best sample based on the scoring test

Treatment	Water content	WAI	WSI	Total
A1B1	2	1	2	5
A1B2	4	2	4	10
A1B3	6	2	6	14
A2B1	1	3	1	5
A2B2	4	6	4	14
A2B3	4	5	3	12
A3B1	5	1	3	9
A3B2	6	3	5	14
A3B3	6	6	6	18

Note: A1B1=ripe banana flour, 15 min; A1B2=ripe banana flour, 20 min; A1B3=ripe banana flour, 25 min; A2B1=unripe banana flour, 15 min; A2B2=unripe banana flour, 20 min; A2B3=unripe banana flour, 25 min; A3B1=mixed banana flour, 15min; A3B2=mixed banana flour, 20 min; A3B3=mixed banana flour, 25 min

3.1. Aroma (Flavor)

The results of organoleptic tests on the aroma of flakes showed that the type of Ambon banana flour and baking time had an effect, this can be seen in Table 3.

Table 3. Results of organoleptic tests on the aroma of banana flakes

Ambon banana flour maturity level (A) and baking time (B)	Average value	Level 5%
A2B2	1,94	a
A3B2	2,21	b
A1B3	2,42	b

Note: The average value marked with the same letter shows no significant difference at the 5% level according to Duncan's further test.

The aroma of food is formed mainly during the baking process, freshly baked food has a very pleasant aroma which quickly disappears upon cooling and storage. The use of ripe Ambon banana flour as raw material for making flakes gives a better smell compared to mixed banana flour and banana flour, where for the treatment of banana flour for 20 minutes based on preference has the lowest average value, while for the treatment of banana flour ripe for 25 minutes has the largest average value. According to [6] aroma and flavor are the main factors that bananas are widely consumed. Chemically, the aroma and flavor of bananas are caused by the presence of volatile components that are received by the alpha receptor. More than 150 volatile components are present in bananas, especially the isoamyl and isobutyl ester groups together with 2-penanone [7]. Various compounds give rise to different aromas, where enzymatic and non-enzymatic browning reactions also produce strong odors, for example the formation of furfural and maltol in the Maillard reaction [8]. Protein in foodstuffs also affects the aroma of foodstuffs. With heating, the protein in food will change and form compounds with other ingredients, for example with amino acids resulting from changes in protein with reducing sugars that form the aroma of food [9].

IOP Conf. Series: Earth and Environmental Science **980** (2022) 012036 doi:10.1088/1755-1315/980/1/012036

3.2. Taste

The results of organoleptic tests on the taste of banana flakes showed that the maturity level of Ambon banana flour and baking time had an effect, this can be seen in Table 4.

Table 4. Results of organoleptic tests on the taste of banana flakes

Ambon banana flour maturity level (A) and baking time (B)	Average value	Level 5%
A2B2	2,01	a
A3B2	2,39	b
A1B3	2,52	b

Note: The average value marked with the same letter shows no significant difference at the 5% level according to Duncan's further test.

Based on the results of Duncan's further test in Table 4, it shows that the type of Ambon banana flour and the baking time on the taste of the A2B2 treatment were significantly different from the A3B2 and A1B3 treatments, the A3B2 treatment was significantly different from the A2B2 treatment, as well as the A1B3 treatment was significantly different from the A2B2 treatment.

Flavor and taste are defined as stimuli evoked by the ingested material, mainly perceived by the senses of taste and smell, as well as other stimuli such as touch and acceptance of the degree of heat in the mouth. Taste is a sensation that is formed from the combination of ingredients and their composition in a food product that is captured by the sense of taste. Taste according to the quality attributes of a product is usually an important factor for consumers in choosing a product [10]. It is known that there are four basic tastes, namely sweet, sour, salty, and bitter. The concept is actually just a simplification, the stimulation received by the brain due to electrical stimulation that is transmitted from the taste cells is actually very complex. It is known that sweetness comes from compounds of 51 sugars such as sucrose, bitter by quinine, salty by salt, and sour by various types of acids. The taste of food products in general does not consist of only one taste but is an integrated combination of various kinds to give rise to the taste of a whole food [11]. Factors that affect taste are chemical compounds, temperature, and interactions with other flavor components. Various chemical compounds give rise to different tastes. The sour taste is caused by proton donors, the salty taste is produced by inorganic salts, the sweet taste is also caused by aliphatic organic compounds and the bitter taste is caused by the alkaloids. Interaction with other components can certainly affect the value of a product's taste [8].

The taste that appears in the product is caused by the presence of ingredients in the product such as banana flour, sugar, eggs, milk and baking powder. It can be seen that the use of ripe banana flour is better than mixed banana flour or mixed banana flour. Where the use of ripe banana flour has a higher level of preference, this shows that ripe banana flour with a baking time of 25 minutes increases the preference for flakes. The more ripe the banana, the sweeter the taste due to the high sucrose content. During the ripening process of bananas, starch is converted into sugar through an enzymatic process where there is a decrease in starch content from 20-30% to 1-2% [12] followed by an increase in the amount of sugar content, especially sucrose to more than 10% by weight of fresh fruit [13].

3.3. Color

The results of the organoleptic test on the color of the flakes showed that the maturity level of Ambon banana flour and baking time had an effect, this can be seen in Table 5.

Table 5. Results of organoleptic tests on the color of banana flakes

Ambon banana flour maturity level (A) and baking time (B)	Average value	Level 5%
A2B2	1,98	a
A3B2	2,39	b
A1B3	2,42	b

Note: The average value marked with the same letter shows no significant difference at the 5% level according to Duncan's further test.

Based on the results of Duncan's further test in Table 5, it shows that the type of Ambon banana flour and the baking time of the color in the A2B2 treatment were significantly different from the A3B2 and A1B3 treatments, while the A3B2 treatment was not significantly different from the A1B3 treatment. But treatment A3B2 significantly different from treatment A2B2.

According to [14], color is the most important quality attribute. Even though a product has high nutritional value, good taste and good texture, if the color is not attractive, it will cause the product to be less attractive. Flakes contain protein and sugar derived from the main raw materials and supporting materials. The possibility of protein and sugar content, the occurrence of a browning reaction during baking. Where for treatment of banana flour mixture for 20 minutes based on preference for flakes has the lowest average value while for treatment of mixed banana flour for 25 minutes has the largest average value. In general, when food is dried, the color of the material will turn brown due to a non-enzymatic browning reaction. This reaction occurs due to baking at high temperatures so that the sugar caramelization reaction occurs into glucose and fructose.

In addition, the color of the flakes is influenced by the Maillard reaction, namely the amino groups of proteins found in ingredients such as banana flour, eggs and milk with the carbonyl groups of reducing sugars found in banana flour and sugar. Browning reaction is defined as the reaction of an amino group on an amino acid, peptide or protein with a hydroxyl group on a sugar, resulting in the formation of a brown nitrogen polymer or melanoidin [10]. The color of the product, apart from being a sufficient factor in determining quality, can also be used as an indicator of whether or not the mixing method or processing method can be marked by the presence of a uniform and even color [8].

3.4. *Crispness*

The results of the organoleptic test on the crispness of the flakes showed that the maturity level of Ambon banana flour and baking time had an effect, this can be seen in Table 6.

Tabel 6. Results of organoleptic tests on the crispness of banana flakes

Ambon banana flour maturity level (A) and baking time (B)	Average value	Level 5%
A2B2	2,17	a
A1B3	2,47	b
A3B2	2,50	b

Note: The average value marked with the same letter shows no significant difference at the 5% level according to Duncan's further test

Based on the results of Duncan's further test in Table 6. shows that the type of Ambon banana flour and baking time on the crispness of the A2B2 treatment were significantly different from the A3B2 and A1B3 treatments, while the A3B2 treatment was not significantly different from the A1B3 treatment, as well as the A1B3 treatment was not significantly different with treatment A3B2 but significantly different from treatment A2B2.

The crispness of breakfast food products is an important factor. In the processing of breakfast foods such as flakes, raw materials are often added, both those that have not been modified or starch that has been modified. According to Muchtadi et al [15], starch has an important role for the manufacture of flakes because it can affect the texture. The effect was mainly due to the ratio of amylose and amylopectin in starch. Amylopectin is known to stimulate the development process (puffin), so that flakes derived from starch with a high enough amylopectin content will be porous, crispy and crunchy, where for the treatment of unripe banana flour to stir for 20 minutes based on preference for flakes in terms of crunchiness has an average value. The lowest average while for the treatment of mixed banana flour for 25 minutes has the largest average value. Changes that occur during the process of making flakes are starch will be gelatinized and mild hydrolysis may occur, browning reactions occur which may be caused by interactions between protein and sugar, caramelization and dextrination processes occur as a result of baking at high temperatures so that crispy flakes are produced due to water reduction which is quite high.

3.5. Hardness

The results of organoleptic tests on the hardness of banana flakes showed that the maturity level of Ambon banana flour and baking time had an effect, this can be seen in Table 7.

Table 7. Organoleptic test results on hardness of banana flakes

Ambon banana flour maturity level (A) and baking time (B)	Average value	Level 5%
A2B2	2,12	a
A1B3	2,44	b
A3B2	2,47	b

Note: The average value marked with the same letter shows no significant difference at the 5% level according to Duncan's further test.

Based on the results of Duncan's further test in Table 7. shows that the type of Ambon banana flour and baking time on the hardness of the A2B2 treatment was significantly different from the A3B2 and A1B3 treatments, while the A3B2 treatment was not significantly different from the A1B3 treatment, as well as the A1B3 treatment was significantly different from the A1B3 treatment but treatment A2B2 not significantly different from treatment A3B2.

According to Muchtadi et al [15], starches with high amylose content, for example starch from tubers, tend to produce hard flakes because the development process occurs in a limited manner, where for the treatment of banana flour it kneads for 20 minutes based on the hardness of the flakes in terms of crispness has the lowest average value while for the mixed banana flour treatment for 25 minutes it has the largest average value.

3.6: Overall

The results of organoleptic tests on overall banana flakes showed that the maturity level of Ambon banana flour and baking time had an effect, this can be seen in Table 8.

Table 8. Results of organoleptic tests on overall of banana flakes

Ambon Banana Flour Maturity Level (A) and Baking Time (B)	Average Value	Level 5%
A2B2	2,17	a
A3B2	2,39	b
A1B3	2,56	b

Note: The average value marked with the same letter shows no significant difference at the 5% level according to Duncan's further test.

Based on the results of Duncan's further test in Table 8. It shows that the type of Ambon banana flour and the baking time of overall in the A2B2 treatment were significantly different from the A3B2 and

A1B3 treatments, while the A3B2 treatment was not significantly different from the A1B3 treatment, as well as the A1B3 treatment was significantly different with A2B2 treatment but not significantly different from A3B2 treatment. This indicates that based on the organoleptic assessment of flakes, the flakes product in treatment A1B3 (25 minutes ripe banana flour) had the highest average value in terms of preference.

4. Conclusion

Based on the results of the study, it can be concluded that the treatment selected according to organoleptic results in each group with attributes of taste, color, aroma, crispness, hardness and overall is a flakes product with treatment A1B3 (ripe banana flour, baking time 25 minutes).

References

[1] Juarez-Garcia E, Religion-Acevedo E, Sayago-Ayerdi S G, Rodriguez-Ambriz S L and Bello-Perez L A 2006 Composition, Digestibility and Application in Breadmaking of Banana Flour *Plant Foods for Human Nutrition* **61** 131-137.
[2] Prahasta A 2009 Agribisnis Pisang CV Pustaka Grafika Bandung (in Indonesian)
[3] Ekafitri R, Surahman D N and Afifah N 2016 Effect of Dextrin and Egg White Addition on Quality of Ripe Banana Flour *Jurnal Litbang Industri* **6**(12) 13-24 (in Indonesian)
[4] Murtiningsih, Suyanti and Muhajir I 1990 Pengaruh Umur petik pisang ambon jepang terhadap mutu tepung *Penelitian Hortikultura* **5**(2): 93-98 (in Indonesian)
[5] Nurjanah E 2000 Analisis karakteristik konsumen dan pola konsumsi pangan sarapan Skripsi Fakultas Pertanian Institut Pertanian Bogor (in Indonesian).
[6] Facundo H V, Garruti D S, Cordenunsi B R and Lajolo F M 2013 Isolation of Volatiles Compounds in Banana by HS-SPME: Optimization for the Whole Fruit and Pulp *International Journal of Bioscience, Biochemistry and Bioinformatics* **3**(2) 110-115
[7] Jordan M J, Tandon K, Shaw P E and Goodner K L 2001 Aromatic Profile of Aqueous Banana Essence and Banana Fruit by Gas Chromatography-Mass Spectrometry (GC-MS) and Gas Chromatography-Olfactometry (GC-O) *J. Agric. Food Chem.* **49** 4813–4817
[8] Winarno F G 1997 Food Chemistry and Nutrition PT. Gramedia Pustaka Utama Jakarta (in Indonesian)
[9] Sudarmadji 1996 Analytical Procedures for Foodstuffs and Agriculture Second Edition, Liberty Yogyakarta (in Indonesian)
[10] DeMan 1997 Kimia Makanan Institut Teknologi Bandung : Bandung (in Indonesian)
[11] Kartika B, Hastuti P and Supartono,W 1988 Pedoman Uji Inderawi Bahan Pangan Universitas Gajah Mada Press Yogyakarta (in Indonesian)
[12] Mohapatra D, Sabuasachi M, and Namrata S 2010 Banana and its By-Product : Overview *Journal of Scientific and Industrial Research* **69** 323-329.
[13] Zhang P, Whistler R L, BeMiller J N and Hamaker B R 2005 Banana starch: production, physical properties and digestibility – a review *Journal of Carbohydrate Polymers* **59** 443-458.
[14] Fenema R O 1985 Food Chemistry 2nd Edition Revised and Expanded Academic Press New York.
[15] Muchtadi D, Koswara S and Dahrul S 1990. Effect of Banana Type and Addition of Anti-browning on Banana Flour Production. Not published.

Hybrid solar fish dryers as a solution for improving the quality of fisheries products to support green technology during the pandemic COVID-19

M Risa[1], A Irawan[2], A Pratomo[3], A Khalid[4] and Y R Fauzi[4]

[1] Department of Business Administration, Politeknik Negeri Banjarmasin, Brigjen H. Hasan Basri, Banjarmasin, South Kalimantan, 70123, Indonesia
[2] Department of Information Management, Politeknik Negeri Banjarmasin, Brigjen H. Hasan Basri, Banjarmasin, South Kalimantan, 70123, Indonesia
[3] Department of Business Digital Department, Politeknik Negeri Banjarmasin, Brigjen H. Hasan Basri, Banjarmasin, South Kalimantan, 70123, Indonesia
[4] Department of Mechanical Engineering, Politeknik Negeri Banjarmasin, Brigjen H. Hasan Basri, Banjarmasin, South Kalimantan, 70123, Indonesia

* email: meyrisa@poliban.ac.id

Abstract. The fishing communities in Sungai Rasau village, South Kalimantan, Indonesia, have several joint business groups that mostly process their fish catch into salted fish products and shrimp paste for sale in the market. The fish drying technique still uses traditional methods that rely on sunny weather, this is not effective because the weather is difficult to predict due to climate change globally. This means that high rainfall affects fish drying production activities. If left unchecked, this certainly results in a decrease in the quantity and quality of salted fish production and affects their income and welfare. To maintain and improve the quality of raw materials for fisheries production, the solution is the use of appropriate technology, namely making salted fish processing equipment or dryers that are energy efficient and not influenced by weather factors such as rain. Using this tool can improve the quality and quantity of production. The form of activity methods carried out includes (1) coordination with related parties to foster local fishing groups; (2) identifying problems and determining solutions; (3) solar fish dryer design; (4) making efficient technology fish dryers that effectively and efficiently utilize solar energy. At the time of implementation of the dryer, the measurement of water content using TDS obtained results that the water content of mackerel has been reduced to about 10%

1. Introduction

As the largest archipelagic country in the world with 2/3 of its territory in the sea, Indonesia has a large area of sea, coast, and small islands that are strategically significant as pillars of national economic development. Specifically for fish catches, Indonesia's potential is very abundant so that it can be expected to be a leading sector of the national economy. The potential of captured fish can be consumed by Indonesian citizens and can even be exported abroad. To get a good level of sales, of course, must be balanced with quality processed products.

IOP Conf. Series: Earth and Environmental Science **980** (2022) 012037 doi:10.1088/1755-1315/980/1/012037

Sungai Rasau Village is one of the coastal and densely populated villages in the prosperous sub-district of Tanah Laut regency with a population of 2,044 inhabitants. About 70% of the population earns a living as fishing and fishing laborers. Sungai Rasau village has the potential to develop fisheries and maritime businesses because it is in the Java seafront position as shown in figure 1.

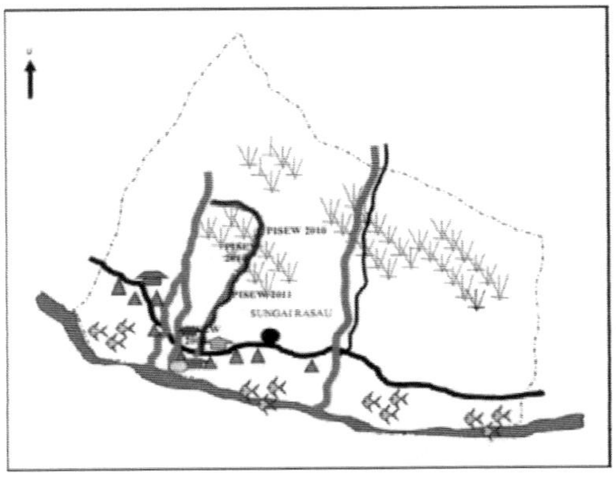

Figure 1. Map location

The headman of Sungai Rasau Village as shown in figure 1 has a plan to solve the problems faced by fish processing groups in tidying up:

- Improving the quality of salted fish and other sea products (quality) that have fallen due to fungal attacks when raw materials have not been dried up.
- Increase salted fish production (quantity) which decreases when rainfall is high.
- Utilization of cheap new renewable energy sources.

Figure 2. Traditional drying method

As shown in figure 2, the fish processing group in Sungai Rasau village still uses traditional methods in which the process of drying salted fish and other sea products is highly dependent on the weather. If the weather is uncertain, the drying process will be disrupted, because the results of dried fish and other marine products are very dependent on the sun's heat. Besides the problem of dependency on the weather, another thing that is a problem in the storage process of the product which is less than the maximum so that sometimes the product becomes damaged. The problems faced by the Fish Processing Joint Business Group (KUB) in Sungai Rasau Village, Tanah Laut South Kalimantan Regency are the inability in mastering the energy-saving salted fish processing technology sector and minimal funding problems.

Fish meat is very good for consumption because it contains a source of about 15-20% protein, vitamins, carbohydrates, and other substances that can dissolve in water (1). Fish is one type of food that contains water, but it would be better if the water content in fish was removed when consumed. (2). Fish require proper handling so that the quality of fish is maintained. There are various methods of preserving fish such as smoking, drying, salting, and freezing. Drying of food products is important to increase resistance to degradation due to decreased water activity (3). Drying is a better choice than refrigeration of food products due to the cost and complexity of the device (4). Fish drying by drying can be done using traditional methods, both open drying and sun drying using hot air. The open drying process still has many weaknesses including the length of time it takes to dry, a large area of land, dust disturbances, animal disturbances and requires a large amount of labor (5). Compared to open sun drying, the use of a greenhouse dryer can reduce drying time by up to 50% and produce a higher quality product in terms of color, texture, and taste (6). Solar dryers for fish products have been developed in several studies. Sengar, Khandetod, and Mohod (2009) investigated the solar dryer with the cost of dried shrimp (Kolambi) (6). Bintang, Pongoh, and Onibala (2013) made a solar fish dryer with a loading and unloading system. Sidhi, Pujianto, Prasetyo, and Muhfizar (2018) conducted an experimental study of drying yellowtail fish under an active greenhouse dryer (2).

2. Method

To achieve its objectives, technology implementation activities in Sungai Rasau village will be carried out through several approaches, including:

a. Data collection
b. Data requirement analysis
c. Design and develop solar powered fish dryer
d. Testing
e. Implementing the fish dryer
f. Evaluation

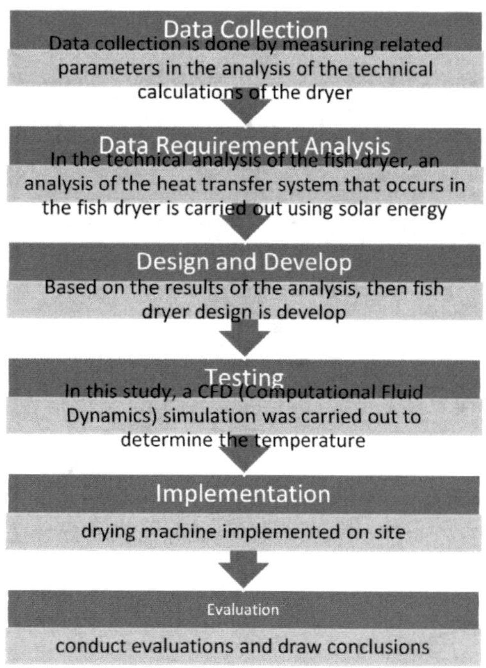

Figure 3. Activity stages

3. Result and Discussion

3.1. Analysis

The heat required for evaporation of water content in fish is:
Heat to heat water + Heat to evaporate water + Heat to raise the temperature of fish
= (389.96 + 6771.6 + 137.17) kJ
= 7298.73 kJ

The result of the calculation of total sun intensity on inclined surface, can be seen in the table 1. While the result of the calculation of can be seen in the table 2 The calculation results of the drying time can be seen in the table 3.

Table 1. Total sun intensity on inclined surface

Hour	Ibt (MJ/m2)	Hour	Ibt (MJ/m2)
08.00	96.005	13.00	183.149
09.00	352.743	14.00	411.159
10.00	137.875	15.00	96.261
11.00	72.716	16.00	68.777
12.00	86.994		

Table 2. Calculation of airflow velocity in the heating room

No	Tb (°C)	Te (°C)	*P*	g (m/s2)	L (m)	Vc (m/s)	G (kg/s)
1	50	33	1.093	9.81	3	0.83	0.031
2	55	33	1.076	9.81	3	0.95	0.032
3	60	33	1.059	9.81	3	1.05	0.034

Table 3. Calculation of drying time

Drying Time	Mt			Removed content	Drying time (hours)		
	50°C	55°C	60°C		50°C	55°C	60°C
1	0.92	1.31	1.69				
2	1.84	2.63	3.38				
3	2.76	3.94	5.07				
4	3.68	5.25	6.75				
5	4.60	6.57	8.44				
6	5.51	7.88	10.13				
7	6.43	9.20	11.82	3 Kg			7.11
8	7.35	10.51	13.51				
9	8.27	11.82	15.20			9.13	
10	9.19	13.14					
11	10.11						
12	11.03						
13	11.95				13.06		
14	12.87						

3.2. Design

The solar fish dryer uses the main material of the Solar Collector which functions to absorb solar radiation energy that falls on the surface of the absorbent plate so that the plate temperature becomes high {Formatting Citation}. Solar cell construction as follows on table 4.

Table 4. Tool specifications

	Tools	Function
1	Solar panel board	Functioning to convert solar energy into electrical energy is used to supply electrical energy in fish drying equipment. The module used is 300 wp, so the power produced is 300 watts/hour.
2	Solar charge controller	Functions as a voltage regulator that comes from the photovoltaic module supply to charge the battery and supply the inverter.
3	Inverter	Functioning as a converter input voltage from the solar regulator in the form of a DC voltage which would then be converted into AC voltage to supply power to the control thermometer, heater, and fan.
4	Battery	serves to store energy from solar photovoltaic modules that can be used as a backup power supply so that the tool is used at night
5	Heater plate	Function as a medium fish dryer
6	Cables	As a power supply

The design of the sea fish dryer into salted fish uses a rack system with the main components in the form of solar panels (9,10), solar charge controller, batteries (batteries), intervers, temperature control, RTD temperature sensors, incandescent lamps, as seen in figure 4 until figure 7.

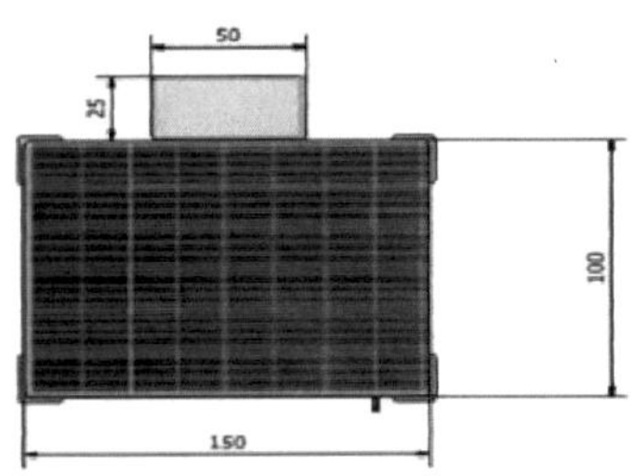

Figure 4. Top view design

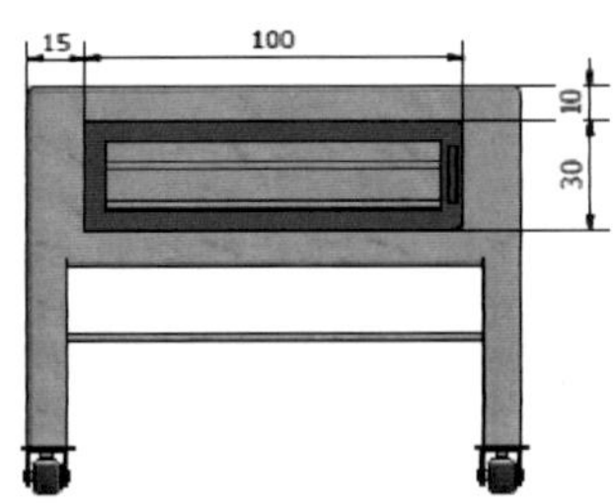

Figure 5. Rear view design

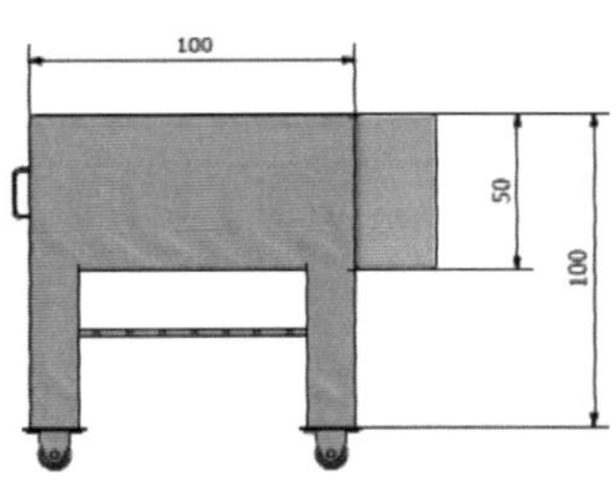

Figure 6. Side view design

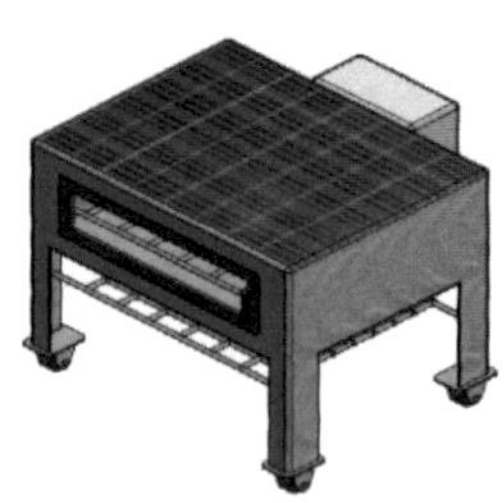

Figure 7. Side view above design

3.3. Implementation

Figure 8. Fish dryer

Figure 9. Dryer fish mat.

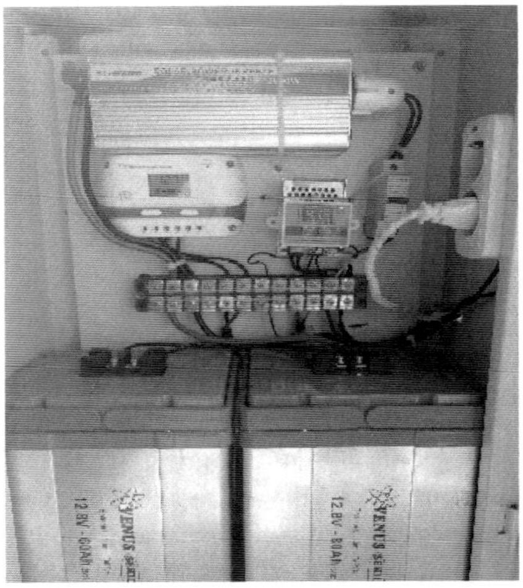

Figure 10. Electronic systems.

Design the size of the Fish drying chamber using data from the calculation of the heat load needed for a fish dryer with a capacity of 5 kg and can be expanded to 7 kg with a tool size of about 1.5 meters, where Qtotal is obtained 12867.212 kJ / the drying cycle and also the collector's need to get heat is 2573.75 kJ / cycle for the needs of the collector used to dry 1 kg of wet fish based on calculations to obtain energy to dry the water fish for 7.11 hours then it is equivalent to the collector of 3.65 m2.

The fish dryer uses a controller that can adjust the temperature automatically, where when the temperature reaches 50° C automatically, the heating machine will turn off. After the temperature drops and reaches 45° C the heater will turn on again. The temperature setting can be adjusted according to the needs of drying raw materials.

The costs incurred for the manufacture of this fish dryer are relatively large, but this is a result of the quality of the durability of the equipment. When compared to a dryer made of wood, of course, the appliance will not last long because it will break quickly. In contrast to tools made of aluminium which have a better level of resistance, termite and fire resistance.

IOP Conf. Series: Earth and Environmental Science **980** (2022) 012037 doi:10.1088/1755-1315/980/1/012037

3.4. Evaluation

For comparison, a load test was carried out. The purpose of load testing is to calculate the time and determine the level of drying efficiency (11). The results of the implementation of the solar powered fish dryer that can be seen are the drying process that is more faster and more production with a better results as seen in the figure 11, figure 12 and table 5.

Figure 11. Drying result using traditional method.

Figure 12. Drying result using solar dryer.

Table 5. Comparison of drying results

	Traditional drying	Using solar dryers
Production / day	0.5 - 1 ton	0.5 - 1 ton
Drying time in hot weather	1 - 2 days	1 day
Drying time on cloudy days	3 - 7 days	1 - 2 days
Production Loss	300 kg	100 kg

Another benefit felt by the community with the solar fish dryer is that more free time can be used to do other jobs such as packing, doing housework, and more time socializing with other parties so that there are more opportunities to expand the market network.

In this activity, a CFD (Computational Fluid Dynamics) simulation is carried out which aims to determine the distribution or temperature distribution of the freshwater fish dryer in the drying chamber. It can be seen that the dryer can spread a uniform temperature, so that the heat is evenly distributed, this can accelerate the drying rate of water in mackerel from 70% to 10% using a fish dryer. At the time of implementation of the dryer, the measurement of water content using TDS obtained results that were not much different from the simulation.

4. Conclusion

Solar fish dryer is a solution to the constraints of the traditional drying process that relies on hot weather, where when the weather is rainy, the resulting product will rotten. The comparison between traditional drying and solar drying shows that solar drying can be done within 1 day while using the traditional method can take up to 2 days. Drying on cloudy days using solar power can be faster with a difference of 2 days. The total production loss for solar drying is 20 percent better than the traditional method, which can be reduced to 10 percent of the total material being dried. At the time of implementation of the dryer, the measurement of water content using TDS obtained results that the water content of mackerel has been reduced to about 10%. During the pandemic, the process of drying fish using a solar powered fish dryer is very useful because it can reduce time outside the home and the drying process can be done at home.

References

[1] Sobukola OP, Olatunde SO. Effect of salting techniques on salt uptake and drying kinetics of African catfish (Clarias gariepinus). Food Bioprod Process. 2011;89(3).
[2] Sidhi SDP, Pujianto A, Prasetyo D, Nurfauzi A, Muhfizar. Experimental Study Of Salted Fish Drying Under Greenhouse Dryer. Russ J Agric Socio-Economic Sci. 2018;77(5).
[3] Bellagha S, Amami E, Farhat A, Kechaou N. Drying kinetics and characteristic drying curve of lightly salted sardine (Sardinella aurita). Dry Technol. 2002;20(7).
[4] Swami VM, Autee AT, T R A. Experimental analysis of solar fish dryer using phase change material. J Energy Storage. 2018 Dec 1;20:310–5.
[5] Setyoko.B., Darmanto S R. Peningkatan Kualitas Pengeringan Ikan Dengan Sistem Tray Drying. Pros SNST ke_3. 2012;
[6] Das T, Tiwari GN. Heat and mass transfer of greenhouse fish drying under forced convection mode. Int J Agric Res. 2008;3(1).
[7] Bintang Y, … JP-MTH, 2013 undefined. Konstruksi dan kapasitas alat pengering ikan tenaga surya sistem bongkar-pasang. ejournal.unsrat.ac.id [Internet]. 2013 [cited 2021 Oct 2];1(2):40. Available from: https://ejournal.unsrat.ac.id/index.php/jmthp/article/view/1632
[8] Handoyo E, Kristanto P, Mesin SA-JT, 2011 undefined. Desain dan pengujian sistem pengering ikan bertenaga surya. fportfolio.petra.ac.id [Internet]. [cited 2021 Oct 2]; Available from: http://fportfolio.petra.ac.id/user_files/91-021/Pengering Ikan.pdf
[9] Nugrahani E, … QA-… and UE, 2018 undefined. Experimental Analysis of Solar Cabinet Dryer for Fish Processing in Gresik, Indonesia. ieeexplore.ieee.org [Internet]. [cited 2021 Oct 2]; Available from: https://ieeexplore.ieee.org/abstract/document/8635737/
[10] Rizal T, engineering ZM-C studies in thermal, 2018 undefined. Fabrication and testing of hybrid solar-biomass dryer for drying fish. Elsevier [Internet]. [cited 2021 Oct 2]; Available from: https://www.sciencedirect.com/science/article/pii/S2214157X18300972
[11] Sengar S, Khandetod Y, Environmental AM-AJ of, 2009 undefined. Low cost solar dryer for fish. ajol.info [Internet]. 2009 [cited 2021 Oct 2];3(9):265–71. Available from: https://www.ajol.info/index.php/ajest/article/view/46076

Accelerated shelf life test method with arrhenius approach for shelf life estimation of tongkol 'euthynnus affinis' balado in cans

A Brilliantina[1], D K Wardani[2*], PT Fadhila[1], B Hariono[1] and R Wijaya[1]

[1]Department of Agriculture Technology, Politeknik Negeri Jember, Indonesia
[2]Department of Agribusiness Management, Politeknik Negeri Jember, Indonesia

dyahwardani@polije.ac.id

Abstract. Many people in Indonesia consumed tongkol fish by processing it into balado tongkol. The process carried out to extend the shelf life of the product is by packing it in cans. Information related to shelf life was needed for canned food, so that it can guarantee that the food was still good for consumption. This study aims to determine the shelf life of balado tongkol in cans from TEFA Fish Canning with the Arrhenius method. Arrhenius is the method of estimating shelf life used was ASLT (Accelerated Shelf Life Test). The ASLT method requires a relatively short testing time with an acceptable level of accuracy. This research was conducted at the TEFA Fish Canning and Food Processing Laboratory of Politeknik Negeri Jember in March – July 2021. Estimating the shelf life used the reaction approach that increases the TBA number as a critical parameter. The kinetic reaction of increasing the TBA number in canned balado tongkol followed the ordo 0 reaction. The results showed that the shelf life of balado tongkol in cans at room temperature was 24,6 weeks.

1. Introduction

Food is the most important basic of human need and its fulfillment is also part of human rights [1]. Food can be interpreted as derived from both treated and untreated biological sources and water. [2]. Public protection from unsafe food circulation is a guarantee that must be obtained by the community as consumer. Shelf life is directly related to the safety of the product and product quality assurance so as not to endanger the health of consumers. The fisheries sub-sector is the main source of food and nutrition for the people of Indonesia. Besides being a source of protein, fish is also recognized as a 'functional food' which has important meaning for health since it contains long-chain unsaturated fatty acids (especially those belonging to omega-3 fatty acids), macro and micro minerals, and vitamins. During the storage process, fish and processed fish products can undergo unwanted changes so as to limit storage time. These changes include protein oxidation ([3]; [4]) and lipid oxidation ([5]; [6]). One of the most commonly consumed fish is tongkol.

Tongkol is a commodity that rapidly declines in quality, in other words, it is easy to decompose because its high level of protein and water is easier to spoilage bacteria to grow [7]. One way to extend the shelf life of tongkol is by the canning process. Canning is one way of preserving food through a processing method that is packaged hermitically or using a very tight, watertight, airtight cover so as to minimize oxidation damage and changes in taste and then sterilize [8]. This study aims to determine the shelf life of a product from TEFA Fish Canning with the Arrhenius method.

2. Research methods

This research done in the Food Processing Laboratory and TEFA Fish Canning, Jember State Polytechnic for 6 months, from February to July 2021. The production process of balado tongkol in can is first carried out by the fish sorting process, followed by the exhausting process. After this, the process of filling the balado seasoning into the tongkol which will be canned is carried out. After the can is closed then the sterilization process is done. The research completed in several stages, (1) testing the production process; (2) identification of critical quality parameters and determination of shelf life.

The expiry date of the Balado Tongkol in can is determined by the accelerated method (Accelerated Shelf-life Testing or ASLT) with the Arrhenius model. Model selection is based on the consideration that the damage of balado seasoning tongkol is suspected to be caused by chemical changes in the food components during storage which could be triggered by storage temperature.

There are 30 cans of Balado tongkol in cans, where each can contains 115gr of Balado tongkol. Storage is carried out at three temperatures, which are 35°C, 45°C, and 55°C. There are 10 cans for each storage temperature. Observations during storage is carried out at the beginning of storage or day 0, within an interval of 1 week in along 1 month. Each observation is conducted on two cans (2 replicates) per each can and storage temperature. Observations are made on the critical parameters of TBA number using the distillation method.

3. Principle of estimating shelf life

The method of determining shelf life can be done by putting away the product until it is damaged under normal storage condition. This method produces the most valid information, but it takes time and impractical for industrial application. Therefore, an accelerated shelf-life testing or ASLT method was developed for estimating shelf life, where the product is kept under extreme storage conditions that can accelerate its deterioration. Shelf life is estimated using a mathematical model, where the factors that can affect product damage are included in the mathematical model [9]. The ASLT method requires a relatively short testing time with an acceptable level of accuracy. The more valid the mathematical model used, the more valid the prediction will be.

ASLT method oftenly used for estimating shelf life are the critical moisture content model and the Arrhenius model. The critical moisture content model is applied to estimate the shelf life of damaged food product by the water absorption of the product. The Arrhenius model is applied to foods that are easily damaged by chemical reactions such as fat oxidation, protein denaturation, Maillard reaction, etc.

The Arrhenius model is carried out by storing food product in final packaging stage at least of three extreme storage temperatures. Experiment with the Arrhenius method aims to determine the reaction rate constant (k) at several extreme storage temperatures, extrapolation then carried out to calculate the reaction rate constant (k) at the desired storage temperature using the Arrhenius equation (equation 1). From this equation, it can be determined the value of k (constant of quality degradation) at the storage temperature of the shelf life. The Arrhenius model equation to determine shelf life is expressed by equation (1).

$$k = ko.\exp\,(Ea/RT) \quad \text{...(1)}$$

where :
K = constant of deterioration rate
Ko = constant (frequency factor of independent temperature)
Ea = activation energy
T = absolute temperature (K)
R = gas constant (1,986 cal/mol K)

4. Results and discussion

4.1 Determination of the critical point

Determination of critical parameters used in estimating shelf life is specific for each product. In canned balado tongkol, the critical quality parameter of TBA number is used as an approach in estimating shelf life. Degradation of polyunsaturated fatty acids (PUFA) during storage leads to the formation of volatile components combined with rancidification [10]. A high percentage of unsaturated fatty acids means that fish tissue is easily pre-oxidized and quickly degraded. Oxidative changes are mainly related to the taste and texture of the fish. Changes in color, nutritional value or secondary product of lipid will be observed at an advanced stage of the lipid peroxidation process [11]. Oxidation of fish lipids is usually based on analysis of refractive index, peroxide value (PV) and reactive 2-thiobarbituric acid (TBARS) as indicators of primary and secondary oxidation products ([12]; [13]; [14]; [15]; [16]).

Table 1. Slope values, interceptions and determination on ordo 0 and ordo 1 reactions increasing of the TBA value in tongkol balado in cans

	Regression Equation		R2	
T (°C)	ordo 0	ordo 1	ordo 0	ordo 1
35	y = 0,1111x + 0,4226	y = 0,182x - 0,8367	**0,9527**	0,9312
45	y = 0,2295x + 0,4208	y = 0,2886x - 0,7866	**0,9763**	0,942
55	y = 0,3248x + 0,4102	y = 0,3439x - 0,7432	**0,9742**	0,9391

4.2 Reaction ordo determination

The speed of quality change of each product parameter is different. If the rate of damage is constant or linear, it follows the ordo 0 reaction. However, if the damage ratio is not constant, logarithmic or exponential, then an ordo 1 reaction is followed. Determination of the reaction ordo is a way to predict deterioration in quality in estimating shelf life. In kinetic reactions, food quality degradation follows ordo 0 and ordo 1. Ordo 0 can be detected by plotting between values of TBA as the Y axis and storage time as the X axis. Ordo 1 can be detected by plotting the value of ln TBA number as the Y axis and storage time as the X axis. The choice of ordo reaction can be seen by plotting the quality degradation data following the ordo 0 and ordo 1, then the linear regression equation is made. The reaction ordo with a larger R2 value is the reaction ordo used [17].

The value of determination (R2) for the increase in TBA number in the orde 0 reaction is higher than the ordo 1 reaction (Table 1). The value of determination (R2) of ordo 0 in canned tongkol balado ranged from 0.9527 to 0.9763. Thus, it can be seen that the reaction kinetics of the increase in the TBA number during storage of Balado tongkol fish product in the package follow the pattern of ordo 0 reaction. The higher the value of determination, the more accurate the result of the data analysis. The positive slope value indicates an increase in the TBA number in canned balado tongkol (Table 1). The value of the gradient or slope (k) states the relationship between the value of quality degradation and storage time. This is in accordance with [18] which showed an increase in the number of TBA (Thiobarbituric Acid) in andaliman sauce during storage. The increase in TBA number in this product is due to the heat-induced oxidation reaction. In the appearance of tomato paste, the brightness of the color also decreased to become more brownish due to lycopene oxidation [19].

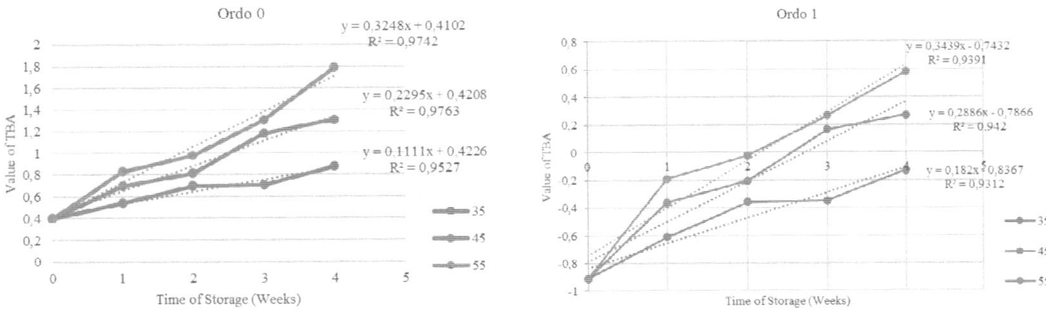

Figure 1. Graphic of relationship of value of TBA degradation to time

4.3 Shelf life estimation

According to [20], The shelf life is the period until the quality of the product deteriorates, and it does not meet the standards specified on the packaging and should not be eaten. (the quality is not yet in the guaranteed quality level), due to the ongoing deterioration reaction. The deterioration reaction leads to poor quality and poor quality of the product, making it unsuitable for consumption. This deterioration reaction will cause changes to the canned balado tongkol fish product. Determination of the critical value of tongkol balado in can is obtained from the TBA number assessment parameter. Response order selection is made by plotting the degradation data after response ordo 0 and response ordo 1, and then writing a linear regression equation.

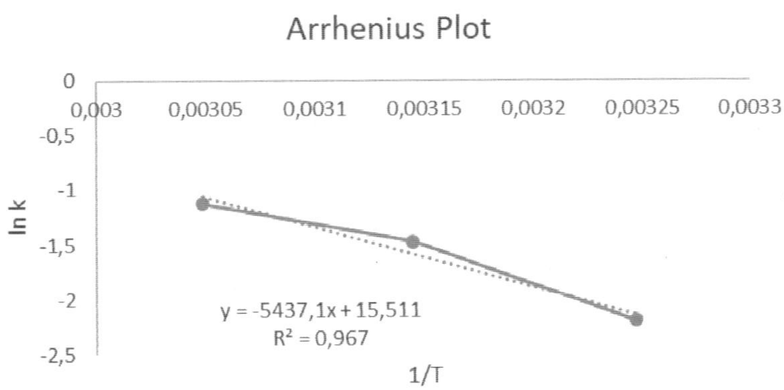

Figure 2. Graphic of TBA value arrhenius plot

The shelf life of tongkol balado fish product in can, assuming the temperature in room temperature of 25°C for 24,6 weeks (24 months 18 days). [21] state that oxidation can lead to the formation of off-flavor compounds and this condition is called rancid. It is the process of breaking down fat. Due to oxidation of unsaturated fatty acids, rancid processed foods lose nutritional value and cause discoloration, which affects quality. Oxidized compounds such as peroxides, ketones, and aldehydes are harmful to human health. While [22] stated that factors affecting the rate of oxidation include the chemical structure of lipids, the amount and type of oxygen, the presence of antioxidant and pro-oxidant compound, the nature of packaging materials, and storage temperature.

5. Conclusion

Estimating the shelf life of tongkol balado in can using the ASLT method by using the reaction approach of increasing the TBA values as a critical parameter. The kinetic reaction of increasing the TBA number in canned balado tongkol followed the ordo 0 reaction. Shelf life of tongkol balado in can using the

ASLT method through an approach of critical parameter of the TBA number at room temperature 25 °C has a shelf life of 24,6 months.

Acknowledgements
This paper is dedicated to Politeknik Negeri Jember. This paper is part of the 2021 Higher Education grant applied research.

References
[1] Hermanuadi, D. Brilliantina, A, and Sari, E.N. 2020. *IOP Conf. Series: Earth and Environmental Science* **411**: 012016
[2] Kurnianto M F, Wibowo M J, Hariono B, Wijaya R, Brilliantina A. 2020. *IOP Conf. Ser.: Earth Environ. Sci.* **411** 012051
[3] Fijuwara, K., Oosawa, T. dan Saeki, H. 2008. Improved thermal stability and emulsifying proper-Ties of carp myofibrillar proteins by conjunction with dextran. *Journal of Agricultural and Food Chemistery* **46**: 1257–1261
[4] Benjakul, S., Viessanguan, W., Thongkaew, C. dan Tanaka, M. 2005. Effect of frozen storage on chemical and gel-forming properties of fish commonly used for surimi production in Thailand. *Food Hydrocolloids* **19**(2): 197–207
[5] Sarma, J., Reddy, G.V.S. dan Srikar, L.N. 2000. Effect of frozen storage on lipids and functional properties of proteins of dressed Indian oil sardine (*Sardinella longiceps*). *Food Research International* **33**: 815–820
[6] Richards, M.P. dan Hultin, H.O. 2012. Contributions of blood and blood components to lipid oxidation in fish muscle. Journal Agriculture Food Chemistry **50**: 555–564
[7] Nabila, L., Tamrin., K.T Isamu. 2017. Karakterisasi organoleptik, kimia, dan mikroba ikan kayu cakalang (katsuwonus pelamis) dan ikan kayu tongkol (euthynnus affinis) yang diproduksi di kota kendari. *Jurusan Sains dan Teknologi Pangan*, Volume **2** (3). P. 530-541
[8] Muntikah dan M. Razak. 2017. Ilmu teknologi pangan. Kemenkes RI: Pusat Pendidikan Sumber daya Manusia Kesehatan
[9] Floros, J.D. dan V. Gnanasekharan. 1993. Shelf Life Prediction of Packaged Foods: Chemical, Biological, Physical, and Nutritional Aspects. Di dalam Chlaralambous, G (Ed.). Instrumental Methods in Food and Beverage Analysis. Elsevier Publ., London.
[10] Pazos, M., Gallardo, J.M., Torres, J.L. dan Medina, I. 2005. Activity of grape polyphenols as inhibitors of the oxidation of fish lipids and frozen fish muscle. *Food Chemistry* **92**: 547–557.
[11] Dragoev, S.G., Kiosev, D.D., Danchev, S.A., Ioncheva, N.I. dan Genov, N.S. 2008. Study on the oxidative processes in frozen fish. *Journal of Agriculture and Science* **4**: 55–65.
[12] Aranda, M., Mendoza, N. dan Villegas, R. 2005. Lipid damage during frozen storage of whole jack mackerel (*Trachurrus symmetricus* Murphyi). *Journal of Food Lipids* **13**: 155–166
[13] Chaijan, M., Benjakul, S., Visessanguan, W. dan Faustman, C. 2006. Changes of lipids in sardine (Sardinella gibbosa) muscle during iced storage. Food Chemistry **99**: 81–83.
[14] Pourashouri, P., Shabanpour, B., Aubourg, S.P., Rohi, J.D. dan Shabani, A. 2009. Aninvestigation of rancidity inhibition during frozen storage of Wels catfish (Silurus glanis) fillet by previous ascorbic and citric acid treatment. International Journal of Food Science and Technology **44**: 1503–1509.
[15] Quitral, V., Donoso, M.L., Ortiz, J., Herrera, M.V., Araya, H. dan Aubourg, S.P. 2009. Chemical changes during the chilled storage of Chilean jack mackerel (Trachurus murphyi): Effect of a plant-extract icing system. LWT–*Food Science and Technology* **42**: 1450–1454.
[16] Yerlikaya P, and Gokogku N. 2010. Inhibition effects of green tea and grape seed extract on lipid oxidation ini bonito fillet during frozen storage. *International Journal of Food Science and Technology* **45**: 252-257.

[17] Arif, A. Bin. 2016. Metode Accelarated Shelf Life Test (ASLT) Dengan Pendekatan Arrhenius Dalam Pendugaan Umur Simpan Sari Buah Nanas, Pepaya Dan Cempedak. *Informatika Pertanian*, **25(2)**, 189–198

[18] Bancin, L. H. 2018. Pengaruh Penambahan Gum Arab Terhadap Mutu Sambal Andaliman dan Penentuan Umur Simpannya. Universitas Sumatera Utara

[19] Sikora, M., Badrie, N., Deisingh, A. K., Kowalski, S. 2008. Sauces and Dressing: A Reviewof Properties and Applications. *Critical Reviews in Food Science and Nutrition*, **1(1),** 50–77

[20] Arpah, M., Syarief, R. 2000. Evaluasi ModelModel Pendugaan Umur Simpan Pangan dari Difusi Hukum Fick Unidireksional. *Buletin Teknologi Dan Industri Pangan*, **XI(I),** 11–16

[21] Secbecic, N. dan Beutelspecher, S.C. 2005. Anti-oxidative vitamins perevent lipid-peroxidation and apoptosisi in corneal endotheliel cell. *Cell Tissue Respirative* **320**: 465–475

[22] McClements, D.J. dan Decker, E.A. 2000. Lipid oxidation in oil-in-water emulsions: Impact of moleculer enviroment on chemical reaction in heterogenous food system. *Journal of Food Science* **65**: 1270–1282.

IOP Conf. Series: Earth and Environmental Science **980** (2022) 012039 doi:10.1088/1755-1315/980/1/012039

Prototype of Integrated Mini Exhausting System For Fish Canning Process

W Suryaningsih[1*], A Bakri[1], S Kautsar[2], B Hariono[1], A Brilliantina[1], R Wijaya[1]

[1]Department of Agriculture Technology, Politeknik Negeri Jember, Indonesia
[2]Department of Engineering, Politeknik Negeri Jember, Indonesia

[*]wahyu_surya@polije.ac.id

Abstract. The pre-cooking process carried out in the fish canning industry still uses the method of steaming on the stove. The difference in the capacity of the pre-cooking and sterilization processes causes a long holding time, which has the potential for microorganism contamination and low productivity. Research on the prototype of an integrated semi-automatic mini exhausting unit with a capacity of up to 300 cans of 115 mL using a hot steam source from a sterilization retort has to increase production capacity and efficiency. The size mini exhausting has a size of 4 meters x 1 meter and is made of 304 stainless material. It has 3 parts, namely the preparation part, the heating part, and the draining section. The drain part was located on the outside of the mini exhausting box. The draining part has a pan that can be rotated 150 degrees to remove the fish liquid in the can before the next process. The results of the exhausting room temperature reached 90°C within 8 minutes at an autoclave pressure of 0.8 bar. Fish meat reaches a temperature of 80 degrees after heating in the 12th minute. In the mini exhausting system, production capacity can be increased up to 8 times, with 25% electrical energy efficiency.

1. Introduction

Precooking is an important to stage in the fish canning process. Precooking aims to reduce the number of microbes, inactivate enzymes, reduce the water content of fish, make meat more compact, reduce fat content, air (oxygen), and reduce the pressure inside the can. The pre-cooking process usually uses hot steam with a minimum temperature of 60°C. The pre-cooking temperature used reached 90°C for 18-20 minutes (Bratt,2010). The initial cooking temperature varies based on the size of the fish to be processed. In general, the initial cooking temperature is around 98°C. This temperature using so that the fishbone temperature reaches a minimum of 60-65°C [1]. The minimum temperature for cooked fish is measure at the deepest part of the fish. According to the official websites of the FDA and USDA, the minimum temperature for cooked fish is 145°F or 62.8 °C.

The precooking process in large-scale fish canning production uses an exhausting device with a steam boiler. Meanwhile, in the small-scale fish canning process, steaming is still used on the stove with a capacity of 40 small 115 mL cans. If the retort used for the sterilization process has the capacity of 200 cans, then the precooking process will cause a long holding time. This condition can lead to contamination and inefficient. This steaming method has many drawbacks, namely uncontrolled temperature, unstable heat, and low production capacity. This condition can cause the fish to be undercooked or overcooked so that it will affect the quality of the fish canning product.

To increase productivity and improve the quality of products. it is necessary to make a prototype of a mini exhausting system integrated with the fish canning process. This concept considers the use of a joint heating system between the pre-cooking and sterilization processes. Water vapor for the pre-cooking process uses steam from the autoclave. After the pre-cooking process is complete, the sterilization process uses the same autoclave, so that there is an efficiency of heating time from the heat stored in the machine from the previous process [2]. The application of this integrated mini exhausting system can shorten the heating and processing time because it uses the heat stored in the retort from the pre-cooking process. The purpose of this research was made and calculated field capacity prototype a mini exhausting system integrated of fish canning process.

2. Related Work

An exhauster is a tool used to create a vacuum condition in the headspace of the can before the can is closed which is called exhausting [3]. This exhausting process aims to reduce the oxygen content in the can (especially when heating in the retort) to reduce corrosion, limit the oxidation process by food, and prevent the growth of aerobic microorganisms that will reduce quality and safety [4].

The exhaust consists of a belt conveyor , a pipe equipped with a spreader, a valve for regulating the flow of hot steam, and an exhaust box. The working principle of the exhauster is to drain hot steam from the boiler through a pipe equipped with a spreader into the exhaust box. This hot steam is used to expel the air in the can headspace running on the rails in the exhauster. The exhausting time is set by adjusting the speed of the belt conveyor. While the exhausting temperature is set by adjusting the steam faucet on the exhauster [5]

The exhausting process has a working principle, namely steam supplied by the boiler, flowing through a pipe into the exhaust box. This hot steam is used to remove the air in the can which runs on the rails in the exhaust box. The speed of the exhausting process is regulated by adjusting the rail speed. Furthermore, the hot steam from the boiler has flowed through the pipe [6]. The exhauster temperature can be adjusted by adjusting the steam valve. The material to be exhausted in the can is placed on a chain on the outside of the exhaust box. The lid of the can container is placed behind the can and passes through the rails together, after exiting the exhaust box the can is immediately closed. After the process is complete the boiler steam flow is stopped and the hot steam flow valve is closed, the tool is turned off and cleaned [3]

3. System Design

3.1 Hardware Design

The Mini Canning Exhausting (MCE) prototype is designed for pre-cooking with a capacity of 154 cans in 115mL packaging. MCE consists of 3 parts: the preparation section, the heating section, and the filter section. The three parts are connected in a conveyor system using stainless ball bearings. The first part is the preparation section, which has a size of 46 cm x 73.5 cm and a capacity of 77 cans of 115 mL which will be used for the precooking preparation process, namely the arrangement of cans and the installation of temperature sensors. The second part is for the process of heating fish or precooking. This section consists of a heating box, steam lines, chimneys, and water reservoirs. The heating box measuring 107 cm x 65.5 cm x 44.5 cm, has a capacity of 154 cans. This section also has a door that can be opened and closed which serves to enter and exit the can during the precooking process. The third part is water and fish filtration. This section is a hollow box made of stainless steel, which can accommodate one pan with a capacity of 77 cans. The box can be rotated up to 150 degrees to remove residual water and oil from the pre-cooking process. The design of the MCE prototype is shown in Figure 1

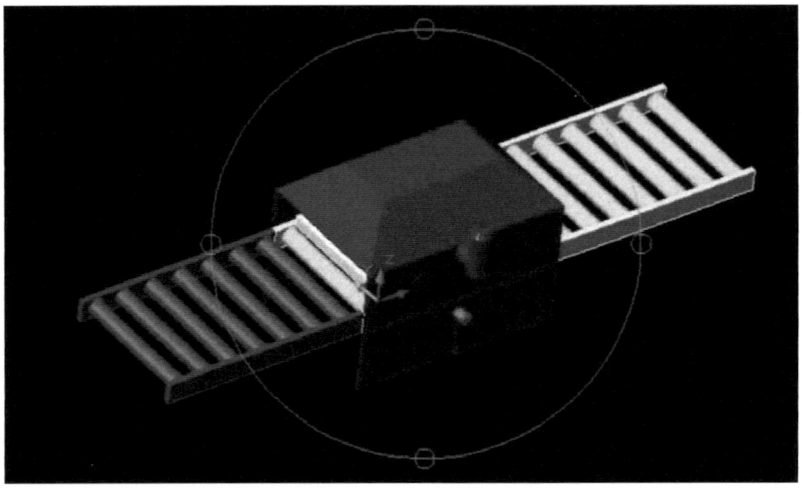

Figure 1. Desain prototype mini canning exhausting

3.2 Software Design

This integrated mini exhausting system equipment is equipped with a microcontroller-based digital monitoring system that functions to measure the quality of the pre-cooking process. This monitoring system consists of 4 DS18B waterproof temperatures sensor and one thermocouple sensor. The DS18B sensor probe is used to measure the temperature in the can, while the thermocouple is used to measure the temperature of the heating box. Sensor values are displayed on the OLED graphic LCD. If the average temperature of the four sensors >= 80 degrees Celsius, the buzzer will sound and the microcontroller will start calculating the heating time. This setting system will make it easier for the operator to set the pre-cooking time to match the desired processing results or set standards. The system can also be connected to a GUI on the computer to store data during the pre-cooking process and display it in graphical form. Figure 2a is a block diagram of the monitoring system on the MCE, and Figure 2b is a programming flow diagram for the ATMega328 microcontroller.

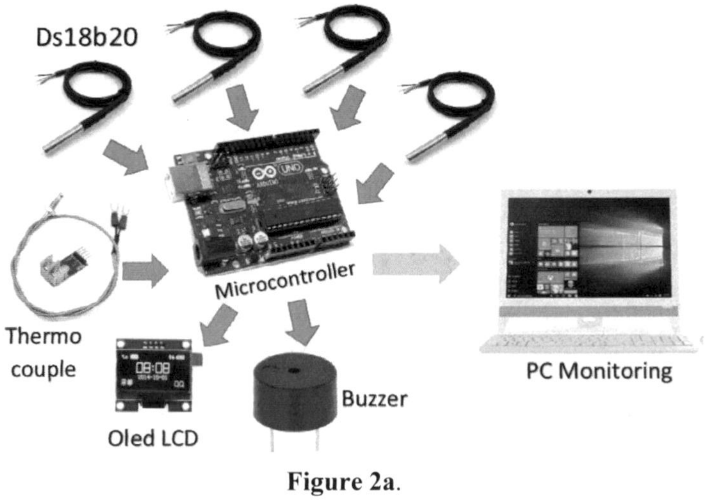

Figure 2a.

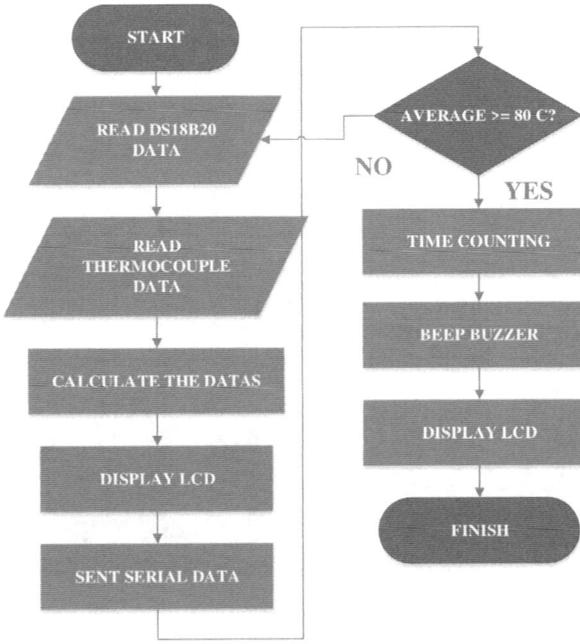

Figure 2b Flowcart system

4. Result and Discussion

4.1. System Realization

The MCE prototype uses 304 stainless steel to be safe from contamination. Figure 3 is the realization of the MCE prototype. The hot steam generated from the retort will be channeled into the heating chamber. The hot steam trapped in the box will circulate throughout the room and heat the fish in the can. There is a chimney at the top of the heating chamber. The chimney is equipped with a valve that can be manually rotated to dissipate hot steam if the room temperature exceeds the pre-cooking threshold. At the bottom of the heating room, there is a container for storing residual water from the heating process. After the heating process is complete, the can is inserted into the canning rotator to remove the remaining water and oil in the can [7].

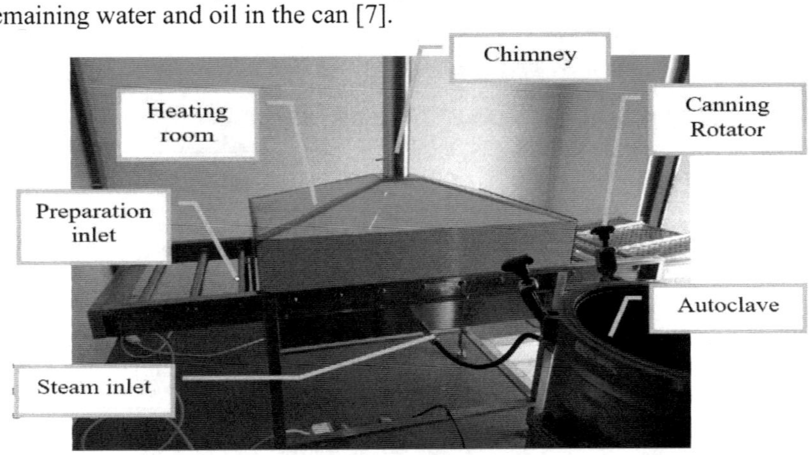

Figure 3. Prototipe MCE

According to the system design in chapter 3, a microcontroller-based MCE temperature monitoring device was created. Figure 4 is the result of the realization of the temperature monitoring device. The monitoring device can be powered in 2 ways: using PLN electricity, and/or using power from the computer's USB port. The temperature data of each sensor, average temperature, and heating time will be displayed on the LCD screen. The sensor used is waterproof with a working capacity of more than 110 degrees Celsius so it is safe to use in the production process that uses hot steam.

Figure 4. MCE

4.2. Testing result

The test was carried out by carrying out 1 production process (154 cans of 115mL size). Fish that have been cut and put in cans are placed on a baking sheet and sensors are mounted on the inside of the meat. Figure 5 is how to place the temperature sensor on the can. The sensor probe is placed inside the fish meat so that the temperature measured during the pre-cooking process is truly accurate. The thermocouple sensor is placed freely on the outside of the can to measure the temperature conditions of the heating room. In the initial process, 20 liters of water are heated in the Autoclave. The heating process to form steam at a pressure of 1.4 Bar takes about 35 minutes. After that, the steam inlet valve is opened to allow hot steam to flow into the heating chamber. From the normal temperature (27 degrees Celsius), the fish reaches an average temperature of 80 degrees after 30 minutes. The heating process data is stored in the application and displayed in the graphic figure 6. The microcontroller will send serial temperature data updates to the computer every 1 second.

Figure 5. Placement of sensors on canned fish

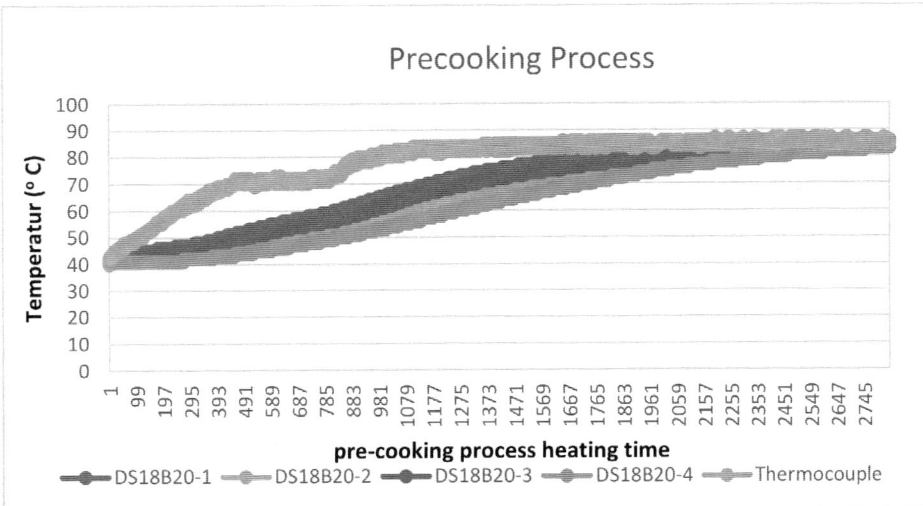

Figure 6. Precooking Graph

Figure 6. The graph of the pre-cooking process shows that the room temperature is around 70 degrees Celsius. This is influenced by the distribution of heat to canned fish. After the average temperature of the fish reaches 70 degrees Celsius, the temperature of the heating room begins to rise again to 85 degrees Celsius. The pre-cooking time starts to count when the average temperature of the four DS18B20 sensors reaches 80 degrees Celsius. According to the pre-cooking procedure of canned fish, the minimum heating time is 12 minutes. The autoclave steam pressure during heating is stable at 1 bar. Figure 7. shows the conditions during the pre-cooking process after the pre-cooking process for 12 minutes.

Figure 7. Precooking Process

Overall, the time for the pre-cooking process is shown in table 1. At the time of the initial heating of the autoclave, the temperature of the heating room was around 31 degrees Celsius. The average temperature of the fish is 27 degrees Celsius because the fish is frozen before slaughter. After 40 minutes, the steam in the autoclave can be used (pressure 1.2bar). The MCE steam inlet valve is then opened about 30% to allow water vapor to flow into the heating chamber. The autoclave pressure will decrease slowly at a pressure of 1 bar. With autoclave steam trapped in the heating chamber, causing heat to be distributed to the fish in the can. 4 sensors are used which are placed evenly to ensure that the heating in the 154 cans is evenly distributed. After the pre-cooking process is complete, the process of seasoning, closing cans, and sterilization is carried out. Cans sterilization is done by placing cans in an autoclave at a pressure of 1 ATM for 20 minutes. Because the autoclave still stores the residual heat from the pre-cooking process, heating for the sterilization process takes 25% faster. So that the sterilization process can save electricity usage of 750Wh [8].

Table 1. Test Data on Mini Exhausting Equipment

Time	Heating Room Temperature	Test Data Fish Average Temperature	Status
0	31 °C	27 °C	Autoclave Heating
40	30 °C	27 °C	Autoclave 1.2 Bar 90 °C
50	70 °C	35 °C	Steam is streamed to MCE
60	72 °C	45 °C	Steam is streamed to MCE
70	85 °C	78 °C	Steam is streamed to MCE
80	86 °C	82 °C	The heating counter starts counting
92	80 °C	78 °C	The heating chamber is opened, the steam stops flowing
100	60 °C	65 °C	The can is removed from the heating chamber

5. Conclusion

This mini exhausting measures 2 meters x 0,75 meter with capacity about 154 cans and is made with 304 stainless materials, and on the outside of the mini exhausting box, there is a rotating pan mechanism to remove the liquid in the can before the next process. The results of the exhausting room temperature reaches 85°C within 12 minutes at autoclave pressure of 1.4 bar. Fish meat reaches a temperature of 80 degrees after heating in the 30th minute. With the use of a mini exhausting system, production capacity can be increased up to 8 times, with an electrical energy efficiency of 25%.

Acknowledgment

The authors would like to acknowledge the financial support of this work by grants from PNBP, State Polytechnic of Jember. The author also thanks to the P3M and Information Technology Department, State Polytechnic of Jember, which has provided support and assistance in completing this research.

References

[1] Fao 2013 *GLOBEFISH RESEARCH PROGRAMME Food and Agriculture Organization of the United Nations Products Technical guide to fish canning*

[2] Val Del Rio A, Pichel A, Fernandez-Gonzalez N, Pedrouso A, Fra-Vázquez A, Morales N, Mendez R, Campos J L and Mosquera-Corral A 2018 Performance and microbial features of the partial nitritation-anammox process treating fish canning wastewater with variable salt concentrations. *Journal of environmental management* 208 112–21

[3] Torregrosa-García A, Ávila-Gandía V, Luque-Rubia A J, Abellán-Ruiz M S, Querol-Calderón M and López-Román F J 2019 Pomegranate extract improves maximal performance of trained cyclists after an exhausting endurance trial: A randomised controlled trial *Nutrients* 11

[4] Miao X, Xiao B, Shui S, Yang J, Huang R and Dong J 2018 Metabolomics analysis of serum reveals the effect of Danggui Buxue Tang on fatigued mice induced by exhausting physical exercise. *Journal of pharmaceutical and biomedical analysis* 151 301–

[5] Girard C, Charette T, Leclerc M, Shapiro B J and Amyot M 2018 Cooking and co-ingested polyphenols reduce in vitro methylmercury bioaccessibility from fish and may alter exposure in humans. *The Science of the total environment* 616–617 863–74

[6] Suryaningsih W, Supriyono S, Hariono B and Kurnianto M F 2020 Improving the Quality of Smoked Shark Meat with Ozone Water Technique *IOP Conference Series: Earth and Environmental Science* 411

[7] Ndahawali D H, Wowiling F, Risnawati, Pongoh S, Kaharu S, Gani S H and Sasara S M 2016 Studi Proses Pengalengan Ikan Di PT . Sinar Pure Foods International Bitung *Buletin Matric* 13 42–53

[8] Zhafirah F and Sipahutar Y H 2021 Proses Pengolahan Ikan Tongkol Abu-abu (Thunnus tonggol) dalam Kaleng dengan Media Air Garam di PT. Jui Fa Interbational Food, Cilcap-Jawa Tengah *in Prosiding Simposium Nasional VIII Kelautan dan Perikanan Fakultas Ilmu Kelautan dan Perikanan, Universitas Hasanuddin, Makassar, 5 Juni 2021* 57–68

IOP Conf. Series: Earth and Environmental Science **980** (2022) 012040 doi:10.1088/1755-1315/980/1/012040

Formulation of Local Food Based Snack Using Linear Program for Nutrisurvey to Prevent Stunting in Under-Five Children

P R Ratri [1*], Sabran [2], M Jannah [1]

[1] Departement of Clinical Nutrition, State Polytechnic of Jember, Jalan Mastrip PO Box 164, Jember 68121, Indonesia.
[2] Department of Health Information Management, State Polytechnic of Jember, Jalan Mastrip PO Box 164, Jember 68121, Indonesia.

* email: putri_ratri@polije.ac.id

Abstract. The prevalence of stunting in Indonesia has decreased from 37.2% in 2013 to 30.8% in 2018. This prevalence is still far from the government target, which is 14%. Immunostimulant prebiotic food formulations in the form of snack bars made from lempeni leaves, edamame, and local herbal plants can be used to increase immunity and prevent stunting among children. However, the development of a local-based diet by considering the nutrient adequacy of the children is lacking. This study aimed to develop an optimized snack bar based on nutrient adequacy using linear programming to meet the nutrient requirement of under-five children. This study used Nutri Survey for Linear Programming 2004 software to develop an optimized snack bar. Nutrition Adequacy Rate among children 1-4 years were used to determine the composition of the snack bar. The composition of the snack bar consists of Simplicia powder of lempeni leaves, edamame, cinnamon powder, ginger powder, oats, almonds, cashew, raisins, honey, peanut butter, and chocolate. The Results show two formulas based on Linear Programming calculation and one formula for control. Snack bar formula fulfilled energy and macronutrient intake requirements within the acceptable macronutrient distribution range. Formula 1 contains 122 kcal energy; 2,9 g protein; 4,9 g fat; 13,4 g carbohydrate, 0,8 mg iron; and 0,4 mg zinc. Formula 2 contains 123 kcal energy; 3,05 g protein; 5,6 g fat; 13,5 g carbohydrate, 0,9 mg iron; and 0,4 mg zinc. Those values are describe the optimization model used in this study provides a mathematical solution for an economical snack bar that approximate the nutrient requirements for under-five children to prevent stunting.

1. Introduction

One of the concerns in the golden age of a child's life is a stunting. Stunting will certainly affect the child's immune system. Stunting is referred to as short stature as a manifestation of chronic malnutrition. Stunting is often not realized at the age of toddlers. Several factors that are thought to be risk factors for stunting include family factors, diet, and socioeconomic conditions [1]. Stunting is also called chronic nutrition which describes a disturbance in height growth that lasts for a long time. Stunting causes poor development in toddlers, impaired cognitive function, metabolism and decreased activity [2].

In addition to maintaining a healthy lifestyle, it can also be done by utilizing local plants as herbal medicines. Herbal plants have been known as local wisdom that have been used since ancient times as alternative medicines. Some of the local plants commonly used include red ginger (*Zingiber officinale*), turmeric (*Curcuma longa* L.), temulawak (*Curcuma xanthorriza* Roxb), green tea (*Camelia sinensis*), meniran (*Phyllantus niruri* L.), salam (*Syzygium polyanthum*), guava (*Psidium guajava*), clove (*Sygizium aromaticum*), and garlic (*Allium Sativum*) [3].

Ginger and cloves are herbal plants that are widely used to be used as immune boosters. In addition to the plants above, there is one local herbal plant that is also efficacious for maintaining endurance, namely the lempeni plant. Lampeni (*Ardisia elliptica*) is one of the plants that exist in Indonesia and is spread in Java, Sumatra, Sulawesi, and Maluku. This plant is in the form of a shrub or small tree. The leaves have a rough and elastic texture. The fruit and leaves of lempeni (*Ardisia elliptica*) are edible and efficacious in treating digestive problems. These plants are certainly very beneficial for the body if they can be consumed [4].

However, for children, consuming pure medicinal plants is a difficult thing to do. Therefore, alternative foods are needed that can also be combined and matched with other plants so that they have benefits that are not only high in nutrition but also can increase the body's immune system. One food source that can be an alternative with nutritional value is edamame. Edamame as soybeans with high production potential has various types of preparations that are easily found in the market due to the high content of phytosterol compounds [5]. Edamame soybean, which is a type of white soybean, has a sweeter taste, stronger nut aroma, softer texture and nutritional content that is easily digested by the body. This soybean has a high content of isoflavones where isoflavones have properties as antioxidants [6]. Edamame has a complete protein content whose quality is equivalent to the protein content in milk, eggs and meat. Edamame is rich in protein, dietary fiber, and micronutrients, especially folate, manganese, phosphorus, and vitamin K. The balance of fatty acids in 100grams of edamame is 361 mg of omega 3 fatty acids-1794 mg of omega 6 fatty acids. In addition, edamame also contains anti-cholesterol substances so it is very good for consumption [1].

This study aims to analyze the potential of snack bars based on local food, edamame, lempeni leaves and herbal active ingredients as snacks for toddlers in order to prevent stunting and increase body immunity.

2. Materials and Methods

This research is an experimental study in a laboratory by applying a linear program consisting of 2 formulation factors and 1 control formulation. The linear program used is the Nutrisurvey for Linear Programming application. The snack bar formulation uses a linear program with the help of chemical analysis data from Lempeni leaf powder, edamame nuts, cinnamon powder, ginger powder, and other supporting ingredients. After obtaining several formulas that are feasible according to a linear program both in terms of quality and cost. Then the linear program result 2 formulations.

The ingredients of snack bar are Lempeni leaf simplicia powder, edamame nuts, almonds, cashews, oats, cinnamon powder, ginger powder, raisins, honey, peanut butter, and chocolate. The tools used in the snack bar manufacturing process include stirrer, scales, aluminium pan, oven, stainless steel knife, stirrer, aluminium container, pan, gas stove, and plastic container.

The snack bar are made by chopped the main raw material for nuts (cashew, almond and edamae). After that it is mixed with other ingredients and then stirred. The mixture is then formed into a rectangular box and baked in the oven in 105°C for 15 minutes. After cooking, wait a few moments for caramelization to form a good *snack bar.*

3. Results and Disscussion

Nutrisurvey for Linear Programming is used to create a formula that is in accordance with the nutritional adequacy rate (RDA) and price (public purchasing power). This snack bar product targets children under five to prevent stunting and increase body immunity. Therefore, the RDA used is the RDA for children aged 1-3 years. The snack bar will be used as a snack for toddlers, where a daily snack contributes 20% of energy needs.

Snack bar is a new product for Indonesian people. Snack bars are solid snacks in the form of sticks and are a mixture of various dry ingredients such as cereals, nuts, dried fruits which are combined into one with the help of a binder. Binders in bars can be in the form of syrup, nougat, caramel, chocolate, and others.[8] Therefore, snack bars have the potential to be a good and effective for children alternative food.

The Control snack bar (Control/F0) are using several food ingredients as composition, including: cashews, almonds, oats, honey, raisins, chocolate, and peanut butter. In the snack bar treatments (F1 and F2), the composition was modified by adding cashew nuts, powdered ginger, cinnamon powder, and powdered lempeni leaves. The product formulations for snack bars F1 and F2 are the results of the analysis using Nutrisurvey for Linear Programming as follows:

Table 1. Result of Snack bar formulation using Nutrisurvey for Linear Programming

Composition	Nutrition	Min – max	Price
		Formula 1 (F1)	
Edamame 10 gram Cashew 5 gram Almond 5 gram Cinnamon 1 gram Ginger powder 1 gram Oats 6 gram Chocolate 2 gram Honey 13 gram Raisins 6 gram Lempeni leaves powder 1 gram Peanut butter 5 gram	Energy = 243,6 kcal Protein = 5,8 g Fat = 9,7 g Carbohydrate = 26,8 g Vitamin C = 1,4 mg Zinc = 0,7 mg Fe = 1,7 mg Calcium = 56,7 mg	Energy = 243 – 297 kcal Protein = 3,6 – 4,4 Fat = 8,1 – 9,9 Carbohydrate = 38,7 – 47,3 Vitamin C = 7,6 – 8,4 Zinc = 0,57 – 0,63 Fe = 1,33 – 1,47 Calcium = 123,5 – 136,5	Rp 5,279
		Formula 2 (F2)	
Edamame 7 gram Cashew 7 gram Almond 7 gram Cinnamon 1 gram Ginger powder 1 gram Oats 6 gram Chocolate 2 gram Honey 13 gram Raisins 6 gram Lempeni leaves powder 1 gram Peanut butter 6 gram	Energy = 245,9 kcal Protein = 6,1 g Fat = 11,2 g Carbohydrate = 27,1 g Vitamin C = 1,3 mg Zinc = 0,8 mg Fe = 1,8 mg Calcium = 65,4 mg	Energy = 243 – 297 kcal Protein = 3,6 – 4,4 Fat = 8,1 – 9,9 Carbohydrate = 38,7 – 47,3 Vitamin C = 7,6 – 8,4 Zinc = 0,57 – 0,63 Fe = 1,33 – 1,47 Calcium = 123,5 – 136,5	Rp 5,878

Source : Primary data

Note: The weight and composition of the snack bar above is the size if it is used for snacks in 1 day (2 times a day). To get 1 time interlude, then the value above is divided into half.

The chemical composition of the raw materials (moisture content, protein content, fat content, and carbohydrate content) is used as the coefficient value of each variable in linear program modeling. In addition to the chemical composition of the raw materials, the data collected also includes the prices of the main ingredients (almond, cashew, edamame, peanut butter, honey, raisin, oats, lempeni leaves powder, ginger powder, and cinnamon powder) as well as the limiting factors that are the nutritional requirements of snack bars. The purpose of using a linear program is to determine the snack bar formula in order to produce a snack bar that is accepted by consumers, and the price is relatively cheaper than the price of snack bars on the market. Based on the calculation results, the first formula is the formula with the lowest price, which is Rp. 5,279 for two snack bar (2,639/one snack bar).

Source: Study documentation

Figure 1. Snack bar and slice of snack bar

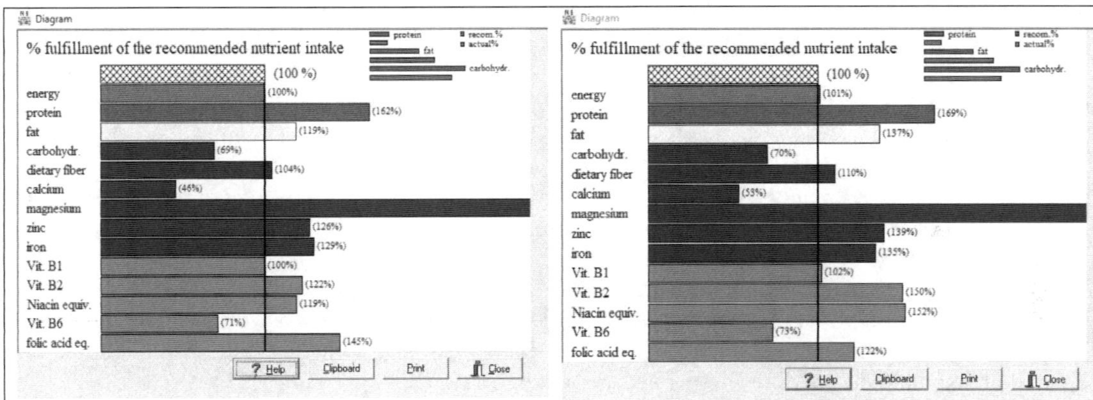

a. Formula 1 b. Formula 2

Figure 2. The fulfilment of recommended nutrient intake of herbal snack bar formula

Based on Figure 2, it can be seen that the two formulas produced by the Linear Program have met the percentage of nutritional needs in accordance with the recommended nutritional intake. In general, formula 2 shows a higher value. This may be influenced by the ratio of the main ingredients used in determining the formula. Formula 1 uses a 2:1 ratio for the main ingredients edamame (double amount) compare to others main formula such as almond, cashew, and oats. While formula 2 uses a 1:1 ratio for all main ingredients.

In addition, calorie content is an important element that must be owned by snack bars as snacks that can delay hunger. Each person's calorie needs are different according to age, height, weight, gender, and level of activity per day. The number of calories in a food product shows the amount of energy contained in an ingredient or food product. The largest source of energy is fat which produces 9 kcal of energy per g, while carbohydrates and protein produce energy of 4 kcal per g. [9] The energy value calculation for this product is 243,6 for formula 1 and 245,9 for formula 2. Based on the results of the study this amount is sufficient to meet the calorie needs for additional food or snacks for toddlers.

4. Conclusion

Snack bar formula fulfilled energy and macronutrient intake requirements within the acceptable macronutrient distribution range. Formula 1 contains 122 kcal energy; 2,9 g protein; 4,9 g fat; 13,4 g carbohydrate, 0,8 mg iron; and 0,4 mg zinc. Formula 2 contains 123 kcal energy; 3,05 g protein; 5,6 g fat; 13,5 g carbohydrate, 0,9 mg iron; and 0,4 mg zinc. Those values are describe the optimization model used in this study provides a mathematical solution for an economical snack bar that approximate the nutrient requirements for under-five children to prevent stunting.

5. Acknowledgment

This study was funded by the DIPA of State Polytechnic of Jember SP DIPA-023.18.2.677607 / 2021, dated 23 November 2021 budgeting year 2021, PNBP Research Grant.

References

[1] Sudiarti D. 2017. The Effectiveness Of Biofertilizer On Plant Growth Soybean Edamame (Glycin max). *SainHealth*. Oct 1;1(2):97–106.
[2] Lestari A, Hanim D. 2020. Edukasi Kader dalam Upaya Pencegahan dan Penanggulangan Stunting di Kecamatan Mondokan Kabupaten Sragen. *AgriHealth: J of Agri-food, Nutrition and Public Health*. Apr 30;1(1):7–13.
[3] Dewi Y K, and Riyandari B A. 2020. Potensi Tanaman Lokal sebagai Tanaman Obat dalam Menghambat Penyebaran COVID-19. *Pharmascience*. Oct 31;7(2):112–28.
[4] Kusumastuti S A, Firdayani F, and Chaidir C. 2015. Potensi Ekstrak Etanol Daun Lampeni (Ardisia Elliptica) Dan Fraksinya Sebagai Agen Antiproliferatif Terhadap Sel Kanker Hati Hepg2. Prosiding SNaPP: Kesehatan (Kedokteran, Kebidanan, Keperawatan, Farmasi, Psikologi). Oct 1;1(1): 389–94.
[5] Kurniawan N D, Setiani B E, and Dwiloka B. 2019. Kadar Lemak, Kadar Air, Kadar Protein, Dan Antioksidan Tempe Edamame (Glycine Max (L) Merrill) Dengan Jenis Pengemas Yang Berbeda. *J Tek Pang*. 3(2).
[6] Nur R, Lioe H N, Palupi N S, Nurtama B. 2018. Optimasi Formula Sari Edamame dengan Proses Pasteurisasi Berdasarkan Karakteristik Kimia dan Sensori. Jurnal Mutu Pangan : *Ind J of Food Qual*. Oct 31;5(2):88–99.
[7] Almatsier S. 2009. Prinsip Dasar Ilmu Gizi. Jakarta: PT Gramedia Pustaka Utama.
[8] Rinda, Ansharullah, and Asyik N. 2018. Pengaruh komposisi snack bar berbasis tepung tempe dan biji lamtoro (Leucaena leucocephala (Lam) de wit) terhadap penilaian organoleptik, proksimat dan kontribusi angka kecukupan gizi. *J Sains & Tek Pang*. 3(3): 1328-1340.
[9] Suloi A N F, Farid J A, Fitriani S N A, and Ramadhani. 2020. Snack bars: Camilan sehat rendah indeks glikemik sebagai alternatif pencegahan penderita diabetes. *J. Abdi*. 2: 118 -125.

Pulsed electric field application on pasteurization of orange milk from low grade orange: study on nutritional, physical, chemical properties, and total microorganism

B Hariono[1*], A Brilliantina[1], E K N Sari[1], MF Kurnianto[1], F Erawantini[2], Supriyono[1], S Kautsar[3]

[1]Departmen of Agriculture Technology, Politeknik Negeri Jember, Indonesia
[2]Department of Health, Politeknik Negeri Jember, Indonesia
[3]Department of Technic, Politeknik Negeri Jember, Indonesia

budi_hariono@polije.ac.id

Abstract. Oranges with low grade categories have a low price, so one of the utilization efforts is combine with milk into orange milk products.. Orange milk, extracted from orange of siam cultivar and added with milk, was pasteurized using a pulsed electric field (PEF) method. The aim of this research was to analyze the changes on nutritional, physical and chemical properties and total microbes from orange milk after pasteurization. The research was conducted from July until September 2021 located in Jember. The PEF treatment was carried out using treatment time variation for 5 minutes (1 cycle), 10 minutes (2 cycles) and 15 minutes (3 cycles). Several parameters of Vitamin C, pH, dissolved solid and total microbes were observed. The result showed that PEF treatment did not significantly change nutritional (vitamin C), physical (dissolved solid), and chemical properties (pH), but total microbes very significantly change after pasteurization with PEF in compared with no treatment pasteurization. The best treatment was found in the variation of 15 minutes (3 cycles) with degradation microbial killing effectiveness reached 94.58%.

1. Introduction

Siam oranges are a type of orange. It is widely developed in Indonesia because the production is high and favored by consumers. Stay at the time of harvest, oranges with low grade categories have a harvest potential of about 15%. Low grade siamese oranges have a low price, thus affecting farmers' incomes[1]. Orange-based combination drinks can be an alternative for orange processed fruit as it has a great public demand. It also be combined with milk. In fresh condition, orange juice and milk cannot stand long at room temperature. For this reason, further processing is needed to make it last longer. Heating is a conventional processing technology to reduce microbial contamination in food. This process ensures product safety, but the possibility of damage to the taste, nutrition and physicochemical properties cannot be avoided [2]. For this reason, the food processing industry continues to develop alternative of preservation technologies to minimize damage caused by excessive heat treatment. Pulsed Electric Fields (PEF) is a non-thermal technology that is possible to inactivate microorganism with no significant damage cause to the taste, color, and nutrients in foods [3].

The PEF process is based on the short pulses application at high voltage (20-80 kV/cm) to a food placed between two electrodes [4]. The PEF application has been widely used by many researchers such as [5], [6] and [7] Research conducted by [8] and [9], shows that the combination of heating and electric field shock was effective to inactivate Pseudomonas and E. coli in milk and coconut water. This study aims to use a pulsed electric field to inactivate total microbes in a combination drink of milk and orange juice with time variations at a constant voltage, so that it is expected to provide knowledge about these applications for food processing.

2. Research methods

2.1 Materials and tools

The material used is siam orange juice obtained from the extraction of siam oranges without the addition of other ingredients. Siam oranges were purchased from Tanjung market-Jember. Other materials are alcohol 95%, cotton and sterile bottles. Pure cow's milk purchased from a cow farmer in Kemuning lor village. The tools used are a series of HPEF tools, high voltage voltmeter, pH meter, petri dish, tds meter, juice extractor, and cool box.

2.2 Research methods

The research was carried out from June to August 2021 at TEFA Canning State Polytechnic of Jember. The method used is descriptive experimental method. Testing is done by varying the processing time. Processing time is 0, 5, 10, and 15 minutes. The volume of apple cider entered in the treatment chamber is 3 liters.

For each processing time with HPEF, 3 (three) replicates of samples is taken and then be analyzed for changes in nutritional value (vitamin C), chemical properties (pH and total dissolved solids) and the number of microbes (total microbes and coliforms) contained in the HPEF will be calculated. Siam orange juice processed with HPEF. The high-voltage pulse generator used in the pasteurization of Siam orange milk with HPEF method consists of several blocks, which are the keypad block, microcontroller, display, flyback converter circuit, high voltage transformer and chamber. The keypad is used to enter the high voltage setting and time for treatment requirement.

The high voltage can be set from 20 kV to 40 kV and the length of treatment can be set in the range of 0 minutes to 15 minutes. The microcontroller is used to display the high voltage and treatment time that is set through the keypad. The flyback converter circuit will receive the output of the microcontroller in the form of square pulses that can be adjusted for the pulse width. The output of flyback converter in the form of a voltage pulse will count the input voltage of the high-voltage transformer so that the output of the transformer will be in the form of a high-voltage pulse. High-voltage transformers can produce a maximum output of 40 kV. Constant frequency is 20 kHz. All PEF components are assembled in a high-voltage generator box made of mica. The high-voltage pulses produced are directly shooted into the chamber (the treatment area) which is coated on the outside by electrodes made of copper. The treatment chamber is made of 4 mm thick stainless steel plate which is safe for all food products. The treatment chamber is in the form of a four-legged cylinder with a volume of 1.7 liters equipped with an exhaust valve. To protect the entire pulsed electric field, the outside is provided with a barrier made of a clear mica box.

3. Results and discussion

The treatment room is a container for orange milk to be pasteurized. High voltage will be fired towards the treatment chamber which is made of stainless steel in the form of a tube. The treatment room is made in a static or batch system with a diameter of 45 cm and a height of 50 cm and is provided with an iron support to keep it upright and a faucet is placed at the bottom of the tube for discharge hole. The walls' thickness of the treatment room is 3 mm. Next to the treatment room, there are three electrodes spaced more than 5 cm apart which conduct electricity to the material. The chamber is entered in an insulator made of mica (15 x 5 x 55 cm). The electrode is mounted on one side of the insulator facing the treatment room. The results of the treatment room design are shown in Figure 1.

3.1 Vitamin C

Vitamin C is a water-soluble nutrient and vitamin that is essential for life and for maintaining health. This vitamin is also known by the chemical name of its main form, namely ascorbic acid. The vitamin C content of Siam orange juice before pasteurization is 9.792 mg/100gr. The highest level of vitamin C after pasteurization in 5 minutes treatment is 9.608 mg/100gr and the lowest in the 15 minutes treatment is 8.714 mg/100gr. The graph of changes in vitamin C levels against treatment time is shown in Figure 2.

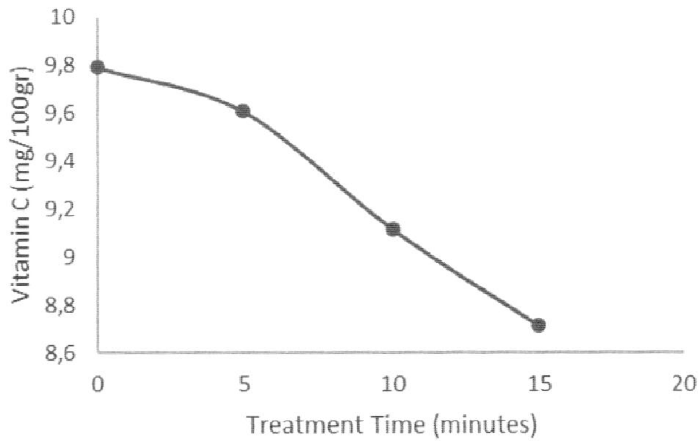

Figure 1. Effect HPEF on the vitamin C

In Figure 2 it can be seen that the presence of nonthermal pasteurization treatment with a high pulsed electric field (HPEF) will cause a decrease in vitamin C level, but this change tends to be insignificant compared to those without HPEF. [10] stated that the higher the temperature and the longer heating, the greater the degradation of vitamin C. Oxidation of vitamin C (ascorbic acid) will convert ascorbic acid into L-dehydroascorbic acid which is chemically very unstable and can undergo further changes to Ldiketogulonic acid which has no vitamin C activity anymore. This is in accordance with [11], in her research which shown that the longer the heating (boiling) time, the lower the vitamin C content of cherry syrup. This is because vitamin C is a vitamin that is easily oxidized, especially by the heating process.

3.2 pH

The pH of Siam orange juice before pasteurization is 4.50. The highest pH after pasteurization in the treatment for 15 minutes of 5.55 and the lowest is in the treatment for 10 minutes that is 5.51. Graphic of changes in pH with treatment time is shown in Figure 3.

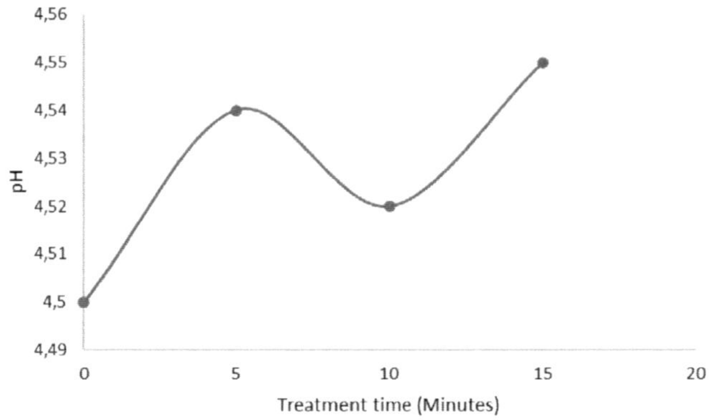

Figure 2. Effect HPEF on the pH

In Figure 3 it can be seen that the pH of non-thermal pasteurized orange juice using a pulsed electric field does not change significantly and is still in the average range with the pH before pasteurization. This is in line with research by [12] that the pH orange juice mixture and carrots treated with PEF did not change significantly. [13] also stated that the HPEF treatment did not significantly change the pH of goat's milk. These changes in the pH value are explained, according with [14], because the migration of ions to the electrodes occurring when an external voltage is applied, which can produce changes of the fluid composition and the chemical properties such as the pH value. The application of HPEF on the orange milk did not show any effect on the pH value.

3.3 Dissolved solid

Dissolved solid total of Siam orange juice before pasteurization was 1303 ppm. The value of Dissolved solid total from orange juice after pasteurization of HPEF tends not to change, except for the 15 minute treatment of 726 ppm. The graphic change in total dissolved solid to treatment time is shown in Figure 7.

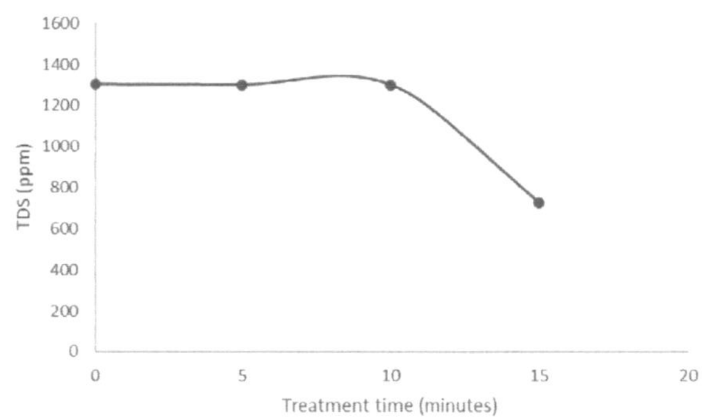

Figure 3. Changes in total dissolved solid with treatment time

The total value of dissolved solid is the sum total content of all substances, both inorganic and organic which dissolved in food. The components in the fruit consist of water-soluble components, such as glucose, fructose sucrose, and water-soluble protein (pectin). The decrease in total dissolved solid value is assumed as the material contained in the siam orange juice is not degraded due to HPEF processing, so that it can reduce the total of dissolved solid in siam orange juice. This is in accordance with the opinion of [15] which states that the decrease in lactose is due to lactose being degraded to glucose and galactose which in turn becomes lactic acid.

r

The total of microbe's milk in siam orange juice before pasteurization is 1.50x105 cfu/ml. The total of these microbes is not in accordance with the Indonesian National Standard (SNI) where the maximum total of microbes contained in fruit juice drinks is 2x102 cfu/ml. The highest microbes total after pasteurization in the 5 minutes treatment is 7.40x104 cfu/ml and the lowest is 8.13x103 cfu/ml after the pasteurization treatment. The graphic of the decrease in microbes total with time changes is shown in Figure 8.

From Figure 8 it can be calculated the effectiveness value of microbial killing which is the difference between the number of microbes before treatment and the number of microbes after treatment divided by the number of microbes before treatment multiplied by 100 percent. The highest killing effectiveness value at 15 minutes is 94.58% and the lowest at 5 minutes is 50.66%. In Figure 8 it can be seen that the longer the pasteurization treatment with PEF, the less the total microbes present in the orange juice milk. In general, the total microbes of orange juice milk after pasteurization have not met the SNI standard, where the maximum limit of total microbes allowed is 2x102 cfu/ml. Therefore, it is necessary to increase the PEF pasteurization time. The longer processing time will result in a greater decrease in total microbes.

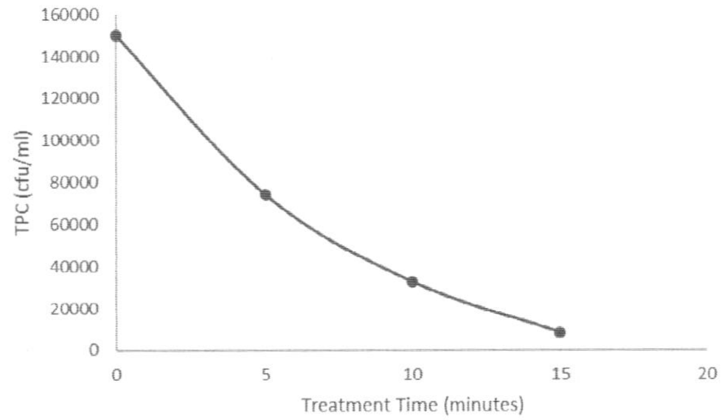

Figure 4. Reduction of total microbes with treatment time

The analysis results showed the presence of coliforms in the fresh siam orange milk. At the time after pasteurization is still be found the presence of coliform. Referring to the Indonesian National Standard (SNI) where the maximum permissible coliform limit is 20 APM/ml, then Siam orange milk juice after HPEF pasteurization treatment is included in SNI, which is still below 20 APM/ml.

The length of treatment also affects the amount of microbial decline, where the longer the treatment time, the greater the microbial mortality. It is suspected that the largest total microbes contained in orange milk juice have a high susceptibility to electric current, causing a high mortality rate. Meanwhile, the remaining total living microbes is assumed to have high resistance to electric shocks, so that the number of subsequent electric shock treatments does not give too much effect in killing the remaining of total microbes. According to [16], spore death by electrocution is limited by the strength electric field generated, the duration of the shock, the number of shocks applied and the cell size. Meanwhile, according to [17], high-voltage electricity for a long time has a greater effect in reducing the number of spores contained in the media.

Non thermal Pulsed Electric Field (PEF) method is one of the non-thermal treatment methods for food preservation, as PEF has the potential to inactivate microbes without changing the taste and nutritional richness of food. The high intensity pulsed electric field process is based on the application of high voltage short pulses (20-80 kV/cm) for a very short time (approximately 1 second) to liquid food placed between two electrodes [18]. PEF technology is more considered than heat treatment of food, because PEF can kill more microbes, avoid or reduce damage in taste, physical properties of food and organoleptic damage [19] and [20]. Microbial inactivation performed with PEF is associated with electro mechanical instability of the cell membrane. The cell membrane protects microbes from environmental condition by working as a semipermeable wall, for example, the membrane regulates the entry of nutrients into the cell and regulates the release of end products of cell metabolic activity [2]. If the cell membrane is ruptured, there is a discharge of fluid from the cell and a loss of cell metabolic activity.

4. Conclusion

The conclusions from this study results are:

1. Chemical properties of pasteurized orange milk juice with PEF, namely pH and dissolved solid dos not change significantly.
2. The vitamin C content of pasteurized orange milk juice with PEF, in which the pH does not change significantly.
3. The highest microbial killing effectiveness reached 94.58%.

Acknowledgements

Authors thanks to Kemenristek BRIN for supporting and financing this program for the 2021 Budget Year

References

[1] Qomariah, R., Hasbianto, A., Lesmayati, S., & Hasan, H. (2013). Kajian prapanen jeruk Siam (Citrus suhuiensis Tan) untuk ekspor. Seminar Nasional Inovasi Teknologi Pertanian. Balai Pengkajian Teknologi Pertanian Kalimantan Selatan (ID), 417-430.

[2] Kurnianto M F, Wibowo M J, Hariono B, Wijaya R, Brilliantina A. 2020. *IOP Conf. Ser.: Earth Environ. Sci.* **411** 012051

[3] Jeyamkondan, S., Jayas, D. S., and Holley, R. A. 1999. Pulsed Electric Field Processing of Foods: A review. Journal of Food Protection. 62: 1088–1096 : New York.

[4] Yeom, H. W., Streaker, C. B., Zhang, Q. H., and Min, D. B. 2000. Effects of pulsed electric fields on the quality of orange juice and comparison with heat pasteurization, Journal of Agricultural Food Chemistry. 48: 4597–4605 : New York.

[5] Barbosa, 2000 dalam Nubhakti O. 2010. Pasteurisasi Non Thermal Teknologi Pulsed Electric Field (PEF) untuk Inaktivasi Staphylococcus aureus pada Sari Buah Tomat. Skrisi. FTP UB : Malang.

[6] Putri, R. I., I. N. Syamsiana., L. C. Hawa. 2010. Design of High Voltage Pulse Generator for Pasteurization by Pulse Electric Field (PEF). International Journal of Computer and Electrical Engineering (IJCEE).Vol.2 No. 5.pp. 916-923. ISSN 1793-8163.

[7] Hawa, L. C., R. I. Putri. 2011. Penerapan Pulsed Electric Field pada Pasteurisasi Sari Buah Apel Varietas Ana: Kajian Karakteristik Nilai Gizi, Sifat Fisik, Sifat Kimiawi dan Mikrobia Total. Agritech. Vol. 31. No. 4: 352 – 358.

[8] Hawa, L. C., B. Susilo., N. E. Jayasari. 2011. Studi Komparasi Inaktivasi Escherichia coli dan Perubahan Sifat Fisik pada Pasteurisasi Susu Sapi Segar Menggunakan Metode Pemanasan dengan Kejut Medan Listrik. Jurnal Teknologi Pertanian Vol. 12 No.1: 31-39.

[9] Wulandari, B. W. 2007. Skripsi Inaktivasi Mikroba dengan Kombinasi Metode Pemanasan dan Kejut Medan Listrik pada Air Kelapa (Cocos nucifera) sebagai Bahan Baku Minuman Isotonik. Jurusan Teknologi Hasil Pertanian FTP UB : Malang.

[10] Craven, H. M., et al. 2000. Evaluation of Pulsed Electric Field and Minimal Heat Treatment for Inactivation of Pseudomonads and Enhancement of Milk ShelfLife. Food Science Australia, Werribee, Victoria 3030 : Australia.

[11] Octaviani LF. 2014. Pengaruh Berbagai Konsentrasi Gula terhadap Aktivitas Antioksidan dan Tingkat Penerimaan Sari Buah Buni (Antidesma bunius). [Skripsi]. Semarang: Universitas Diponogoro.

[12] Ameliya R, Nazaruddin, Handito D. 2018. Pengaruh lama pemanasan terhadap vitamin C, aktivitas antioksidan dan sifat sensoris sirup kersen (Muntingia calabura L.). Pro Food (Jurnal Ilmu dan Teknologi Pangan 4 (1).

[13] Rivas, A., D. Rodrigo, A. Martínez, G.V. Barbosa-Cánovas, dan M. Rodrigo. (2006). Effect of PEF and heat pasteurization on the physical-chemical characteristics of blended orange and carrot juice. LebensmittelWissenschaft und Technologie 39: 1163-1170.

[14] Hariono B, Wijaya R, Kurnianto MF, Sutrisno, Seminar KB, Brilliantina A. 2020. Quality of goats milk exposed ultraviolet and high pulsed electric field. IOP Conf. Series: Earth and Environmental Science **411** (2020) 012052

[15] Meneses C H S G, R L A Bruni, P D Fernandes, W E Pereira, L H G M Lima, M M A Lima, and M S Vidal. 2011. Germination of cotton cultivar seeds under water stress induced by polyethyleneglycol-6000. Crop Science. 68(2):131-138.

[16] Zakaria, Y. 2009. Pengaruh Jenis Susu dan Persentase Starter yang Berbeda terhadap Kualitas Kefir. Agripet. J. 9 (1) : 26—30.

[17] Sitzmann, V. 1995. High Voltage Pulse Technique for Food Preservatives. *In* : New Method for Food Preservation : 236-252. Ed. G.W Gould. London, UK : Blackie Academic and Professional.

[18] Gould, G.W. (1995). New Methods Food Preservatives. Chapman Hall. New York.

[19] Cueva, Olga A. 2003. Pulsed Electric Field Influences on Acid Tolerance, Bile Tolerance, Protease Activity and Growth Characteristics of Lactobacillus Acidophilus La-K. Escuela Agrícola Panamericana Zamorano. Honduras

[20] Quass, D. W. 1997. Pulsed Electric Field Processing In The Food Industry. A Status Report On Pulsed Electric Field. Palo Alto, CA. Electric Power Research Institute. Page 23-35.CR-109742.

[21] Andreea-Manuela C., C. Csatlós, C. Bica. 2007. Pulsed Electric Field Processing of Liquid Foods. Simulation of the electric field distribution between electrodes. University of Transilvania, Brasov Romania

The analysis of phosphor content and sensory quality of dragon fruit and moringa oleifera soft candy

A D Elisanti [1]*, E T Ardianto [2] and R Rindiani [1]

[1] Clinical Nutrition Study Program, Health Department, Politeknik Negeri Jember, Jl. Mastrip 164 Jember, East Java, Indonesia
[2] Health Information Management Study Program, Health Departement, Politeknik Negeri Jember, Jl. Mastrip 164 Jember, East Java, Indonesia

E-mail : *alinea@polije.ac.id

Abstract. This study aims to determine the effect of adding dragon fruit and moringa oleifera to the phosphor content and acceptability of soft candy. This study is an experimental study using a completely randomized design (CRD) with the concentration of adding dragon fruit 10%, 12%, 14%, and 16% Moringa Oleifera 1%, 1%, 2%, and 3% and fresh milk 50 ml. Observation parameters: Phosphor, with the panelists' preference test for colour, texture, taste, and smell. The results showed that the best treatment for adding 12% dragon fruit and 1% moringa leaves was preferred by panelists' in terms of colour, taste, texture, and smell based on analysis using the Mann Whitney test showed a p-value of 0,010 with a mean difference of 53.75, phosphor content 0,288% and meets the quality standard requirement (SNI 2803: 2010). In 100 grams soft candy contains 576 mg of phosphor. Thus, it is considered as an alternative to meet the phosphor daily intake of children under five in Indonesia.

1. Introduction

Phosphor is one of the minerals that affect the incidence of stunting. This is exactly associated with the process of bone growth in the child's body. The skeleton, which contains over 90 percent of the phosphor, provides a reserve for regulation and the maintenance of these elements. Both have important functions in cells and body fluids [1].

The results showed that phosphor intake in stunted children aged 24-59 months was significantly lower than in non-stunted children. The group of children who had a low phosphor intake had a 2.29 times greater risk of experiencing stunting than the group of children who had sufficient phosphorus intake. The lowest known calcium and phosphorus intake is phosphorus intake, which is 68.75% [2], [3].

A study in Thailand reported that there were differences in the average intake of energy, protein, and nutrition in school children related to the incidence of stunting. Significantly more boys are stunted than girls. Stunting men had a low average intake of energy, protein, calcium, phosphor, and zinc. Also has a lower average arm muscle area compared to non-stunted children. Stunting at school age can have an impact on decreasing adult size, thereby reducing work capacity and adverse reproductive systems. Stunting is associated with anorexia and hypogeusia caused by insufficient intake of zinc and phosphorus [4], [5].

The impact of stunting is quite complex, including difficulties in learning to read, impaired intellectual development during childhood, the potential for imperfect growth and development and even failure to thrive, low motor skills and productivity, and a higher risk for suffering from non-communicable diseases. Meanwhile, the impact of stunting on women is the potential to contribute to perinatal mortality and give birth to stunted children, all of which will lead to low work productivity in adulthood [6]. Therefore, it is necessary to conduct a study related to functional foods to support mineral intake in stunting toddlers.

There have been many studies related to functional foods that have mineral content, such as the research we have done, producing biscuits made from dragon fruit and Moringa leaves that contain iron. Thus, it is possible to be an alternative to prevent nutritional problems in children, adolescents, and adults [7]. Similar research was also carried out to identify the physical and chemical quality of functional food cheese sticks from red bean flour and beetroot flour, the chemical quality test being tested was the zinc content [8]. The study of functional food in the form of biscuits has also tested the carbohydrate content of jackfruit seeds that have been fortified with iron, the carbohydrate content of the waste biscuits of jackfruit seed fortified with Fe (% bb) was 73.9139% and carbohydrate content (% wt) was 75.8999% bk [9].

Although many functional foods have been created to prevent and treat stunting, functional foods in the form of sweets are still rarely studied. In addition, the results of research aimed at identifying phosphorus in functional foods have never been carried out. Thus, it is necessary to innovate functional food products in the form of candy and identify phosphorus content, to support mineral intake in stunting toddlers.

2. Material and Methods

This study used an experimental method with a factorial Completely Randomized Design (CRD) through two repetitions.

2.1. Material

The raw materials of this research; sugar, gelatin, pure cow's milk, dragon fruit pulp 10%, 12%, 14%, 16%, and Moringa Oleifera leaf flour 1%, 1%, 2%, and 3%. Sample 1 includes : 10% dragon fruit pulp; 1% Moringa Oleifera leaf powder; 50 ml of factory fresh milk); sample 2 (12% dragon fruit pulp; 1% moringa leaf powder; 50 ml pure cow's milk), sample 3 (14% dragon fruit pulp; 2% moringa leaf powder; 50 ml pure cows milk), sample (16% dragon fruit pulp; 3% Moringa leaf flour; 50 ml pure cow's milk).

2.2. Methods

The research stages started from candy formulation. Followed by the hedonic test which was analyzed by the Kruskal Wallis and Mann-Whitney test with an alpha of 5%. Hedonic test data were analyzed using excel and SPSS 19 programs. The phosphor content was tested using Atomic Absorption Spectrophotometry (AAS). The quality of the phosphor content test was assessed using the SNI standard number 2803:2010. The hedonic test was carried out on 30 semi-trained panelists who were carried out in limited off-line. The instrument used was a hedonic test form including color, taste, aroma, and texture with a rating range of 1-7. The criteria for assessing the quality of candy products were 1. Very bad, 2. Bad, 3. Slightly bad, 4. Moderate/enough, 5. Fairly good, 6. Good, 7. Very good. This criterion was a modification of the hedonic scale presented by Peryam & Girardot, 1952, which consisted of 9 hedonic scales; like extremely, like very much, like moderately, like slightly, neither like nor dislike, dislike slightly, dislike moderately, dislike very much, dislike extremely [10]. The research was carried out at the Politeknik Negeri Jember, and the phosphor content test was carried out at the Bioscience Laboratory Politeknik Negeri Jember.

3. Result and Discussion

This research was conducted for 6 months with the results including hedonic tests and phosphor content. The results were presented in univariate and multivariate.

3.1. Univariate Analysis

3.1.1. Quality of Hedonic Test (Overall)

The description of the quality of hedonic test can be seen in Figure 1. The picture is a description of the mean (average) value of the hedonic test including color, taste, aroma, texture, and overall soft candy quality. Descriptively, sample 1 has the highest average value of hedonic quality test results on all test items (color, taste, aroma, texture, overall), followed by sample 2. sample 4 has the lowest color hedonic quality compared to other samples.

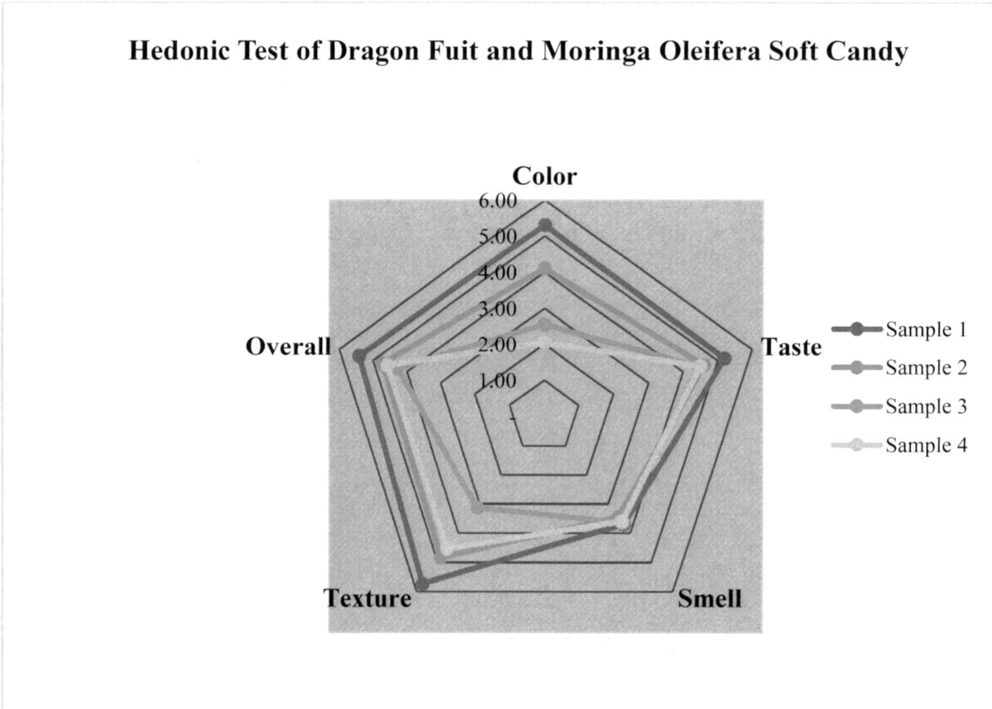

Source: Hedonic test, 2021

Figure 1. Hedonic Test of Dragon Fruit, Moringa Oleifera Sof Candy

Overall, the lowest hedonic quality test is the smell. This smell is influenced by Moringa Oleifera leaf flour which does have a distinctive and unpleasant smell. Even so, Moringa leaves have a complete nutritional content including minerals, vitamins, and high protein [11]. In addition, Moringa contains calories (cal), protein (g), fat (g), carbohydrates (g), fiber (g), calcium (mg). magnesium (mg), phosphorus (mg), potassium (mg), copper (mg), iron (mg), sulfur (mg), vitamin B1 (mg), B2 (mg), B3 (mg), C (mg) , E [12].

3.1.2. Descriptive explanation of Hedonic Test

Descriptive explanation based on hedonic quality test items can be seen in Figures 2,3,4 and 5

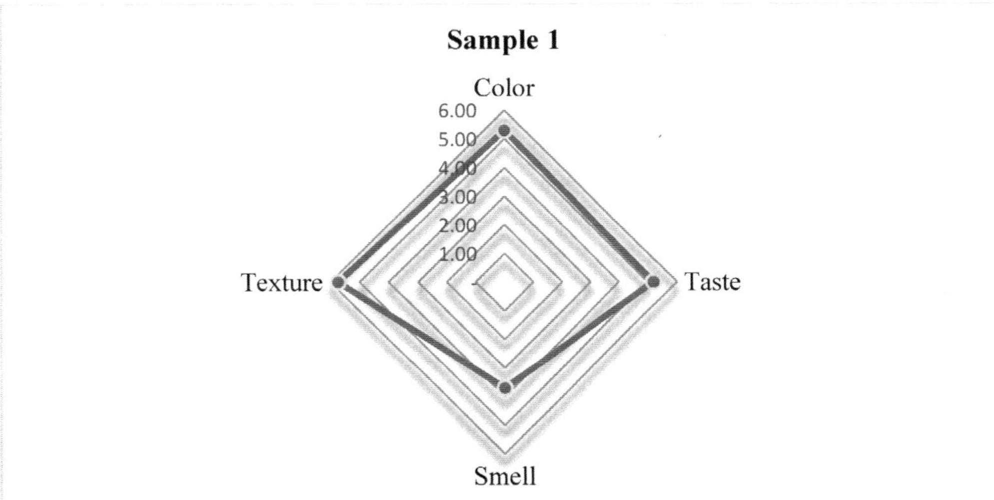

Source: Hedonic test, 2021

Figure 2. The results of the analysis of hedonic differences (color quality, taste, smell, texture) of dragon fruit and Moringa Oleifera soft candy (sample 1)

Sample 1 has a lower hedonic quality of smell compared to the quality of color, taste and texture. Moringa leaves have a distinctive unpleasant odor. There are several processing techniques that can minimize the unpleasant smell, one of which is through blanching. Moringa leaf flour does not go through the blanching process to avoid the loss of minerals and vitamins through the heating process. Generally, blanching produces a decrease in the nutritional value of foods. Nutrients leach out from the product especially during water blanching [13].

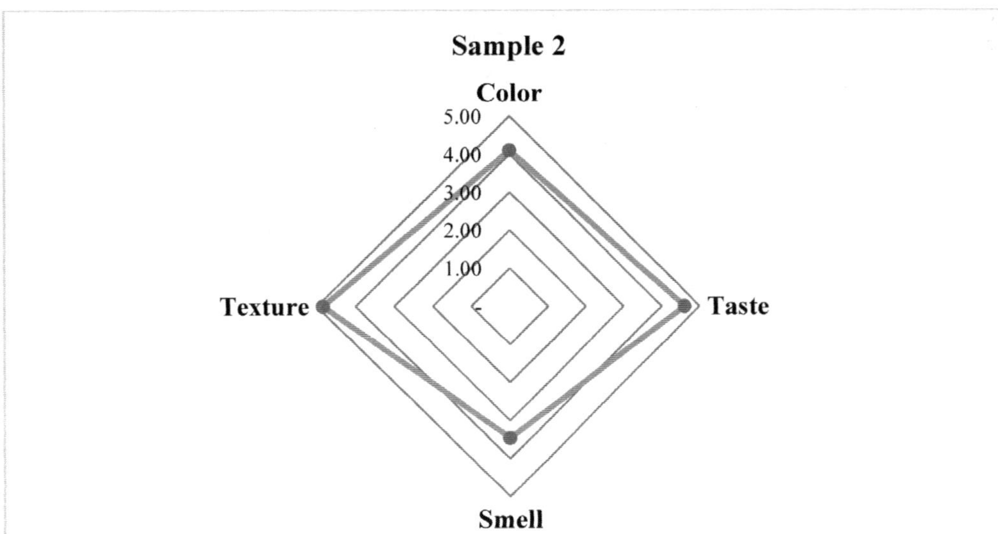

Source: Hedonic test, 2021

Figure 3. The results of the analysis of hedonic differences (color quality, taste, smell, texture) of dragon fruit and Moringa Oleifera soft candy (Sample 2)

Sample 2 has the lowest hedonic quality of smell compared to the quality of color, taste, and texture. Nonetheless, the color was medium. The smell of soft candy is on slightly bad category, color and taste were Moderate/enough, and texture was Fairly good.

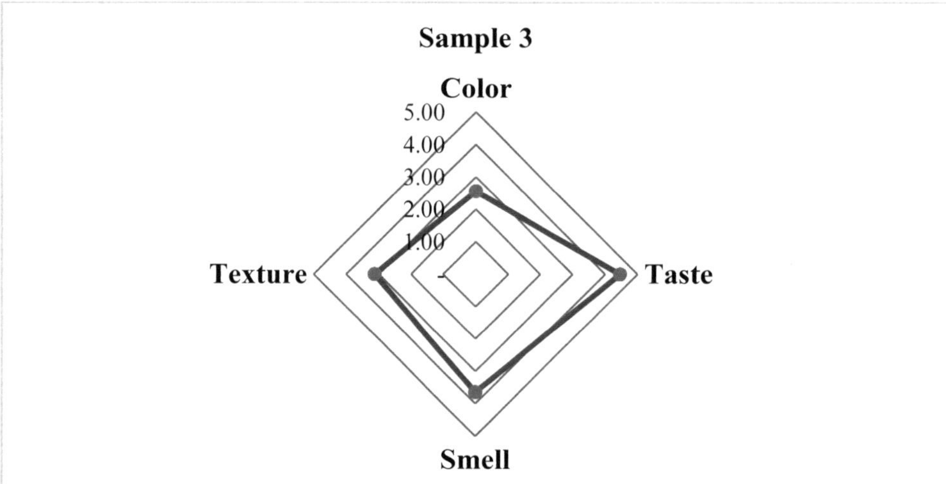

Source: Hedonic test, 2021

Figure 4. The results of the analysis of hedonic differences (color quality, taste, smell, texture) of dragon fruit and Moringa Oleifera soft candy (Sample 3)

Sample 3 has moderate/ enough taste hedonic quality, close sufficient smell but less or Slightly bad texture and color.

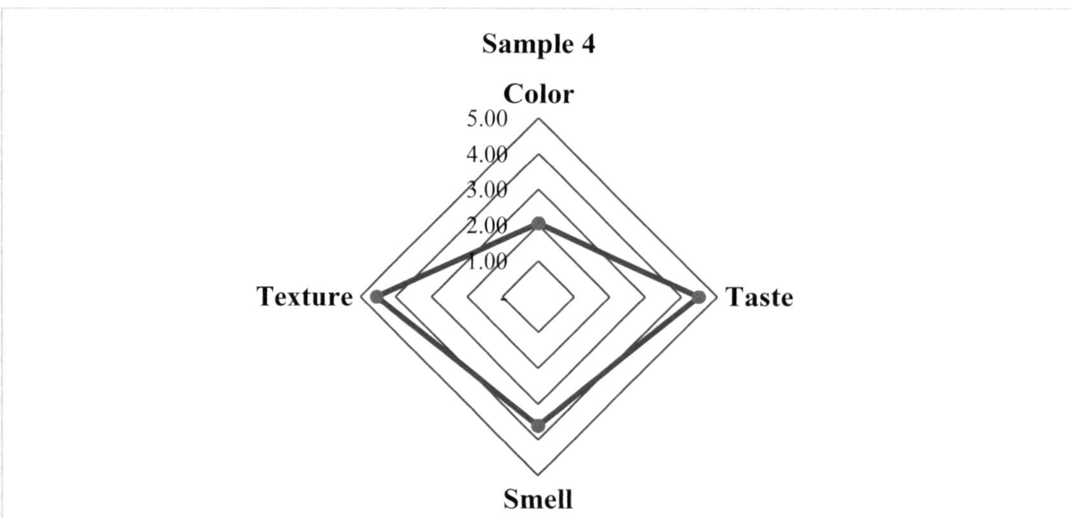

Source: Hedonic test, 2021

Figure 5. The results of the analysis of hedonic differences (color quality, taste, smell, texture) of dragon fruit and Moringa Oleifera soft candy (Sample 4)

Sample 4 has a fairly good hedonic quality of texture and taste, sufficient smell and poor (bad) color. Based on the results of the study, the color hedonic quality test is included in the less preferred category. This is due to the content of raw materials for making candy. The results shows that there is gelling effect on the color characteristics, but do not affect the aroma, texture, and taste [14]. The color of the soft candy is largely determined by the natural color of the fruit juice and the result of browning during the process [15].

3.2. Multivariate Analysis

A multivariate test was used to look for differences in treatment groups and to determine interactions between groups. The results of the Kruskal Wallis and Mann Whitney test analysis of dragon fruit powder candy, moringa leaf flour, cow's milk covering color, taste, smell, and texture are shown in table 1.

Table 1. The hedonic test analysis (quality) of dragon fruit powder and moringa leaf flour soft candy

Variable	Sample	Hedonic average	Significance (α 0.05)
Colour	1	78.20	0.010
Taste	2	59.28	
Smell	3	53.75	
Texture	4	50.77	

Source: Hedonic test, 2021

The hedonic test (quality) of dragon fruit powder, moringa leaf flour, cow's milk candy which include color, taste, smell and texture have differences between samples with a sig value of 0.010. The different samples were analyzed using the Mann-Whitney test (Table 2).

Table 2. The hedonic difference test (quality) of dragon fruit, cow's milk and moringa leaves soft candy

Sample	Sample 1	Sample 2	Sample 3	Sample 4
1	-	0.023	0.004	0.007
2	0.023	-	0.299	0.490
3	0.004	0.299	-	0.711
4	0.007	0.490	0.711	-

Source: Mann whitney test, 2021

Sample 1 was significantly different with sample 2 ($p= 0.023$), sample 3 ($p= 0.004$) and sample 4 ($p= 0.007$). Sample 2 was not different with sample 3 ($p= 0.299$) and sample 4 ($p= 0.490$), as well as sample 3 was not different with sample 4 ($p= 0.711$). The most preferred soft candy sample was sample 1 (sample composition of 10 % of dragon fruit: 1 % of *Moringa Oleifera* leaf flour: 50 ml of factory cow's milk (mean: 78.20). Nonetheless, sample 1 is a control (comparison) to sample 2 with a composition of 12% dragon fruit powder: 1% *Moringa oleifera* leaf flour: 50 ml of pure cow's milk with an average: 59.28 is the sample with the best quality.

3.3. Phosphor Content of Dragon Fruit and Moringa Oleifera's Soft Candy

The resulting soft candy contains several minerals, but in this study, an identification was made on the phosphor content. The reason for choosing phosphor is because there are less studies that discuss phosphor, even though phosphor plays a very important role in increasing appetite. Although in small amounts this type of phosphor should be met in number in the body. The content of the phosphor is described in table 3.

Table 3. Phosphor content of dragon fruit powder and moringa Oleifera soft candy (SNI 2803:2010)

Parameters	**Sample 1**	**Sample 2**	**Sample 3**	**Sample 4**
Phosphor (%)	0,217	0,169	0,288	0,100

Source:Atomic Absorption Spectrophotometry test

Phosphor content was measured in 0.5-gram sample based on SNI 2803:2010. Sample 3 with the addition of 14% dragon fruit powder and 2% Moringa oleifera leaf powder has the highest phosphor content, which is 0.288 %. This is equivalent to 2,88 mg/kg. Phosphor needs for children aged 1-3 years are 460 mg/day and for children aged 4-5 years is 500 mg/day [16]. In 100 grams soft candy contains 576 mg of phosphor. This means candy can be considered as an alternative to meet the phosphor daily intake of children under five in Indonesia. Phosphor becomes a major ingredient of not only bones and teeth, but also the brain. Phosphor facilitates the permeability of cell membranes that helps nutrients pass in and out of cells. Phosphor also mediates energy and maintains consistent blood sugar levels [17]. Phosphor is also required for metabolic processes, coagulation of phospholipids, DNA, RNA metabolism of fats, proteins, and carbohydrates [18].

4. Conclusion

The best phosphor content in dragon fruit and *Moringa Oleifera* soft candy was found in sample 3 with the addition of 14% dragon fruit powder and 2% *Moringa Oleifera* leaf flour with a content of 0.288%. In 100 grams soft candy contains 576 mg of phosphor and meets the quality standard requirement (SNI 2803: 2010). The hedonic test showed the best treatment for adding 12% dragon fruit powder and 1% moringa oleifera leaf flour base on (color, taste, texture, and smell).

5. Acknowledgment

We gratefully acknowledge the funding from the Ministry Of National Education, Culture, Research and Technology of The Republic Of Indonesia through the PNBP Competitive Research Grant Politeknik Negeri Jember Number 354/PL17.4/PG/2021.

References

[1] Copp D H 1957 Calcium and phosphorus metabolism *Am. J. Med.* **22** 275–85
[2] Sari E M, Juffrie M, Nurani N and Sitaresmi M N 2016 Asupan protein, kalsium dan fosfor pada anak stunting dan tidak stunting usia 24-59 bulan *J. Gizi Klin. Indones.* **12** 152
[3] Widiyaningsih E N and Wahyuningtyas A P 2017 Protein, Magnesium and Phosphorus Intake at Stunted and Non-Stunted Children in Panularan Surakarta *3rd Int. Conf. Sci. Technol. Humanit.* 27–34
[4] Chairunnisa E 2017 *Inadekuat Asupan Vitamin D , Kalsium Dan Fosfor Pada Anak Stunting Usia 12-24 Bulan Program Studi Ilmu Gizi*
[5] Gibson R S, Manger M S, Krittaphol W, Pongcharoen T, Gowachirapant S, Bailey K B and Winichagoon P 2007 Does zinc deficiency play a role in stunting among primary school children in NE Thailand? *Br. J. Nutr.* **97** 167–75
[6] Elisanti A D 2021 Bab 4. Stunting, determinan dan kebijakan pencegahan pada masa pandemi COVID-19 *Pena Persada* (Banyumas Jawa Tengah: Pena Persada) pp 46–68
[7] Tri Ardianto E, Subaktilah Y and Elisanti A D 2020 Formulasi Biskuit Buah Naga dan Daun Kelor untuk Mencegah Anemia *J. Kesehat.* **8** 10–5
[8] Harahap R A 2020 Uji Mutu Fisik Dan Uji Mutu Zinc Cheese Stick Tepung Kacang Merah Dan Tepung Bit Sebagai Pangan Fungsional
[9] Hidayat A 2021 Kadar karbohidrat : fortifikasi fe pada biskuit limbah biji nangka sebagai cemilan fungsional bagi penderita stunting *J. Pertan. dan Pangan* **3** 28–35
[10] Lim J 2011 Hedonic scaling: A review of methods and theory *Food Qual. Prefer.* **22** 733–47
[11] Muawanah A, Djajanegara I, Sa'duddin A, Sukandar D and Radiastuti N 2012 Penggunaan Bunga Kecombrang (Etlingera Elatior) Dalam Proses Formulasi Permen Jelly *J. Kim. Val.* **2**
[12] Wahyudi I and Nurhaedah M 2017 Ragam Manfaat Tanaman Kelor (Moringa oleifera Lamk)

Bagi Masyarakat *Info Tek. EBONI* **14** 63–75
[13] Powers, Joseph R, Jose´I.ReyesDeCorcuera R P C 2004 Blanching of Foods *Encycl. Agric. ,Food, Biol. Eng.* 1–5
[14] Ahmad D and Mujdalipah S 2017 Karakteristik Organoleptik Permen Jelly Ubi Akibat Pengaruh Jenis Bahan Pembentuk Gel *Edufortech* **2** 52–8
[15] Harijono J K dan S A M 2001 Pengaruh Kadar Karaginan dan Total Padatan Terlarut Sari Buah Apel Muda terhadap Aspek Kualitas Permen Jelly *J. Teknol. Pertan.* **2** 110-116.
[16] Kemenkes RI 2019 Angka Kecukupan Gizi Mayarakat Indonesia *Menteri Kesehat. Republik Indones. Peratur. Menteri Kesehat. Republik Indones.* **Nomor 65** 2004–6
[17] Sawyer P 2004 Bearer of Light Homeopathic Phosphorus *MIDWIFERY TODAY* 64–9
[18] Putri Rahmi 2019 Peran Nutrisi Bagi Tumbuh dan Kembang Anak Usia Dini *Bunayya J. Pendidik. Anak* **5**

IOP Conf. Series: Earth and Environmental Science **980** (2022) 012043 doi:10.1088/1755-1315/980/1/012043

Effect of banana flour and tempeh flour on sensory characteristics of prebiotic snack bar

G Purnasari[1], I R A Sasmita[1], Y Susindra[1]
[1] Clinical Nutrition Department, Politeknik Negeri Jember, Jl. Mastrip PO. Box 164 Jember, Indonesia
*galih_purnasari@polije.ac.id

Abstract. Prebiotic snack bar is a food innovation in the form of a bar that contains a source of functional nutrients that can improve chronic small intestinal inflammation syndrome (pediatric environmental enteropathy) in stunting children, increase the absorption of nutrition, and then improve the nutritional status of stunting children. The aim of this research was to determine the sensory characteristics of snack bar with any level of banana flour and tempeh flour addition. Sensory analyses were carried out on flavor, color, taste, texture and overall of the snack bar. The results showed that the ratio of banana flour : tempeh flour (95% : 5%) was more acceptable in terms of flavor (3,89), color (3,60), taste (3,91), texture (3,77) and overall (4,00). The result also showed that the addition of tempeh flour and banana flour to the formulation snack bar had a significant effect ($p \leq 0{,}05$) on the sensory quality of prebiotic snack bar

1. Introduction

Stunting or short based on height compared to age index of the standard population (<-2SD) is still a health problem in Jember Regency. The 2016 Nutrition Status Monitoring survey data stated that Jember is the district with the second highest stunting incidence in East Java (percentage stunting 39.2%) [1]. Stunting infant is at risk of growing short in adolescence. Children who are stunted at an early age (0-2 years) and remain short at 4-6 years of age have a 27 times risk of staying short before entering puberty [2].

Stunting children suffer a chronic inflammatory syndrome in the small intestine, which is called pediatric environmental enteropathy (PEE). This chronic inflammation causes a characteristic shortening of the villi and reduces the absorptive surface of the intestine. The result is a reallocation of nutrients needed for growth and development of children as well as disruption of hormonal pathways that regulate bone growth activity. Chronic inflammation is hypothesized to affect brain development and then induce negative cognitive effects [3].

Risk factors for stunting children are lack of nutritional intake, incidence of diarrhea and infectious diseases such as cough and flu [4]. Stunting children generally experience frequent episodes of diarrhea [5,6] Diarrhea is associated with high levels of pathogenic bacteria in the gastrointestinal tract. The composition of the gastrointestinal microbiota during diarrhea changes to a higher composition of pathogenic bacteria compared to probiotics in the gastrointestinal tract [7,8].

The right intervention for stunting children is to provide functional foods that contain prebiotics so that it can improve intestinal function in the absorption of nutrients and can improve the immune system [9]. Functional foods can be made using local food sources of prebiotics that are easily available. These local foods include *kepok* banana (*Musa paradisiaca*) and tempeh soy bean.

Tempeh from yellow soybean has been known as a functional food. Yellow soybean being fermented using fungus *Rhizopus sp* (*R. oryzae, R. stolonifer, R. Oligosporus*) [10]. Tempeh can stimulate the growth of digestive bacteria (microflora), especially from the genus *Bifidobacterium* and non-pathogenic *E. coli*. The growth of microflora bacteria that colonize in the digestive tract can recreate the balance of microflora so that it becomes one of the mechanisms to prevent diarrhea and relieve diarrhea symptoms [11]. Banana kepok (*Musa paradisiaca*) is a natural prebiotic that contains inulin and fructooligosaccharides (FOS) compounds. The properties of inulin as a prebiotic are also considered as "colonic foods" for the intestinal microflora [12].

Interventions are still needed even after passing the First 1000 Days of Life (HPK) to prevent the increase in stunting rates. This study aims to determine the sensory characteristics of snack bar with any level of banana flour and tempeh flour addition.

1.1. Material

The main ingredients used were banana kepok flour and tempeh flour obtained from "Kusuka Ubiku" Jogjakarta. The ingredients for making the snackbar were sugar, margarine, egg yolk, crispy rice, almonds, raisins, vanilla, and salt. The tools used were digital scales, bowls, spatulas, rolling pins, baking trays, cutters, and oven.

1.2. Methods

The snack bars making process refers to Indrawan's research (2018) which has been modified [13]. The stages of making snack bars were weighing the materials, mixing process stage I and stage II. Stage I were mixing banana kepok flour, tempeh flour, and salt. Stage II were mixing margarine and sugar, added the egg yolks then mixed all them. Then, mixed stages I and II, then added almonds, rice crispies, and raisins. The dough is molded and then baked in an oven at 150°C for 40 minutes. After 40 minutes, the snack bars were cooled at room temperature at ± 28-30°C for 20 minutes.

2. Analysis

2.1. Sensory Evaluation

Sensory tests performed include hedonic tests and hedonic quality tests. The panelist was asked to evaluate according to five sensory attributes, consisting of flavor, colour, taste, texture, and overall. The hedonic test rating scale ranges from 1 to 5 (criteria from dislike to like very much). Sensory testing using 35 panelists.

2.2. Statistical Analysis

Statistical analysis was performed using SPSS (version 16.0).One way ANOVA test was used to obtain the significant difference between each factor. Duncans multiple range test (DMRT) at p ,0.05 was performed to compare means between sample.

3. Result and Discussion

3.1. Hedonic Test

The hedonic test is a testing activity carried out with several panelists aiming to determine the value of people's preferences for snack bars. The hedonic test was carried out with a scoring scale preference test by 35 untrained panelists. Parameters used for hedonic test are aroma, color, taste, texture and overall snack bar. Panelis preference is shown in **Table 1**. Based on the ANOVA analysis, it was shown that there was no significant difference in hedonic values ($P>0.05$) between F1, F2, F3 and F4 on the flavor parameter. However, in the parameters of color, taste, texture, and overall there were significant differences (P<0.05) between F1, F2, F3, and F4.

Table 1. Hedonic test of prebiotic snack bar

	F1	F2	F3	F4
Flavor	3.89 ± 0.87^{a}	3.43 ± 1.01^{a}	3.69 ± 0.90^{a}	3.60 ± 0.97^{a}
Color	3.60 ± 0.60^{b}	2.97 ± 0.86^{a}	3.60 ± 0.88^{b}	3.23 ± 0.94^{ab}
Taste	3.91 ± 0.85^{b}	3.23 ± 1.14^{a}	3.86 ± 0.97^{b}	3.11 ± 1.11^{a}
Texture	3.77 ± 0.87^{b}	3.17 ± 0.98^{a}	3.40 ± 0.88^{ab}	3.14 ± 1.09^{a}
Over all	4.00 ± 0.77^{b}	3.34 ± 0.97^{a}	3.89 ± 0.90^{b}	3.26 ± 0.95^{a}

Mean value ± standard deviation. Means within a column sharing followed by the same letters do not differ statistically from each other ($P < 0.05$)

Note : F1(95% banana flour : 5% tempeh flour), F2(90% banana flour : 10% tempeh flour), F3(85% banana flour : 15% tempeh flour), F4 (80% banana flour : 20% tempeh flour)

Based on **Table 1**, it is shown on the flavor parameter, panelists indicate the level of likes. snackbar with a composition of 95% banana flour and 5% tempeh flour was preferred by panelists to other compositions. This is due to the greater the addition of banana flour, the resulting snackbar has a banana flavor. On the color parameter, the composition of the snackbar preferred by the panelists was the composition of 95% banana flour and 5% tempeh flour. The color of the snackbar that the panelists like is pale brown. This is because a little tempeh flour is added. The more tempe flour that is added, the more the color of the snackbar gets brown. The color difference is caused by a non-enzymatic browning reaction due to heating [14]. The acceptance of the taste on the snackbar that the panelists liked was indicated by the composition of 95% banana flour and 5% tempeh flour with the highest score of 3.91. This composition gives a sweet taste because more banana flour is added than tempeh flour. The acceptance of the texture and overall on the snackbar that the panelists liked was indicated by the composition of 95% banana flour and 5% tempeh flour with the highest score of 3.77 and 4.00. The resulting texture is soft. Based [15] The most preferred snackbar texture is that it is soft, easy to bite and chew.

3.2. Hedonic Quality Test

Hedonic quality test is an organoleptic test that can be used to measure the level of preference for a food product. The level of preference measured is a preference for a particular attribute or preference for a food product as a whole [16]. The results of the hedonic quality test are shown in **Table 2.**

Table 2. Hedonic quality test of prebiotic snack bar

	F1	F2	F3	F4
Flavor	1.66 ± 0.76^{a}	1.91 ± 0.89^{a}	1.71 ± 0.79^{a}	1.94 ± 0.91^{a}
Color	2.80 ± 1.05^{a}	2.37 ± 1.11^{a}	2.66 ± 1.08^{a}	2.83 ± 1.22^{a}
Taste	1.77 ± 0.91^{a}	2.29 ± 1.07^{b}	2.00 ± 0.77^{ab}	2.34 ± 0.99^{b}
Texture	3.74 ± 0.89^{b}	3.49 ± 0.85^{ab}	3.23 ± 0.97^{a}	3.06 ± 0.97^{a}

Mean value ± standard deviation. Means within a column sharing followed by the same letters do not differ statistically from each other ($P < 0.05$)

Note : F1(95% banana flour : 5% tempeh flour), F2(90% banana flour : 10% tempeh flour), F3(85% banana flour : 15% tempeh flour), F4 (80% banana flour : 20% tempeh flour)

Based on **Table 2**, it is shown that the addition of tempe flour and banana flour did not have a significant effect on the aroma and color attributes. Meanwhile, the taste and texture attributes have a significantly different effect. The highest value representing the taste of the prebiotic snackbar at the composition of 80% banana flour and 20% tempeh flour. The higher the concentration level of the addition of tempeh flour, the more bitter the resulting taste. This is due to changes in the content of isoflavone glycoside compounds to aglycones in tempe during the drying process [17].

Based on [15], Snackbar texture is soft, easy to bite and chew. This is obtained by F1 which is the composition of 95% banana flour and 5% tempeh flour. The higher the concentration of addition of tempe flour, the harder of texture. Tempe flour has a protein content of 46.1% [18]. Enhancement protein content increases the hardness of the snackbar. Protein has hydration properties that can increase the water absorption of the snackbar. The increased water absorption causes the water to evaporate when roasting leaving an empty space so that the snackbar becomes more crispy [19].

4. Conclusion

Based on the results of the study, the addition of banana flour and tempeh flour had a significant effect on the taste and texture of the prebiotic snackbar. In the hedonic test, it was found that the results of adding banana flour and tempeh flour had a significant effect on the level of preference for color, taste, texture, and overall. Meanwhile, the aroma attribute does not have a significant effect. Snackbar products with a composition of 95% banana flour and 5% tempeh flour have the highest panelist acceptance rate compared to other compositions. This is due to the soft texture of the snackbar, easy to bite and chew, the sweet taste and the pale brown color.

5. References

1. Badan Penelitian dan Pengembangan Daerah Jawa Timur (Bappeda Jatim). 2017. Kasus Gizi Buruk Turun. [online] http://bappeda.jatimprov.go.id/2017/07/17/kasus-gizi-buruk-jatim-turun.
2. Aryastami, N. K. & Tarigan, I. 2017 Kajian Kebijakan dan Penanggulangan Masalah Gizi Stunting di Indonesia. Bul. Penelit. Kesehat. 45, 233–240.
3. Vonaesch P, Morien E, Andrianonimiadana L, Sanke H, Mbecko J. 2018. Stunted childhood growth is associated with decompartmentalization of the gastrointestinal tract and overgrowth of oropharyngeal taxa. *Proceedings of the National Academy of Sciences*. 115(36):8489-8498.
4. Mwaniki dan Makokha. 2013. Nutrition status and associated factors among children in public primary schools in Dagoretti, Nairobi, Kenya. African Health Sciences 13(1):39-46.
5. Checkley W, Buckley G, Gilman RH, Assis AMO, Guerrant RL, Valentiner-branth P, Lanata CF, Black RE, Morris SS. 2008. Multi-country analysis of the effects of diarrhoea on childhood stunting. International Journal of Epidemiology 37:816-830. http://doi.org/10.1093/ije/dyn099.
6. Pop M, Walker AW, Paulson J, Lindsay B, Antonio M, Hossain MA, Stine OC. 2014. Diarrhea in young children from low-income countries leads to large-scale alterations in intestinal microbiota composition. Genome Biol 15(6):1-12. http://doi.org/10.1186/gb-2014-15-6-r76.
7. Dinh DM, Ramadass B, Kattula D, Sarkar R, Naumova EN, Kang G, Ward HD. 2016. Longitudinal analysis of the intestinal microbiota in persistently stunted young children in South India. Plos One 11(5):1-17.http://doi.org/10.1371/journal.pone.0155405.
8. Gough EK, Stephens DA, Moodie EEM, Prendergast AJ, Stoltzfus RJ, Humphrey JH, & Manges AR. 2015. Linear growth faltering in infants is associated with Acidaminococcus sp . and community- level changes in the gut microbiota. Microbiome 3(24): 1-10.doi:10.1186/s40168-015-0089-2.
9. Waliyo E, Agusanty SF, Hariyadi D. 2020. Formula Prebiotik Berbasis Pangan Lokal Dapat Meningkatkan Z-Skor Pb/U pada Anak Stunting. *AcTion: Aceh Nutrition Journal, November 2020 (5)2: 130-139.*
10. Astawan M, Wresdiyati T, Maknun L. 2017. Tempe Sumber Zat Gizi dan Komponen Bioaktif untuk Kesehatan. Bogor: PT Penerbit IPB Press.
11. Kuligowski M, Kuligowska IJ, Nowak J. 2013. Evaluation of Bean and Soy Tempeh Influence on Intestinal Bacteria and Estimation of Antibacterial Properties of Bean Tempeh. Polish Journal of Microbiology. 62(2): 189-194
12. Hardisari R dan Amaliawati N. 2016. Manfaat Prebiotik Tepung Pisang Kepok (Musa paradisiaca formatypica) terhadap Pertumbuhan Probiotik Lactobacillus casei secara In Vitro. Jurnal Teknologi Laboratorium. Vol.5, No.2, September 2016, pp. 64 ~67.

13. Indrawan I, Seveline, Ningrum RIK. 2018. Jurnal Ilmiah Respati. Vol 9, No 2. DOI: https://doi.org/10.52643/jir.v9i2
14. Winarno, F.G. 2004. Kimia Pangan dan Gizi. Jakarta: Gramedia Pustaka Utama.
15. Ho H, Tang H, Mazaitul A S, Mohd A H, Roslan A. 2016. Development of novel energy snack bar by utilizing local Malaysian ingredient. International Food Research Journal, 23: 5, 2280-2285
16. Septianingrum, Enny, R. 2009. Identifikasi Penyebab dan Upaya Pengurangan Aftertaste Pahit pada Cookies Ubi Jalar (Ipomea batatas L.) dengan Karakteristik Tekstur Menyerupai Cookies Keladi. Skripsi. Institut Pertanian Bogor. Bogor.
17. Suprijono, M.M dan Sutedja, A.M. 2008. Efek Metode Blanching Uap dalam Pembuatan Biskuit Tepung Tempe terhadap Penerimaan Konsumen. Prosiding Teknologi Proses 152-159.
18. Astuti, Mary dan Hardiman. 1983. Tepung Tempe, Pemahaman Sifat-Sifat dan Penyesuaian Penggunaannya. Laporan Penelitian Fakultas Teknologi Pertanian Universitas Gajah Mada.
19. Rauf R. 2015. Kimia Pangan. Yogyakarta. Andi

IOP Conf. Series: Earth and Environmental Science **980** (2022) 012044 doi:10.1088/1755-1315/980/1/012044

Antioxidant activity and hedonic evaluation of melinjo peel with variations of sodium metabisulfite concentration and soaking time

MW Apriliyanti[1], M Ardiyansyah[2], DE Wulandari[3], P Arum[4], Jayus[5], A Sjaifullah[6]

[1,2,3]Food Industrial Technology Study Program, Agricultural Technology Departement, Politeknik Negeri Jember, Indonesia

[4]Clinical Nutrition Study Program, Health Departement, Politeknik Negeri Jember, Indonesia

[5]Agricultural Products Technology Department, Faculty of Agricultural Technology, Universitas Jember, Indonesia

[6]Chemistry Department, Faculty of Mathematics and Science, Universitas Jember, Indonesia

Email : mulia_apriliyanti@polije.ac.id

Abstract. Melinjo peel is part of the melinjo plant which contains many compounds that are beneficial for the health of the body. This research was conducted to obtain a treatment that has optimal antioxidant activity and hedonic evaluation from melinjo peel steeping which has been pretreated with sodium metabisulfite in concentrations of 0.2%, 0.4%, and 0.6% with a long soaking time of 20 and 30 minutes. Based on the results of testing the antioxidant activity of steeping melinjo peel, the average value ranged from 37.2%-43.6%. n the hedonic organoleptic test, the concentration of sodium metabisulfite and the length of soaking time were significantly different for color with the highest average value (3.56) and taste with the highest average value (3.16) while the aroma was not significantly different from the average value. the highest average (2.96).The best treatment was the concentration of sodium metabisulfite 0.2%, immersion time of 20 minutes with antioxidant activity of 37.2%; color 2.92 (clear yellow); Aroma 4.28 (scented melinjo) and taste 3.96 (slightly taste melinjo peel).

1. Introduction

Melinjo plant (*Gnetum gnemo L*) is a native Indo-Malay plant which has a size of approximately 15.24 m which is commonly found in Indonesia such as in Sumatra and Java. According to [1] the seeds of the melinjo plant are used as chips, while the peel has not been used optimally even though the potential is quite large. According to [2] showed that melinjo plants, both leaves and seed coats, contain antioxidant compounds such as lycopene and carotenoids. Antioxidant activity of the phenolic content of melinjo was comparable to that of butylated hydroxytolune (BHT).

Antioxidants are compounds that can absorb or neutralize free radicals so that they can prevent degenerative diseases such as cardiovascular, carcinogenesis and other diseases. In the research of [3] melinjo peel tea which was processed without enzymatic oxidation (non-oxidation) by withering at room temperature for 8 hours then dried in an oven at 105°C for 25 minutes, the yield was 6.48%.

water, ash content 3, ash content 5%, phenol content 34.8% and yield 56.9% with 37.93% brightness. The drying process can cause a decrease in natural antioxidant activity [4] and the organoleptic of melinjo peel. The drying temperature correlates with the degradation process of phenolic compounds so that it has become decreased antioxidant activity. According to [5] that melinjo peel contains polyphenols, tocopherols, and ascorbic acid which have antioxidant activity and have potential as xanthine oxidase inhibitors. Xanthine oxidase plays a role in the formation of uric acid by catalyzing hypoxanthine to xanthine and then uric acid. Several antioxidant compounds have potential as xanthine oxidase inhibitors because they are able to capture electrons.

Based on research [6] which discusses the optimum conditions for pretreatment and drying of dried dragon fruit peel, namely soaking with sodium metabisulfite on dragon fruit peel can increase antioxidant activity. Sodium metabisulfite can help retain bioactive compounds. [7] explained that the antioxidant activity of dragon fruit peel [7] can be increased through oven drying treatment and sodium metabisulfite soaking. Concentration factors and soaking time in metabisulfite solution affect the optimal antioxidant activity and organoleptic.

This study aims to determine the effect of sodium metabisulfite concentration and soaking time on antioxidant activity and organoleptic evaluation of brewing melinjo peel. The benefits of this research is expected to optimize the utilization of melinjo peel as a raw material for functional beverage products.

2. Material and Method

2.1. Tools and Material
The equipment used includes analytical balance, drying oven (Model R-5A Harvest Saver, USA), and spectrophotometer (Ultrospec 2100 Pro UV-Visible). The ingredients used are orange to red melinjo peel (Magetan Regency, East Java), sodium metabisulfite, aquades, methanol (pa, QREAC), and DPPH (2,2-diphenyl-1-pikridirazil) (sigma Aldrich).

2.2. Experimental Design
This study used a factorial Completely Randomized Design (CRD) consisting of 2 factors with 3 replications. The first factor was the concentration of sodium metabisulfite with 3 levels (0.2; 0.4; and 0.6%). The second factor was the soaking time of the sodium metabisulfite solution with 2 levels (20 and 30 minutes).

2.3. Raw material preparation
The melinjo peel was sorted and selected from orange to red. After that, washing was carried out to remove dirt.

2.4. Soaking with Sodium Metabisulfite
The melinjo peel was weighed as much as 100 g for each treatment. After that, it was soaked with a concentration of 0.2%, 0.4% and 0.6% sodium metabisulfite solution until all parts of the melinjo peel were submerged. Soaking was carried out for 20 and 30 minutes. Next process was draining. Modification method of [7].

2.5. Melinjo Peel Drying
Drying was carried out in an drying oven at 50°C for 12 hours.

2.6. Brewing
The brewing process was carried out to facilitate the analysis of antioxidant activity and organoleptic (hedonic).

2.7. Determination of Antioxidant Activity
Analysis of antioxidant activity based on [8] on melinjo peel that has been soaked with various concentrations of sodium metabisulfite and variations in soaking time using the DPPH (2,2-diphenyl 1-pikrihidrazi) method. Preparation of the DPPH solution is to weigh 0.0040 g of DPPH dissolved in methanol until the volume is 100 ml. For blank preparation, 0.5 ml of methanol is added, 3.5 ml of DPPH is added, then is mixed, left it for 15 minutes and the absorbance measured at a wavelength of 517 nm using a spectrophotometer. Further testing of antioxidant activity, 0.1 ml of sample is put into a test tube and then 0.4 ml of methanol is added. The suspension was then added with 3.5 ml of 1.25 mM DPPH solution which is then homogenized using a vortex. The series of reaction activities are carried out in a dark room and at room temperature. Next, the absorbance is measured at a wavelength of 517 nm. Capacity is expressed in terms of percentage inhibition against DPPH radicals by calculating:

$$\% \text{ Antioxidant Activity} = \frac{(\text{Blank Abs} - \text{Sample Abs}) \times 100\%}{\text{Abs blank}}$$

Abs blank = absorbance of the blank before being reacted with the sample.
Abs sample = Absorbance of the sample after being reacted with the sample

2.8. Hedonic Evaluation
Acceptance test or preference test used hedonic test with 25 panelists. Hedonic evaluation showed the level of preference of the panelists. Panelists fill out the questionnaire by giving an assessment on a scale of 1-5, namely from dislike very much, dislike, slightly like, like, and like very much.

2.9. Data Analysis
Data were analyzed using analysis of variance or ANOVA (Analysis of Variance) with Microsoft Excel 2010. If significant, followed by the Least Significant Difference (LSD) with a significant level of 5%. Determination of the best treatment with the De Garmo method.

3. Results and Discussion

3.1. Antioxidant Activity
Antioxidants are compounds that have a molecular structure that can donate electrons to free radical molecules without being disturbed at all by their function and can break the chain reaction of these free radicals. Epidemiological studies stated that there is an increase in natural antioxidants found in fruits, vegetables, flowers and other parts of plants that can prevent diseases caused by oxidative stress such as cancer, heart disease, kidney and liver inflammation. One of them is melinjo peel which has antioxidant activity.

The antioxidant activity test used the DPPH method (2,2-diphenyl 1-picrylhydrazyl), because this method is quite simple, easy to do and does not take much time. The presence of antioxidant compounds in plant extracts can change the color of the DPPH solution from purple to yellow. The results of the antioxidant activity of melinjo peel steeping can be seen in table 1.

Table 1. Average of Antioxidant Activity in Melinjo Peel

Sodium Metabisulfite Consentration (%)	**Soaking Time (Minutes)**	**Average of Antioxidant Activity (%)**
0,2	20	37.2
0,2	30	39.4
0,4	20	38.9
0,4	30	40.3
0,6	20	41.7
0,6	30	43.6

Table 2. Average of Antioxidant Activity by Treatment with Sodium Metabisulfite Consentration

Sodium Metabisulfite Consentration (%)	Average of Antioxidant Activity (%)
0,2	38.3[a]
0,4	39.6[b]
0,6	42.65[c]

Note: The numbers followed by different lowercase letters show that they are significantly different based on the LSD follow-up test at the 5% level

Table 3. Average of Antioxidant Activity by Treatment with Soaking Time

Soaking Time (Minutes)	**Average of Antioxidant Activity (%)**
20	39.2[a]
30	41.1[b]

Note: The numbers followed by different lowercase letters show that they are significantly different based on the LSD follow-up test at the 5% level

Based on the results of analysis of variance (ANOVA) on the antioxidant activity of melinjo peel, it shows that the F count < F table is 1.16 < 6.92 (1% level) and 1.16 < 3.88 (5% level) so that it shows the concentration sodium metabisulfite and soaking time were not significantly different on antioxidant activity. The results of testing the antioxidant activity of steeping melinjo peel obtained an average value ranging from 37.2%-43.6%. The treatment that produced the highest antioxidant activity (43.6%) was 0.6% sodium metabisulfite concentration with 30 minutes of soaking while the treatment that produced the lowest antioxidant activity (37.2%) was 0.2% sodium metabisulfite concentration with 20 minutes of soaking. In the study of [6] the antioxidant activity of melinjo peel tea without pretreatment was lower (32.37%). Antioxidant activity are also affected by the processing through the heating and storage process. [9] explained that storage at room temperature will also reduce the color intensity and antioxidant activity of the natural dye solution of dragon fruit flesh. Meanwhile, the storage treatment at refrigerator temperature also decreased the pH value, color intensity, and antioxidant activity but the decrease was not as large as that in room temperature storage.

Table 2. and 3 shows that the treatment of sodium metabisulfite concentration and soaking time, each showed a very significant effect. The higher concentration of sodium metabisulfite, the higher the antioxidant activity of steeping melinjo peel. The longer of drying process caused lower the antioxidant activity. The antioxidant activity in avocado leaf tea decreased from 85.11% to 73.13% with a drying time of 1 hour at 50^0C. The longer of drying time caused lower the antioxidant activity of melinjo peel tea. Therefore, pre-treatment with sodium metabisulfite soaking can optimize antioxidant activity in steeping melinjo peel. Increasing the concentration of citric acid and sodium metabisulfite solutions will increase the antioxidant activity of dry matter [10].

The treatment soaking time, 20 minutes treatment was significantly different from the 30 minutes treatment. The length of soaking time in sodium metabisulfite solution will affect the antioxidant activity of steeping melinjo peel, the longer the soaking time, the higher the antioxidant activity. This is in accordance with the statement of [7] that the length of soaking time and the addition of sodium metabisulfite concentration can increase the compounds contained in food ingredients.

3.2 Hedonic Evaluation
The organoleptic test is carried out with a hedonic evaluatin or what is called a preference test, which is a test based on personal responses from panelists about the product or sample presented, whether they like it or not. The scale used starts from very like, like, somewhat like, dislike and dislike very much. Based on this scale, it can be seen that there are differences in preferences for the product or sample presented.

3.2.1 Color Color is one of the sensory properties that can be assessed by the sense of sight (eyes). The first impression that someone is interested or not with a product or food is influenced by its attractive color. The results of the hedonic test of steeping melinjo peel can be seen in table 4.

Table 4. Average of The Hedonic Evaluation

Sodium Metabisulfite Consentration (%)	Soaking Time (Minutes)	Color	Aroma	Taste
0,2	20	3.4^{ab}	2.56	3.16^{c}
0,2	30	3.56^{b}	2.76	3.08^{c}
0,4	20	3.24^{ab}	2.72	2.64^{a}
0,4	30	3.12^{ab}	2.96	2.8^{ab}
0,6	20	3^{ab}	2.8	2.8^{ab}
0,6	30	2.84^{a}	2.72	2.68^{a}

Note: The numbers followed by different lowercase letters show that they are significantly different based on the LSD follow-up test at the 5% level

Based on the results of analysis of variance (ANOVA) of color hedonic organoleptic evaluation data, it is found that if F count > F table is 3.02 > 2.28 (5% level) and F count < F table is 3.02 < 3.17 (1% level) then the concentration of sodium metabisulfite and the duration of soaking significantly affected the color of the melinjo peel brewing. The results of the hedonic evaluation on the color of brewing melinjo peel obtained an average value ranging from 2.84-3.56. The treatment with the highest value of the color hedonic evaluation was found at 0.2% sodium metabisulfite concentration and 30 minutes soaking time, which was 3.56 (lemon yellow) while the treatment with the lowest value was 0.6% sodium metabisulfite concentration and 30 minutes immersion time, which was 2.84 (clear yellow). Panelists prefer the color of melinjo peel with the treatment of lower sodium metabisulfite concentration and fast soaking time because the higher the addition of sodium metabisulfite, the resulting color will dissolve and decrease. [11] stated that the color of melinjo peel steeping water is influenced by carotenoid pigments. Panelists like green melinjo peel tea with a greenish yellow color. This is because the existing chlorophyll compounds are more easily dissolved in water and the cell walls of the melinjo peel are exposed due to heating.

3.2.2 Aroma Aroma is a sensory property that can be assessed using the sense of smell. Everyone has a different level of smell based on their respective preferences. The results of the hedonic test on the aroma of steeping melinjo peel can be seen in table 4.

The results of analysis of variance analysis (ANOVA) obtained that hedonic evaluation on aroma showed that the concentration of sodium metabisulfite and the length of soaking time did not significantly affect the aroma of brewing melinjo peel. Hedonic aroma evaluation of steeping melinjo peel obtained an average value ranging from 2.56-2.96. The treatment with the highest average value from the hedonic evaluation was found at 0.4% sodium metabisulfite concentration for 30 minutes soaking time, which was 2.96 (slightly like) while the lowest average value was 0.2% sodium metabisulfite concentration treatment 20 minutes soaking peel that has been soaked with various concentrations of sodium metabisulfite and the difference in soaking time in the heating (drying) process will cause different aroma changes. The higher the concentration of sodium metabisulfite used, the panelists preferred the aroma of steeping melinjo peel. The processing process can trigger the emergence of a natural aroma, such as due to the roasting process, roasting and other processes. During the heating process there will be a change in aroma caused by the evaporation of volatile compounds, caramelization of carbohydrates, decomposition of protein and fat and protein coagulation [14].

3.2.3 Taste Taste is one of the sensory properties that can be assessed using the sense of taste. Compared to other sensory properties, taste is an important factor to accept or reject a food to be consumed. The results of the hedonic evaluation on the taste of steeping melinjo peel can be seen in the table 4.

Based on the results of analysis of variance (ANOVA) hedonic test data on the taste of steeping melinjo peel shows if F count > F table, namely 2.36 > 2.28 (5% level) and F count < F table 2.36 < 3.17 (1% level) then the concentration of sodium metabisulfite and the duration of immersion significantly affect the taste of steeping the peel. Hedonic test on the taste of steeping melinjo peel obtained an average value ranging from 2.64-3.16 (dislike to slightly like). The treatment with the highest average value from the hedonic taste test was at 0.2% sodium metabisulfite concentration, 20 minutes soaking time, which was 3.16 (slightly like) while the treatment with the lowest average value was 0.4% sodium metabisulfite concentration treatment 20 minutes soaking time, which is equal to 2.64 (dislike). Panelists preferred the soaking treatment with a concentration of 0.2% sodium metabisulfite with an soaking time of 20 minutes, which had a melinjo peel taste.

Melinjo peel which contains phenolic compounds from the flavonoid group can bring bitterness to steeping. The higher the temperature used, the faster the taste of the tea produced. Polyphenol compounds and essential oils contained in melinjo peel contribute to the taste. The taste contribution given is in the form of a bitter and astringent taste which can be reduced if using a high temperature and drying time. If the drying time is too long and the drying temperature is too high, the taste of Torbangun leaf tea will become less bitter because the essential oil evaporated [12].

3.3 Determination of Best Treatment

Based on the results of De Garmo's analysis to determine the best treatment with hedonic evaluation index and antioxidant activity, it was obtained if the best treatment was from steeping melinjo peel, namely the treatment with 0.2% sodium metabisulfite concentration, 20 minutes soaking time with 37.2% antioxidant activity; color 2.92 (clear yellow); Aroma 4.28 (scented melinjo) and taste 3.96 (slightly taste melinjo peel).

4. Conclusion

Based on the results of this study, the following conclusions were obtained:

1. Interaction concentration of sodium metabisulfite and the length of soaking time were not significantly different from the antioxidant activity. However, the treatment of sodium metabisulfite concentration and immersion time each showed a significant effect.
2. In the hedonic evaluation, the concentration of sodium metabisulfite and the length of immersion time were significantly different for color with the highest average value (3.56) and taste with the highest average value (3.16) while the aroma was not significantly different from the average value highest average (2.96).
3. The best treatment is the concentration of sodium metabisulfite 0.2%, soaking time 20 minutes with antioxidant activity 37.2%; color 2.92 (clear yellow); Aroma 4.28 (scented melinjo) and taste 3.96 (slightly taste melinjo peel).

5. Acknowledgments
Thanks are conveyed to the Directorate of Research and Community Service (DRPM) of the Directorate General of Strengthening Research and Development of the Ministry of Education, Culture, Research, and Technology for Funding Higher Education Cooperation Research in the 2021 budget.

References

[1] Siregar, T , Cornelia M, Ermiziar R 2009 Studi Kandungan Karotenoid, Vitamin C dan Aktivitas Antioksidan Kulit Melinio (Gnetum gnemon L) *Pros. Semin. Nas. PATPI*

[2] Cornelia M, Siregar T E 2010 The Study of Carotenoid And Antioxidant Activity Of Melinjo Peel (Gnetum ghemon L) *Proc. Nat. Pigment. Conf. South East Asia Malang, 20-21 March 2020 20-21 March 2020*

[3] Ardiyansyah - and Apriliyanti M 2016 Karakteristik Kimia Teh Kulit Melinjo *J. Ilm. Inov.* **16** 89–92

[4] Hossain M B, Barry-Ryan C, Martin-Diana A B and Brunton N P 2010 Effect of drying method on the antioxidant capacity of six Lamiaceae herbs *Food Chem.* **123** 85–91

[5] Santoso M, Naka Y, Angkawidjaja C, Yamaguchi T, Matoba T and Takamura H 2010 Antioxidant and DNA damage prevention activities of the edible parts of gnetum gnemon and their changes upon heat treatment *Food Sci. Technol. Res.* **16** 549–56

[6] Apriliyanti M W, Handayani A M and Gangsar A I 2020 Optimum response of melinjo peel (gnetum gnemon) antioxidant activity using response surface methodology (RSM) *IOP Conf. Ser. Earth Environ. Sci.* **411** 0–8

[7] Apriliyanti M W, Suryanegara M A, Wahyono A and Djamila S 2020 Kondisi Optimum Perlakuan Awal dan Pengeringan Kulit Buah Naga Kering *J. Teknol. dan Ind. Pangan* **31** 155–63

[8] Molyneux P 2004 The Use of the Stable Free Radical Diphenylpicryl-hydrazyl (DPPH) for Estimating Antioxidant Activity *Songklanakarin J. Sci. Technol.* **26** 211–9

[9] Nataliani M M, Kosala K, Fikriah I, Isnuwardana R and Paramita S 2018 The Effect of Storage and Heating on Physical Stability and Antioxidant Activity of Natural Dye Solution from Dragon Fruit Flesh (Hylocereus costaricensis *J. Tumbuh. Obat Indones.* **11** 1–10

[10] Chaethong K and Pongsawatmanit R 2015 Influence of Sodium Metabisulfite and Citric Acid in Soaking Process after Blanching on Quality and Storage Stability of Dried Chili *J. Food Process. Preserv.* **39** 2161–70

[11] Tarmizi E, Putri A Z E, Nusandari M A, Husnil Y A, Saragih R and Lalasari L H 2019 Identification of chemical structure of anthocyanin and other active substances of red color melinjo peels by FTIR and LC-MC analysis *IOP Conf. Ser. Mater. Sci. Eng.* **578** 4–8

[12] Saragih R 2014 Uji Kesukaan Panelis Pada Teh Daun Torbangun (Coleus amboinicus) *J. Kesehat. dan Lingkung.* **1** 46–52

Effect of time and temperature on vitamin c contents and antioxidant activity in the hot water extraction of melinjo peels

MW Apriliyanti[1]*, M Ardiyansyah[2], W Febrianti[3], P Arum[4], Jayus[5], A Sjaifullah[6]

[1,2,3]Food Industrial Technology Study Program, Agricultural Technology Departement, Politeknik Negeri Jember, Indonesia

[4]Clinical Nutrition Study Program, Health Departement, Politeknik Negeri Jember, Indonesia

[5]Agricultural Products Technology Department, Faculty of Agricultural Technology, Universitas Jember, Indonesia

[6]Chemistry Department, Faculty of Mathematics and Science, Universitas Jember, Indonesia

*mulia_apriliyanti@polije.ac.id

Abstract. Melinjo peel contains ascorbic acid, tocopherol, polyphenols and has the potential as a xanthine oxidase inhibitor. Melinjo peel also has the potential as an antioxidant naturally good for the body. The purpose of this studies to determine effect of variation of time and temperature extractions on antioxidant activity and vitamin C of melinjo peel. This study used a completely randomized design with two factors. firstly, 75 and 100°C for temperature and then secondary factors is different extraction times are 30, 45, and 60 minutes. The highest value of vitamin C content is 313,30 mg/L at 100°C for 60 minutes. Although, the highest value was 30,89% at 100oC for 30 minutes. Consequently, the best treatments at 100 °C and 30 minutes would be appropriate extraction operating conditions.

1.Introduction

Melinjo (Gnetum gnemon L.) is a commodity with various benefits. Melinjo fruit is commonly consumed as the additional ingredients in making dishes and the raw material of melinjo crackers (emping) as traditional foods in Indonesia. During the process of making emping, several of peel from melinjo is wasted. Red melinjo peel has bioactive compounds such as lycopene, flavonoid, phenolic, B-carotene, and vitamin C which also can be used as natural antioxidant compounds for human body [1] [2].

Antioxidant is several compounds which provides one or more electron to counter the free radicals. Free radicals itself are reactive compounds which generally have unpaired electrons at its outer layer [3]. Those free radicals could bring negative effect to the body such as destroying the protein, DNA, and cell membrane. This will cause degenerative disease that the body needs a certain compound namely antioxidant to inhibit the free radicals. Vitamin C or ascorbic acid is an essential vitamin for human body. It is organic substance needed for the body in a small amount to maintain the body metabolism system, repair the body tissue, accelerate collagen formation to heal the wound, decrease cholesterols, and become an aqueous antioxidant which effectively inhibit the free radicals. Vitamin C is able to react to free radicals and oxidants within human body.

This research is aimed to find out the effect of extraction duration and temperature on vitamin C and antioxidant activity of melinjo peel. Normally, extraction process is carried out through maceration method with room temperature. However, the method is not yet considered as the effective one since the less soluble compound will not be perfectly dissolved with room temperature. Therefore, we added variety of temperature and duration to the maceration process to optimize the extraction process.

2. Material and Method

2.1 Tools and Material

Tools used during the melinjo peel extraction process were drying oven (Model R-5A Harvest Saver, USA), hot magnetic stirrer (SH2), thermometer, and spectrophotometer (UV-1800). Ingredient used in this research were orange to red melinjo peel (Magetan Regency), sodium metabisulfite, aquades, methanol 96% (pa, QREAC), DPPH (2,2-diphenyl-1-pikridirazil) (sigma Aldrich), starch indicator, and iodine.

2.2 Experimental Design

This research was carried out through Completely Randomized Design (CRD) with 2 factors and 3 replications.

Factor 1: Extraction Time Duration (Minutes)
T1 = 30
T2 = 45
T3 = 60
Factor 2: Extraction Temperature (°C)
S1 = 75
S2 = 100
From the factors above, we could make some treatment combinations as follows:
S1T1 = Temperature 75°C : Duration 30 minutes
S1T2 = Temperature 75°C : Duration 45 minutes
S1T3 = Temperature 75°C : Duration 60 minutes
S2T1 = Temperature 100°C : Duration 30 minutes
S2T2 = Temperature 100°C : Duration 45 minutes
S2T3 = Temperature 100°C : Duration 60 minutes

2.3 Soaking with Sodium Metabisulfite

Initialy, before starting the soaking process, the melinjo peel were sorted and cleaned to remove the dirt on the peel. Next, the peel were immersed in the 0,6% sodium meta-bisulfite solution for 30 minutes. After that, the peel was dried by using dryer oven at 50°C temperature for 12 hours.

2.4. Extraction

The melinjo peel was then extracted by using maceration method with aquadest. We put 5 grams of melinjo peel into the beaker glass 1000 ml and added 500 ml aquadest. Next, we carried out the maceration process for 30, 45, and 60 minutes at the temperature of 75°C and 100°C. Then, we strained the result and obtained the melinjo peel extract with aquadest solvent.

2.5 Determination of Vitamin C Contents

Vitamin C contents on melinjo peel extracts was tested through titration methods from AOAC Official Method 967.21. Sample 5 ml was put into the Erlenmeyer 250 ml, then we added 10 drops of starch indicator 1%, 20 ml aquadest, then titrated with Iodine 0,01 N solution until the color changed from yellow to green and blue. The calculation was applied to the amount of iodine solution spent with the following formula:

$$\text{mg/g} = \frac{Vol\ iodine\ titration\ x\ 0.7832\ x\ total\ volume}{g\ sample}$$

Where:

Vol iodine titration : iodine volume of the titration result
Total volume : 500 ml
g sample : weighed sample (in gram)

2.6. Determination of Antioxidant Activity

Antioxidant activity on melinjo peel extract was determined by DPPH methods [4]. 2 ml of DPPH solution mixed with 800 µl ethanol 96%. Then, 200 µl of sample added to DPPH-etanol 96 % mixture solutions homogenized with vortex for 15 minutes. The mixture solution was incubated at room temperature in the dark for 30 minutes. The sample absorbance was measured at 517 nm wave length. Antioxidant activity was calculated through the following formula:

$$\%\ \text{inhibition} = \frac{Blank\ Absorbency - Sample\ Absorbency}{Blank\ Absorbency} \text{X } 100\%$$

2.7. Data Analysis

Observation result data was analyzed through analysis of variance or ANOVA (Analysis of Variance) factorial Completely Randomized Design (CRD) methods followed by tukey's test with a significant level of 5%.

3. Results and Discussion

Melinjo peel contains vitamin C, tocopherol and polyphenol as the antioxidants which could potentially become the xanthine oxidase inhibitor [5]. The ANOVA statistical result of vitamin C content and antioxidant activity within the melinjo peel extracts with variation of time and temperature extractions is significantly ($p<0.005$) was shown at the Table 1.

Table 1. Result of Vitamin C Content and Antioxidant Activity on Melinjo Peel

Treatment		Vitamin C (mg/100mL)	Antioxidant Activity (%)
Temperature (°C)	Duration (minutes)		
75	30	254,50^{a}	29,32^{c}
	45	250,60^{a}	28,36^{b}
	60	274,10^{b}	27,00^{a}
100	30	293,70^{c}	30,89^{f}
	45	303,50^{c}	30,06^{d}
	60	313,30^{c}	30,48^{e}

The average value of vitamin C content obtained were varied from 254,50 mg/L – 313,30 mg/L. the highest vitamin C contents at 100°C temperature until 60 minutes while the lowest was on the one with 75°C temperature for 30 minutes. Decreasing of vitamin C can inhibit with sodium metabisulphite until dryng and extraction operating conditions. This result is similar with [6] where vitamin C and phenolic compounds was found in dried mango slices.

Variation of temperatures and times extraction gave different effect on the antioxidant activity. The highest antioxidant activity was 30,89% at 100°C temperature for 30 minutes while the lowest was 27,00% at 75°C for 60 minutes. Increasing temperatures in a short span of time in line with antioxidant activity with increased. An extraction process with high temperature and short times could accelerate the drawing process of a compound from a material and increase the solubility of the active components on a material since the material cells were damaged by the heating process. Besides, increasing of antioxidant activity in this research was caused the water solvents which had polar properties. The antioxidant properties on the melinjo peels are supposed to have hydrophilic properties which was a compound bound to water due to its polar shape, that it extracted quite much antioxidant bioactive compounds.

According to [7] the effect of using solvent with polar properties would result on a polar extract which indicated an effective antioxidant activity. Meanwhile, [8] stated that temperature had a significant effect on extracted compounds, either the bioactive or anti-nutritional compounds. [9] concluded that antioxidant on several food materials were increased along with the increasing temperature at 45-100°C and it would decrease if the extraction temperature up to 120°C. Then, the effect of higher temperatures and longer exposure times make polyphenol compound degradations [6].

Based on the obtained result of antioxidant activity, the best result was 30,89% with 100°C temperature and 30 minutes extraction times. On the others hand [2] antioxidant activity from red melinjo peel used ultrasonic-assisted extraction with 30 minutes and 20 °C was 37,35%. In contrast, seed extracts using hot water showed the antioxidant activity at 41.86 ± 4.22 % [10].

4. Conclusion

The highest value of vitamin C content is 313,30 mg/L at 100°C for 60 minutes. Although, the highest antioxidant activity was 30,89% at 100°C for 30 minutes. It can be concluded, the optimum treatments at 100 °C and 30 minutes with vitamin C content and antioxidant activity is 293,70 mg/L and 30,89 %.

5. Acknowledgments

Thanks you to the Directorate of Research and Community Service (DRPM) of the Directorate General of Strengthening Research and Development of the Ministry of Education, Culture, Research, and Technology for Funding Higher Education Cooperation Research in the 2021 budget.

References

[1] Siregar, T , Cornelia M, Ermiziar R 2009 Studi Kandungan Karotenoid, Vitamin C dan Aktivitas Antioksidan Kulit Melinjo (Gnetum gnemon L) *Pros. Semin. Nas. PATPI*

[2] Fatimatuzzahroh D, Kunarto B and Pratiwi E 2020 Lama Ekstraksi Kulit Melinjo Merah (Gnetum gnemon L.) Berbantu Gelombang Ultrasonik Menggunakan Pelarut Etil Asetat terhadap Likopen, β-Karoten dan Aktivitas Antioksidan *J. Teknol. Pangan dan Has. Pertan.* **15** 41

[3] Rahmi H 2017 Review: Aktivitas Antioksidan dari Berbagai Sumber Buah-buahan di Indonesia *J. Agrotek Indones.* **2** 34–8

[4] Brand-Williams W, Cuvelier M E and Berset C 1995 Use of a free radical method to evaluate antioxidant activity *LWT - Food Sci. Technol.* **28** 25–30

[5] Santoso M, Naka Y, Angkawidjaja C, Yamaguchi T, Matoba T and Takamura H 2010 Antioxidant and DNA damage prevention activities of the edible parts of gnetum gnemon and their changes upon heat treatment *Food Sci. Technol. Res.* **16** 549–56

[6] Jiménez-Durán A, Santos-Sánchez N F, Hernández-Carlos B, Juliani H R and Salas-Coronado R 2020 Mango-seed extract and sulphites as promoters of color and bioactive compounds retention during tray drying of mango slices *Ital. J. Food Sci.* **32** 893–911

[7] Wangensteen H, Samuelsen A B and Malterud K E 2004 Antioxidant activity in extracts from coriander *Food Chem.* **88** 293–7

[8] Qu W, Pan Z and Ma H 2010 Extraction modeling and activities of antioxidants from pomegranate marc *J. Food Eng.* **99** 16–23

[9] Azman Abdul Rahim M S, Salihon J, Yusoff M M, Bakar I A and Damanik M R M 2010 Effect of temperature and time to the antioxidant activity in Plecranthus amboinicus Lour *Am. J. Appl. Sci.* **7** 1195–9

[10] Wazir D, Ahmad S, Muse R, Mahmood M and Shukor M Y 2011 Antioxidant activities of different parts of Gnetum gnemon L. *J. Plant Biochem. Biotechnol.* **20** 234–40

Edible coating development to extend shelf life of mangoes (*Mangivera indica* L.)

L A Handojo, D Shofinita, G Evelina, A N Nasution

Food Engineering Department, Bandung Institute of Technology, Let. Jend. Purn. Dr. (HC) Mashudi Road no.1, Jatinangor, Sumedang District, West Java, Indonesia

email: lienda@che.itb.ac.id

Abstract. Mango is one of the most favored tropical fruits which potentially as an export commodity in Indonesia. However, this fruit is easily damaged and has a short shelf life. One way to extend the shelf life of fruits is to use edible coatings. This study aimed to determine the effect of storage temperature and edible coatings on the shelf life of mangoes. Using 6.4% beeswax and a composite of 10% Arabic gum and 1% chitosan as edible coatings, these results were compared to control. Several parameters were used to evaluate the effectiveness of edible coatings, specifically weight loss, pH, fruit skin color, fruit flesh color, hardness, decay incidence (DI), and soluble solids concentration (SSC). Mangoes were stored in a variety of storage conditions, namely room temperature (26.5°C), showcase (10.2°C), and wine cooler (12.4°C). Experimental results indicated that applicating edible coatings and storing fruits at lower temperatures generally could reduce weight loss above 20% and DI by 85%, slowing changes in SSC, pH, and color of the fruits' skin and flesh, as well as maintaining hardness. Moreover, 6.4% beeswax was more effective than the composite of 10% Arabic gum and 1% chitosan in extending the shelf life of mangoes.

1. Introductions

Fruits are rich in carbohydrates, vitamins, and minerals, as well as low protein content with a pH value ranging from 7 to slightly acidic and high water content. They are easily damaged because they contain 80-90% water of their total weight [1]. These conditions are suitable for the growth of bacteria, fungi, and molds [7, 10].

Indonesia has very diverse types of tropical fruits which have the potential to become an export commodity. The main of Indonesia's annual fruits are bananas, mangoes, oranges/tangerines, pineapples, and durians which almost of all are produced in every province in Indonesia. Since the climacteric fruit ripening process will continue after being picked from the tree, mangoes should be harvested when they reach maximum growth but are not yet ripe [11]. For mangoes that will be transported out of town or for export, the ripeness of the mango harvested is usually 60-70%, thus the mangoes are not damaged during the transportation and are 90-100% ripe until their destination.

Using edible coatings to coat the fruits is one of the methods developed to extend the shelf life of fruits. A thin layer is formed by the edible coating that is safe for consumption as well as protects the fruits from oxygen, moisture, and microbes. Edible coatings have a high potential to control browning activity, foul-smelling activity, unpleasant taste, and microbial activity of fruits and vegetables [9]. The main benefit of using edible coatings is that some of the active ingredients can be incorporated into the polymer matrix and consumed with food, thereby increasing safety or even nutritional and sensory attributes [1].

Coated mangoes using Arabic gum 10% and chitosan 1% could reduce mass loss, respiration rate, ethylene production, change in soluble solid concentration (SSC), decay incidence (DI), and maintain firmness. This proves that their application as natural preservative are a simple and effective technique to delay ripening and maintain mango fruit quality during cold storage (13°C; 80% RH for 28 days) without using fungicides [2]. In addition, several external factors also affect the fruit shelf life, such as temperature which will be investigated further in this study. Arumanis mango (Mangifera indica L. var. arum manis) is used then it is coated with Arabic gum and chitosan as composite and beeswax. The fruits are stored at room temperature (26.5°C), wine cooler (12.4°C), and showcase (10.2°C). This study aimed to determine the effect of storage temperature and edible coatings on the shelf life of mangoes.

2. Material and Methods

The fruits used in this research were Arumanis mangoes (Mangifera indica L. var. arum manis) with a maturity level of 70% which were characterized by a slight aroma, hard texture when pressed, and the absence of black spots on the mangoes. While the mangoes were obtained from Griya supermarket which is located in Jatinangor, West Java, Indonesia, the ingredients for making edible coatings bought from Cipta Kimia, Jakarta, Indonesia for Arabic gum, Monodon Group, Lampung, Indonesia for chitosan, and farmer in Berau Barat, Indonesia for beeswax.

The research method consisted of several stages, namely preparation and washing, coating the fruits with edible coatings, then fruits storage and testing. The edible coatings treatment on mangoes were (1) aqua dm as control, (2) Arabic gum 10% + chitosan 1% as composite, and (3) beeswax 6.4%. After the fruits had been coated, they were stored at several temperatures such as room temperature (26.5°C), 12.4°C using a wine cooler (Polytron SCN 233D), and 10.2°C using showcase (GEA XW-85).

The fruits were tested by measuring weight loss, hardness, skin and flesh color, decay incidence (DI), soluble solids concentration (SSC), and pH. To measure the weight loss, two mangoes from each group were marked separately before being stored and weighed using a scale (Ningbo ETDZ Hansen Measurement Co., Ltd) at the beginning of the experiment then these same fruits were weighed consistently every 7 days. The results were then calculated as a percentage by comparing the mass of the sample in a certain time range with the mass of the initial sample.

Fruit hardness was measured using a texture analyzer (Stable Micro Systems serial number 41141) with blade probe, while the color of the skin and flesh was measured by colorimeter (3NH Technology Co., Ltd type NH310). These measurements were carried out at several different locations such as the top, middle, and bottom of the fruits. The color measurement results from the colorimeter are then converted to the total color change (ΔE) by the following formula:

$$\Delta E = (\Delta L^2 + \Delta a^{*2} + \Delta b^{*2})^{0.5} \qquad (1)$$

These results were compared with the results of measuring ΔE of the flesh and skin color on day 0 of the experiment. DI and SSC measurements were carried out using the method described by Khaliq et al (2016). Samples preparation for pH testing is the same as that used for SSC testing. The fruits' pH was measured by using a pH meter (pH buffer groups USA model 220).

3. Results and Discussion

3.1. Effect of Storage Temperature on Mangoes

Figure 1 represents that the fruit weight loss increased along with the rise in storage temperature and time for each variation of edible coatings. Therefore, the smaller fruit weight loss indicated better results, as a consequence of reducing metabolic activity as well as slowing the chemical changes at the lower storage temperature. While the smallest weight loss from each type of coating variation was the mango that was stored in a showcase (10.2°C), the largest was found in those stored at room temperature.

The optimum temperature for storing mangoes is at 13°C with an RH of 85-95% which can provide a shelf life of 2-3 weeks for mature green mangoes [6], almost similar to the wine cooler's temperature (12.4°C), thus storage mangoes in a wine cooler should be able to minimize the weight loss. However, this research results showed that storing mangoes at a showcase (10.2°C) retained their moisture better than in wine cooler (12.4°C) because the storage relative humidity level was too low (RH 65% and 71%)

compared to the optimum (RH 85-95%) as well as unstable storage conditions in the wine cooler due to too many mangoes stored in the wine cooler resulted with the slow and uneven of the cooling process.

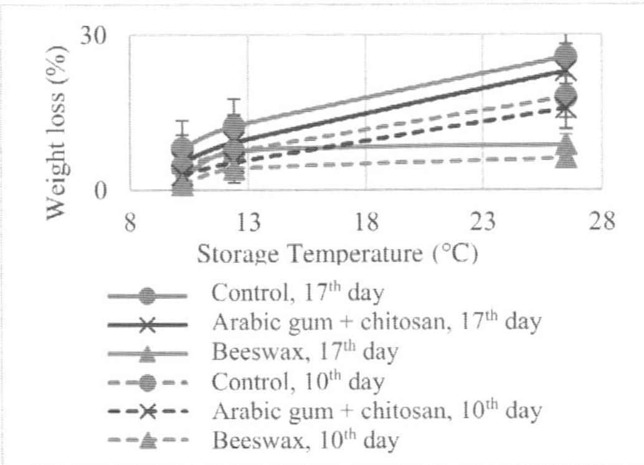

Figure 1. Comparison of weight loss at various storage temperatures on the 10th and 17th day.

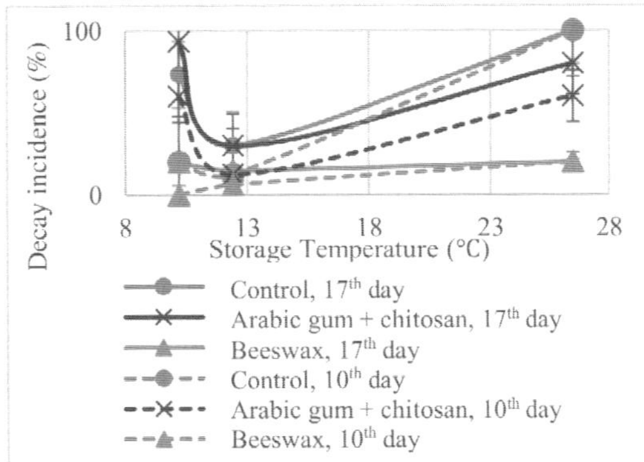

Figure 2. Comparison of DI at various storage temperatures on the 10th and 17th day.

Based on figure 2, decay incidence (DI) as an indicator of the level of fruit rot continued to rise along the period. The lower value of DI implied that the fruit underwent a slower decay process. On the 17th day of storage, for control and beeswax coating treatments, a lower storage temperature could reduce the DI. The largest DI occurred on the control stored at room temperature compared to mango coated with beeswax stored in the wine cooler (12.4°C) which was the least.

3.2. Effect of Edible Coatings on Mangoes

3.2.1. Weight loss

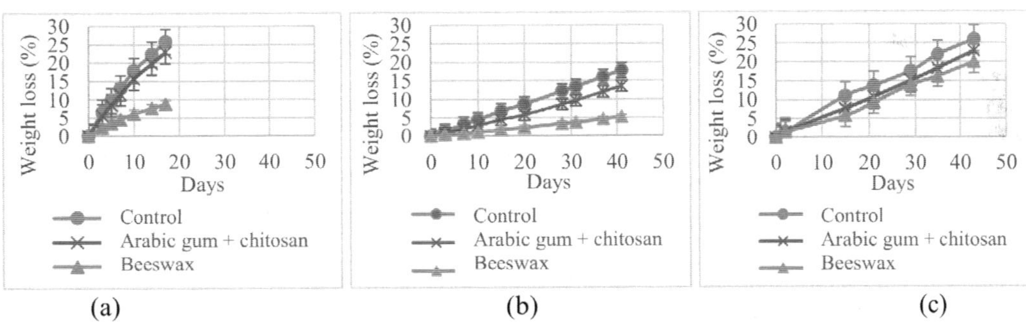

Figure 3. Comparison of weight loss on storage at (a) room temperature 26.5°C, (b) showcase 10.2°C, and (c) wine cooler 12.4°C over storage time.

Weight loss was used as an index of fruit quality during storage whose value would continue to rise over the time. The control had the largest weight loss at the end of the period, which were 25.9%, 25.74%, and 17.75% for mangoes that used storage temperature at room temperature, showcase, and wine cooler, respectively. On the contrary, the smallest weight loss was found in mangoes that used beeswax as edible coatings, which was only 5.19% for showcase storage and 8.72% for room temperature storage. In addition, the lowest weight loss of mangoes for storage at wine cooler with a value of just under 20% was experienced by mangoes coated with Arabic gum and chitosan.

Based on figure 3, beeswax had a better impact in reducing fruit weight loss compared to Arabic gum and chitosan. Due to Arabic gum and chitosan are carbohydrate-based edible coatings that have strong hydrophilic properties, they produce water vapor and do not work well at blocking gas formation.

IOP Conf. Series: Earth and Environmental Science **980** (2022) 012046 doi:10.1088/1755-1315/980/1/012046

Meanwhile, beeswax is a lipid-based edible coating that has hydrophobic properties thus it is better at preventing moisture loss. The weight loss of fruits coated with edible coatings was lower than the control because of they form a barrier to mass transfer such as water, oxygen, and solutes, as well as slow down transpiration (the difference of water pressure at the outside and inside the fruit) and respiration.

3.2.2. Decay Incidence (DI)

Decay incidence (DI) of mangoes from each treatment rose over time which the highest was occurred in control stored at room temperature, as shown in figure 4. However, mangoes coated with Arabic gum and chitosan had a higher DI value than two other variations at showcase storage and it had almost the same value as the control in wine cooler storage. Whereas, Arabic gum and chitosan should reduce the DI value because of their film-foaming properties that can inhibit the microbial process in the fruit [2]. As mentioned in Chapter 3.1, this can occur due to the low relative humidity of storage compared to its optimum storage at 85-90% also the storage conditions in the wine cooler are less stable, and too many mangoes were stored resulting in the slow and uneven of the cooling process.

The DI value of mangoes coated with beeswax was the lowest compared to the others for each temperature storage. This is supported by the result of the mangoes weight loss testing, which was also the smallest than others due to the hydrophilic nature of beeswax. Using edible coating should be able to inhibit the respiration and transpiration process then reduce changes in physiochemical properties.

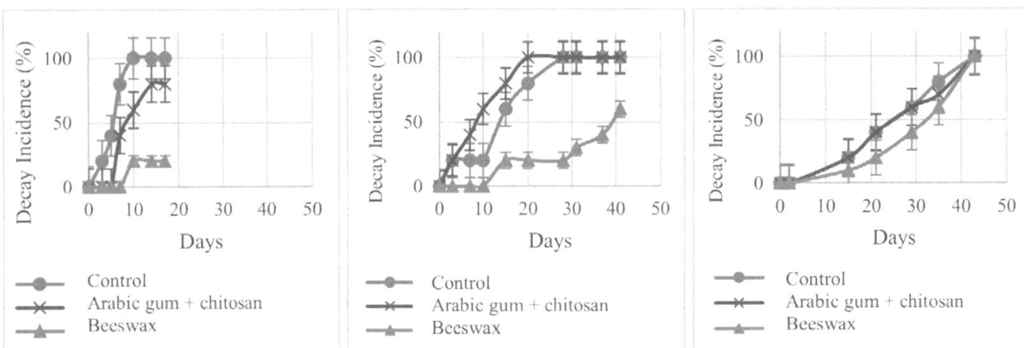

(a) (b) (c)

Figure 4. Changes in the DI of mangoes on storage at (a) room temperature 26.5°C, (b) showcase 10.2°C, and (c) wine cooler 12.4°C over storage time.

3.2.3. Soluble Solids Concentration (SSC)

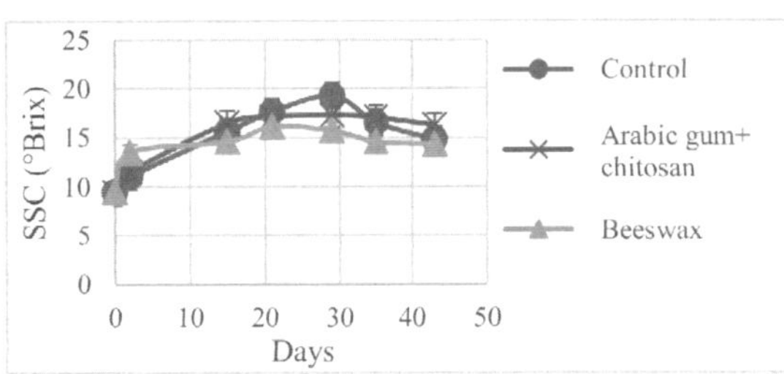

Figure 5. SSC changes of mangoes on storage in the *wine cooler* 12.4 °C over time.

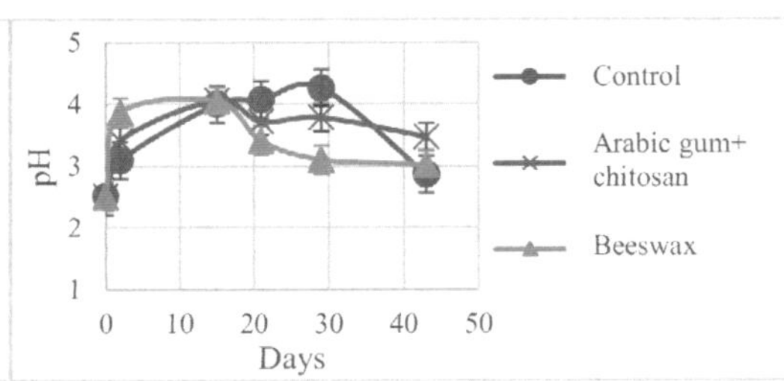

Figure 6. pH changes of mangoes on storage in the *wine cooler* 12.4 °C to storage time.

SSC is an index of fruit maturity and is often used to measure fruit quality since the ratio of sugar to acid has a significant role in determining the level of ripeness and taste of the fruit. SSC increased gradually until the mango fruit was ripe due to the increase in carbohydrates into simple sugars and glucose. During the ripening process, starch is hydrolyzed into simple sugars, such as glucose, fructose, and sucrose which are predominant in ripe fruit. The increasing amount of enzymes sucrose synthase, invertase, and amylase hydrolyzed starch to sucrose [2].

The SSC of mango beeswax variation on storage in the wine cooler at 12.4°C tends to be the lowest among others, as shown in figure 5. The fruit coated with edible coating had a lower SSC value than the control because edible coatings provide an excellent semipermeable barrier around the fruit that can modify the fruit's internal atmosphere by reducing oxygen, increasing carbon dioxide, and suppressing ethylene production. The decrease in respiration rate will slow down synthesis and utilization of metabolites, as well as conversion of starch to sugar, thereby reducing the SSC value [5].

3.2.4. pH

The pH test is carried out to measure the strength of the acid contained in the fruit. Its value of the fruit would continue to increase until a certain day, then decreased over the storage period, caused by a decrease and increase in the level of acid content in the fruit.

As is presented in figure 6, mangoes in each variation experienced an increase in pH up to a certain day, then the pH turned to decrease. For control, pH experienced an upward tendency over a-29-days of periods, then it declined until the end of the period. Meanwhile, both mangoes coated with Arabic gum and chitosan and beeswax experienced an upward trend and reached a peak at the 15th day, then it decreased. The increasing acidity of stored fruits occurs due to the formation of carbonic acid or the production of acid from the fermentation of sugar. Moreover, the decrease in acidity occurs when the fruit is ripe because the levels of malic acid and citric acid are reduced supports the formation of sugar. Malic acid is considered the main substrate for the respiration process so that a decrease in acidity is expected to occur in fruits that have a high respiration rate [3].

Figure 6 reveals that the lowest change in pH occurred in the beeswax variety mango. The use of beeswax in mangoes can reduce the respiration rate of the fruit so that the utilization of organic acids can be delayed. In addition, the change in pH of the control variation mango had the highest value. This is because the edible coating layer acts as a semipermeable layer that functions in delaying fruit ripening by slowing the respiration rate of the fruit so that it affects the formation of acid [5].

3.2.5. Skin color

Based on figure 7, the ΔE value of mango skin tended to increase. On the last day of fruit storage, the ΔE value of mango peel was in the range 11 - 49, which indicated that the skin color was more likely to be different. The color changed from green to yellowish-green in mango was caused by the degradation of chlorophyll and the appearance of carotenoid yellow pigments. Carotenoids are stable compounds synthesized during the developmental stage but are masked by the presence of chlorophyll.

After 15 days, the mangoes began to show blackish spots which were getting bigger due to the high natural sugar content in the fruit and indicating the maximum maturity level of the mangoes. The mangoes which were threatened by edible coatings had a smaller chance of ΔE compared to the control. This depicts that edible coating provided a protective layer for the fruit, thereby reducing the rate of damage that caused changes in the skin color of the fruit due to metabolic processes and microorganism activity. Treating mango fruit using edible coating could also slow down the respiration rate of the fruit and inhibit the rate of ethylene production to create a modified atmosphere around the fruits. In addition, the delay of mangoes skins' color can be caused by the decreased rate of chlorophyll degradation in the mango skin tissues [8].

3.2.6. Flesh color

Based on Figure 8, the value of ΔE of mango flesh tends to increase over storage time. On the end of the storage period, the ΔE value of mango flesh was in the range 11 - 49, which signified that the flesh color was more likely to be different, because of the yellow color of the mango flesh changes to an orange color enzymatically. At the end of the storage period, mangoes showed a reduction in the total amount of carotenoids which could be associated with the oxidation reaction of pigments or phenolic compounds causing darkening of the mango flesh [4].Treated mango fruit with beeswax experienced the smallest change in the value of ΔE of the flesh compared to others. Therefore, the edible coating provided a protective layer for the mangoes, thereby reducing the rate of damage that led discoloration of the fruit flesh due to metabolic processes and microorganism activity.

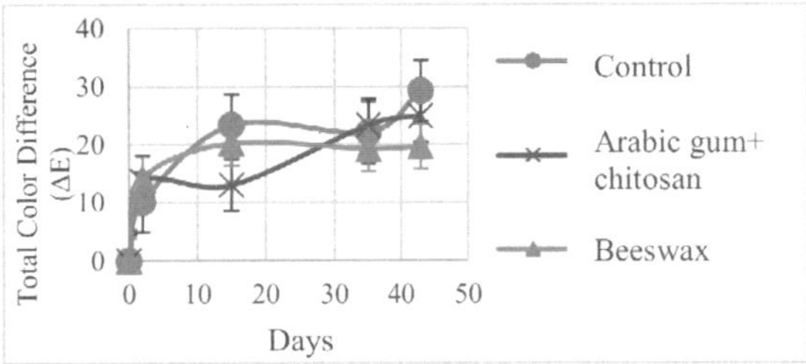

Figure 7. The ΔE of mangoes skin on storage in the wine cooler 12.4°C over storage time.

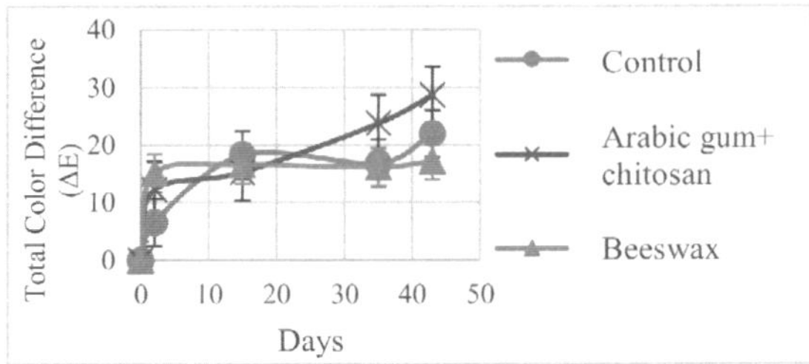

Figure 8. The ΔE of mangoes flesh on storage in the wine cooler 12.4°C over storage time.

3.2.7. Texture (Hardness)

The level of fruit maturity can be determined by reducing the level of fruit hardness because its flesh will be more tender and juicy during the fruit ripening process. According to figure 9, the hardness value of beeswax variation is the greatest among others. However, mangoes coated with edible coating had a firmer fruit texture than control. As a result of a modified atmosphere that was formed by the edible coatings around the surface of the fruit, thus affected the delay in changing the texture of the fruit [2]. In addition, the edible coating was able to prevent water loss and maintain fruit hardness, as the fruit softening is influenced by the loss of water content in the fruit and associated with enzymatic hydrolysis reactions of cell wall components and pectin degradation. Moreover, the use of edible coatings inhibited the respiration process as the breakdown of carbohydrates into water-soluble compounds was reduced.

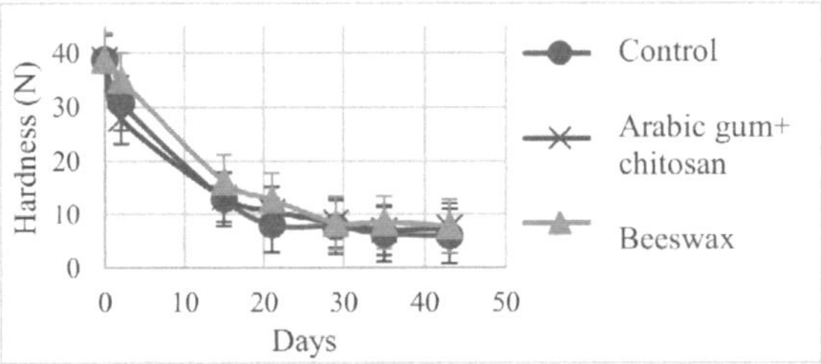

Figure 9. The changes of hardness on storage in the wine cooler 12.4°C over storage time.

4. Conclusions

Based on this research, it could be concluded that stored mangoes at lower temperatures and coated with edible coatings in general, could reduce weight loss and decay incidence (DI), slow down soluble solids concentration (SSC) changes, pH, and color on the skin and flesh of the fruit, and maintain fruit hardness. However, the application of beeswax was more recommended than Arabic gum and chitosan.

References

[1] Dhall R K 2013 Advances in Edible Coatings for Fresh Fruits and Vegetables: A Review *Critical Reviews in Food in Food Science and Nutrition* **53** 435-50

[2] Khaliq G, Mohamed M, Ding P, Ghazali H, Ali A 2016 *Storage behaviour and quality responses of mango (Mangifera indica L.) fruit treated with chitosan and gum arabic coatings during cold storage conditions International Food Research Journal* **23** 141-48.

[3] Kittur F S, Saroja N R, Habibunnisa, Tharanathan R N 2001 Polysaccharide-based composite coating formulations for shelf-extension of fresh banana and mango *European Food Research and Technology* **213** 306-11

[4] Lima A M, Cerqueira M A, Souza B W S, Santos E C M, Teixeira J A, Moreira R A,Vicente A A 2010 New edible coatings composed of galactomannans and collagen blends to improve the postharvest quality of fruits – influence on fruits gas transfer rate *Journal of Food Engineering* **97** 101–9

[5] Maqbool M, Ali A, Alderson P G, Zahid N, Siddiqui Y 2011 Effect of a novel edible composite coating based on gum arabic and chitosan on biochemical and physiological responses of banana fruits during cold storage *Journal of Agricultural and Food Chemistry* **59** 5474-5482

[6] McGregor, Brian M 1987 *Tropical products transport handbook* U.S. Department of Agriculture, Agriculture Handbook No. 668.

[7] Mohapatra D, Mishra S, Giri S, Kar A 2013 Application of hurdles for extending the shelf life of fresh fruits *Trends in Post Harvest Technology* vol 1 (Maharashtra :Jakraya Publications (P) Ltd) p 37-54

[8] Pelayo C, Vilas-Boas E V B, Benichou M, Kader A A 2003 Variability in responses of partially ripe bananas to 1-methylecyclopropene *Postharvest Biology and Technology* **28** 75–85

[9] Raghav P, Agarwal N, Saini M 2016 Edible coating of fruits and vegetable : a review *International Journal of Scientific and Modern Education* **1** 188-204.

[10] Ramos B, Miller F A, Brandão T R S, Teixeira P, Silva C L M 2013 Fresh fruits and vegetables- An overview on applied methodologies to improve its quality and safety *Innovative Food Science and Emerging Technologies* **20** 1-15

[11] Santosa 2006 Panen dan Pascapanen Buah Mangga vol 5 (Limapuluh Kota: Jurnal Penelitian Lumbung) p 558-564

IOP Conf. Series: Earth and Environmental Science **980** (2022) 012047 doi:10.1088/1755-1315/980/1/012047

Organoleptic and chemical properties test on cookies made from Mocaf and oyster mushroom flour

M F Kurnianto[1], R Wijaya [1], A M Handayani [1], B Hariono [1], and A Brilliantina [1]

[1] Department of Agricultural Technology, Politeknik Negeri Jember, Jl. Mastrip 164 Jember, East Java, Indonesia

[*] email: rizza.wijaya@polije.ac.id

Abstract. The purpose of our study was to characterize the quality of oyster mushroom flour dried using the hybrid fluidized bad dryer with a UV lamp. Making cookies was implemented by oyster mushroom flour and modified starch flour called mocaf. Methods in our study using Completely Randomized Design with comparision modified starch flour (mocaf) and oyster mushroom flour (100% of mocaf flour; 95% of mocaf flour and 5% of oyster mushroom flour; 90% of mocaf flour and 10% of oyster mushroom flour; 85% of mocaf flour and 15% of oyster mushroom flour; 80% of mocaf flour and 20% of oyster mushroom flour; 75% of mocaf flour and 25% of oyster mushroom flour. Water content, fat content and fiber content were analyzed as chemical properties of cookies. Organoleptic test was designed by 25 panelists using hedonic test to characterize sensory properties. The results showed that chemical properties consist of water content about 3.3 - 7.1%, fat content about 20.65 – 28.9%, fiber content about 3.03 – 6.55%. Observations on aspects of physical properties carried out by organoleptic tests on the parameters of taste, color and aroma through the Kruskal Wallis test showed no effect. In the texture parameter of cookies products, the variation in the amount of mocaf flour and oyster mushroom flour treatment has an effect.

1. Introduction

Indonesia has the potential for abundant and diverse food sources, but in reality, staple food production in Indonesia almost always relies on imports to meet the food needs. Indonesia's natural wealth has not been utilized and managed optimally. This can be seen from the dependence on various imported food ingredients which is still very high. People's dependence on imported food is wheat flour or wheat flour [1]. Wheat flour/wheat flour as a food ingredient has entered all aspects of life of every level of society in Indonesia. The food diversification program based on local products has not been successful, so the consumption of wheat-based food continues to increase. Currently, Indonesia's wheat consumption per year reaches 21 kg/capita, the second largest after rice. All wheat needs in Indonesia are still 100 percent imported. The imports of wheat seeds reached 6.3 million tons with a value of 2.3 billion USD. Some people have used cassava as an ingredient substitute for rice because of the economic inability to buy rice. Cassava has been developed into a modified form of flour [2]. Cassava flour that has been modified by microbial fermentation is usually referred to as mocaf flour (Modified Cassava Flour). Cassava flour is limited to food ingredients, such as flour substitution by 5% in instant noodles and pastries. Mocaf is used as raw material for noodles, bakery, cookies and semi-wet food. Bowness, steamed cakes, sponge cake, made from mocaf mixture up to 80%. Mocaf produces good quality when using the sponge dough method, which is the use of a dough starter. Mocaf dough is better done with warm water (40-60°C). the process of fermenting cassava into mocaf flour using a fermentation process using Lactobacillus

Plantarum can increase protein levels up to 3.39% [3]. The cassava fermentation process causes changes in the characteristics of the flour which increases viscosity, rehydration, and ease of dissolving and the taste of mocaf flour become neutral by covering the cassava taste up to 70% and has characteristics similar to flour so that it can be used as a substitute for flour or flour mixture [4].

Cookies are a type of product made from a soft dough that uses flour-based, has a heavy texture, tends to be dense and melts in the mouth. Cookies are generally favored by many people, ranging from children to even adults, this product is preferred because it is practical and easy to serve, has a slightly sweet and savory taste and has a soft texture and has a long shelf life of 3-6 months. The main parameters of snack products in the form of physical properties that refer to texture/crispy as well as taste and aroma can be obtained from organoleptic testing of these products so that the innovation of snack products made from corn and cassava in the form of mocatilla chips can be accepted by consumers. This study aims to determine the characteristics of making cookies with mocaf flour and mushroom flour as raw materials. The observed characteristics are in the form of the physical and chemical properties of the resulting product. Cookies are made from a mixture of mocaf flour and oyster mushroom flour with various compositions. This research will produce a recipe or product development formula cookies using a mixture of mocaf flour and oyster mushroom flour [5].

1.1. Mocaf Flour

Mocaf flour has a competitive edge that has high added value by using the biotechnology principle of LAB (*Lactic Acid Bacteria*) fermentation technique. This technology is inspired by the original technology of cassava and cassava sour starch from Brazil. Mocaf is an edible cassava flour product. The color of mocaf is whiter than the color of ordinary cassava flour. The use of mocaf flour can reduce the consumption of wheat flour. Besides being able to be produced from local agricultural products, the use of mocaf flour has many advantages over wheat flour. In mocaf flour there is no gluten protein which often has to be avoided by certain people who have allergies to gluten, children with autism, and people with celiac disease. MOCAF (Modified Cassava Flour) is modified cassava flour through a fermentation process by lactic acid bacteria so that it changes its functional properties and can be used to replace wheat in the manufacture of food products made from wheat [6]. MOCAF does not contain gluten so MOCAF cookies can't expand so Therefore, to help its development, soluble flour is used. Soluble flour has high starch digestibility of 84.35%, low amylose content of 29.67-31.34%, and low amylose content. The high flower content of 54% makes cookies softer, crunchy, and easy to digest. MOCAF and this soluble flour do not have a binding power and crunchiness, therefore maltodextrin is added [7].

1.2. Oyster Mushroom Flour

White oyster mushroom (Pleurotus ostreatus) is currently quite popular and much loved by people in the world, besides being delicious, it is also full of nutrients, high in protein, and low in fat. Each 100 g of dried mushrooms contains 7.8-17.72 g protein, 1-2.3 g fat, 5.6-8.7 g crude fiber, Ca 21 mg, Fe 32 mg, thiamin 0.21 mg, riboflavin 7.09 mg, and 57.6-81.8 g carbohydrates, with 328 -367 kcal of energy. This mushroom can increase metabolism and regulate autonomic nervous function. It is also for the treatment of hepatitis, digestion, duodenum, and stomach. Oyster mushrooms are very good for cardiovascular heart sufferers and cholesterol control. Oyster mushrooms contain mevinolin and similar compounds that are potent inhibitors of HMG CoA (3-hydroxy-3-methylglutaryl coenzyme A reductase), the main enzyme in cholesterol biosynthesis. Beta Glucan Health Center states that oyster mushrooms contain pleural compounds, protein (19-30%), carbohydrates (50-60%), amino acids, vitamins B1, B2, B3 (Niacin), B5 (pantothenic acid), B7 (biotin).), Vitamin C, minerals calcium, iron, Mg, phosphorus, K, P, S, and Zn. It also acts as an anti-tumor, antioxidant and lowers cholesterol. Consumption of this food mushroom can be done in various ways, depending on the taste and purpose of consuming the oyster mushroom in question. Some are consumed fresh, usually for side dishes mixed with meat, fish, or other vegetables [8]. Some are dried, usually, if you want to cook mushrooms at any time, dried mushrooms are poured with hot water. Another way is in the form of powder or flour, usually for making food.

Oyster mushrooms are perishable foodstuffs, like other types of vegetables. A few days after harvesting, the quality of the oyster mushroom drops rapidly until it is unfit for consumption. Changes in the quality of oyster mushrooms include wilting, the color becomes brown, soft and the taste changes, in Indonesia there is not much preservation of commercial food mushrooms, in supermarkets, mushrooms are usually stored at a cold temperature of 15-200C. At this temperature, mushrooms can only survive (still fit for consumption) for 3-5 days, even though they have been packaged in polyethylene plastic [9]. To overcome this problem, further processing is needed so that the shelf life of oyster mushrooms can be extended. One way to extend the shelf life of oyster mushrooms is to process oyster mushrooms into oyster mushroom flour. Making oyster mushroom flour is one of the efforts to extend the shelf life, improve the quality of food, provide convenience in handling, and expand the application of oyster mushrooms in a variety of products. The process of making oyster mushroom powder is considered quite important in its optimization because oyster mushrooms have the potential as a dietary nutrient (nutriceutical), which is currently popular and quite attractive to the public. The optimization here is related to how the process of processing oyster mushrooms into flour, without or minimally eliminating the nutrients contained when in the form of fresh oyster mushrooms. To produce mushroom flour, in general, there are 7 stages, namely (1). Selection of good, healthy mushrooms, uniform in size, and the attached dirt removed, (2). Washing with clean running water, then drained until the water is gone. (3). Mushrooms were sliced thinly (shredded), then blanched with hot water containing 2000 ppm sodium bisulfite (0.2%) for 5 minutes, then drained. (4). The mushrooms were dried in a drying oven by setting the initial temperature to 30oC and gradually increasing to 60oC for 13 hours. Drying can also be done directly by drying in the sun for 3-5 days depending on the weather until the mushrooms dry brownish. (5). Dried mushrooms are stored in tightly closed containers. (6). For the manufacture of mushroom flour, dried oyster mushrooms are then mashed with a food grinder. (7). Mushroom flour is packed in clean and food-grade packaging [10].

1.3. Cookies
Cookies (pastries) are one type of snack that is very popular with people both in urban and rural areas. Cookies can be produced using a variety of flours including flour that does not contain gluten because cookies do not require development. Therefore, cookies have the potential to be made with various local flours. Cookies are one of the most popular dry food products in the market. Various studies have carried out the substitution of wheat flour with local ingredients. The use of local materials is in line with the food diversification program to realize food security in Indonesia. The cookies used in this studywere made from mocaf flour substituted with oyster mushroom flour.

2. Material and Methods
This study used a RAL research design (Completely Randomized Design) to determine the differences in each cookie formula made from mocaf flour and oyster mushroom flour based on chemical properties and level of preference for color, aroma, taste, and texture categories. The research design was chosen because the research was conducted at the Jember State Polytechnic's food analysis laboratory so that the homogeneity of the units can be guaranteed. In addition, the experiment involved a small experimental unit [11]. Levels of pleasure categories of color, aroma, taste, and texture form the basis for analyzing the acceptability of cookies products made from mocaf flour and oyster mushrooms [12].

2.1. Research Procedure
The first stage of research is to conduct product development experiments. The formula determination development process did not go through the QDA (Quantitative Descriptive Analysis) stage because the product to be made already had a standard recipe from the home industry, therefore direct experiments were carried out by providing 5 different formula comparisons between mocaf flour and oyster mushrooms.

2.2. Research Location
This research was conducted at the TEFA Bakery Laboratory and the Food Analysis Laboratory of Politeknik Negeri Jember.

2.3. Tool and Material
The main ingredients used in this study were mocaf flour and oyster mushroom flour which were produced by the team for the basic ingredients for making cookies. The tools used are proximate analysis equipment, Atomic Absorption Spectrophotometer (AAS) Shimadzu ASC-7000, High Performance Liquid Chromatography (HPLC) Agilent 1200 series with Agilent multi-wavelength detector (MWD) type detector, millipore, 0.45 m filter, Texture Analyzer, organoleptic testing equipment, incubators and desiccators.

2.4. Research Methodology
The research was carried out in several stages, namely: The first stage of making cookies using raw materials in the form of mocaf flour and oyster mushrooms as a mixture with various variations of additions. The variations are A1 = 100% Mocaf Flour (Control), A2 = 95% Mocaf Flour : 5% Oyster Mushroom Flour, A3 = 90% Mocaf Flour : 10% Oyster Mushroom Flour, A4 = 85% Mocaf Flour : 15 Oyster Mushroom Flour , A5 = 80% Mocaf Flour: 20% Oyster Mushroom Flour, A6 = 75% Mocaf Flour: 25% Oyster Mushroom Flour. Respondents include the parameters of taste, color, aroma and texture of the product. Each experiment and test was carried out 2 times [12].

2.5. Proximate Analysis
The analyzes carried out where water and ash content using the oven method, fat using the Soxhlet method, and crude fiber. The analysis followed the guidelines of the Association of Official Analytical Chemistry (AOAC 2012).

2.6. Organoleptic Test
Tests on the acceptance of cookies were carried out by organoleptic tests including taste, color, texture/crisp, and aroma. The evaluation was carried out using a semi-trained panel of 25 people. The assessment is based on the level of preference by giving a scoring based on the criteria tested: score 5 = very much like; score 4 = like; score 3 = somewhat like, score 2 = dislike and score 1 = strongly dislike [14].

3. Result and Discussion
Research observations on the response ratio of mocaf flour (modified cassava flour) and oyster mushroom flour in the manufacture of cookies include the chemical and organoleptic properties of the product. The chemical properties observed included water content, fat content and crude fiber. While the organoleptic parameters include taste, color, aroma and taste texture. Based on data analysis on chemical properties according to ANOVA, it showed that the ratio of mocaf flour and oyster mushroom had no significant effect on water content but had a very significant effect on crude fiber content.

3.1. Chemical Properties
3.1.1. Water Content
The average moisture content of cookies with various ratios of mocaf flour and oyster mushrooms ranged from 3.3% - 7.1% (Table 1). The highest water content was obtained from the A4 treatment, the ratio of mocaf flour and oyster mushroom flour was 85:15. While the lowest level was found in the A5 treatment, the ratio of mocaf flour and oyster mushroom flour was 80:20.

The ratio of mocaf flour and oyster mushroom flour had no significant effect on the moisture content of cookies (Table 1). Based on the control of pure mocaf flour (100%) it produces the average water content is 3.7%, which means there is no difference between treatments A2, A3, A5 and A6. In the treatment with variation, A4 obtained a large water content with a value of 7.1%.

IOP Conf. Series: Earth and Environmental Science **980** (2022) 012047 doi:10.1088/1755-1315/980/1/012047

Table 1. Water Content (%)

Treatment	Water Content (%)			
	Repetition 1	**Repetition 2**	**Total**	**Average**
A1(CONTROL)	3,6	3,8	7,4	3,7
A2 (95%:5%)	3,4	3,5	6,9	3,45
A3(90%:10%)	4,7	4,9	9,6	4,8
A4(85%:15%)	7	7,2	14,2	7,1
A5(80%:20%)	3,3	3,3	6,6	3,3
A6(75%:25%)	5	5	10	5

3.1.2. Crude Fiber

The average crude fiber content of various ratios of mocaf flour: oyster mushroom flour ranged from 3.02% - 6.55% as shown in Table 2. The highest crude fiber content was obtained from the ratio of mocaf flour: oyster mushroom flour as much as 75: 25, namely the treatment A6 by 6.55%. While the lowest crude fiber content was found in the treatment ratio of 20: 80, namely in treatment A2 which had a crude fiber content value of 3.55%.

Table 2. Crude Fiber Value

Treatment	Crude Fiber (%)			
	Repetition 1	**Repetition 2**	**Total**	**Average**
A1(CONTROL)	3,045	3,01	6,055	3,0275
A2 (95%:5%)	3,61	3,5	7,11	3,555
A3(90%:10%)	4,53	4,6	9,13	4,565
A4(85%:15%)	5,21	5,15	10,36	5,18
A5(80%:20%)	5,88	5,73	11,61	5,805
A6(75%:25%)	6,59	6,51	13,1	6,55

The crude fiber content of cookies with the composition of mocaf flour: oyster mushroom flour 100:0 (A1) was significantly different from treatments A2, A3, A4, A5 and A6 all treatments gave very significantly different results. This difference is due to the different compositions of fiber in mocaf flour and oyster mushroom flour. The higher the ratio of mocaf flour with oyster mushroom flour, the higher the crude fiber content produced. The crude fiber content in mocaf is around 1.9-3.4%, the results of observations of steamed brownies states that the crude fiber content is 2.27-6.93%. This difference is very real in the steamed brownie cake which has been substituted with mocaf flour. The amount of crude fiber increased with increasing concentration of mocaf flour. The increase in crude fiber was also due to an increase in mocaf flour which contains fiber. The manufacture of dry noodles with mocaf substituted with edamame produces a fiber content of 3.85%, this is almost equivalent to the proportion of mocaf and flour in the manufacture of steamed brownies with a proportion of 40:60. This is influenced by each fiber in different composition and comes from the materials used [13].

3.1.3. Fat level

The average fat content of various ratios of mocaf flour: oyster mushroom flour ranged from 20.65% - 28.9% as shown in Table 3. The highest fat content was obtained from the ratio of mocaf flour: oyster mushroom flour as much as 85: 5, namely treatment A2 of 28.9%. While the lowest fat content is found in the treatment ratio of 85: 15, namely in the A4 treatment which has a fat content value of 25.15%.

Table 3. Fat Level

Treatment	Fat Level (%)			
	Repetition 1	Repetition 2	Total	Average
A1(CONTROL)	20,8	20,5	41,3	20,65
A2 (95%:5%)	29	28,8	57,8	28,90
A3(90%:10%)	26	26,1	52,1	26,05
A4(85%:15%)	25	25,3	50,3	25,15
A5(80%:20%)	27	27,4	54,4	27,20
A6(75%:25%)	27	27,2	54,2	27,10

3.2. Organoleptic Test
The organoleptic test is also called sensory assessment because it involves the senses in humans. This assessment involves panelists. The panelists taken were untrained panelists. The parameters used in the organoleptic assessment are aroma, taste, color and texture. The assessment criteria used started from liking, liking, neutral, disliking and very disliking. The range of values that must be entered also varies from 5 – 1 (very much like to very dislike). The results of the organoleptic test can be seen in Figure 1 with 4 test parameters.

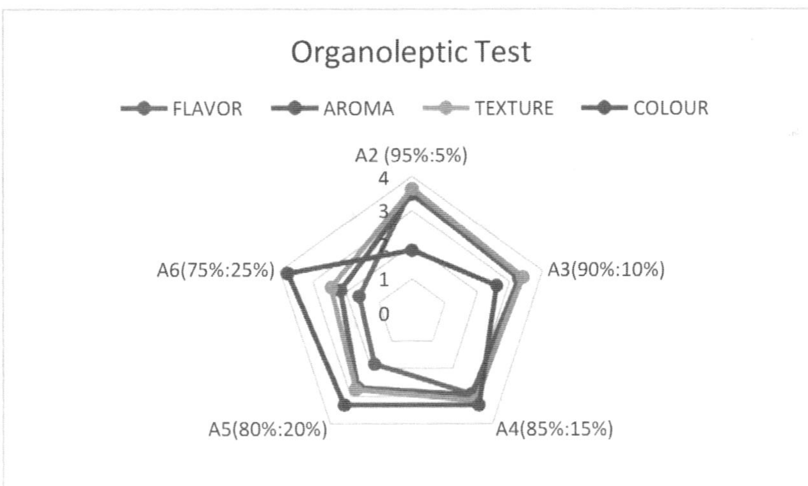

Figure 1. Organoleptic Test

3.2.1. Aroma
Aroma is one of the parameters of the level of consumer preference for a food ingredient. The results of the research on aroma cookies that appear during steaming are accepted by the panelists at different scales. The higher the ratio of mocaf flour, the intensity of the aroma of cassava flour is getting stronger and conversely the lower the ratio of mocaf flour to oyster mushroom flour, the intensity of the aroma of cassava flour is getting weaker. In contrast to the results that have been cooked and cooled, the aroma that appears has a neutral smell. In this study the samples were given in cold conditions.

Table 4. Aroma Organoleptic Test Results

Sample	Aroma
A2 (95%:5%)	3,48
A3(90%:10%)	3,28
A4(85%:15%)	2,96
A5(80%:20%)	2,68
A6(75%:25%)	2,16

The percentage of the lowest level of preference for the aroma of cookies was obtained from the ratio of mocaf flour to oyster mushroom flour (75:25), which has an average value of 2.16. While the highest level of preference in the treatment of mocaf flour: oyster mushroom flour (95:5), which is 3.48. The existence of an average value of 2.68 indicates that the panelists' preference for the aroma of cookie products is neutral. Test Kruskal Wallis obtained p value = 0.406 >α = 0.05, this indicates the ratio of mocaf flour and oyster mushroom flour has no significant effect on cookie products.

3.2.2. Flavor
Flavor is one of the factors determining the level of consumer acceptance of the product. Differences in taste that arise in a test of a product can be caused by subjective traits in humans, such as lack of concentration, feeling tired or tired. Panelists' assessments tend to be individual subjective. It is also influenced by individual sensitivity and preference for objects (cookies). Therefore, the panelist's assessment is carried out by trained consumers in the market group. The scale used in the hedonic test is 5: very much like, 4: like, 3: neutral, 2: dislike, 1: very dislike. The results of the organoleptic test for the taste of cookies had the highest average value of 3.64% in the ratio of mocaf flour: oyster mushroom flour (95:5). While the lowest average is 1.6% in the ratio of mocaf flour: oyster mushroom flour (75:25) as in Table 5. This is due to the presence of a distinctive mushroom taste in oyster mushroom flour, so that with the high ratio of oyster mushroom flour to mocaf flour, a distinctive mushroom taste appears in cookie products.

Table 5. Flavor Organoleptic Test Results

Sample	Flavor
A2 (95%:5%)	3,64
A3(90%:10%)	3,40
A4(85%:15%)	2,92
A5(80%:20%)	1,84
A6(75%:25%)	1,60

The results of the Kruskal Wallis test showed that the p value = 0.372 > = 0.05, this indicates that the flavor has no significant effect on the taste of the cookie product.

3.2.3. Texture
The average level of panelists' preference for texture cookies ranged from 3.44 to 3.64. The lowest percentage of preference level is found in the ratio of mocaf flour and oyster mushroom flour as much as 80:20 with a value of 2.76. While the highest level of texture preference is in the ratio of mocaf flour and oyster mushroom flour 95:5 with a value of 3.64. The panelists' organoleptic test results on the texture of these cookies can be seen in Table 6.

Table 6. Texture Organoleptic Test Results

Sample	Texture
A2 (95%:5%)	3,64
A3(90%:10%)	3,40
A4(85%:15%)	3,16
A5(80%:20%)	2,76
A6(75%:25%)	2,44

The decreasing proportion of flour causes the gluten content in flour to decrease. This gluten functions as an air trap. As a result, the less the number of cavities in the cookies, this is what causes the cookies to be harder and less desirable. The results of the observations showed that the value of P = 0.002 < = 0.05, this means that this texture has a significant effect on the quality of cookies.

3.2.4. Color

Color is the first thing or sensory thing that is seen which also determines the acceptability of the product. If the appearance is less attractive, it is usually not liked by consumers even though sometimes the product has a good taste and has high nutritional value. The results of this color test use a scale hedonic and are assessed by trained panelists. The hedonic scale used is 5: very much like, 4: like, 3: neutral, 2: dislike, 1: very dislike. The average organoleptic color can be seen in Table 7.

The results showed that the best color according to the panelist test was the ratio of mocaf flour: oyster mushroom flour as much as 75: 25 with an average of 3.76. While the color that has the lowest value in the ratio of mocaf flour: oyster mushroom flour is 95: 5 with an average of 1.84. The results of the Kruskal Wallis test showed that the value of p = 0.524 > = 0.05 this means that the ratio between mocaf and mushroom flour has no significant effect on the color of cookies.

Table 7. Color Organoleptic Test Results

SAMPLE	COLOR
A2 (95%:5%)	1,84
A3(90%:10%)	2,6
A4(85%:15%)	3,32
A5(80%:20%)	3,32
A6(75%:25%)	3,76

4. Conclusion

From the results of this study it can be concluded The treatment on various variations did not cause significant differences related to the amount of mocaf flour and wheat flour on the chemical properties of cookie products. In crude fiber parameters, variations in the amount of mocaf flour and oyster mushrooms can affect it. Observations on aspects of physical properties carried out by organoleptic tests on the parameters of taste, color and aroma through the Kruskal Wallis test showed no effect. In the texture parameter of cookies products, the variation in the amount of mocaf flour and oyster mushroom flour treatment has an effect.

5. Acknowledgments

Our thank goes to Politeknik Negeri Jember, who has helped support for this research.

References

[1] Wijaya R and Hariono B, 2020 The Mathematical Analysis of the Drying of Cassava Grater by Using Pneumantic (flash) Dryer with Heat Recirculation Method *J. Phys. Conf. Ser.* **1569**, 4.
[2] Brilliantina A Wijaya R and Hariono B, 2021 the Effect of Natrium Metabisulfite Immersion and Dryng Temperature for Tapai Flour Production *Food Sci. J.* **3**, 1 p. 41.
[3] Pornpraipech P Khusakul M Singklin R Sarabhorn P and Areeprasert C, 2017 Effect of temperature and shape on drying performance of cassava chips *Agric. Nat. Resour.* **51**, 5 p. 402–409.
[4] Hariono B Kurnianto M F Bakri A Ardiansyah M and Wijaya R, 2018 Improvement of Sensory and Chemistry Quality of Fried Edamame by Freezing *IOP Conf. Ser. Earth Environ. Sci.* **207**, 1 p. 1–7.
[5] Díaz A Dini C Viña S Z and García M A, 2018 Technological properties of sour cassava starches: Effect of fermentation and drying processes *Lwt* **93**, December 2017 p. 116–123.
[6] Liu W Zhang M Adhikari B and Chen J, 2020 A novel strategy for improving drying efficiency and quality of cream mushroom soup based on microwave pre-gelatinization and infrared freeze-drying *Innov. Food Sci. Emerg. Technol.* **66**, June p. 102516.
[7] Hariono B Djamila S Bahariawan A Wijaya R and Pambudi S, 2020 Energy study of edamame freezing process at PT Mitratani Dua Tujuh Jember *J. Phys. Conf. Ser.* **1450**, 1.
[8] Veeramanipriya E and Umayal Sundari A R, 2021 Performance evaluation of hybrid photovoltaic thermal (PVT) solar dryer for drying of cassava *Sol. Energy* **215**, January p. 240– 251.
[9] Kosasih E A Zikri A and Dzaky M I, 2020 Effects of drying temperature, airflow, and cut segment on drying rate and activation energy of elephant cassava *Case Stud. Therm. Eng.* **19**, February p. 100633.
[10] Argo B D Sandra S and Ubaidillah U, 2018 Mathematical modeling on the thin layer drying kinetics of cassava chips in a multipurpose convective-type tray dryer heated by a gas burner *J. Mech. Sci. Technol.* **32**, 7 p. 3427–3435.
[11] Kurnianto M F Wibowo M J Hariono B Wijaya R and Brilliantina A, 2020 The Analysis of Consumer Perception on Quality of Soybean Milk Used Importance Performance Analysis Method *IOP Conf. Ser. Earth Environ. Sci.* **411**, 1.
[12] Suratno Kurnianto M F Bachri S Hariono B and Brilliantina A, 2021 Optimization of making white oyster mushroom with the effect of immersion time and concentration of natrium metabisulphite *IOP Conf. Ser. Earth Environ. Sci.* **672**, 1.
[13] Hermanuadi D Brilliantina A and Novitasari E K, 2021 Value added analysis of cassava "tape" supply chain in Bondowoso regency *IOP Conf. Ser. Earth Environ. Sci.* **672**, 1 p. 1–6.

Risk analysis of Musculoskeletal Disorder (MSDs) on corn noodles production

A Indriati[1*], N K I Mayasti[1], T Rahman[1], C Litaay[1], C E W Anggara[1], H M Astro[1], D N Surahman[1], F Novianti[1], and Y Andriana[2]

[1] Research Center for Appropriate Technology, Subang, Indonesia
[2] Research Unit for Natural Product Technology, Yogyakarta, Indonesia

[*] email: ashriindriati@gmail.com

Abstract. Corn noodles are one of the non-wheat food diversification products to reduce dependence on wheat flour. In making corn noodles, it is suspected that there is a risk of Musculoskeletal Disorders (MSDs). This study aimed to identify the level of risk of MSDs in workers while making corn noodles. Data were collected using questionnaires, interviews, and observations to determine the risk level of MSDs using the Rapid Entire Body Assessment (REBA) method. The results showed that when workers did the corn noodle-making process, several work activities showed the risk of MSDs. Activities that pose a high risk of MSDs are when using an extruder; medium risk during mixing and steaming, while weighing ingredients and drying noodles had a low risk of MSDs. Work postures that cause complaints include bending over, looking down, and when lifting. The MSDs complaints were in the neck, shoulders, back, waist, arms, and fingers.

1. Introduction

Corn noodles are a diversified product of non-wheat processed food. Corn noodles are made from corn kernels milled into flour, which can be done by wet and dry milling [1]. The sheeting slitting method in making corn noodles is done by compacting the dough and then making sheets and strands [2]. Corn noodles contain carotenoids as a source of vitamin A with a low glycemic index and can increase the added value of corn and are suitable for business. [2,3].

In a business, an ergonomic work system is one thing that is quite important because it can increase company revenues and reduce expenses [4]. Ergonomic work systems are designed to eliminate potential hazards at work by providing a comfortable and efficient work station design, equipment, and work equipment according to the needs of workers to produce a healthy work environment [5]. One of the impacts of a system that is not ergonomic is the risk of Musculoskeletal disorders (MSDs); for example, if someone is accustomed to working in a bent position, they will feel complaints in the form of pain or soreness in the back area [6].

MSDs are a disorder of the musculoskeletal system with symptoms such as pain due to damage to nerves and blood vessels in various body parts such as the neck, shoulders, wrists, hips, knees, and heels [7]. MSDs appear when workers perform repetitive movements exacerbated by various risk factors at work [8].

In member countries of the European Union (EU), MSDs are the main causes of work disability, sickness absence, presenteeism, and loss of productivity up to 2% of gross domestic product (GDP) [9]. In addition, in the United States from 1992-2010, MSDs accounted for 29-35% of all occupational injuries [10]. Other research that has been conducted shows that in Colombia during 2005, there were 23,477 cases of MSDs with a total cost of handling the US $ 171.7 million or 0.2% of Colombia's Gross Domestic Product in that year [11]. Some work activities can cause the occurrence of MSDs, such as lifting objects, which has a very significant effect on lower back pain, and the movement of pulling important things makes wrist pain [12]. Therefore, it is necessary to conduct an ergonomic study on the corn noodle business to minimize the occurrence of MSDs during production.

The scope of the research is ergonomic testing in the aspect of motion analysis and work posture when doing noodle production. The purpose of this study is to identify the level of MSDs risk when workers produce corn noodles so that it can be a recommendation for improvement.

2. Material and Methods

2.1. Tools and materials

The equipment used in this research includes the manufacture of corn noodles such as extruders, mixers, steamers, scales, and other supports. In addition, the equipment for data collection includes a camera, stationery, hygrometer, infrared thermometer, sound level meter, and lux meter. The ingredients for making corn noodles include corn flour, mocaf flour, rice flour, salt, and water.

2.2. Data Collection Method

The research was conducted from March to September 2021 at Research Center for Appropriate Technology. The object of research is corn noodle SMEs. The data used are primary and secondary. Data were collected using questionnaires, interviews, and observations.

Information about MSDs complaints that were felt when making corn noodles was obtained through questionnaires and interviews. Meanwhile, direct observation was carried out to find out the description of the working posture of each stage of making corn noodles. The risk level for MSDs uses the Rapid Entire Body Assessment (REBA) method.

2.3. Data analysis

The questionnaire and interview data analysis results were to find out the complaints of MSDs that were felt during making corn noodles. In contrast, the observation data were analyzed using the REBA method.

3. Results and Discussion

3.1. Respondent Characteristics

Research respondents are corn noodle SMEs in Subang and have been producing noodles for 2-3 years. The number of respondents was three people, with an age range of 18-24 years, unmarried and not smoking. Respondents generally work for 8 hours a day for the noodle-making process with a rest time of about 1 hour.

3.2. Workplace Characteristics

One that affects the comfort in working is the condition of the work environment. Working environment factors that influence include light intensity, air temperature, humidity, noise, and others. The average condition of the room at the time of corn noodle production is shown in table 1, and measurements were made when the noodle-making process was carried out.

Table 1. Workplace Characteristics

Activity Type	Noise level (dB)	Light intensity (Lux)	Room temperature (°C)	Relative Humidity (%)
Weigh	56.13	326	22.5	72.5
Mixer	69.1	375.33	23.1	74.17
Steaming	62.77	848	25	83
Extrusion	79.17	364.78	26.67	75.56
Noodle decomposition and drying	53.54	562	24	67

Based on Table 1, the noise level generated when work activities are carried out is in the Low-Intensity category (<85dB) and does not cause hearing damage [13]. The intensity of light in the workspace has met the requirements for the health of the work environment in offices and industry, which is at least 100 lux in the room [14]. The room temperature based on the health requirements of the work environment in offices and industry is in the range of 18-28°C so that it is included in the applicable prerequisites. The humidity in the room exceeds the prerequisite range (40%-60%), so it is recommended to use a dehumidifier so that the existing conditions follow the health requirements in the work environment [14].

3.3. Corn noodle making

Corn noodles can be made using calendaring and extrusion techniques [1]. In this study, the manufacture of corn noodles using the extrusion technique. This process requires an extruder. The steps for making noodles with the extrusion technique are shown in Figure 1.

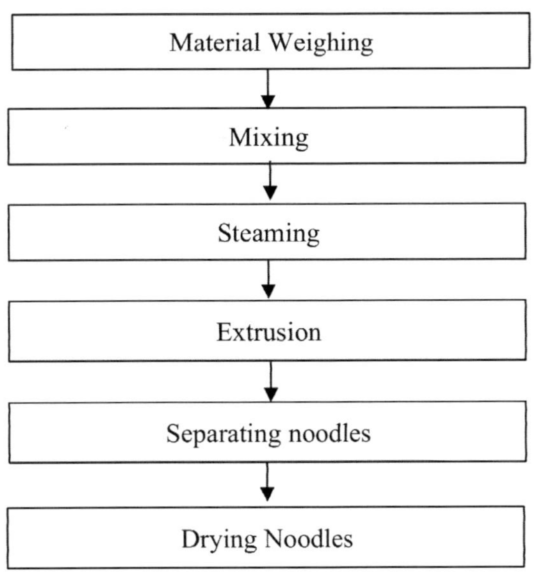

Figure 1. Corn noodle making

In the extrusion process, the equipment used is a single screw-type extruder (Figure 2). The capacity of the extruder used is 12 kg/hour. The extruder specifications are listed in Table 2.

Figure 2. extruder single screw type

Table 2. Specification of Extruders

Specifications	Extruder
Machine type	Single Screw
Capacity	12 kg/h
The product	Gluten-free noodle
Length of the screw	502 mm
Screw diameter	62 mm
Screw pitch	43 mm
Screw rotation	40-50 RPM
Transmission	Gearbox
Heater	Electric Elemen 500 W
Motor	Electric motor 3,77 kW
Materials (screw and barrel)	Stainless Steel 304

3.4. REBA Analysis

There are several stages of activities in the manufacture of corn noodles, including:

a. Material weighing

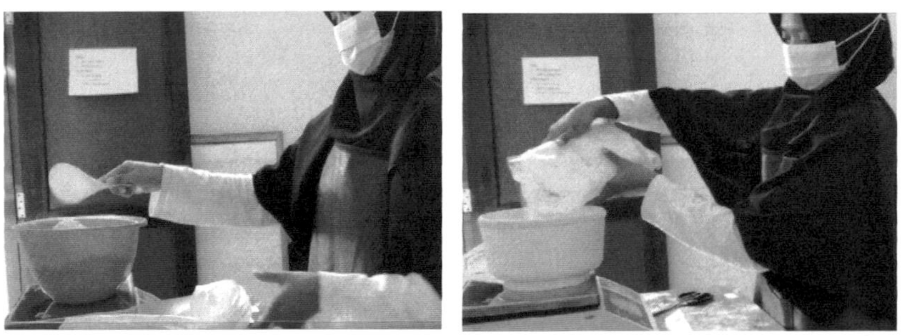

Figure 3. Material weighing process

At this stage, the material is weighed according to the required formulation. The back is straight; the neck is slightly bent; the feet are resting on both feet and, the wrist is twisted.

b. Mixing and steaming

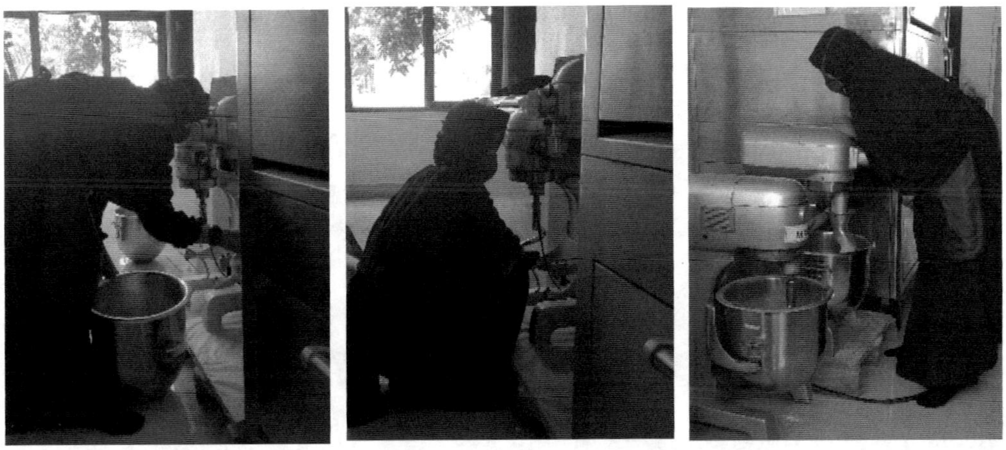

Figure 4. Mixing and steaming

At this stage, all the ingredients are mixed using a mixer. When mixing the ingredients, the body position is resting on both feet, the body posture is flexion 20-60°, the neck is bent down, and the wrist is rotated when preparing for mixing the ingredients and changing the ingredients to be mixed.

c. Material steaming
The mixed ingredients are then steamed. At this stage, the neck is slightly bent, the body is resting on both legs, the arms are above the arm line, and the wrists are twisted and crossed.

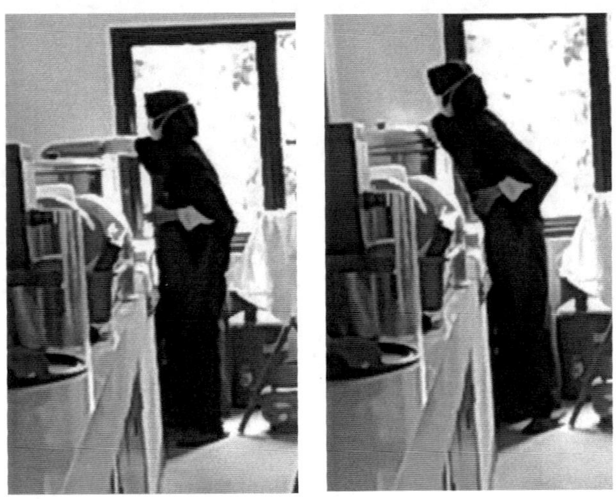

Figure 5. Steaming corn noodles

IOP Conf. Series: Earth and Environmental Science **980** (2022) 012048 doi:10.1088/1755-1315/980/1/012048

d. Extrusion
At this stage, there are several activities, including entering the material in the extruder and cutting noodles. When inserting the material and ensuring that the material enters the tool, the back is bent, the neck is bent, the body is resting on both legs, and the arms are stretched out to spill the material onto the tool. The position of cutting noodles, the body position is in a sitting position, the back is bent, the neck is slightly bent, the arm position is sometimes above the upper arm line, and the wrist is sometimes twisted and crossed. Another position is to untangle the noodles and bring the pan containing the noodles to the drying.

Figure 6. Extrusion process

e. Decomposition and drying

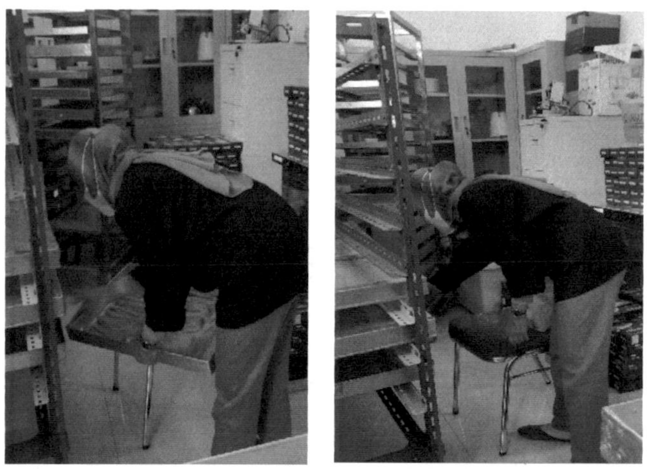

Figure 7. Decomposition and drying

Before the drying process, the noodles are separated, and then the noodles are arranged on a baking sheet. At this stage, the position of the back and neck is bent, the body position is resting on both feet and sitting, and the wrists sometimes twist or cross when storing the pan on the shelf.

Table 3. Score A, Score B, and Score C in the process of making corn noodles

Activity	Awkward posture	Score A	Score B		Score C	
			Right	Left	Right	Left
Material weighing	Neck down	1	5	5	3	3
Material mixer	bend back and Neck down	4	5	5	5	5
Steaming	Material weighing	3	9	9	7	7
Extrusion						
- Enter material	bend back and Neck down	4	9	9	9*	9*
- noodle cutting	bend back and Neck down	5	7	6	9*	8*
Decomposition and drying	bend back and Neck down	3	3	3	3	3

*addition of 1 due to repetition

Table 4. REBA Score and Action Level in making corn noodles

Activity	REBA Score		Action level	Risk Level	Corrective action
	Right	Left			
Material weighing	3	3	1	Low	May be necessary
Material mixer	5	5	2	Medium	Necessary
Steaming	7	7	2	Medium	Necessary
Extrusion					
- Enter material	9*	9*	3	High	Necessary soon
- noodle cutting	9*	8*	3	High	Necessary soon
Decomposition and drying	3	3	1	Low	May be necessary

*addition of 1 due to repetition

The following is an analysis of the REBA score at the stages of making corn noodles in Table 3 and Table 4. Activities that have MSDs risk are when mixing materials which can be caused by an inappropriate mixer storage position so that they have a moderate MSD risk, the use of an extruder in a position to enter the material is above so that it requires a footrest, lowering the neck and slightly bending the body in addition to That's the position of the noodles coming out of the tool in a position that causes the body to be in a sitting position and bent down to facilitate cutting. In addition to weighing materials and drying have a low risk of MSDs.

3.5. MSDs Complaints
Based on the questionnaire results, several complaints were felt at the time of making corn noodles. The perceived complaints are listed in Table 5. Activities in the manufacture of corn noodles that cause pain include: Insert dough into the extruder; Material Mixing; Turning the dough over; Cutting noodles and, Storing noodles on racks.

The pain has been felt for 1-3 days and, complaints of pain have been handled since using the extruder. When compared with the results of the REBA analysis, it is seen that there is a relationship between body posture and complaints of MSDs. Activities in making corn noodles that are felt to cause pain include entering the dough into the extruder, mixing ingredients, turning the dough over, cutting noodles, and storing noodles. The use of an extruder has a high risk of MSDs complaints, and according to the results of interviews, the pain is felt every time you use the tool, so it is necessary to take control measures to minimize the risk of MSDs. Control measures that can be taken include avoiding the same position simultaneously, and modifications can be made to the place and work pattern [15].

Table 5. MSDs Complaints on Corn Noodle Making

Complaint type	Felt complaint
Pain or stiffness in the upper neck	Sore
Pain or stiffness in the lower neck	Sore
Pain in the left shoulder	Sore
Pain in the right shoulder	Sore
Pain in the left upper arm	sore; fever
Pain in the upper back	Sore
Pain in the upper right arm	sore; fever
Pain in the lower back	Sore
Pain in the waist	Sore
Pain in the buttocks	Sore
Pain in the left elbow	Sore
Pain in the lower left arm	sore; fever
Pain in the lower right arm	sore; fever
Pain in the fingers of the left hand	sore; fever
Pain in the fingers of the right hand	sore; fever
Pain in the left calf	Sore
Pain in the right calf	Sore
Pain in the left foot	Sore
Pain in the right foot	Sore

4. Conclusion

Corn noodle-making activities have a risk of MSDs complaints. Activities that have a high risk of MSDs are at the time of extrusion, so improvements are needed on the tool to reduce the risk of MSDs. Improvements can be made by redesigning the extruder; for example, the insertion section of the material can be made lower so that it is easier to insert the material into the tool and does not require additional footrests. Other; In addition, the noodle outlet can adjust the user's target anthropometry to reduce the slouching position, and it is also possible to manufacture an automatic noodle cutter. The process of mixing ingredients and steaming has a moderate risk, and the process of weighing and drying has a low risk for repairs on both risks can be done by improving the layout of the tool so that the user is more comfortable when using the tool. The MSDs complaints felt during the corn noodle manufacturing process were proportional to the MSDs risk level of each method.

References

[1] Ekafitri R 2010 *J Pangan* **19**(3) 283–93
[2] Kusuma P and Mayasti N 2014 *Agritech.* **34**(2) 194–202
[3] Putra GB 2019 *Analysis of Baso Noodle Merchants and Consumers Preferences on Corn Wet Noodles with Extrusion Technology* (Bogor: Institut Pertanian Bogor)
[4] Widodo L, Pramudya B, Herodian S and Syu'aib MF 2009 *Study of Ergonomic Aspects in Agro-Industry Work Systems*. In: Seminar Nasional Dan Gelar Teknologi PERTETA. (Mataram:

IOP Conf. Series: Earth and Environmental Science **980** (2022) 012048 doi:10.1088/1755-1315/980/1/012048

UNRAM) p 37–46.
[5] Mayasari D, Saftarina F 2016 *J Kedokt Dan Kesehat Univ Lampung* **1**(2) 369–79.
[6] Setyaningsih A 2015 *Assessment of Work Posture and Complaints of Musculoskeletal Disorders (Msds) on Micro-Scale Wood Frame and Door Industrial Workers* (Bogor: Institut Pertanian Bogor)
[7] Cho K, Cho H-Y, and Han G-S 2016 *J Phys Ther Sci.* **28**(1) 56–62.
[8] Sari D 2015 *Analysis of Musculoskeletal Disorders (MSDs) Risk on Logging and Skidding Workers at IUPHHK-Ha PT Sari Bumi Kusuma, Central Kalimantan* (Bogor: Institut Pertanian Bogor)
[9] Bevan S 2015 *Best Pract Res Clin Rheumatol* **29**(3) 356–73
[10] Bhattacharya A 2014 *Int J Ind Ergon.* **44**(3) 448–54
[11] Piedrahita H 2015 *International Journal of Occupational Safety and Ergonomics* **2**(4) 379-386
[12] Basahel AM 2015 *Procedia Manufacturing* **3** 4643-4649
[13] Sihombing TN and Arvianto A 2018 *Industrial Engineering Online Journal* **7**(4)
[14] Minister of Health 2002 *Keputusan Menteri Kesehatan Republik Indonesia Nomor 1405/MENKES/SK/XI/2002 tentang Persyaratan Kesehatan Lingkungan Kerja Perkantoran Dan Industri* (Jakarta: Ministry of Health)
[15] Kurniawidjaja. Theory and Application of Occupational Health. Jakarta: Universitas Indonesia; 2011.

IOP Conf. Series: Earth and Environmental Science **980** (2022) 012049 doi:10.1088/1755-1315/980/1/012049

Rapid and non-labelling detection method in food safety and adulteration

B B Andriana [1*], P N Paramitha [1,3], R Zakaria [1,2], P Parikesit [2,3], K Hashimoto [1], H Sato [1]

[1] School of Biological & Environmental Sciences, Kwansei Gakuin University, 2-1 Gakuen, Sanda, Hyogo-Ken, 669-1337, Japan
[2] Department of Biology, Universitas Padjadjaran, Jl. Raya Bandung - Sumedang Km 21, Jatinangor, Sumedang 45363, Indonesia.
[3] Centre for Environmental & Sustainability Science (CESS), Universitas Padjadjaran, Jl. Raya Bandung - Sumedang Km 21, Jatinangor, Sumedang 45363, Indonesia.

[*] email: ftc87897@kwansei.ac.jp

Abstract. One of the goals from The United Nations' 17 Sustainable Development Goals (SDGs) is good health and well-being has, directly and indirectly, absolutely has relation with the safety and fraud of food. In the field of food sciences, the availability of analyzing tools for both safety and fraud of food is a must for producing the sustainable of good health and well-being of consumers, as well as a need early detection of contaminants in the food supply that be able to be serious to certify the safe food. Raman spectroscopy is one of the unconventional techniques that can detect and indicate those problems classes rapidly, sensitively, noninvasively, and relatively inexpensive method. The quantitative analysis and modeling of biomaterials based on empirical data are useful for connecting molecular biology to new areas, particularly in Industry 4.0 era, such as computational of food science. Raman spectroscopy is considered one of the possible methods by which we can observe the biological sample in rapid and non- labeling methods. This could improve to check up the quality and safety of food in the field of food science rapidly and safety. Furthermore, it will largely contribute to society in the future. The present article introduces some preliminary techniques based on Raman spectroscopy and evaluates their applications in intact food ingredients, such as lipid and Stingless Bees Honey.

1. Introduction

From several hundred years ago until the present, the problem of adulteration persists at everylevel of food from food preparation to consumption. Particularly in 20 centuries, most of the food manufacturers, processors, and restaurant owners are all elaborating in one way in this unethical practice of adulteration. Foods are adulterated by using various harmful chemicals, toxic artificial colors, rotten foods, perishables foods, and unexpected foods are stored then sold and served to buyers. Meanwhile, many buyers do not have the skills to make some distinguishing. Just a few buyers are still concerned about distinct it due to they have a sensitivity sense.

Furthermore, in the digital era, many people are busy with virtual and real activities wherethere is no relation to the knowledge of food adulteration. Uses of un-expecting foods in replacement the original foods are evident in highest degrees which are making the more profitfor the producers. Such as lipid where there is the possibility to blend one species' lipid with the other species lipid that is cheaper and abundant compared with the expected lipid. This case is a serious threat to public trust in any country for the last couple of decades. Focusing on sheep's lipid and Goat's lipid, there is still confusion to the market whether mutton has meaning for the meat of sheep or goat. Meanwhile, lamb is the meat of the young sheep and has a high price compared with adult sheep and goat meat due to the lamb containing a smallamount of lipid in its body. The goat has a small amount of lipid comparing the sheep's lipid.If lamb's lipid is compared to goat's lipid just a special customer or buyer can directly make difference between lamb`s and goat`s lipid.

Not only ruminant's lipid, but the same adulteration case also occurs with honey bees. There is the other honey are produced by stingless bees. In this preliminary study of stinglessbees, the report is just focusing to test whether Raman spectroscopy can distinguish the honey produced by two colonies of stingless bees that are harvested from the wild and farmed stingless bees. Stingless bees will produce honey influenced by the availability of nectar surrounding their colonies or their foraging ecology. The kind of nectar depends on the flower itself. Conventional methods (causing damage to the sample, labeling,and taking time) and unconventional methods (Rapid, non-destructive, and non-labeling) are potential for studying the adulteration of animal products. Some conventional methods with different techniques have been used for the detection of adulteration of cheap meat species andproducts using electrophoresis [1], enzyme-linked immunosorbent assay ELISA [2], and the mass spectrometric methods [3]. The classification of meat species using their DNA markers,namely PCR or Polymerase chain reaction (PCR) has been used [4], gas chromatography-massspectrometry [5], and high-pressure liquid chromatography [6]. Although, DNA and protein-based methods such as PCR and ELISA are the most specific and sensitive techniquesfor the detection of adulteration of meat species, however, these methods require expensive equipment, highly skilled technical expertise, damaging the sample, labeling. Moreover, chromatographic methods are a low rate of reproducibility.

Particularly in the application of technology for identification of the adulteration havealready one by many researchers where nearly their methods are still categorized thedestruction methods. The sample will be used for analyzing and the damage of samples cannotbe avoided. Raman spectroscopy, its wavelength of 785 nm, has been used by our laboratoryfor the determination of the chemical composition of any kind of biological samples [7], meatspecies [8][9][10]. Raman spectroscopy and chemometric data analysis are potential todetermine food adulteration in several seconds duration [11]. The advantage of Ramanspectroscopy is the capability to convert the biochemical structure of vibration molecules tobe digital data without any changing and have need of small amount of the sample [12], thenit is high potential for the evaluation of food security and quality during harvesting, handling, processing, packaging, and storage [13]. Most lipids initiated in adipose tissue are in the formof triacylglycerol that the lipid metabolic system in various ways can be affected by the typeand amount of fatty acids and depending on their chain length and degree of unsaturation [14].In these preliminary studies, first Raman spectroscopy has been used as a non-destructive and non-labeling method for detecting the distinct of lamb's lipid and goat's lipid where theband at 1085 cm^{-1} might be as a marker of lamb's lipid and the band at 1105 cm^{-1} might be asa marker for goat's lipid. Secondly, Raman spectroscopy can distinguish two colonies of stingless bees from the forest and farmer.

2. Materials and Methods

2.1. Materials

The sheep's or lamb's lipid (n = 5 samples) and goat's lipid (n = 5 samples; no age information)were purchased from retailer halal meat shop in Kobe Japan, which in the form of frozen meat.Lambs were imported from Australia and repackaged in Japan, meanwhile, the goats were produced by farmers in Japan. As the preliminary research, the materials research wasdesigned by collecting the frozen samples without the information of the age, strain, sex, feedtype, and the duration of frozen storage. The lipid sample was just from the subcutaneous lipidof sheep (lamb) and goats. We did not buy the mutton when

IOP Conf. Series: Earth and Environmental Science **980** (2022) 012049 doi:10.1088/1755-1315/980/1/012049

the seller was not quite sure aboutdifferent mutton meat. Some of the sellers have some understanding that the mutton can be the meat of the mature sheep or goat.

After purchasing, the frozen lipid was thawed for 1 hour and directly conducted measurement. There was no specific preparation before measurement. The lamb's lipid and goat's lipid samples were thawed at room temperature prior to the collection of Raman spectra.Raman spectra were obtained directly from lipid samples by placing them under the microscope of Raman spectroscopy. Clean aluminum sample holders were used to hold lipid samples upon it. Figure 1 shows the performance of sheep - lamb (see fig. 1A) and goat (seefig. 1B) wherein some part of society or customer cannot make a difference which one sheep and goat is.

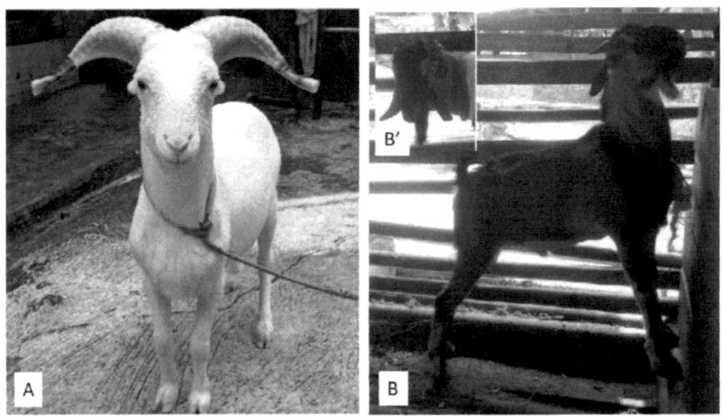

Figure 1. Male sheep - Lamb (A) and male goat (B) have some differences in theirperformance, sheep (*Ovis aries*) and goat (*Capra aegagrus hircus*). Goat has a long ear and beard (B').

Figure 2. Stingless bees on the flower (A), the distance of Batu Ampar to Hovat are 350km where Batu Ampar is at low land and Hovat is at high land – Forest (B), Stingless bees are on the pealed orange fruit as the human effort to prepare the alternative feeding (C). Red arrows are stingless bees.

The honey of stingless bees was collected from forests and farming. For preliminary research, the number of colonies was decided to one colony from the forest (at Hovat, Kalimantan, Indonesia) as the wild stingless bees and the farmed stingless bees (at Batu Ampar, Kalimantan, Indonesia). The method of honey measurement did not also need some preparation and directly we drop the honey into the sample folder at room temperature. Figure 2 depicting the stingless bees (see fig. 2A), the place where the samples were collected (see fig. 2B), and several stingless bees on the pealed fresh orange (see fig. 2C).

2.2. Raman Spectroscopy Information

The Raman spectra from the lipid and Stingless bee samples were obtained by using single excitation lasers at 785 nm. The Raman system is microscopy Raman spectroscopy [15] usedfor obtaining the spectra from lipid and stingless bee honey samples by using laser diode at 785 nm, which has a working distance of 200 – 800 μm. The laser was focused on the lipid surface with a 20x objective lens and Raman spectra were measured in the backscattering configuration. The laser power on sample points was 50 mW. Each Raman spectrum was acquired from 500 to 1800 cm^{-1} with an exposure time of 30s x 2. To exclude any experimental artifact, six Raman spectra were recorded from each fat sample, and the number of samples was 5 samples. From each sample, six Raman spectra were recorded from the sample surface by shining a laser power of 90 mW. To record Raman spectra, a microscopic objective lens of 20x magnification was used to focus the laser on the upper surface of lipid tissue samples for acquiring an optimal Raman signal. The calibration of Raman spectrometerswas conducted via recording the Raman spectrum of a silica crystal wafer at 521 cm^{-1}. Figure3 shows the performance of Raman spectroscopy that we used for these preliminary studies.

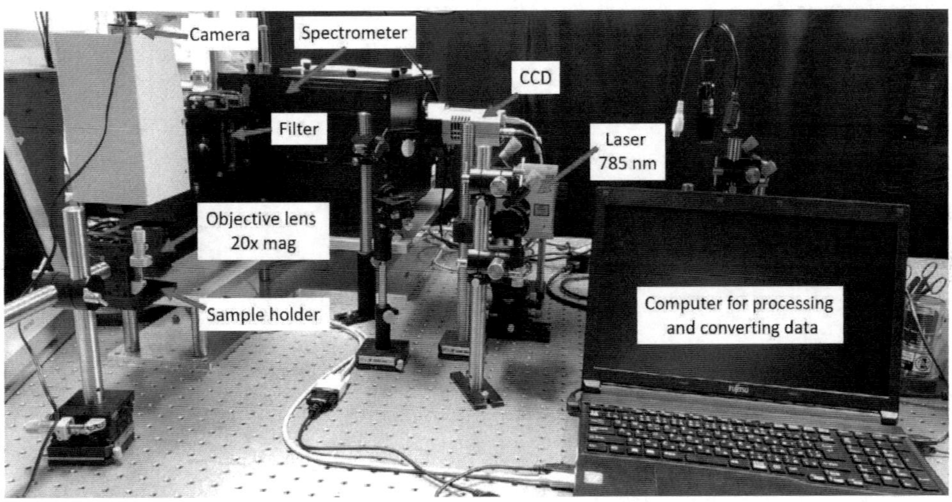

Figure 3. Microscopy Raman spectroscopy with the magnification of lens 20x and laser wavelength 785 nm

2.3. The Preprocessing of Raman spectra

All the raw Raman spectra (Such as fluorescence background, noise, and spectral overlappingof nearby bands) were processed to be easily understood for visually identifying them. Therefore,getting rid of these unwanted components in the Raman spectra is very important. All Raman spectra were preprocessed in MatLab (MathWorks) for removing the noise, making a baseline correction, and determining the normalization (at the band of 1440 cm^{-1} [14] for studying sheep's and goat's lipid)) so that molecular bands and functional groups can be identified correctly. Normalization of stingless bee honey is at the band of 890 cm^{-1}. All the Raman spectra were smoothed by the smoothing method of

Savitzky-Golay. For the spectral analysisof data, the fingerprint with the range of Raman spectra was from 500 to 1800 cm^{-1} that contained prominent spectral signatures. The concurrent determination of specific analyses incomplex tissue is difficult and requires handling huge amounts of data. Principal Component Analysis (PCA) was applied on the pre-processed Raman spectra of lipid samples only for their classification based on their spectral fingerprints. Furthermore, PCA is for the elimination of repetitious data, then the multivariate data analysis techniques were applied for thedetermination of specific fatty acids (Palmitic acid, etc.) in sheep's lipid and goat's lipid. Meanwhile, the stingless bees were just used to demonstrate that not only from livestock products can adopt the Raman spectroscopy, but the stingless bees honey from the forest (wild) and farming can be distinct.

The Raman spectra of two colonies of stingless bees were analyzed by making the subtraction of their spectra only due to this measurement is just verification effort for this preliminary study of adulteration determination by our microscopy Raman spectroscopy. Processing the data with simple methods (subtraction) were done after spectrum normalizationbased on the broadband at 1406 cm^{-1} was applied [16] to recompense the gross distinguish in the spectral response that was caused by physical effects. Discriminations of fatty acids content using a chemometric method (PCA) were performed by Unscrambler X (Service pack - USA).

3. Result and Discussion

The Raman spectra of sheep's lipid and goat's lipid encompassed molecular signatures that provided the biomolecule changing of the sample after lighting by laser. This study's goal isto investigate whether our microscopy Raman spectroscopy can differ lamb's lipid and goat's lipid from subcutaneous lipid or not. And the testing was also conducted on two colonies of stingless bees for detecting the difference. The Raman scattering is independent of exciting laser wavelength to the molecules, where the wavelength of 785 nm laser has been testified for strong Raman scattering of proteins, lipids, and fatty acids [17]. Thus, the laser has been used as the excitation source for Raman scattering from lipid and honey samples of Sheep/goatand stingless bees, respectively, to investigate their molecular composition and the differencesamong them.

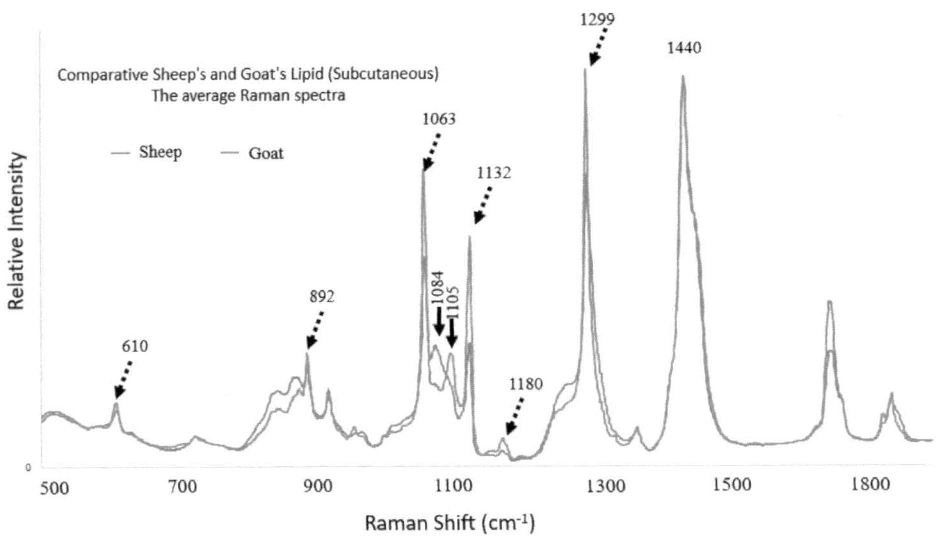

Figure 4. The average of Raman spectra for sheep's lipid (lamb) and goat's lipid in subcutaneous part, the black arrows indicate the main difference of Ramanspectra where correspondence with C N stretching mode of proteins and lipid mode of lesser degree (1084 cm^{-1}) and phenylalanine or protein (1105cm^{-1}) for Lamb's lipid and goat's lipid, respectively.

The mean of Raman spectra of 5 lamb's lipid and 5 goat's lipid is shown in figure 4, where6 Raman spectra of lipid (n = 5 heads) were acquired from each sample using a 785 nm laser.The region along the spectrum is average that potential presented to visualize spectra among them. In the case of Raman spectra lipid, the band at 1084 cm^{-1} just appears at the lamb's lipid and it is correspondence with C N stretching mode of proteins and lipid mode of lesser degree [18]. Goat's lipid has a tendency to represent more peak intensity at the band of 1105 cm^{-1} or protein [19].

Principal Component Analysis (PCA) discriminates more detail regarding the fatty acid content where the peak intensity of Palmitic acid (at the bands of 1065, 1131, 1298 cm^{-1}) is higher to goat's lipid than lamb's lipid. The age of the goat might have a relation with the domination of palmitic acid in the goat's lipid compared to the lamb. Cholesterol is detected at the band of 610, 965, 1674 cm-1. The other peak intensity shows the assignment of CH2 stretch of lipids and Phospholipid. In addition, the score and loading plots of both species are shown in Fig. 5 thatmaking it is possible to analyze not only specific bands (such as at 1084 and 1105 cm^{-1}) but theother fatty acid and ester groups belonging to both species have different peak intensities.

Goat's lipid has a higher intensity of palmitic acid compared with the lamb's lipid. Ramanstudies of adipocytes tissue have been examined and produced the predominant bands of palmitic acid, cholesterol, cholesterol ester, CH, and CH2 stretch of lipids [20]. This lipid spectrum dominantly performs with the palmitic acid spectrum as one of the most frequent saturated fatty acids in the animal kingdom [21]. However, these preliminary studies are still needed verification and further studies due to the status of samples were the frozen lipid and no detailed information concerning the feed and recording management of them. Therefore, itis necessary to design more detail of methodology, and also the fresh meat is recommended.

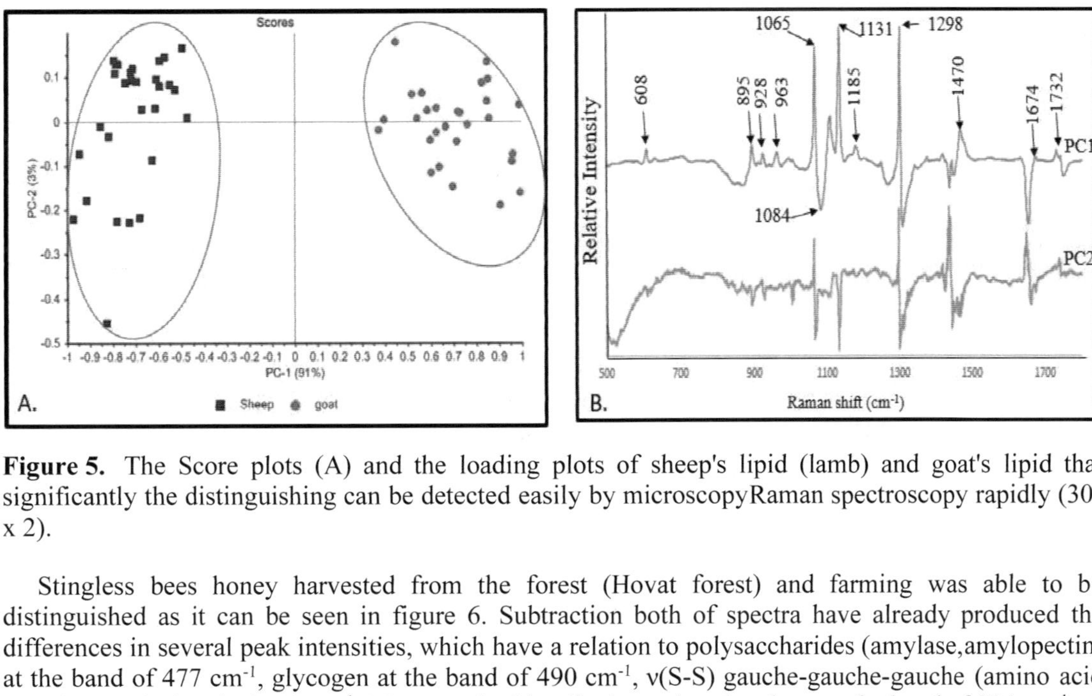

Figure 5. The Score plots (A) and the loading plots of sheep's lipid (lamb) and goat's lipid that significantly the distinguishing can be detected easily by microscopyRaman spectroscopy rapidly (30s x 2).

Stingless bees honey harvested from the forest (Hovat forest) and farming was able to be distinguished as it can be seen in figure 6. Subtraction both of spectra have already produced the differences in several peak intensities, which have a relation to polysaccharides (amylase,amylopectin) at the band of 477 cm^{-1}, glycogen at the band of 490 cm^{-1}, ν(S-S) gauche-gauche-gauche (amino acid cysteine) at the band of 510 cm^{-1}, Monosaccharides (β-glucose) and maltose at the band of 894 cm^{-1}.

The differences might have a correspondence with the diversity of trees in the forest, in other words depending on the foraging ecology of stingless bees [22], in producing the nectar. The honey harvested from wild stingless bees might be made from the nectar where there is no human involvement or interference. Meanwhile, the farmed stingless bees might have a relation with the human interference [23] for the availability of tree and plantation (related tothe availability of natural nectar) and the fresh sweetie fruit pealed by the owner, such as orange, guava, rambutan, mangosteen fruit, etc. (see fig. 2C). For further studies, we recommend analyzing more about the phytochemical of this stingless bee honey related to the diversity of fruit and flower surrounding the colonies (foraging ecology), including the differences in the harvesting of stingless bees honey itself. And Raman spectroscopy can coverto explore these studies in the near future.

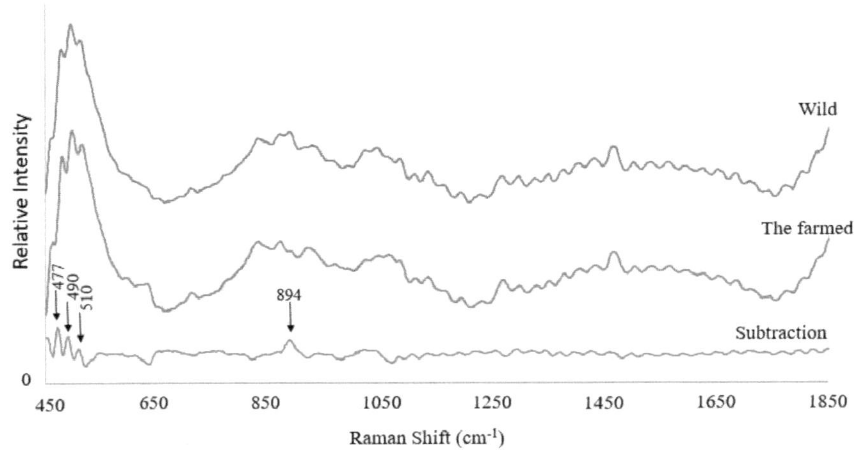

Figure 6. Raman spectroscopy demonstrates its ability to make differences betweenthe honey produced by the wild and the farmed Stingless bees.

4. Conclusion
Raman spectroscopy has a powerful in identification sheep's lipid (lamb) and Goat's lipid where the Raman peak at 1084 and 1105 cm^{-1} appearing just in Lamb's lipid and Goat's lipid,respectively, rapidly it was just one minute. However, it must be more detail to be studied using the fresh lipid and good information of their food and development management. Ramanspectroscopy also has potential and can be promoted to be the faster method (one minute) to distinct the honey produced by the wild and the farmed stingless bees in near future.

5. Acknowledgment
Authors are obliged for providing the Raman spectra of two lipid types by using a 785 nm laser. We would like also to thank Drs. Nurdiana (from Pusdiklat TeuweulSumedang, West Java, Indonesia) has many helpful in collecting stingless bees honey forthis preliminary research, including the information of their foraging ecology in Kalimantan, Indonesia.

References
[1] Belinda V-C, Gonza´lez-Co´rdova AF, Mazorra-Manzano MA Rodrı´-guez-Ramı´rezR 2005. *Journal of Separation Science* 28:826–836.
[2] Asensio L, Gonza´lez I, Garcı´a T and Martı´n R 2008. *Food Control* 19: 1–8.
[3] Balog J, Perenyi D, Guallar-Hoyas C, Egri A, Pringle SD, Stead S, Chevallier OP, Elliott CT and Takats Z 2016. *Journal of Agricultural and Food Chemistry*64:4793–4800.
[4] Di Pinto A, Bottaro M, Bonerba E, Bozzo G, Ceci E, Marchetti P, Mottola A and Tantillo G 2015. *Journal of Food Science and Technology*. 52:2479–2484.
[5] Quehenberger O, Armando AM and Dennis EA 2011. *Biochimica et BiophysicaActa*. 1811:648–656.
[6] Boselli E, Pacetti D, Curzi F and Frega NG 2008. *Meat Science*. 78:305–313.

[7] Sato H, Ishigaki M, Taketani A and Andriana BB 2018. *Biomedical Spectroscopy andImaging.* 7: 97-104.
[8] Fowler SM, Schmidt H, van de Ven R and Hopkins DL 2018. Meat Science. 138:53–58.
[9] Beganovic A, Hawthorne LM, Bach K and Huck CW 2019. *Foods.* 8:1–18.
[10] Ostovar Pour S, Fowler SM, Hopkins DL, Torley PJ, Gill H and Blanch EW 2019. *Analyst* 144:2618–2627.
[11] Schmidt H, Sowoidnich K and Kronfeldt HD 2010. *Applied Spectroscopy.* 64:888 –894.
[12] Marquardt BJ and Wold JP. 2004. *LWT Food Science and Technology.* 37:1–8.
[13] Herrero AM 2008. *Food Chemistry.* 107:1642–1651.
[14] Meksiarum P, Andriana BB, Matsuyoshi H and Sato H 2016. *Scientific Report.* 6(1):37068.
[15] Sato H, Suzuki T, Andriana BB, Morita S, Maruyama A, Shinzawa H, Komachi Y, Kanai G, Ura N, Masutani K, Matsura Y, Toi M, Shimosegawa T and Ozaki Y. 2009. *Proc. SPIE.* 7169, 1.
[16] Ozbalci B, Boyaci IH, Topcu A, Kadilar C and Tamer U 2013. *Food Chemistry.* 136: 1444-1452.
[17] Casella M, Lucotti A, Tommasini M, Bedoni M, Forvi E, Gramatica F and Zerbi G. 2011 *Spectrochimica Acta - Part A Molecular and Biomolecular Spectroscopy.* 79:915–919.
[18] Stone N, Kendall, C., Shepherd, N, Crow, P and Barr H 2002. J. Raman Spectroscopy.33: 564–573.
[19] Lakshmi RJ, Kartha VB.,\Krishna CM, Solomon JGR, Ullas G and Devi PU 2002. *Radiation Research.* 157 (2): 175–182.
[20] Huang NY, Short M, Zhao JH, Wang,HQ, Lui H, Korbelik, M and Zeng, H.S 2011.Optics Express, 19 (23): 22892–22909.
[21] Saleem M, Amin A and Irfan M 2021. *Journal of Food Science and Technol.* 58(1):234-243.
[22] Hall MA, Brettell LE, Liu H, Nacko S, Spooner-Hart R and Riegler M 2020. *FEMS Microbiology Ecology.* 30; 97(1): fiaa236.
[23] Trinkl M, Kaluza BF, Wallace H, Heard TA, Keller A and Leonhardt SD. 2020. *Insects* .15; 11(2):125.

The Antibiotic resistance of lactic acid bacteria isolated from kefir made from Etawah goat milk

T Budiati [1*], W Suryaningsih [1] and S O N Yudiastuti [1]

[1] Food Engineering Department, Politeknik Negeri Jember, Mastrip PO Box 164, Jember 68101 East Java, Indonesia

* email: titik_budiati@polije.ac.id

Abstract. The objectives of the study were to determine the antibiotic resistance of Lactic Acid Bacteria (LAB) isolated from kefir made from Etawah goat milk. A total of 25 isolates were isolated from etawah goat milk, yogurt and kefir. LAB isolates were resistant to Ceftazidime (Caz, 100%), Ceftriaxone (Cro, 100%), Clindamycin (Da, 100%), Lincomycin (Li, 100%), Rifampicin (Rd, 100%) and Tobramycin (Tob, 100%). The bacteria were susceptible to Sulphamethoxazole-Trimethoprim (Sxt, 60%) and Tetracycline (Te, 40%). The percentage of antibiotic resistance of LAB was found to be in the range between 40 to 100%. The presence of antibiotic resistance in good bacteria may introduce the horizontal genetic transfer to bad bacteria which has the potential to disrupt human health.

1. Introduction

The COVID-19 pandemic that has hit the world since 2020 until now has killed more than 4 million people with the number of positive cases of COVID 19 reaching more than 200 million people [1]. The strategy used to reduce cases is with 3M, namely wearing masks, maintaining distance and washing hands [2]. Preventive efforts that are currently being intensively carried out are the provision of vaccines for the entire human population on earth to be able to bring up herb immunity [3]. Another effort made to increase immunity is by consuming foods that contain probiotics [4][5].

Various studies have examined the possibility of giving probiotics to prevent and reduce the risk caused by COVID 19. This happens because probiotics contain good live microorganisms so that they are able to fight bad microorganisms that live in the human digestive area. The competition between the good and bad microorganisms may resulted the reduction in the population of bad microorganisms. This is expected to increase human immunity and reduce the risk of COVID 19 [2, 4, 5].

However, there is another concern that arises in the natural conditions of probiotics, namely the existence of resistance to antibiotics. This may be due to the over-administration of antibiotics in animals [6]. The dairy products are a good natural medium for the growth of probiotics. This has an impact on various products produced from milk processing. Antibiotics carried in the animal's metabolic system will affect the genetic properties of the probiotics that live in it [6]. These resistance properties can be transmitted vertically or horizontally between microorganisms [7][8][9][10]. If these properties are transmitted to bad microorganisms, it will cause big problems for healing a disease in humans [11]. Etawah goat's milk and various processed products are products that are known to contain high nutrition and can increase human immunity and can cure various digestive diseases in humans [12]. One of the processed goat milk products is kefir [13]. Kefir is a fermented product that contains probiotics with a very large variety of microorganisms such as lactic acid bacteria [14]. The concern that arises is the existence of resistance properties in lactic acid bacteria which can be transferred to bad microorganisms such as pathogenic bacteria and viruses [6]. Until now there have been limited studies that provide information about the nature of antibiotic resistance in kefir made from Etawah goat milk. Therefore, it is necessary to do this research. The objectives of the study were to determine the antibiotic resistance of Lactic Acid Bacteria (LAB) isolated from kefir made from Etawah goat milk

2. Material and methods

2.1. Sample

A total of 25 samples of Lactic acid bacteria sp. isolated from kefir made from etawah goat milk were obtained from 7 local markets in Jember and Surabaya (May 2020 to October 2021).

2.2. Disc diffusion assay

By using disc diffusion assay, the antimicrobial activity was measured as the method of Bauer **[15]**[16] for Ceftazidime (10 μg), Clindamycin (2 μg), Ceftriaxone (5 μg), Lincomycin (2 μg), Rifampicin (30 μg), Spectinomycin (10 μg), Sulphamethoxazole-Trimethoprim 19:1 (25 μg), Tetracycline (30 μg). The antibiotics discs were obtained from Oxoid (UK). Fresh cultures were prepared in Luria Broth medium (LB, Himedia, India) which were incubated in a shaker incubator for 16 h at 37 °C. Prior to the observation of the antibiotic resistance test, the turbidity of the diluted culture was equivalent to 0.5 on the McFarland scale. For disc diffusion observations, cultures were grown on Mueller Hinton Agar (Oxoid, England) media by swabbing method. The discs of antibiotics were placed on top of the culture. This observation was made after the sample was incubated at 37°C for 18-24 hours and repeated for 3 times. The inhibition zone measurement is expressed in millimeters. These experiments use negative control by dripping 15 μL of 10% DMSO onto sterile disc paper. Penicillin G (10 units/ml) was used as a positive control.

2.3. Statistical Analysis

The antimicrobial activity of antibiotics to inhibit the growth of lactic acid bacteria was carried out by using one way ANOVA at a significant level of $P < 0.05$.

3. Result and Discussion

The antibiotic resistance among lactic acid bacteria isolated from kefir was resistant to Ceftazidime, Ceftriaxone (25/25), Clindamycin (25/25), Lincomycin (25/25), Rifampicin (25/25) and Tobramycin (25/25). Some isolates were susceptible to Sulphamethoxazole-Trimethoprim (15/25) and Tetracycline (10/25)

Ceftazidime and ceftriaxone interferes with inhibition of bacterial cell wall synthesis [17]. In other antimicrobials, this study found that most of *Lactic acid bacteria* isolates were resistant to Lincomycin which inhibit protein synthesis that used for cell structures or ribosomes. This is not surprising that lincomycin might be used heavily in Etawah goat farming system. Sharma *et al.* [18] revealed that clindamycin was used to treat the respiratory tract infections in goat.

Up to 60% of *Lactic acid bacteria* isolated from kefir made from Etawah goat milk were resistantto tetracycline. Acquired tetracycline resistance was attributed to the extensive use of antibiotics in etawah goat farming in some countries to treat the disease [18]. All isolates from kefir made from etawah goat milk were resistant to rifampicin. The action of rifampicin was binding to the specific site on the DNA-dependent RNA polymerase [19]. Other study revealed that the resistance to this antimicrobial was caused by mutation on the β subunit of RNA polymerase [20]. Therefore, the RNA chain initiation and elongation were inhibited [19] and the isolates could decrease the affinity of rifampicin [21].

In this study, the majority of LAB isolated from kefir were susceptible to sulphamethoxazole-trimethoprim (Table 1). These may be initiated by the absence of genes in the chromosomes and plasmids. Therefore, it becomes an option to avoid this antibiotic as a cure for a disease to allow lactic acid bacteria to grow optimally. However these bacteria were resistant to Ceftazidime, Ceftriaxone, Clindamycin, Lincomycin, Rifampicin, Tobramycin and may increase the risk of horizontal transfer gene to pathogenic bacteria.

Table 1. Number (%) of *Lactic Acid Bacteria* (LAB) isolated from kefir made from Etawah goat milk resistant to antibiotics

Antibacterial agents	Total isolates n = 25
Ceftazidime (Caz)	25 (100%)[a]
Cefriaxone (Cro)	25 (100%)[a]
Clindamycin (Da)	25 (100%)[a]
Lincomycin (Lin)	25 (100%)[a]
Rifampicin (Rd)	25 (100%)[a]
Sulphamethoxazole-Trimethoprim (Sxt)	10 (40%)[c]
Tetracycline (Te)	15 (60%)[b]
Tobramycin (Tob)	25 (100%)[a]

Note : [a,b] = different alphabet means significant difference at P<0.05 in the same column

4. Conclusion

The present work highlights on antibiotic resistance of Lactic acid bacteria sp. in kefir made from Etawah goat milk which was resistant to Ceftazidime, Ceftriaxone, Clindamycin, Lincomycin, Rifampicin, Tobramycin and susceptible to Sulphamethoxazole-Trimethoprim and Tetracycline. The antibiotic resistance percentages ranged from 40 to 100%.

References

[1] Worldometer 2020 Covid-19 coronavirus pandemic: Worldometer
[2] Baud D, Dimopoulou Agri V, Gibson G R, Reid G and Giannoni E 2020 Using Probiotics to Flatten the Curve of Coronavirus Disease COVID-2019 Pandemic Frontiers in Public Health **8**
[3] Jeyanathan M, Afkhami S, Smaill F, Miller M S, Lichty B D and Xing Z 2020 Immunological considerations for COVID-19 vaccine strategies Nature Reviews Immunology **20** 615
[4] Olaimat A N, Aolymat I, Al-Holy M, Ayyash M, Abu Ghoush M, Al-Nabulsi A A, Osaili T, Apostolopoulos V, Liu S Q and Shah N P 2020 The potential application of probiotics and prebiotics for the prevention and treatment of COVID-19 npj Science of Food **4**
[5] Kurian S J, Unnikrishnan M K, Miraj S S, Bagchi D, Banerjee M, Reddy B S, Rodrigues G S, Manu M K, Saravu K, Mukhopadhyay C and Rao M 2021 Probiotics in Prevention and Treatment of COVID-19: Current Perspective and Future Prospects Archives of Medical Research **52** 582
[6] Mathur S and Singh R 2005 Antibiotic resistance in food lactic acid bacteria - A review International Journal of Food Microbiology **105** 281
[7] Polz M F, Alm E J and Hanage W P 2013 Horizontal gene transfer and the evolution of bacterial and archaeal population structure Trends in Genetics **29** 170
[8] Scott J M, Li S, Harper S Q, Welikson R, Bourque D, Dellorusso C, Hauschka S D and Chamberlain J S Viral vectors for gene transfer of micro-, mini-, or full-length dystrophin Neumuscular disorder **12** S23
[9] Budiati T, Suryaningsih W, Umaroh S, Poerwanto B, Bakri A and Kurniawati E 2018 Antimicrobial activity of essential oil from Indonesian medicinal plants against food-borne pathogens IOP Conference Series: Earth and Environmental Science **207**

[10] Budiati T, Suryaningsih W, Biyanto T R, Pangestika N P, Pangestu M T, Saputra F, Hidayat A, Widyawati A, Firdaus F N and Sabilla D V 2020 Finite Impulse Response Type Multilayer Perceptron Artificial Neural Network Model for Bacteria Growth Modeling Inhibited by Lemon Basil Waste Extract IOP Conference Series: Earth and Environmental Science **411**

[11] Kothari D, Patel S and Kim S K 2019 Probiotic supplements might not be universally-effective and safe: A review Biomedicine and Pharmacotherapy **111** 537

[12] Sumarmono J, Sulistyowati M and Soenarto 2015 Fatty Acids Profiles of Fresh Milk, Yogurt and Concentrated Yogurt from Peranakan Etawah Goat Milk Procedia Food Science **3** 216

[13] Purnomo H and Muslimin L D 2012 Chemical characteristics of pasteurised goat milk and goat milk kefir prepared using different amount of indonesian kefir grains and incubation times International Food Research Journal **19** 2

[14] Otles S and Cagindi O 2003 Kefir: A Probiotic Dairy-Composition, Nutritional and Therapeutic Aspects Pakistan Journal of Nutrition **2** 2

[15] Budiati T, Wibisono Y, Pambayun R A, Fahrezy M F, Ariyani R, Kurniawati E, Suryaningsih W, Yudiastuti S O N and Bakri A 2020 Inhibition of Listeria monocytogenes by natural antimicrobial IOP Conference Series: Earth and Environmental Science **411**

[16] Budiati T, Suryaningsih W, Yudistira H and Azhar S W 2021 Antimicrobial activity of jengkol and petai peel extract to inhibit listeria monocytogenes IOP Conference Series: Earth and Environmental Science **672**

[17] Serrano P H 2005 Responsible Use of Antibiotics in Aquaculture FAO Fisheries Technical paper **469** (Rome : Food and Agriculture Organization of the United Nations)

[18] Sharma M, Vemu B and Dumka V K 2017 Pharmacokinetics of Lincomycin Following Single Intramuscular Administration in Goats International Journal of Agricultural **7** 555

[19] Wehrli W 1983 Rifampin: Mechanisms of Action and Resistance Clinical Infectious Diseases **5** 3

[20] Brandis G, Wrande M, Liljas L and Hughes D 2012 Fitness-compensatory mutations in rifampicin-resistant RNA polymerase Molecular Microbiology **85** 142

[21] Campbell E A, Korzheva N, Mustaev A, Murakami K, Nair S, Goldfarb A and Darst S A 2001 Structural Mechanism for Rifampicin Inhibition of Bacterial RNA Polymerase Cell **104** 6

Interaction of parasitoids associated with fruit flies attacking star fruit *(Averrhoa carambolae)* in Denpasar City, Bali Province, Indonesia

K A Yuliadhi[1], I W Susila[1], I W Supartha[1,*], A Sultan[2], I K W Yudha[3], I W E K Utama[3], P A Wiradana[4]

[1] Laboratory of Integrated Pest Management (IPMLab), Faculty of Agriculture, Udayana University, Jalan P.B. Sudirman, Denpasar City 80231, Bali-Indonesia
[2] Study Program of Agro-Ecotechnology, Faculty of Agriculture, Udayana University, Jalan P.B. Sudirman, Denpasar City 80231, Bali-Indonesia
[3] Graduate Program of Dry Land Agriculture, Faculty of Agriculture, Udayana University, Jalan P.B. Sudirman, Denpasar City 80231, Bali-Indonesia
[4] Study Program of Biology, Faculty of Health, Science, and Technology, Universitas Dhyana Pura, Jl. Raya Padang Luwih, Dalung, North Kuta, Badung Regency 80351, Bali-Indonesia

* email: yansupartha@yahoo.com

Abstract. Star fruit (Averrhoa carambola) is a fruit plant that has a high consumption rate, especially in the Central/South America and Asia-Pacific regions including Indonesia. The fruit fly, Bactrocera spp. become a major pest on star fruit commodities in Indonesia which can harm quantity and quality which in turn hampers export activities. The aims of this study was to identify the type of pest, the level of attack, and the type of parasitoid associated with star fruit flies in the Denpasar City area. Fruit samples were collected using a purposive sampling method, namely collecting symptomatic fruit planted in four sub-districts in Denpasar City. Identification was carried out on the types of fruit flies and parasitoids that emerged after the rearing period. The percentage of fruit damage and the degree of parasitization was also determined. The fruit flies identified were Bactrocera carambolae and B. dorsalis and two parasitoids, namely Fopius arisanus and Diachasmimorpha longicaudata. The percentage (%) of fruit damage was 83.10%. The interaction that occurs between the fruit fly population and the level of damage is also very close. In all observation sites in Denpasar City, the parasitization rate of F. arisanus was higher than that of D. longicaudata. Our findings indicate that fruit flies are one of the main causes of star fruit damage in Denpasar City and should be integrated into a more comprehensive Integrated Pest Management (IPM) program by the relevant authorities.

1. Introduction

The starfruit, Averrhoa carambola L., is one of the most significant fruit commodities found in Southeast Asia [1]. The nutritional composition of star fruit contains vitamin A, vitamin C, vitamin B1 & B2, citric acid, ketoglutaric acid, oxalic acid, tartaric acid, and carotene with mineral content in every 100 grams of fruit [1]. In Indonesia, starfruit plants are commonly planted in communal settlements and are utilized for different household requirements until they are sold to traditional markets. However, although

Indonesia is geographically located in a tropical climatic condition, several invasive insects and parasites that target star fruit have been claimed to have led to star fruit output declines and poor quality [2].

Fruit flies (Diptera: Tephritidae) are the most severe pests damaging fruit commodities in Indonesia, particularly during the rainy season. These pests affect star fruit (A. carambola L.), mango (Mangifera indica L.), and guava (Psidium guajava L.) commodities. The action of female fruit flies, for example, when laying eggs on fruit, can cause harm to fruit flesh. Furthermore, the eggs that are put in the flesh of the fruit that has developed into larvae will devour the fruit, therefore amplifying the amount of harm. Attack symptoms, such as black spots on the fruit, lead the fruit to fall before maturity. Secondary infection with fruit rot fungus and bacteria was also linked to qualitative damage. The incidence of crop failure causes quantitative harm in the form of a significant reduction in the number of harvests. According to some reports, fruit fly infestations can result in production losses of up to 50-75 %. Fruit fly infestations cause a 40-70 % reduction in fruit production in other nations, such as Bangladesh [3].

This is critical because each country must consider biosecurity problems such as the presence of invasive pests when exporting and importing fruit commodities. This practice is carried done to prevent pests from spreading to other nations via fruit flesh. In relation to this, agricultural product regulation in Indonesia is governed by Regulation of the Minister of Agriculture Number 61/PERMENTAN/OT.140/10/2010. The legislation explicitly specifies that it is required to develop and maintain technical plant quarantine policies, as well as to conduct plant biosafety surveillance against Invasive Alien Species (IAS), which has been the object of monitoring for agricultural commodities to date.

Farmers in the field, on the other hand, continue to employ chemical insecticides to control fruit fly infestations, and the application of these pesticides is massively utilized to mix various types of insecticides and spray them with a high frequency. Poor agronomic practices can lead to increased heavy metal pollution as well as other contaminants, such as the development of foodborne pathogens and pesticides that accumulate in fruit [4,5]. In terms of agroecology, pesticide application damages the variety of pollinating insects, predators, and parasitoids, leading to the establishment of new populations of resistant invading pests.

Fruit fly pests must be managed to enhance star fruit output. Integrated Pest Management (IPM) approaches include the employment of biological organisms like parasitoids, which are natural adversaries of fruit pests and can be employed as an alternative since they show promising possibilities. In nature, parasitoids can continue to develop indefinitely and do not require significant expenditures to use their services. Such control techniques are critical in the field to optimize the function of natural enemies of agricultural pests [6,7]. According to reports, the spontaneous administration of parasitoid agents can protect grapevines and is a low-cost method of controlling plant pests [8]. The parasitoids' mode of action is to parasitize the body of the host (insect) in order to weaken and harm it [9]. Fruit fly infestations have been extensively observed to be controlled by parasitoids from the Order Hymenoptera, Family Braconidae. Suputa et al. [10] report how the parasitoids F. arisanus, Agasnapis sp., and Asobara sp. successfully parasitize star fruit fly in Yogyakarta's Special Region (DIY). Putra et al. [11] reported an average of 11.12 % parasitoid parasitization on star fruit in Gianyar Regency, Bali.

There is currently no information available concerning fruit flies and the types of parasitoids found on star fruit in the Denpasar City region of Bali. This is significant since Denpasar City is the capital and largest urban region in Indonesia, with a population density of 10,798 people / km2. Fruit flies can be transmitted by population mobility activities that occur across cities/districts by transporting fruit commodities given to Traditional Markets (Badung Market) to supermarkets. The aim of the research is to species identification of fruit flies and their parasitoids mainly feed on star fruit in the Denpasar City region. These findings will be used to supplement prior data from a star fruit plantation in Gianyar, Bali [11]. This is, to the state of the art, the first survey of the impact of pest damage and parasitoids on star fruit in Denpasar.

2. Materials and methods

2.1. Study sites

The research was carried out between November 2020 - March 2021. In the Denpasar City region, field research is being conducted in residential areas with starfruit trees. The Integrated Pest Management Laboratory (IPMLab), Faculty of Agriculture, Udayana University, Bali, conducted laboratory-scale research.

2.2. Determination of location and samples

Purposive sampling was used to collect data in four districts of Denpasar City: Denpasar, West, South Denpasar, East Denpasar, and North Denpasar. At each location in the sub-district, 5 fruit trees were chosen, resulting in a total of 21 fruit trees as examples (see Figure 1). Up to ten-star fruits that exhibited symptoms of an assault were placed in a sterile plastic bag with a label on each tree. The sample period was repeated 5 times every 4 days.

Figure 1. Study sites in Denpasar City, Bali Province

2.3. Maintenance method

The star fruit samples were placed in a clear plastic bottle that was used as a storage container. The bottle has a height of 23 cm and a diameter of 8.5 cm, and it has been filled to a height of 10 cm with sand. One symptomatic star fruit was placed in one bottle. Gauze was used to seal the bottle (Figure 2). Every day, observations on the appearance of parasitoids and fruit flies were made.

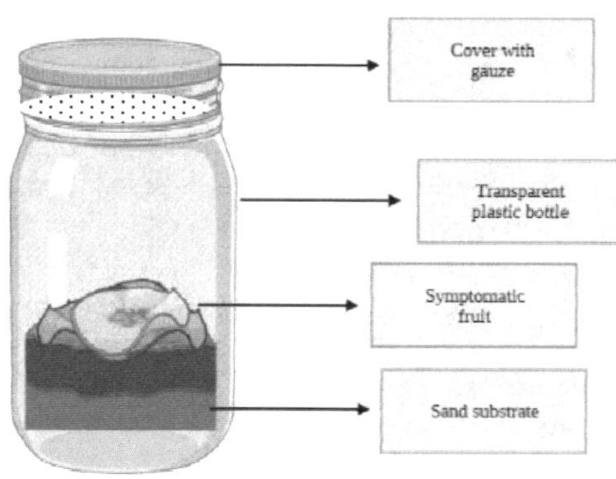

Figure 2. Fruit fly breeding container (Created with BioRender.com)

2.4. Research procedure

2.4.1. Identification of fruit flies and parasitoids
Adult flies in each rearing container were identified using a microscope based on the morphological characteristics of each species. Schutze et al. were cited as identification guidelines [12], International Center for the Management of Pest Fruit Flies (ICMPFF) & Ministry of Agriculture Republic of Indonesia (KEMENTAN) [13], and Suputa et al. [10]. As for the type of parasitoid identified referring to Sharkey and Wahl [14] and Carmichael et al. [15]. The findings of the identification are recorded and reported.

2.4.2. Relative abundance (%)
The relative abundance of fruit fly populations was determined using the equation of McCarthy and Magurran [16] namely:

$$Relative\ abundance = \frac{species\ number\ at\ x\ location}{total\ number\ of\ species\ at\ x\ location}\ x\ 100\% \quad (1)$$

2.4.3. Attack percentage (%)
The percentage of fruit fly attacks was determined by randomly selecting up to 50 fruits from each tree. The star fruit included in the assessment has a yellowish-green to totally yellowish color, including the fruit that has falling and those that are still on the tree. The following is the attack percentage equation:

$$Attack\ percentage = \frac{number\ of\ infected\ fruit}{total\ number\ of\ fruit}\ x\ 100\% \quad (2)$$

2.4.4. Parasitzation rate
The parasitization rate possessed by parasitoids was determined using the following equation:

$$Parasitization\ rate = \frac{\Sigma\ number\ of\ parasitoids}{\Sigma\ adult\ fruit\ fly + \Sigma\ adult\ parasitoid\ (a+b+c)}\ x\ 100\% \quad (3)$$

2.5. Data analysis
The collected data was examined descriptively, that is, by methodically, factually, and properly summarizing the facts and characteristics of the people in specific locations. Ms software was used to tabulate the data on relative abundance, attack percentage, and parasitization rate. Microsoft Excel 2019 (Microsoft, USA). Tables, graphs, and images are used to show information [17].

3. Results and discussion

3.1. Fruit flies
According to the identification findings, there are two species of fruit flies that attack star fruit in the Denpasar City area: *Bactrocera carambolae* and *Bactrocera dorsalis*. Table 1 shows the image of the identification result.

Table 1. Types of star fruit flies in the Denpasar City area

Bactrocera carambolae			
No	**Morphology**	**Figure**	**Description**
1	Whole body		• Body dark brown to black. • The caput is dark orange. • Legs light brown.
2	Head	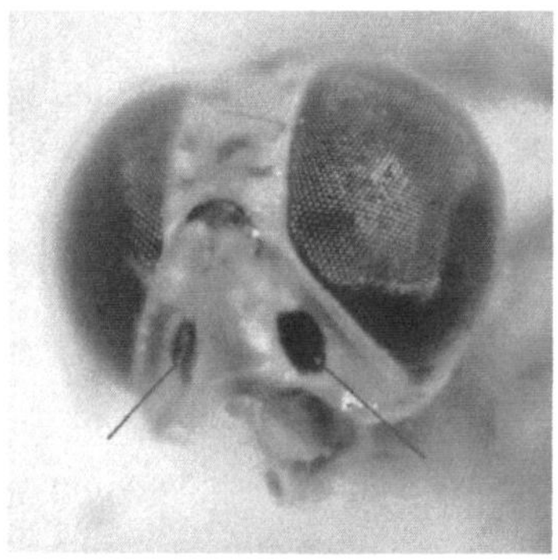	The head is equipped with a pair of medium-sized black spots
3	Thorax	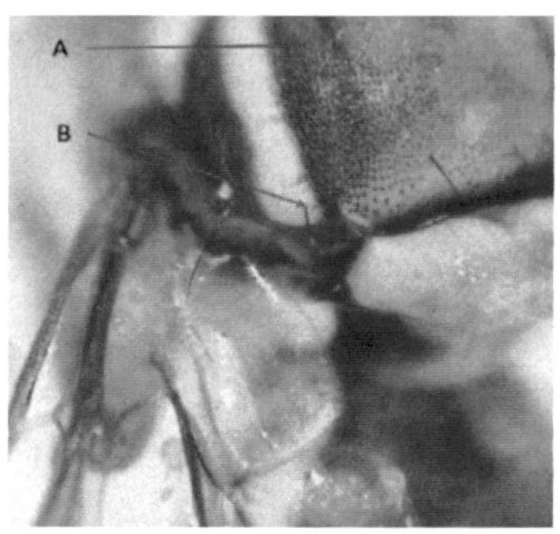	The dark color on the scutum (A), setae present before the tip of the yellow band on the lateral side (B).
4	Wings	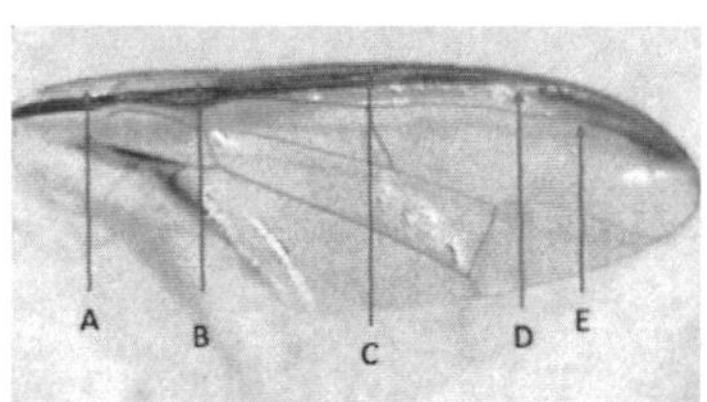	Wings, Costal band 1 (A), Costal band 2 (B), R1 (C), R2+3 (D), R4+5 (E). The wing pattern at the end (apex) is shaped like a fishing rod and overlaps at R2+3

IOP Conf. Series: Earth and Environmental Science **980** (2022) 012051 doi:10.1088/1755-1315/980/1/012051

5	Insect limbs	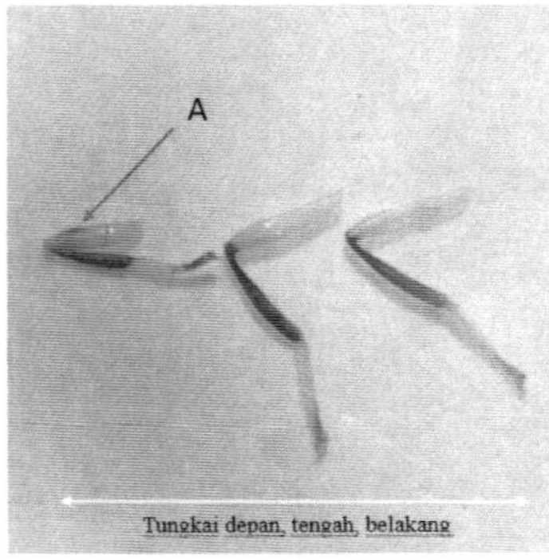	The limbs of the front, middle and hind femurs are pale. dark front, middle, and back tibias. There is a dark spot on the apical femur (A).
6	Abdomen	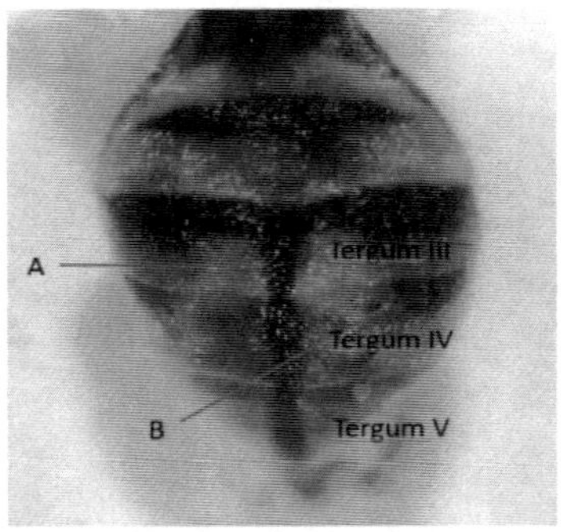	The abdomen has a medial longitudinal line on tergum III-V (A), and there is a black pattern with a rectangular shape on terga IV (B).

Bactrocera dorsalis

No	Morphology	Figure	Description
1	Whole body		The whole body is brownish orange.

2	Head	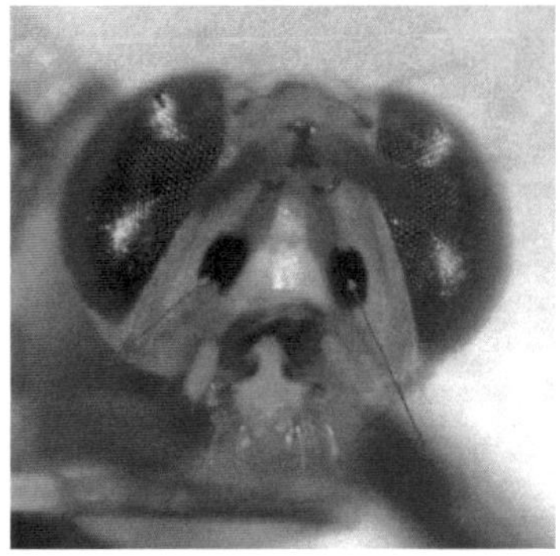	On the head, there is a large black spot.
3	Thorax	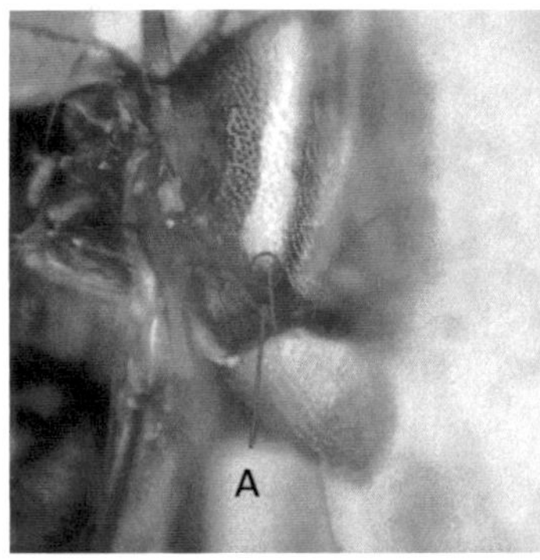	The setae of the thorax are located at the tip and outside of the yellow band (A).
4	Wings	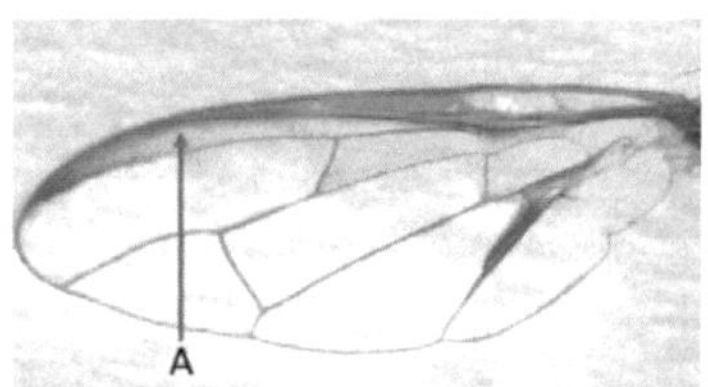	On the wing, the costal band is confluent with R2+3 (A).
5	Insect limbs	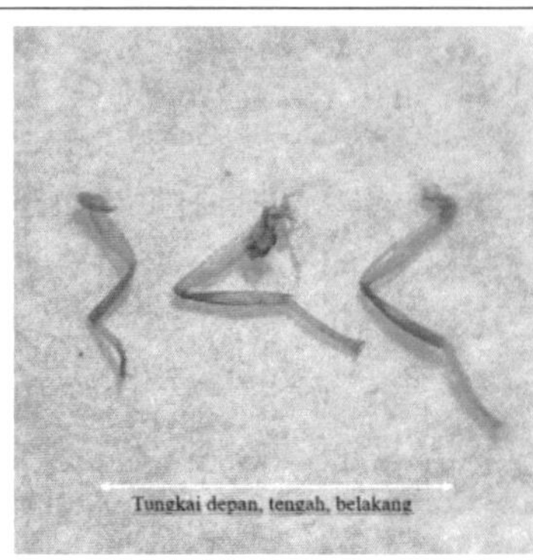	The color of the front, middle, and hind femurs are pale. The color of the front tibia is dark, middle dark, back dark to black.

6	Abdomen	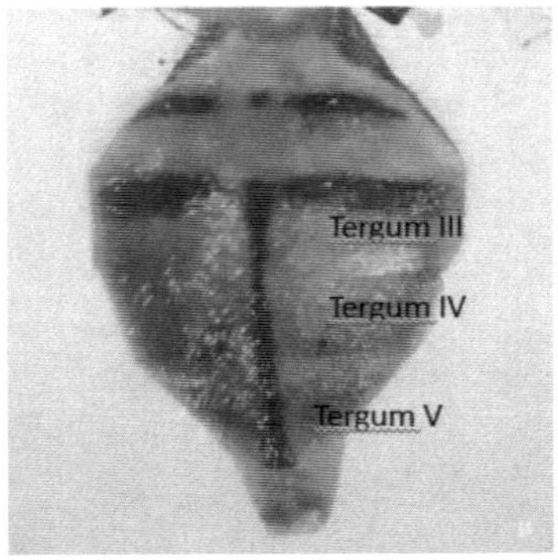	Abdomen with well-defined segments, at tergum III –V orange-brown with T. pattern

3.2. *Relative abundance*
The proportional quantity of fruit flies that attack star fruit varies by observation region. The relative abundance of fruit fly population in East Denpasar District with *B. carambolae* species was determined to be 62.89 % (1,822 adults), followed by B. dorsalis species at 37.11 % (1,075 adults). The relative abundances of *B. carambolae* and *B. dorsalis* in South Denpasar District were 66.74 % (1,782 adults) and 33.26 % (888 adults) respectively. Furthermore, the relative abundance of *B. carambolae* fruit flies in the West and North Denpasar sub-districts was 63.48 % (1.884 adults) and 60.94 % (1.884 adults), respectively (1.743 adults, respectively). Meanwhile, *B. dorsalis* was identified in 36.52 % (1,084 adults) and 39.06 % (1,084 adults) of the last two sub-districts (1,117 adults). Figure 3 depicts the relative features of fruit flies.

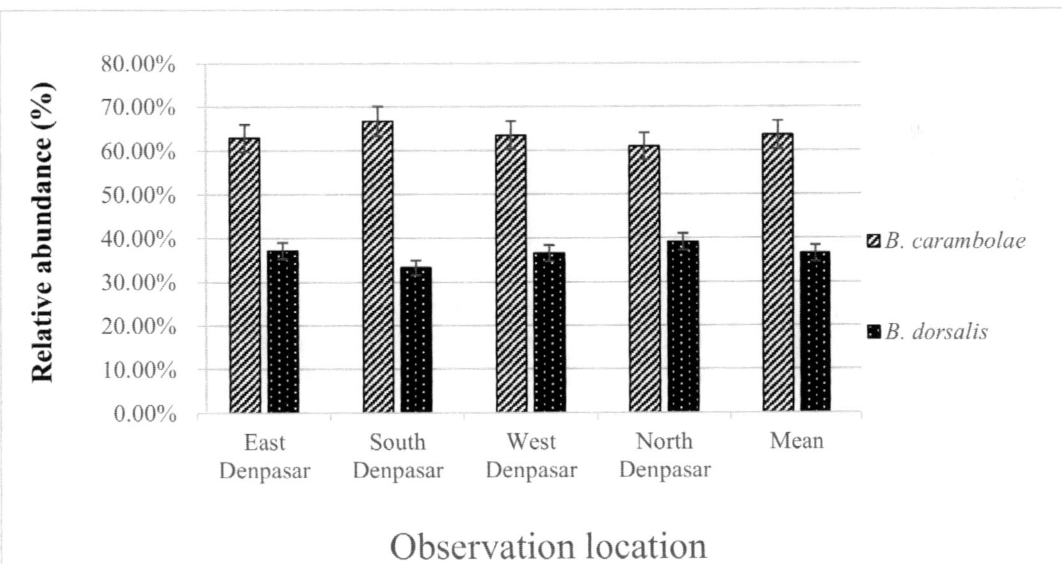

Figure 3. The relative abundance of fruit flies found on star fruit in the Denpasar City area

The variation in the number of adult flies in each sub-district can be attributed to a variety of factors, including the availability of fruit in the field, the level of fruit ripeness, and alternative hosts. The West Denpasar District has a disproportionately large amount of star fruit trees, making them simple to locate. When compared to other sub-districts, almost every house in the West Denpasar region has a starfruit tree with dense fruiting circumstances, and there are alternate hosts such as mango, guava, and water guava that are adjacent to the starfruit tree. The presence of diverse varieties of fruit in a region provides a favorable environment for the growth of fruit fly pests [18]. Fruit maturity is dominated by yellowish-green to yellow in the West Denpasar District. Fruit flies choose these hues to deposit their eggs. This supports the results of Sahetapy et al. [19], who found that fruit flies prefer white and yellow over other colors. The golden color of ripe fruit has been identified as an appropriate host for female fruit flies to deposit eggs [20]. Susila and Supartha [21] have reported that the color and scent of the host fruit influence fruit fly activity in locating hosts.

3.3. Attack rate

In Denpasar City, the percentage of fruit fly attacks on star fruit varies from 83.10 percent to 83.10 %. According to our data, the West Denpasar area had the greatest percentage of attacks (85.20 %), while the South Denpasar area had the lowest (80.80 %). The outcome of the attack is depicted in Figure 4 below.

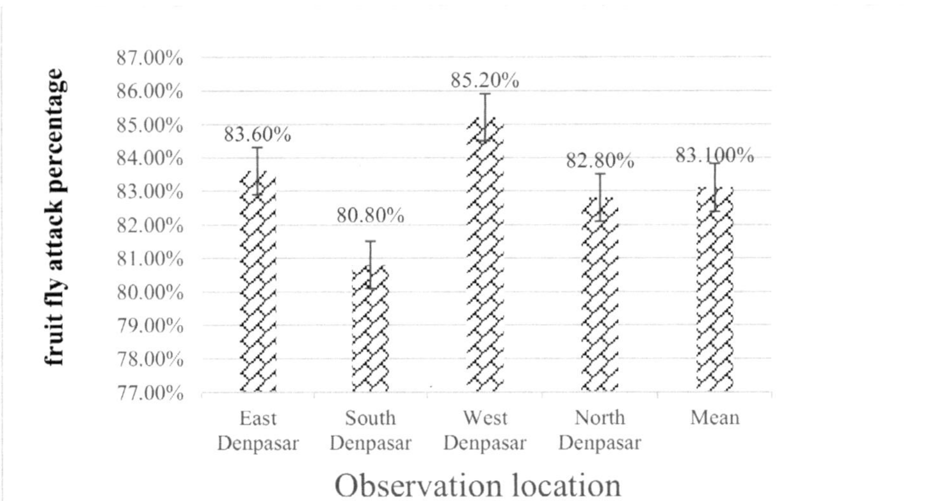

Figure 4. Percentage of fruit fly attacks on star fruit in Denpasar City

The proportion of attacks carried out is determined by the number of adult flies in the field; as the number of adults increases, so does the percentage of attacks. Furthermore, control measures like as trapping, cleanliness, and pesticide usage could all have an impact on the proportion of attacks [22].

3.4. Parasitism rate

Diachasmimorpha longicaudata and *Fopius arisanus* are two parasitoids that interact with fruit flies, according to our results. The interaction is in the form of parasitism that uses fruit flies as its host to continue its life development.The level of parasitoid parasitization in the four sub-districts in Denpasar City was ranging from 8.22 to 9.30 percent. The incidence found in East Denpasar sub-district, the parasitization level of *F. arisanus* was *(8.89%; 286* adults) and *D. longicaudata* (1.09%; 35 adults), while in South Denpasar sub-district, the parasitization level of *F. arisanus* was 8.22% with 241 adults and *D. longicaudata* was 1.09% with 35 adults). Parasitization rates of parasitoid *F. arisanus* were (9.30%; 308 adults) and (8.42%; 266 adults) in the Districts of West Denpasar and North Denpasar, while the parasitization rate of *D. longicaudata* was 1.06% with 35 adults and 1.04% with 33 adults) in West Denpasar (Figure 5).

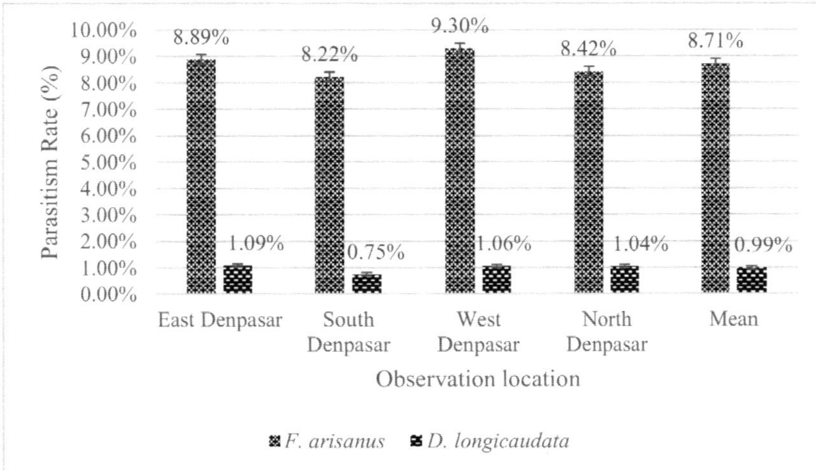

Figure 5. Parasitism rate of parasitoids against fruit flies on star fruit in Denpasar City area

The data revealed that the West Denpasar District has the greatest degree of parasitization. We believe this occurrence was caused by the sub-high district's quantity of fruit flies in comparison to other sub-districts in the Denpasar City region. According to Herlinda et al. [23] the parasitoid population can be influenced by the variety of the host plant and the host population (fruit fly). This proved that the rise in the number of parasitoids was exactly proportionate to the increase in the number of fruit flies in the field, and opposite. Several studies recommend attempts to improve parasitoids' efficacy as natural enemies, which may be accomplished by supplying increased food supplies such as sugar, which has a role in parasitoids' survival and reproduction [24,25]. Host quality, on the other hand, can influence the fitness and effectiveness of progeny parasites [26]. Different sizes of the same parasitoid affect various biological features, showing that larger parasitoids have longer life spans and are more competitive [27].

F. arisanus dominated the parasitization rate in all observation regions. *F. arisanus* is a single koinobiont parasitoid found in a variety of fruit species, particularly those of the genus Bactrocera. Researchers in Africa introduced *F. arisanus* for biological control of *Bactrocera dorsalis* in a variety of fruits and crops [28]. Similarly, prior research reported *F. arisanus* in *Silba adipata*, which attacked the white chili pepper *(Capsicum frutescens L.)* in Bali Province [29]. The findings also revealed that *D. longicaudata* had the highest parasitization rate in *S. adipata*, which attacked white chili peppers in Bali Province.

Based on these data, we hypothesized that the difference in parasitization rate of parasitoids was also highly impacted by the type and quantity of the host in the field. However, the parasitoid rate of parasitoids in star fruit flies in this study was high due to the high abundance of fruit flies in general, and the presence of star fruit cultivated in residential areas was not interfered with by the use of synthetic pesticides, so it became one of the indications that parasitoids still survive in plant hosts.

4. Conclusions

Parasitoids that interact with fruit flies on star fruit in Denpasar are *D. longicaudata* and *F. arisanus* with *B. carambolae* and *B. dorsalis*. The form of interaction is parasitism by utilizing fruit flies as a host for the survival of each parasitoid. The more dominant parasitoid parasitizing fruit flies was *F. arisanus* with parasitization levels ranging from 8.22 to 9.30 percent in the field. *B.carambolae* had a higher population abundance in all locations compared to *B. dorsalis* with an attack rate ranging from 80 - 85 percent in the field.

5. Acknowledgment
The authors would like to thank the Head of the Integrated Pest Management Laboratory at Udayana University, Bali, Indonesia, for providing facilities and funding assistance.

References
[1] Muthu N, Lee S Y, Phua K K and Bhore S J 2016 Nutritional, Medicinal and Toxicological Attributes of Star-Fruits (Averrhoa carambola L.): A Review *Bioinformation* **12** 420–4
[2] Badan Karantina Pertanian 2017 *Deskripsi dan visualisasi jenis asing invasif (JAI)/Invasive Alien Species (IAS) kelompok tumbuhan dan organisme yang berasosiasi dengan tumbuhan* (Jakarta)
[3] Hossain M S, Sarkar B C, Hossain M M, Mian M Y, Rajotte E G, Muniappan R and O'Rourke M E 2020 Comparison of biorational management approaches against mango fruit fly (Bactrocera dorsalis Hendel) in Bangladesh *Crop Prot.* **135** 104807
[4] Drechsel P and Keraita B 2014 *Urban Vegetable Farming Sites, Crops and Cropping Practices*
[5] Lente I, Keraita B, Drechsel P, Ofosu-Anim J and Brimah A K 2012 Risk Assessment of Heavy-Metal Contamination on Vegetables Grown in Long-Term Wastewater Irrigated Urban Farming Sites in Accra, Ghana *Water Qual. Expo. Heal.* **4** 179–86
[6] Begg G S, Cook S M, Dye R, Ferrante M, Franck P, Lavigne C, Lövei G L, Mansion-Vaquie A, Pell J K, Petit S, Quesada N, Ricci B, Wratten S D and Birch A N E 2017 A functional overview of conservation biological control *Crop Prot.* **97** 145–58
[7] Mansion-Vaquié A, Ferrante M, Cook S M, Pell J K and Lövei G L 2017 Manipulating field margins to increase predation intensity in fields of winter wheat (Triticum aestivum) *J. Appl. Entomol.* **141** 600–11
[8] Möller G, Keasar T, Shapira I, Möller D, Ferrante M and Segoli M 2020 Effect of Weed Management on the Parasitoid Community in Mediterranean Vineyards *Biology (Basel).* **10** 7
[9] Badriasih K, Supartha I W and Susila I W 2019 Population Density and Distribution Patterns Fruit Flies (Bactrocera spp.) (Diptera: Tephritidae) that Attacking Mangoes (Mangifera indica L.) in Buleleng Regency *J. Trop. Agroecotechnology* **8** 294–301
[10] Siwi S, Hidayat P and Suputa 2007 Taksonomi dan Bioekologi Lalat Buah Penting di Indonesia (Diptera: Tephritidae) *J. Perlindungan Tanam. Indones.* **13** 106–14
[11] Putra I N W, Susila I W and Bagus I G N 2019 Abundance of Fruit Fly Species (Diptera: Tephritidae) and Its Parasitoid Associated on Star Fruit Plants (Averrhoa carambola L.) in Gianyar regency. *Agrotrop J. Agric. Sci.* **9** 1
[12] Schutze, M., McMahon, J., Krosch, M., Strutt, F., Royer, J., Bottrill, M., Woods, N., Cameron, S., Woods, B., Blacket M (Eds.). 2018 *The Australian Handbook for the Identification of Fruit Flies (Version 3.1)* (Canberra, Autralia: Plant Health Australia)
[13] ICMPFF and KEMENTAN 2006 *Training Workshop on Fruit Flies of Indonesia: Their Identification and Pest Status* (Darwin, Australia: Griffith University and Ministry of Agriculture Republic of Indonesia)
[14] Sharkey M J 1992 Cladistics and tribal classification of the Agathidinae (Hymenoptera: Braconidae) *J. Nat. Hist.* **26** 425–47
[15] Carmichael A E, Wharton R A and Clarke A R 2005 Opiine parasitoids (Hymenoptera: Braconidae) of tropical fruit flies (Diptera: Tephritidae) of the Australian and South Pacific region *Bull. Entomol. Res.* **95** 545–69
[16] McCarthy B C and Magurran A E 2004 Measuring Biological Diversity *J. Torrey Bot. Soc.* **131** 277
[17] Supartha I W, Yudha I K W, Wiradana P A and Susila I W 2020 Response of parasitoids to invasive pest Phenacoccus manihoti Matile-Ferrero (Hemiptera: Pseudococcidae) on cassava crop in Bali, Indonesia *Biodiversitas J. Biol. Divers.* **21**
[18] Reyes J and Ortiz G 2019 *Fruit Sampling Guidelines for Area-Wide Fruit Fly Programmes Fruit Sampling Guidelines for Area-Wide Fruit Fly Programmes Edited by Joint FAO / IAEA*

Programme of Nuclear Techniques in Food and Agriculture Joint FAO / IAEA Programme of Nuclear Techniques i

[19] Sahetapy B, Uluputty M R and Naibu L 2019 Identifikasi Lalat Buah (Bactrocera spp), pada Tanaman Cabai (Capsicum Annum L.) dan Belimbing (Averrhoa Carambola L.) dikecamatan Salahutu kabupaten Maluku Tengah. *Agrikultura* **30** 63

[20] Rattanapun W, Amornsak W and Clarke A R 2010 Is a mango just a mango? Testing within-fruit oviposition site choice and larval performance of a highly polyphagous fruit fly *Arthropod. Plant. Interact.* **4** 35–44

[21] Susila I W and Supartha I W 2020 Jenis dan Peranan Parasitoid dalam Mengendalikan Populasi Lalat Buah (Bactrocera dorsalis Complex.) (Diptera: Tephritidae) yang Menyerang Buah Mangga (Mangifera indica L) di Kabupaten Buleleng *Agrotrop J. Agric. Sci.* **10** 29

[22] Wahyuni S, Supartha I W, Ubaidillah R and Wijaya I N 2017 Parasitoid community structure of leaf miner Liriomyza spp. (Diptera: Agromyzidae) and the rate of parasitization on vegetable crops in Lesser Sunda Islands, Indonesia *Biodiversitas* **18** 593–600

[23] Herlinda S, Noni O, Suwandi S and Hasbi H 2020 Exploring entomopathogenic fungi from South Sumatra (Indonesia) soil and their pathogenicity against a new invasive maize pest, Spodoptera frugiperda *Biodiversitas J. Biol. Divers.* **21**

[24] Tian J-C, Wang G-W, Romeis J, Zheng X-S, Xu H-X, Zang L-S and Lu Z-X 2016 Different Performance of Two Trichogramma (Hymenoptera: Trichogrammatidae) Species Feeding on Sugars *Environ. Entomol.* **45** 1316–21

[25] Harvey J A, Cloutier J, Visser B, Ellers J, Wäckers F L and Gols R 2012 The effect of different dietary sugars and honey on longevity and fecundity in two hyperparasitoid wasps *J. Insect Physiol.* **58** 816–23

[26] Liu Z, Xu B, Li L and Sun J 2011 Host-Size Mediated Trade-Off in a Parasitoid Sclerodermus harmandi ed G Smagghe *PLoS One* **6** e23260

[27] Lampson L J, Morse J G and Luck R F 1996 Host Selection, Sex Allocation, and Host Feeding by Metaphycus helvolus (Hymenoptera: Encyrtidae) on Saissetia oleae (Homoptera: Coccidae) and Its Effect on Parasitoid Size, Sex, and Quality *Environ. Entomol.* **25** 283–94

[28] Nanga Nanga S, Hanna R, Gnanvossou D, Fotso Kuate A, Fiaboe K K M and Djieto-Lordon C 2019 Fruit Preference, Parasitism, and Offspring Fitness of Fopius arisanus (Hymenoptera: Braconidae) Exposed to Bactrocera dorsalis' (Diptera: Tephritidae) Infested Fruit Species ed R Schmidt-Jeffris *Environ. Entomol.* **48** 1286–96

[29] Yuliadhi K A, Supartha I W, Darmiati N N, Bangun A, Yudha I K W, Utama I W E K and Wiradana P A 2021 Short Communication: Silba adipata (Diptera: Lonchaeidae) parasitoids on cayenne pepper (Capsicum frutescens) in Bali, Indonesia *Biodiversitas J. Biol. Divers.* **22**

Increasing maize yield and farming benefit through integrated plant management in Madura Island East Java

W Handayati, E Latifah, Sugiono, H Subagio, D Sihombing and R U Fitria
Assessment Institute of Agriculture Technology East Java, Jl. Raya Karangploso KM 4 Malang

e-mail: wahyuhandayati@yahoo.com

Abstract. Maize is one of the primary plant cultivations on Madura Island and is usually cultivated by farmers with traditional technology and using open-pollinated local varieties. According to this condition, an experiment and a demo plot for local farmers were conducted in Larangan Slampar village Tlanakan District Pamekasan Regency. The experiment was arranged in randomized block design. The treatment was integrated plant management (IPM) using HJ21 and Bima14 hybrid and as control was conventional technology applying local Madura variety. The analyzed data indicated that IPM technology could increase plant yield and farming benefits. The highest harvest yield was shown by IPM1 with Bima 14 variety up to 6.35 ton ha-1, meanwhile existing technology (local Madura variety) only 1.22 ton ha-1 or the productivity was increased up to 422,40 %. The results of the farming analysis showed that benefit with IPM1 reached Rp 15,520,500 and B/C ratio 1,35; while existing technology only Rp 2,168,000 and B/C ratio of 0.55.

1. Introduction

Madura Island is one of the maize cultivation centers in Indonesia. The harvest area is quite wide, but the productivity still very low and far below the national production average. According to BPS 2021, the average productivity was only 2,187 tons (Bangkalan), 1,729 tons (Sampang), 2,258 tons (Pamekasan), and 2,278 tons ha-1 (Sumenep) [1], meanwhile the national productivity up to 5,474 tons ha-1 [2]. Supposedly, this condition is affected by conventional technology applied by farmers, especially the local varieties usage. In addition, knowledge of farmers is still lacking about Integrated Plant Management (IPM) and its application in the field. About 88% of farmers in Madura are cultivating the local varieties of maize to fulfill the family's food supplies. Some farmers reason to use local varieties, namely seed availability, lower farm costs, longer shelf life, lower post-harvest cost, and higher selling prices [3][5].

Integrated maize crop management aimed to increase maize productivity, maintain sustainability, and improve production efficiency. This development must consider the local resource conditions [6] [7]. In this regard, the technological components that have been obtained from various studies are assembled into one technology package that can provide a synergistic effect. And the application must be made with an integrated maize management system. Implementation of the components of production technology such as superior varieties, quality seeds, optimal plant populations, and efficient fertilization, must be adjusted to the land conditions and socio-economic farmer communities [8]. On marginal lands, such as on Madura Island, maize productivity can be significantly increased with the support of IPM technology innovation.

To get maximum productivity, then the superior varieties must be combined with others planting techniques. Because the highest yields can be resulted by plants if the superior varieties get optimal accessibility to the water, fertilizer, and other planting techniques combination. So far, the cultivation technology components of maize have been used partly by the farmer, especially in the low productivity field. Therefore, there is no significant effect on farming production. A previous study on IPM technology application with new hybrid varieties showed the increasing yield of up to 8.8 tons on Bima 4, 8.3 ton Bima 5, and 8.4 tons on Bisi 2, meanwhile the existing only 4.90 – 5.80 tons ha^{-1} [9].

To accelerate the IPM technology adoption by farmers, it is necessary to introduce this technology through field activities involving farmers and farmers' groups as the main actors, and researchers or extension workers as facilitators and supervisors in the application of technology. These activities must be designed in such a way that farmers are interested in replicating them in subsequent farming activities. In this regard, the research activities have been carried out as well as demonstration plots for farmers. The study was aimed to validate the increase of maize productivity, farming benefits, or feasibility by the determination of cultivation technology with an integrated plant management approach.

2. Materal and Method

The study was done in the dryland of Larangan Slampar village Tlanakan District Pamekasan Regency. This assessment was arranged in a randomized block design with six replications, The treatments were the cultivation technology applied i.e., integrated plant management (IPM) approach using HJ 21 and Bima 14 varieties, and existing technology using local variety (Madura) as control. Details of each treatment in the field were assembled in Table 1.

Table 1. Treatment arrangement of IPM technoloy

Activity	Treatments		
	IPM1	IPM2	Existing (control)
1. Land cleaning	yes	yes	yes
2. Land processing	yes	yes	No
3. Varieties	Bima 14	HJ 21	Local variety (cv Madura)
4. Seed (kg/ha)	20	20	15
5. Plant space (cm)	70 x 20	70 x 20	40 x 40
6. Organic manure (ton/ha)	1	1	0
7. An-organic manure (kg/ha)	Urea 300 kg; NPK Ponska 200 kg	Urea 300 kg; NPK Ponska 200 kg	Urea 100 kg
8. Irigation	6 x	6 x	0
9. Weed control	Herbicide 1 application in 2 week after palnting	Herbicide 1 application in 2 weeks after palnting	No
10. Pest and disease control	Pesticide 1 application	Pesticide 1 application	No

The land has tilled by hand tractor and flattened with a hoe. The plant hole was made with a wood digger. After seed planting, the hole was covered by organic fertilizer. Inorganic fertilizer dose was based on the Dry Soil Test Kit (DSTK) and Leaf Color Chart (LCC) analyses. Application of fertilizer was carried out twice, as much as 1/3 dose at 14 days after planting, and the remaining at 30 days after planting. Plant maintenance was done according to the treatment arrangement. The observed variables consist of plant growth, yield components, yield, and analyses of farming.

Analyzed of variance was done for tabulated data and continued with the LSD test at a level of 0.05 if between the treatments had a significant difference. All agronomic variables observed was The farmer income or financial benefit by IPM technology treatment was estimate through a simple farming analysis.

To measure the feasibility of improvement technology (IPM) used MBCR (Margin Benefit Cost Ratio) analysis. If the value of MBCR > 1 then the improvement technology (IPM) must be replacing the technology applied by farmers. It can be simply derived with a formula:

$$MBCR = \frac{Total\ Benefit\ of\ IPM - Total\ Benefit\ of\ exising}{Total\ cost\ of\ IPM - Total\ cost\ of\ existing}$$

3. Result and discussion

3.1. The field agroecoclimate conditions

The assessment was placed in the dry lowland with dry climate conditions. The land was a clay texture and was classified as alluvial soil. DSTK analyses indicated that the land contained a low P, K, and C-organic content and neutral pH. A different study reported that the soil in this district was very poor in nutrients element i.e., N 0.01%, P205 1.29 mg kg-1, K 0.16 m/100 g, C-organic 0.46, and pH 6.2 [10].

Generally, the rainy period in this location was only from November to April. In the assessment, there was no rain from the beginning of planting until harvesting, Therefore, the water necessity was fulfilled only from watering. Daily temperatures range from 28-32°C and humidity 65-70%.

3.2. Analysis of variance for parameters observed

Anova for all parameters was presented in Table 1. There was no significant effect (P > .05) by the blocking data to all parameters observed. Then, the blocking effect was ignored, so that the degree of freedom Error (df) can be increased. All agronomic characters were significantly influenced by the treatments, except for the weight of net kernels (Table 2).

Table 2. Anova for maize agronomic characters

Parameter	Source of Variation	Degree of freedom (Df)	Sum of squares	Means squares	F	Sig.
Plant height (cm)	Between groups	2	25,111.80	12555.90	80.08	7.01E-07
	Within groups	15	1,808.73	204.95		
	Total	17	26,920.53	1583.56		
Stalk diameter (cm)	Between groups	2	10.31	5.16	51.29	5.53E-06
	Within groups	15	1.87	0.27		
	Total	17	12.18	0.72		
Leaf length (cm)	Between groups	2	8066.09	4033.05	155.72	2.91E-08
	Within groups	15	367.03	47.51		
	Total	17	8433.12	496.07		
Leaf width (cm)	Between groups	2	110.22	55.11	102.80	2.15E-07
	Within groups	15	11.92	1.85		
	Total	17	122.14	7.18		
Cob length (cm)	Between groups	2	227.94	113.97	26.97	9.36E-05
	Within groups	15	76.59	11.09		
	Total	17	304.53	17.91		
Cob diameter (cm)	Between groups	2	26.71	13.35	87.11	4.71E-07
	Within groups	15	3.35	0.52		
	Total	17	30.06	1.77		
Cob weight (g)	Between groups	2	168409.71	84204.86	156.70	2.83E-08
	Within groups	15	9392.18	1341.06		
	Total	17	177801.89	10458.93		
Kernel weight/cob (g)	Between groups	2	106192.71	53096.36	188.62	1.15E-08
	Within groups	15	3361.560	390.82		
	Total	17	109554.27	6444.37		
Net kernel weight (%)	Between groups	2	0.01	0.00	1.55	0.26
	Within groups	15	0.08	0.01		
	Total	17	0.09	0.01		
100 kernel weights (g)	Between groups	2	2906.34	1453.17	191.52	1.07E-08
	Within groups	15	209.12	34.24		
	Total	17	3115.47	183.26		
Yield (ton)	Between groups	2	82.46	41.23	208.48	7.05E-09
	Within groups	15	3.16	0.43		
	Total	17	85.62	5.04		

3.3. Performance of plant growth

The data showed that plant growth was significantly affected by introduced cultivation technology (IPM). The highest plants were found in IPM1 and IPM2, and it was significantly different from the existing treatment. Stalk diameter was also found in a significant difference between treatments tested. The biggest stalk diameter was found in IPM2 (Table 3).

IOP Conf. Series: Earth and Environmental Science **980** (2022) 012052 doi:10.1088/1755-1315/980/1/012052

Observation of leaf showed that there was a significant difference between treatments both on the leaf length and leaf width, especially to existing (Table 3). This finding was similar to the present report that all local Madura varieties had a small plant posture and cob. [11]

Table 3. Effect of IPM technology to plant growth of maize in dry season at Larangan Slampar village Tlanakan district Pamekasan regency.

Treatments (Cultivation technology)	Plant height (cm)	Stalk diameter (cm)	Leaf length (cm)	Leaf width (cm)
IPM1 (Bima 14)	174.83 a	2.81 a	85.53 a	9.59 a
IPM2 (HJ 21)	179.66 a	2.88 a	89.90 a	9.55 a
Existing (local variety)	98.12 b	1.24 b	42.97 b	4.32 b
CV	8.30	13.74	6.99	9.36
SEM	5.11	0.13	2.08	0.30

3.4. Plant yield

The data analyses of the harvest yield component indicated that the cob length was significantly affected by the treatments. The longest cob was found in IPM1 treatment (18.66 cm), and its significant differences from the existing (10.94 cm) (Table 4).

The cob diameter showed that the largest cob was resulted by IPM1, and it has a significant difference to existing. IPM1 also produced the heaviest cobs up to 267.07 g and had a significant difference with the existing (Table 4). This result indicates that varieties had the highest effect on the harvest yield component.

Table 4. Effect of IPM technology to maize yield in dry season at Larangan Slampar Tlanakan Pamekasan

Treatments (Cultivation technology)	Cob length (cm)	Cob diamete r (cm)	Cob weight (g)	Kernel weight/ cob (g)	Net kernel weight (%)	100 kernel weights (g)	Yield (ton)	Increased yield of IPM to existing (%)
IPM1 (Bima14)	18.66 a	4.82 a	267.39 a	216.45 a	0.81 a	44.22 a	6.35 a	422.40
IPM2 (HJ 21)	18.30 a	5.13 a	221.81 b	170.69 b	0.78 a	41.64 a	4.70 b	286.38
Existing (local variety)	10.94 b	2.40 b	43.24 c	35.53 c	0.82 a	16.07 b	1.22 c	
CV	12.88	9.51	13.06	11.91	5.13	8.11	10.88	
SEM	0.84	0.16	9.46	6.85	1.69	1.25	0.18	

The heaviest kernel per cob has resulted from IPM1. It was significantly different from IPM2 and existing. The weight per 100 kernels was significantly different from existing (Table 4).

The net weight of the kernel is a comparison of the kernel weight to the cob ear weight before threshed. There was no significant difference between the treatment examined. The heaviest net kernel was obtained by existing (Table 4). This fact indicated that this variety could produce kernels with an optimum weight, even though the maize cobs size of local Madura was a lot smaller than other varieties.

The yield data in Table 4 showed that IPM1 produces the highest yielding 6.35 tons ha-1, and it was a significant difference to IPM2 and a very significant difference to existing which just produced 1.22 tons ha-1. Furtherly, this treatment could enhance the yield by 422.40% than existing technology. Another study also reported the same result that yields on IPM technology (cooperative farmers) were higher than conventional technology (non-cooperative farmers). Srikandi Kuning and Sukmaraga varieties gave yields of 7.45 and 7.90 tons ha-1 respectively or 119.28% higher than farmers' technology with the local variety

Manado Kuning (only 3.50 t ha-1) [12]. The character expression will be optimum if the crop was planted in a suitable environment, as in the IPM treatment condition. Inversely, if an environment of growth was inappropriate, the character performance will not be optimum. A character performance with low heritability has a significant environmental effect. Hence, its appearance will be easy to change. However, the high heritability character has a little environmental effect. Therefore, the performance will be quite stable [13].

Table 5. The correlation coefficients of harvested yield and agronomic parameters

Parameters	Plant height (cm)	Stalk diameter (cm)	Leaf length (cm)	Leaf width (cm)	Cob length (cm)	Cob diameter (cm)	Cob weight (g)	Seed weight/ cob (g)	Net kernel weight (%)	100 kernel weight (g)	Sampling yield (kg)/5,46 m2
Plant height (cm)	1										
Stalk diameter (cm)	0.916	1									
Leaf length (cm)	0.911	0.889	1								
Leaf width (cm)	0.903	0.803	0.921	1							
Cob length (cm)	0.860	0.864	0.840	0.763	1						
Cob diameter (cm)	0.918	0.869	0.904	0.825	0.816	1					
Cob weight (g)	0.920	0.897	0.934	0.865	0.860	0.885	1				
Seed weight/ cob (g)	0.897	0.864	0.919	0.909	0.815	0.852	0.977	1			
Net kernel weight	-0.207	-0.275	-0.261	0.041	-0.359	-0.266	-0.277	-0.073	1		
100 kernel weight (g)	0.928	0.931	0.953	0.856	0.922	0.912	0.982	0.947	-0.340	1	
Sampling yield (kg/5,46 m2)	0.909	0.912	0.921	0.902	0.861	0.897	0.963	0.977	-0.108	0.958	1
Harvest yield (ton/ha)	0.890	0.865	0.878	0.893	0.855	0.816	0.932	0.959	-0.034	0.918	0.959

Correlation analyses among agronomic variables indicated that all the parameters were correlated to the yield, except the weight of the net kernel). The plant height, the cob length, diameter, and weight; and kernel weight/cob, and 100 kernel weights had the highest coefficients correlation to the harvest yield (Table 5. The same result in earlier research showed that harvest yield was positively correlated with plant height, ear height, and field weight and with high genotypic and phenotypic correlation coefficients [14]. A different report quoted that kernel was positively correlated and had a significant correlation with plant and flag leaf height, 300-kernel weight, total kernel per row, and per ear [15].

3.5. Farming Analysis

The results of the farming analysis showed that the IPM1 assembly technology provided the highest income or revenue up to Rp 26,987,500, reaching the R/C ratio of 2.35 while in existing only 6,100,000 with an R/C ratio of 1.55. Although the production costs used in IPM1 increase, it can provide higher profits than existing. It looks at the B/C ratio of 1.5 in IPM1 and 0.55 in existing technology with a benefit difference reaching Rp 13,352,000 for IPM1.

Based on feasibility analysis, the highest MCBR value showed by technology introduced IPM1 (Bima 14 variety) reached 1.77. Although the costs incurred were higher than existing technology, the IPM1 technology assembly is suitable for development because it can provide higher profits and incomes to farmers. Thus, the application of IPM1 with Bima 14 can be recommended as a location-specific technology.

This result is in line with previous research that the gaining of income or profits in IPM technology (Rp 8,687,500) increased 197 % than existing technology (Rp4,390,000) [12]. Another study showed that IPM with different varieties also affects the cost of production and maize yields. Farming cost about Rp 6,013,140 for BISI and Rp 3,826,620 for Lamuru varieties with productivity of 4,785 kg and 4,607 kg ha-1 respectively [16]. The study in Sumenep showed that maize yields 4.93 tons with a profit up to Rp 6,944,099 [11]. Another study showed that 3.90 tons ha-1 of corn yields resulted in a profit of Rp 2,104,950 [17].

Table 6. Farming analisys of IPM technology application on maize cultivation in dry season at Larangan Slampar village Tlanakan district Pamekasan regency

Description	Units	IPM1 (Bima 14)			IPM2 (HJ 21)			Existing (lokal Madura)		
		Volume	Unit price (Rp)	Total (Rp)	Volume	Unit price (Rp)	Total (Rp)	Volume	Unit price (Rp)	Total (Rp)
Cost										
1. Seed	kg	22	56,000	1,232,000	22	56,000	1,232,000	-	-	-
2. Organic fertilizer	kg	1,000	500	500,000	1,000	500	500,000		500	-
3. NPK Ponska fertiizer	kg	200	2,300	460,000	200	2,300	460,000	100	2,300	230,000
4. Urea fertilizer	kg	300	1,800	540,000	300	1,800	540,000	100	1,800	180,000
5. Pesticide	package	1	500,000	500,000	1	500,000	500,000	-	500,000	-
6. Land processing (hand tractor)	package	1	1,000,000	1,000,000	1	1,000,000	1,000,000	1	1,000,000	1,000,000
7. Planting	worker	20	40,000	800,000	20	40,000	800,000	20	40,000	800,000
8. Plant maintanance (fertilizer application, weeding, pest/disease	worker	50	40,000	2,000,000	50	40,000	2,000,000	20	40,000	800,000
9. Irrigation	package	6	500,000	3,000,000	6	500,000	3,000,000		500,000	-
8. Harvesting	worker	20	40,000	800,000	20	40,000	800,000	20	40,000	800,000
9. Thresher	kg	6,350	100	635,000	4,700	100	470,000	1,220	100	122,000
A. Total cost	Rp			**11,467,000**			**11,302,000**			**3,932,000**
B. Revenue	kg	6,350	4,250	**26,987,500**	4,700	4,250	**19,975,000**	1,220	5,000	**6,100,000**
C. Benefit	Rp			**15,520,500**			**8,673,000**			**2,168,000**
D. Δ Benefit of IPM to existing technology	Rp			**13,352,500**			**6,505,000**			
E. R/C ratio				**2.35**			**1.77**			**1.55**
F. B/C ratio				**1.35**			**0.77**			**0.55**
G. MCBR				**1.77**			**0.88**			

4. Conclusions

Application of IPM technology, especially IPM1 with Bima 14 variety could increase the harvesting yield and farming benefit or farmer income. The harvesting yield obtained up to 6,35 ton ha-1, and it increased up to 422.40% compared to existing. The revenue or income increased up to Rp 26,987,500 with an R/C ratio of 2,35. The farming benefits about Rp 15,520,500 with a B/C ratio of 1,35. This benefit increases up to Rp 13,352,500 than existing. The highest MCBR value in technology introduced IMP1 (Bima 14 variety) reached 1.77.

This IPM technology assembly using Bima 14 and others superior hybrid varieties was recomendated to replace the conventional tehnology

References

[1] BPS_Indonesia Statsitic 2021 Hasvested area, production and productivity of maize in Madura.

[2] Astuti K, O R Prasetyo and Khasanah I N 2021 *The 2020 Analysis of Maize and Soybean Productivity in Indonesia. (The Results of Crop Cutting Survey)*. BPS-Statistics Indonesia. 110p

[3] Sugiarti T and Hayati M 2009 Persepsi petani Madura dalam menolak komoditas jagung varietas baru. Embryo **6** 1 : 35 -46.

[4] Purwandari S E, A Anto and Suriansyah 2014 *Teknologi Budidaya Jagung dengan Pendekatan PTT*. Balai Pengkajian Teknologi Pertanian (BPTP) Kalimantan Tengah. 28p

[5] Zubachtirodin, Saenong S, Pabbage M S, Azrai M, Setyorini D, Kartaatmadja S and Kasim F 2016 *Pedoman Umum PTT Jagung*. Badan Penelitian dan Pengembangan Pertanian. Kementerian Pertanian. Jakarta. 20p

[6] Badan Litbang Pertanian. 2007. Pengelolaan Tanaman Terpadu (PTT) Jagung Sawah Irigasi. Petunjuk Teknis Lapang. Badan Penelitian dan Pengembangan Pertanian. Departemen Pertanian.

Jakarta.
[7] Suyamto, Zubachtirodin, Pabbage M S, Saenong S and Widiarta I N 2008 *Panduan Pelaksanaan Sekolah Lapang Pengelolaan Tanaman Terpadu (SL-PTT) Jagung*. Departemen Pertanian 43p
[8] Syuryawati dan Faesal. 2015. Pengaruh penerapan teknologi PTT jagung terhadap pendapatan usahatani jagung. Prosiding Seminar Nasional Serealia, 2015. p642 – 648
[9] Soehendi R dan Syahri B 2013 Potensi Pengembangan Jagung di Sumatera Selatan J. Lahan Suboptimal **2** 1 81-92
[10] Cahyani N K M D, Nurhatika S and Muhibuddin A 2014 Eksplorasi Mikoriza Vesikular Arbuskular (MVA) Indigenous pada Tanah Aluvial di Kabupaten Pamekasan Madura. J. Sain dan Seni POMITS **3** 1 E 22-25 http://dx.doi.org/10.12962/j23373520.v3i1.5525
[11] Santoso, R, H. Sudarmadji, dan Awiyanto. 2013. Faktor-faktor yang mempengaruhi produksi jagung di Kabupaten Sumenep. Cemara **10** 110-17.
[12] Tamburian Y, W. Rembang dan Bahtiar 2011. Kajian usahatani jagung di lahan sawah setelah padi melalui pendekatan PTT di kabupaten Bolmong Suawesi Utara. Seminar Nasional Serealia 2011. http://balitsereal.litbang.pertanian.go.id/wp-content/uploads/2016/12/30bpros11.pdf
[13] Haryati Y dan A Sinaga 2016 Pengujian adaptasi beberapa varietas jagung hibrida spesifik lokasi di Kabuaten Majalengka. J Agrotek Lestari **2** 1 5 –58
[14] Nzuvel F, Githiri S, Mukunya D M and Gethi J 2014 Genetic Variability and Correlation Studies of Grain Yield and Related Agronomic Traits in Maize J Agricultural Sci **6** 9 166-176 http://dx.doi.org/10.5539/jas.v6n9p166
[15] Moradi H,. Akbari G A, Khorasani S K and Ramshini H A 2012 Evaluation of drought tolerance in corn (*Zea mays* L.) new hybrids with using stress tolerance indices European J. Sustainable Development **1** 3 543-560
[16] Margaretha SL. dan Syuryawati 2017 Adopsi Teknologi Produksi Jagung dengan Pendekatan Pengelolaan Tanaman Terpadu pada Lahan Sawah Tadah Hujan. Penelitian Pertanian Tanaman Pangan **1** 1 53 – 64
[17] Mahdiana, M, S. Sulastri, dan H.S. Handayawati 2010 Analisis pengaruh penggunaan factor produksi usahatani jagung (Zea mays L). Wacana **13** 4 684-688.

IOP Conf. Series: Earth and Environmental Science **980** (2022) 012053 doi:10.1088/1755-1315/980/1/012053

Determinants and strategies of social forestry development for integrated coffee agribusiness in the foot of mount Argopuro area, Jember regency

R A Djamali[1), D T Cahyaningrum[2), and D E Putra[3)

[1, 2,3)] Agribusiness Management Department, Politeknik Negeri Jember, Jember 68101 East Java, Indonesia. Politeknik Negeri Jember

abdoel_djamali@polije.ac.id

Abstract. LMDH "Rengganis" Pakis Village, District, Jember Regency has received the Recognition and the Forestry Partnership ("KULIN-KK") from the Ministry of Environment and Forestry, covering an area of ± 1,032.19 hectares. The forest land is cultivated by 536 families by planting robusta coffee. The research objective is to describe the determinants and strategies for developing an integrated coffee agribusiness area through the social forestry program. The methodology used is descriptive exploratory with an expert system approach. The analysis technique used is cognitive mapping analysis and hierarchy process analysis. Conclusions: (a) Five priority factors are the main determinants of coffee agribusiness development in social forestry programs, namely: legal certainty in forest use, dynamics and cooperation in forest village farme business groups, downstream processing of smallholder coffee products, shortening the trade system chain, support for facilities and infrastructure from the government; (b) Strategy for developing an integrated coffee agribusiness area based on priorities, namely (1) Objectives: developing an integrated coffee agribusiness area in a forest village; (2) Stakeholders, namely: forest farmers/members of LMDH, LMDH Management, LMDH Business Unit, Perhutani, Government, (2) Integrated Coffee Agribusiness Area Development Program: social engineering to increase cooperation in forest village farmer groups, engineering technology to increase coffee productivity, engineering technology downstream processing ofpeople's coffee, economic engineering to design an efficient coffee marketing chain, development of integrated agro-tourism based on agribusiness and processing of products from social forestry programs, namely coffee and durian commodities.

1. Introduction

Social forestry is defined as a form of industrial forestry (conventional) that is modified to allow distribution of benefits to local communities (Gilmour and Fisher, 1991 *in* Suharjito and Darusman, 1998)[10]. The concept of social forestry can be implemented on traditional forest lands, namely state forest areas and other lands, such as yards, fields, or gardens. Therefore, first of all, we view that the purpose of developing social forestry is to involve the communities living around and within forest areas to participate in empowering existing forest resources. The concept of "social forestry" is more accurately stated by Tiwari (1983) *in* Hendarto , Kresno Agus (2003) and Alfitri (2005) that social forestry is the science and art of planting trees and or other plants on land that is possible for certain purposes, within and outside the forest area, and manage it intensively by involving the community [1]. This management is integrated with other activities, so that there is a balance and complement each other's land use with a view to providing goods and services widely to both individual cultivators and the community. Wiersum (1984) views that Social Forestry should be an activity related to the professionalism of foresters whose specific goal is to increase local community participation in forest management to meet family needs and accommodate their aspirations for forestry development [13].

Social forestry is a sustainable forest management system implemented in state forest areas or private forest/customary forests carried out by local communities to improve their welfare, environmental balance and socio-cultural dynamics. For the 2015-2019 period, the government allocated 12.7 million hectares for Social Forestry, through the following schemes: (1) Village Forest (VF) with VFMR tenure or Village Forest Management Rights, (2) Community Forest (CF), the permit granted is CFUBP or Community Forest Utilization Business Permit, (3) Community Plantation Forest (CPF), the permit granted is TFPUBP-CPF or Timber Forest Product Utilization Business Permit - Community Plantation Forest, (4) Customary Forest (CF), the tenure is Determination of Inclusion Customary Forests, (5) Forestry Partnerships (FP) in the form of RPFP or Recognition of Protection of Forestry Partnerships and SFFUP or Social Forestry Forest Utilization Permits on Java Island (http://pkps.menlhk.go.id/). Forest management by involving community participation has actually existed in several other concepts. Among them are community forestry (CF) and community forest or community foresty programs. Social forestry can be carried out on privately owned land ("Usaha Hutan Rakyat") or in state forest areas ("Community Forest") [2].

In accordance with the confirmation by the President of the Republic of Indonesia in April 2017 regarding the declaration of a government program for economic equality, the Ministry of Environment and Forestry ("KLHK") issued Minister of Environment and Forestry Regulation Number 39 of 2017 namely Social Forestry in the working area of Perum Perhutani. Based on data from Perhutani Forest Management Unit (KPH) Jember, there are 59 villages spread across 16 sub-districts in the forest area.

One of the Forest Village Community Institutions ("LMDH") in Jember Regency which has received a Decree from the Indonesian Ministry of Environment and Forestry concerning the Recognition and Protection of Forestry Partnerships ("KULIN KK") Number 4298/MENLHK/-PSKL/PKPS/PSL.0/4/2019 including LMDH "Rengganis" Pakis Village, Panti District, Jember Regency with the Jember Forest Management Unit ("KPH") covering an area of ± 1,032.19 hectares. The forest area managed by LMDH "Renggaris" is cultivated by 536 families or an average of 1.93 hectares per household. What is meant by the issuance of the KULIN KK is for: (a) area utilization business, (b) timber forest product utilization business in permanent production forest area, (c) non-timber forest product utilization business in the form of utilization of intercropping plants, use of land under stands, (d) business of utilizing environmental services, and (e) rehabilitation of protected areas or protected forests.

Considering that the social forestry program carried out by LMDH Rengganis is very strategic which includes 3 main things, namely: (a) economic aspects, namely the area of the land has been planted with several types of plants, namely Robusta coffee/Coffea canephora (85%) and durian (7%) , avocado (3%), banana (3%), and banana (2%), which of course have made a real contribution to regional economic growth and have a multiplier effect on other development fields, (b) social aspects, from 536 households as farmers The perhutani land cultivators have been able to improve the welfare of forest village farmers who have usually been indicated as being one of the poverty enclaves in Jember Regency, and (c) the legal aspect, the issuance of the KULIN KK is of course very important as a strong legal basis for carrying out cultivating plants in the utilization of production forest areas and protected forests.

The existing conditions of the coffee agribusiness of farmers working on this social forestry program are: (a) the low productivity of robusta coffee plantations for under industrial forest stands at 0.8 quintals of Oose coffee per hectare and for robusta coffee under protected forests, 1.3-1.5 quintals of ose coffee per hectare, (b) types of robusta coffee varieties grown are not superior varieties, (c) traditional coffee cultivation methods, (d) low level of knowledge of farmers in mastering harvest technology and post-harvest handling which is still low which results in low coffee quality, (e) 100% of coffee products are marketed in ose form to middlemen at the village level, (f) there is no roasted coffee and ground coffee processing agroindustry at the village level, (g) there is no coffee farmer cooperative that acts as a joint economic support institution for coffee farmers. So it is hoped that through the Social Forestry program in Pakis Village, Panti District, it will later become an integrated and sustainable robusta coffee agribusiness area.

Seeing the gap between the potential business opportunities and the existing conditions of coffee farmers in the LMDH "Rengganis", it is necessary to conduct in-depth research on the determinants and strategies for developing an integrated coffee agribusiness area through a social forestry program at the foot of Mount Argopuro, Pakis-Panti Village.

2. Methodology

The selection of the research location was carried out with a purposive (deliberate) approach at LMDH "Rengganis" Pakis Village, Panti District, Jember Regency. The main considerations for choosing the location at the "Rengganis" LMDH are: (a) it has been successful and has been used as a role model in the social forestry partnership pattern of Perum Perhutani East Java, (b) able to map the potential and utilize under forest plantation stands for coffee agribusiness activities. (main commodities), durian, jackfruit, petai, and avocado, (c) has developed the concept of an integrated ecotourism area, namely integrating the concept of developing agribusiness and agro-industry processed products of superior commodities with forest tourism.

The research method used is a combination of quantitative and qualitative research. Mixed methods' is a research approach whereby researchers collect and analyze both quantitative and qualitative data within the same study [9]. Mixed methods research requires a purposeful mixing of methods in data collection, data analysis and interpretation of the evidence. The key word is 'mixed', as an essential step in the mixed methods approach is data linkage, or integration at an appropriate stage in the research process. Purposeful data integration enables researchers to seek a more panoramic view of their research landscape, viewing phenomena from different viewpoints and through diverse research lenses [4] This research also uses an expert system approach, which is a computer program that imitates the reasoning of an expert with expertise in a particular area and science [5].

The technique of collecting data is through in-depth interviews, namely the process of obtaining information for research purposes by means of question and answer while face to face between the interviewer and the respondent or the person being interviewed, with or without using an interview guide in which the interviewer and informant are involved. in a relatively long social life [12]. The research instrument used in this study was a questionnaire containing a set of combinations of open and closed questions [11]

To discuss the purpose of this research, there are two types of appropriate analytical techniques, namely:

1. To describe the determinants of the development of an integrated coffee agribusiness area through the Social Forestry program at the foot of Mount Argopuro, Pakis Village, Panti District, Jember Regency using a Cognitive Mapping Analysis approach. Cognitive mapping is categorized as a soft methodology and is different from the traditional formal methodology which consists of several types of analysis and the results are still general. Cognitive mapping uses a map technique base that is able to represent elements of complex problems, which are organized and arranged using arrow diagrams. The direction of the arrow shows the connection and relationship between indicators. In this study, two main variables in cognitive mapping are used, namely domain and centrality. Domain is an important factor in cognitive maps because it describes the density or number of indicators that are directly related to certain indicators regardless of their direction. The higher the domain value of an indicator indicates the large number of indicators that

affect/influence the indicator. This shows that downstream indicator effects are not only directly related to other indicators but also are not directly related. The meaning of centrality shows strategic meaning because it describes the cumulative impact of a number of indicators outside of direct influence. The higher the centrality score, the more significant all indicators of the development of an integrated coffee agribusiness area as a system. Domain and centrality are the ideas of Eden and Akerman (1998) *in* [6] , where both are the main tools in compiling cognitive mapping. To calculate the centrality score using the formula:

$$C_j = \frac{S_m}{m} + \cdots + \frac{S_n}{n}, j = 1, 2, 3 \ldots n \quad (1)$$

where :
Cj : the central score of the j-th level of the m indicator
Sj : number of m level indicators

The central score can show the strategic value of an indicator/attribute because it reflects not only the number of indicators that have a direct impact but also all the effects of indirect relationships with other indicators.

2. To formulate the objectives in formulating a strategy for developing an integrated coffee agribusiness area using the Analytical Hierarchy Process (AHP) approach. According to [5] that the stages of AHP analysis are as follows:

a. Hierarchy

The problem to be solved is then broken down into its elements, namely criteria and alternatives, followed by compiling a hierarchical structure, namely goals, criteria and then alternatives.

b. Criteria and Alternative Assessment

Criteria and alternatives are assessed through pairwise comparisons using a scale of 1 to 9, where 9 is the best scale for expressing opinions. The value and definition of qualitative opinion from the comparison scale can be seen in the following table [3].

Table 1. Values and Definitions of Comparison Scale

Value	Description
1	A is as important as B
3	A is slightly more important than B
5	A is clearly more important than B
7	A is much more important than B
9	A is absolutely more important than B
2, 4, 6, 8	Both elements have values between two adjacent consideration values

c. Prioritization

Priority determination can be obtained by conducting pairwise comparisons for each level of the hierarchy. The relative comparison values are then processed to determine the relative rank of all alternatives. The compared qualitative and quantitative criteria will produce weights and priorities from the calculation of matrix manipulation or through solving mathematical equations.

d. Logical consistency

All elements are grouped logically and ranked consistently according to a logical criterion. [8] explains the steps of the Analytical Hierarchy Process (AHP) calculation process which is written in his journal as follows:

1. Define the problem and determine the desired solution.
2. Create a hierarchy that begins with a general goal, followed by criteria and sub-criteria.
3. Create a pairwise comparison matrix that shows the relative contribution or influence of each element to each goal or criterion at the level above it.
4. Perform pairwise comparisons so that a total score of n x n or n x {(n-1)/2} is obtained with the value of n being the number of elements being compared.

5. Create eigenvalues and test their consistency, if they are not consistent then the data collection is repeated.
6. Repeat steps c and d for all levels of the hierarchy.
7. Create eigenvectors from each matrix with pairwise comparisons. The eigenvector value is the weight of each element. This seventh step aims to synthesize judgments in determining the priority of elements at the lowest hierarchical level until the achievement of goals.
8. Checking the consistency of the hierarchy, if the value is more than 10% then the assessment must be corrected.

3. Result and Discussion

The community of Pakis Village on the slopes of the Hyang Argopuro Foot Mountains has long used the forest as a source of livelihood which is incorporated in the Forest Village Community Institution (LMDH) "Rengganis". The concept of independent forest management as Rimba Sosial has been used as a pilot project by the East Java Provincial Office, LMDH Rengganis has partnered with Perum Perhutani since 2005 in the form of Community Forest Management. Then it was used as a pilot project since the issuance of the SK KULIN KK in April 2019 with an area of 1,032.29 Ha. Currently, the status of forest use is stronger in protection for managing forests because it has received recognition from the government. LMDH Rengganis, which has more than 700 members, has become a pilot model because of its strong partnership with Perum Perhutani. The East Java Provincial Government provides a number of facilities and assistance, including the ease of disbursing People's Business Credit (KUR) worth IDR 17 billion in 2017 which is the basis for the development of the coffee business. In addition, they also received assistance from the East Java Fisheries Service in the form of local fish seeds. Not only coffee commodities, after obtaining the Social Forestry Decree, currently LMDH Rengganis can form a Social Forestry Business Group (KUPS) for several other commodities such as durian, jackfruit, petai, and avocado.

A. Determinants of the Development of an Integrated Coffee Agribusiness Area at the Foot of Mount Argopuro, Pakis Village, Jember Regency

Based on the results of in-depth interviews with experts, 11 determining factors were obtained in the effort to develop an integrated coffee agribusiness area in Pakis Village (see Figure 1). In mapping cognitive (cognitive mapping), each factor has the same opportunity to influence and be influenced by other determinants (concepts). This is shown in the form of arrows in either one or two directions from each factor. This soft methodology is able to translate and describe the direct and indirect influence relationships between these factors.

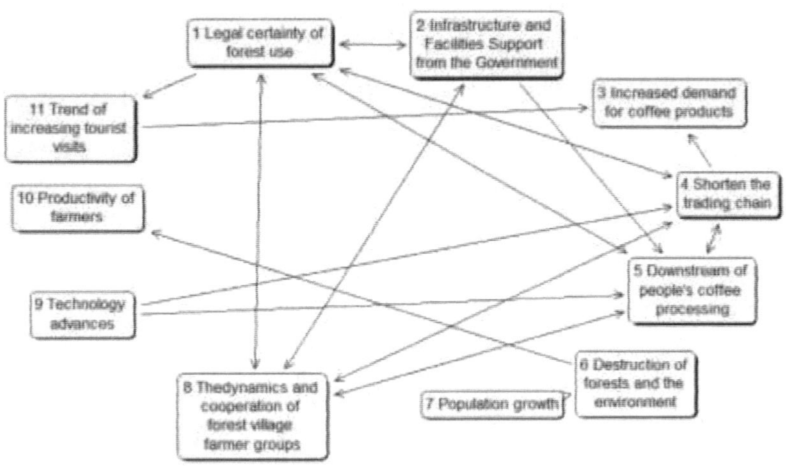

Figure 1. Determinants of the development of an integrated coffee area

Table 2. Centrality Score and Domain of Coffee Agribusiness Development at the Foot of Mount Argopuro-Pakis Village, Jember Regency

No	Determinants	Centrality Score	Domain (link around)
1	Legal certainty of forest use	6 from 7 concepts	9
2	Support facilities and infrastructure from the government	4 from 7 concepts	5
3	Increased demand for coffee products	4 from 7 concepts	2
4	Shorten the marketing chain	6 from 7 concepts	8
5	Downstreaming of people's coffee processing	6 from 7 concepts	8
6	Forest and environmental damage	2 from 2 concepts	2
7	Population growth	1 from 2 concepts	1
8	Group dynamics and collaboration	5 from 7 concepts	8
9	Technological advances	4 from 7 concepts	2
10	Low productivity of farmers	1 from 2 concepts	1
11	Trend of increasing tourist visits	4 from 7 concepts	2

Source: Primary Data Processed, 2021

B. Integrated Coffee Agribusiness Area Development Strategy

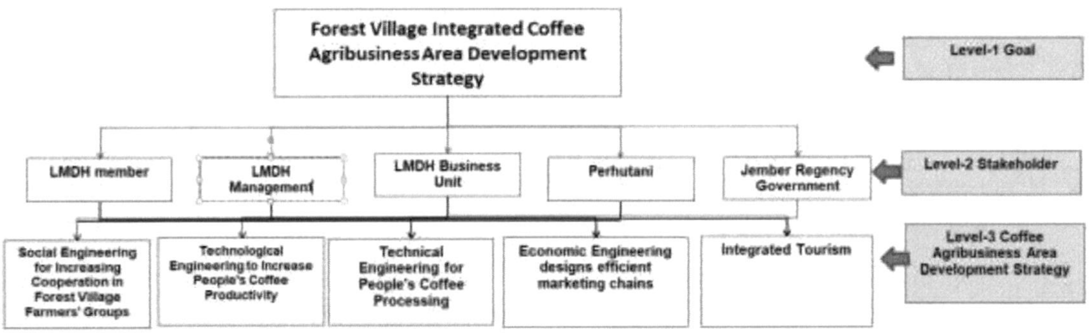

Figure 2. Strategy for the Development of an Integrated Coffee Agribusiness Area at the Foot of Mount Argopuro, Pakis Village, Jember Regency

The results of in-depth interviews with selected experts resulted in a strategic structure for developing an integrated coffee agribusiness area at the foot of Mount Argopuro, Pakis Village (see Figure 2).) Level-2 Stakeholders consisting of LMDH Members (MEMBERS), LMDH Management (MENAGEMENT), LMDH Business Units (BUSINESS), Perhutani (PERHUTANI), and Jember Regency Government (GOV), and (3) Level-3, namely the Integrated Coffee Agribusiness Area Development Strategy which consists of: (a) Social engineering to increase cooperation in village farmer groups (SOS-ENG) (b) Technological Engineering to increase people's coffee productivity (TECH-ENG 1), (c) Technological engineering for the downstream processing of people's coffee (TECH-ENG 2, (d) Economic engineering in designing efficient marketing chains (EC-ENG), and (e) Integrated Tourism Area (TURISM)

Furthermore, the AHP analysis was carried out at level-2 where successively as stakeholders in the development of an integrated coffee agribusiness area, the average value for LMDH Members (0.404), LMDH Management (0.263), LMDH business units (0.182), Perhutani (0.100), and Jember Regency Government (0.052). This means that as the main driving force for achieving the development of an integrated coffee agribusiness area, LMDH internal stakeholders are prioritized over stakeholders outside LMDH. The active role and involvement of each member, management, and business unit of LMDH "Rengganis" will determine the success and achievement in making the Foot of Mount Argopuro into an integrated agribusiness area based on social forestry. The social dynamics that are formed based on internal power will be stronger and more sustainable in carrying out program activities that lead to changes in the social status of a society. The planning of activity programs based on needs and desires driven by internal stimuli will be better and more implementable because they are actually in accordance with the real conditions of the community.

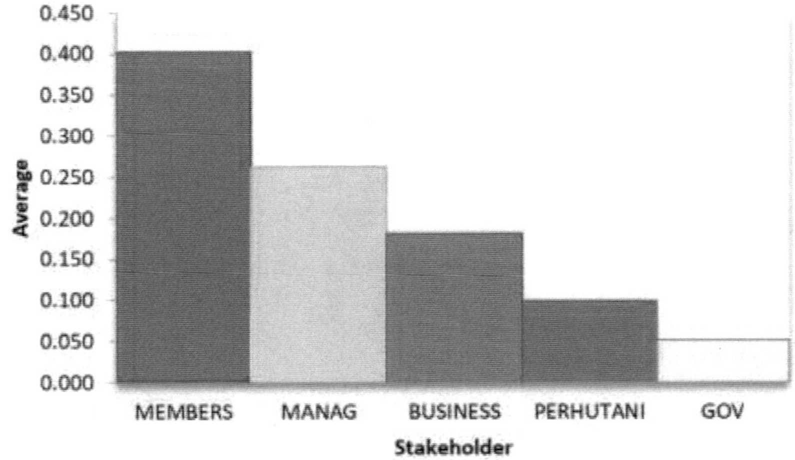

Figure 3. Stakeholder Level Priorities in the Integrated Coffee Area Development Strategy

For the determination of level-3 priority in determining the strategy of developing agribusiness areas after the AHP analysis in the combined matrix stage shows the following priorities: SOS-ENG (0.295), ECO-ENG (0.277), TECH-ENG 2 (0.247), TURISM (0.099), TECH-ENG 1 (0.081). The strategy in carrying out social engineering is intended to increase the cooperation of farmer groups under LMDH Reggganis. The roles and activities of members are to actively participate in group activities in planning activities, solving problems, implementing technological innovations in the cultivation and processing of people's coffee. Economic engineering, namely structuring the trade system for green bean coffee products, which so far have only been sold to middlemen at the village level. Of course, the bargaining power of farmers is very weak with prices that are very detrimental. The main obstacles experienced by farmers are: bonded system, limited knowledge and market information, weak business capital, and limited knowledge and facilities for harvesting and post-harvesting. Engineering technology-2, namely downstream processing of people's coffee, means that there are concrete steps to process some coffee products into roasted coffee and ground coffee. Product marketing can be developed through cooperative networks and social media-based

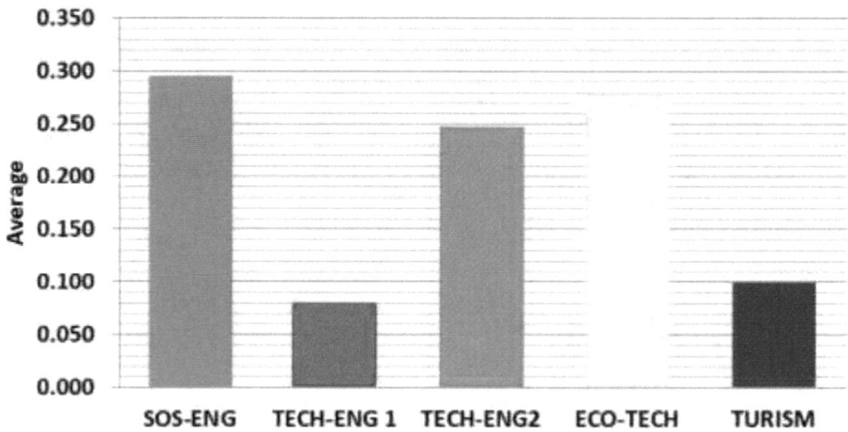

Figure 4. Strategy for the Development of an Integrated Coffee Agribusiness Area

The priority strategies for developing an integrated coffee agribusiness area are:

a. Level-1 Objectives: development of an integrated forest village coffee agribusiness area;
b. Level-2 Stakeholders are: forest farmers/members of LMDH, LMDH Management, LMDH Business Unit, Perhutani, Local Government,
c. Level-3 Integrated Coffee Agribusiness Area Development Program: social engineering to increase collaboration between forest village farmer groups, technological engineering to increase coffee productivity, engineering downstream technology for people's coffee processing, economic engineering to design efficient coffee marketing chains, development of integrated agro-tourism based on agribusiness and processing of products from social forestry program, namely coffee and durian commodities.

4. Conclusion

(1). Five priority determinants of coffee agribusiness development in social forestry programs, namely: (a) legal certainty in forest use, (b) dynamics and cooperation in forest village farmer business groups, (c) downstream processing of smallholder coffee products, (d) shorten the chain of commerce, (e) support for facilities and infrastructure from the government;

(2). The strategic priorities for developing an integrated coffee agribusiness area are:

   a. Level-1 Objectives: development of integrated forest village coffee agribusiness area;
   b. Level-2 Stakeholders are: forest farmers/members of LMDH, LMDH Management, LMDH Business Unit, Perhutani, Local Government,
   c. Level-3 Integrated Coffee Agribusiness Area Development Program: social engineering to increase collaboration between forest village farmer groups, technological engineering to increase coffee productivity, engineering downstream technology for people's coffee processing, economic engineering to design efficient coffee marketing chains, development of integrated agro-tourism based on agribusiness and processing of products from social forestry program, namely coffee and durian commodities.

5. References

[1] Alfitri. 2005. Model perhutanan sosial berbasis partisipasi masyarakat pada program konservasi hutan taman nasional kerinci seblat (tnks). Indonesia journal for sustainable future, 1 (2). Pp. 29-42.

[2] Ditjen rehabilitasi lahan dan perhutanan sosial. 2002. Buku panduan rehabilitasi lahan dan perhutanan.

[3] Falatehan, A. F. 2016. Hendarto, kresno agus. 2003. Perhutanan sosial dan peubah-peubah sosiologi yang harus diperhatikan: sebuah tinjauan teoritis. Jurnal manajemen hutan tropika Vol. IX No.1

[4] Ivankova nv. 2006.. Using mixed methods sequential explanatory design: from theory to practice. Field methods.18:3–20. doi: 10.1177/1525822x05282260.

[5] Marimin. 2004. Teknik dan aplikasi pengambilan keputusan kriteria majemuk. Grasindo.

[6] Mendoza GA, Prabhu R. 2002. Qualitative multi-criteria approaches to assesing indicators of sustainable forest resources management. Forest ecology and management. 5913: 1-5.

[7] National collaborating centre for methods and tools. 2015. Appraising qualitative, quantitative, and mixed methods studies included in mixed studies reviews: The MMAT. Hamilton, on: BMJ publishing group. Http://www.nccmt.ca/resources/search/232 (accessed may 2021).

[8] Noor, s. 2014. Penerapan analisis swot dalam menentukan strategi pemasaran daihatsu luxio di malang. Jurnal intekna. 14(2): 102-209.

[9] Shorten A , Shorten b , Kennedy hp. 2014. Complexities of choice after prior cesarean: a narrative analysis. Birth ;41:178–84.doi: 10.1111/birt.12082

[10] Suharjito, D. and D. Darusman. 1998. Kehutanan masyarakat; beragam pola partisipasi masyarakat dalam pengelolaan hutan. Diterbitkan atas kerjasama IPB dan the ford foundation, Bogor.

[11] Sugiyono. 2014. Metode penelitian pendidikan pendekatan kuantitatif,. kualitatif dan R&D. Bandung

[12] Sutopo. 2006. Metodologi penelitian kualitatif: dasar teori dan terapannya dalam penelitian. Universitas sebelas maret.

[13] Wiersum. 1984. Developing strategies for social forestry: a conceptual approach. Honolulu (US): Environment and policy institute east west center.

Effect of Soilless Media with Alternate Wetting-Drying (AWD) as Basic Irrigation on the Growth of Two Varieties of Rice

Damanhuri [1*], T W Widodo[1], I Muhklisin[1]

[1] Department of Agriculture Production, Politeknik Negeri Jember, Indonesia 68124

*email: damanhuri@polije.ac.id

Abstract. The decreasing area of agricultural land and declining water sources are forcing rice farmers to adopt various farming technology. Soil-less farming and alternate wetting-drying (AWD) irrigation can alternatively fix the problems (water substitutes soil as a proper growing media, while in AWD irrigation systems water can be saved), especially in the urban area. This study aimed to observe the growth of two rice varieties grown under soil-less media with AWD irrigation systems. The study was conducted in Sumberjeruk village, Jember city of Indonesia, from June-September 2021. A completely randomized design (CRD) within two factors and three replications was used in the experiment. The first factor was growing media, namely control (soil) + non AWD, water media + rice husk + non AWD, water media + non AWD, and water media + 1 day (once) AWD. The second factor was rice variety specifically Mapan 05 and Ciherang. The result showed that water media + rice husk gave the highest performance at plant height (73,88 cm), the number of the stem (35,25 stems), and the number of leaves (108,92 leaves). Moreover, there was a significant difference in stem diameter number between Mapan 05 and Ciherang (6,95 mm and 5,80 mm respectively).

1. Introduction

Rice is the most popular staple food in Indonesia as well as one of the three main staple foods in the world. Currently, various issues such as low rice productivity, reduced land area, climate change, and drought in rice cultivation threaten the growth and production of rice plants. The area of rice fields is decreasing over time due to land conversion as projected in 2015 there were 8.087 million hectares (a decrease of 0.3% compared to the previous year) [1].

The decreasing area of agricultural land, expensive irrigation costs, and difficulties in obtaining capital for agricultural businesses [2] ultimately force farmers to apply various alternatives and innovations in rice cultivation, and the application of subsistence farming systems is an example. In the subsistence farming system, there are technologies such as soil-less farming and AWD irrigation systems that can be a way to overcome the problems above. In soil-less farming, water can replace the role of soil as a growth medium, while in AWD irrigation systems water use can be saved. The combination of soil-less farming (water media) and AWD can be a promising method for rice cultivation, especially in urban areas.

Soil-less farming is a method to grow a crop without soil as root growth media [3]. It utilizes artificial media that play a role in providing nutrients, water, and growing space [4]. The soil-less farming system and the AWD irrigation system will provide two advantages: optimizing the function of vacant land in urban areas and providing food for households; not only fruits and vegetables but also staple foods such as rice. Generally, lowland rice requires 2500 liters of water to produce 1 kg of rice seeds [5] This is an unusually high amount for agricultural irrigation. The availability of water will commonly be more limited in urban areas than in rural areas, therefore, cultivating lowland rice in urban areas will be a tremendous challenge. However, the wet-dry irrigation method (AWD) in paddy fields is claimed to save irrigation water up to 44% [6]. With the presence of soil-less farming and AWD irrigation, rice cultivation in urban areas which are limited in land and water sources will hopefully be possible.

The use of soil-less farming and AWD irrigation systems are used to maximize plant growth and yield through the use of alternative growing media and saving irrigation water. Until now, the use of alternative media, especially water, which is combined with the AWD irrigation system has not been widely informed. Therefore, further research to observe AWD irrigation and soil-less farming specifically on rice cultivation is needed.

2. Material and Methods

2.1. Material

The research material consisted of two rice varieties namely Mapan 05 and Ciherang varieties grown on several planting media (soil, topsoil, husks, and water) in a special planting tub with dimensions of 60 cm x 40 cm x 20 cm which had a drainage channel using a tap installed on one side of the planting tub to adjust drying-wetting irrigation. Various sources of nutrients were used (AB Mix, urea fertilizer, and NPK fertilizer) to ensure the nutrient needs of rice plants were fulfilled.

2.2. Methods

This research was conducted in Sumberjeruk Village, Kalisat District, Jember city of East Java province of Indonesia from June to September 2021. This study used a completely randomized design (CRD) with two factors. The first factor was planting media, consisting of four media types namely control (soil) + non AWD, water media + rice husk + non AWD, water media + non AWD, and water media + AWD once a day (water drained from 5 pm to 7 am daily, conducted for 30 days from 15 to 45 days after planting (DAP)). The second factor was rice varieties consisting of Mapan 05 and Ciherang varieties. Based on the treatment design, there were 8 treatments with three repetitions. Consequently, there were 24 experimental units applied. Observation parameters consisted of plant height, number of tillers, stem diameter, and number of leaves. All data obtained were then processed using analysis of variance (ANOVA) and continued with Duncan's Multiple Range Test (DMRT) if the results showed significant differences.

3. Result and Discussion

The results showed that the types of soil-less planting media and rice varieties were significantly different for all observed variables, but there was no significant difference in the interaction between soil-less media and rice varieties on the observed variables.

3.1. Effect of soil-less planting media

The results of this study showed that soilless planting media treatment significantly affected plant height, number of tillers, and number of leaves, as can be seen in Table 1. Table 1 shows that the highest plant growth was found in water + rice husk + non-AWD media. The addition of rice husks to water media seemed to be able to store nutrients thereby increasing the availability of nutrients for a longer period (slow-released). This is similar to Hayati (2006) who reported that a mixture of rice husk and sand media had a high ability to absorb and store nutrients so that the availability of nutrients for plants increased [7]. Moreover, rice husk has properties that are easy to bind water, lightweight, weather-resistant, and provide potassium nutrients for plants.

Table 1. Average Plant Height, Number of Tillers, and Number of Leaves in Several Types of Soil-less Planting Media

Type of Soil-less Media	Plant Height (cm)	Number of Tillers	Number of Leaves
Water + rice husk + non AWD	73,88 a	35,25 a	108,92 a
Control (soil) + non AWD	68,18 ab	25,67 b	87,58 a
Water + non AWD	63,73 b	16,00 c	49,08 b
Water + AWD once a day	63,11 b	18,17 bc	54,33 b

Notes: Different letters in columns indicate a significant difference between treatments at 5% DMRT

The utilization of water + rice husk media has the potential to be developed in urban farming. Compared to water media, the use of water + rice husk media can trigger plant growth even though the EC (electric conductivity) value is quite high. A high EC value indicates an increased TDS (Total Dissolved Solids) which will hinder the absorption of nutrients and water. Until now, research on the use of husk + water as a growing medium has not been widely reported. However, soilless cultivation systems are currently one of the fastest-growing cultures in urban farming [8]. Rusli *et al.* (2018) stated that rice husk application as growing media worked as a pH buffer in soil-less media (the pH media were steadily maintained at 6.5), therefore, increasing *Cucumis sativus* yield due to increasing nutrients uptake [9]. Frasetya (2019) suggested that the use of rice husk silicate extract may increase rice plant height, EC, and pH in a hydroponic system [10].

3.2. Effect of Rice Variety

The results showed that the types of rice varieties had a significant effect on stem diameter. The average stem diameter in various treatments of rice varieties is shown in Table 2.

Table 2. Average Stem Diameter in Two Rice Varieties

Rice Variety	Stem Diameter (mm)
Mapan 05	6,95 a
Ciherang	5,80 b

Notes: Different letters in columns indicate a significant difference between treatments at 5% DMRT

Table 2 shows that the treatment of the Mapan 05 variety was significantly different from the Ciherang variety. The genetic differences between the two varieties resulted in different nutrient absorption abilities that affected plant growth, especially stem diameter. Thus, the stem diameter of each rice variety will vary. This is following Purwansyah *et al.* (2021) informed that both environmental and genetic factors significantly affect the length and diameter of rice stems [11]. Research that has been done by Djaman *et al.* (2018) regarding AWD in rice plants in the Sahel area shows a synergistic relationship between rice genotypes and nitrogen levels [12]. They also stated that AWD affects soil biochemical and physical processes (such as nitrification, denitrification, mineralization, percolation, and leaching) due to water and air equilibrium changes that contribute to rice growth and development. Differences in rice genotypes will affect differences in Nitrogen Use Efficiency (NUE), which is then seen from rice growth and yields. Yang *et al.* (2017) stated that AWD irrigation is an effective irrigation technique to increase rice yields by reducing redundant rice growth, increasing root systems, and improving canopy structure due to increased NUE [13]. The diameter of the stem is closely related to the process of cell division and enlargement. Cell enlargement is supported by the large turgidity of cells. Lack of water will reduce cell turgor, thereby suppressing plant growth and development and cell wall synthesis [14].

4. Conclusion

a. Water + rice husk media gave the best results on plant height, the number of tillers, and the number of leaves.
b. There was a significant difference in stem diameter between Mapan 05 variety and Ciherang variety.

5. Acknowledgment

The authors would like to acknowledge the support from the Research and Development Agency of State Polytechnic of Jember, East Java, Indonesia in 2021.

References

[1] Badan Pusat Statistik. Luas Lahan Sawah Menurut Provinsi Tahun 2020. [serial online]. https://www.bps.go.id/. Diakses 04 April 2020.
[2] Damanhuri, Setyohadi DPS, Utami MMD, Kurnianto MF, Hariono, B 2017 Capital Strategy in Diversification Farming Efforts Using SWOT Analysis *Journal of Physics. Conf. Series* 953.
[3] Savvas D, Gianquinto G, Tuzel Y, and Gruda N 2013 Soilless culture Good agricultural practices for greenhouse vegetable crops, principles for mediterranean climate areas FAO *Plant Production and Protection Paper* 303–354.
[4] Samsuddin MH, Saud HM, Ismail MR, Omar MH, Habib SH, Bhuiyan MSH, and Kausar H 2014 Effect of different combinations of coconut coir dust and compost on rice grown under soilless culture *Journal of Food, Agriculture & Environment* **12** 1280-1283.
[5] Bouman B 2009 How much water does rice use *Management* **69** 115-133.
[6] Rejesus RM, Pallis FG, Rodriguez DGP, Lampayan RM, and Bouman BAM 2011 Impact of alternate wetting and drying (AWD) water-saving irrigation technique: evidence from rice producers in the Philippines *Food Policy* **36** 280-288.
[7] Hayati M 2006 Penggunaan Sekam Padi Sebagai Media Alternatif dan Pengujian Efektifitas Penggunaan Media Pupuk Daun Terhadap Pertumbuhan dan Hasil Tanaman Tomat secara Hidroponik *J. Floratek* **2** 63-68.
[8] Balliu A, Youbin Z, Glenda S, Juan AF, Nazim SG, Yuksel T 2021 Environmental and Cultivation Factors Affect the Morphology, Architecture and Performance of Root Systems in Soilless Grown Plants *Horticulturae* **7** 1-25.
[9] Rusli SH, Salehuddin SMF, Affandi NFL, Man SHC, Mohamad Z, Yunus A, Suhaini AM, Baharulrazi N 2018 Effect of Indigenous Microorganisms (IMO) and Rice Husk on pH of Soilless Media and Yield of Cucumis sativus *Chemical Engineering Transactions* **63** 367-372.
[10] Frasetya B, Subandi M, Ginandjar S, Setiati Y 2019 The Effect of Rice Husk Silicate Extract on Plant Height, Electrical Conductivity and pH of Paddy Hydroponics *IOP Conf. Series: Earth and Environmental Science* **334** 1-6.
[11] Purwansyah TS, Rosanti D, dan Kartika T 2021 Morfometri Beberapa Varietas Tanaman Padi (*Oryza sativa* L.) di Kecamatan Pulau Rimau Banyuasin *Jurnal Indobiosains* **3** 28-38.
[12] Djaman K, Mel VC , Diop L, Sow A , El-Namaky R, Manneh B, Saito K, Futakuchi K, Irmak S 2018 Effects of Alternate Wetting and Drying Irrigation Regime and Nitrogen Fertilizer on Yield and Nitrogen Use Efficiency of Irrigated Rice in the Sahel *Water* **10** 1-20.
[13] Yang J, Zhoua Q, Zhang J 2017 Moderate wetting and drying increases rice yield and reduces water use, grain arsenic level, and methane emission *The Crop Journal* **5** 151-158.
[14] Solichatun, Anggarwulan E, Mudyantini W 2005 Pengaruh ketersediaan air terhadap pertumbuhan dan kandungan bahan aktif saponin tanaman ginseng Jawa (*Talinum paniculatum* Gaertn.) *Biofarmasi* **3** 47-51.

Design of the Needs Model for the Development of Young Generation Interests in the Agricultural Sector in Banyuwangi Regency

Muksin [1*], I Juanda [1], N D Wahyono [1], E Eliyatiningsih [2], I Harlianingtyas [2], D Purwoko [3]

[1] Agribusiness Departement, Politeknik Negeri Jember, Indonesia
[2] Department of Agriculture Production, Politeknik Negeri Jember, Indonesia
[3] The Indonesian Institute of Science (LIPI)

[*] muksin@polije.ac.id

Abstract. The agricultural sector is the leading sector of the Banyuwangi Regency. However, the contribution of the farming sector to the Gross Regional Domestic Product (GRDP) of Banyuwangi Regency continues to decline by up to 30% in 2019. The decline in the agricultural sector's contribution to the GRDP of Banyuwangi Regency is partly due to the declining interest of the younger generation in the farming industry so that the regeneration of agricultural, human resources continues very slow. This study aims to analyze the strategy of developing the younger generation's interest in the Banyuwangi Regency. The analytical method used is Interpretative Structural Modeling (ISM) to produce designs according to conditions. The results of the analysis show that the agricultural preference of the younger generation is digital-based agricultural cultivation. The strategy that needs to be developed is local government policy. Local Government policy is a critical factor in developing young people's interest in the agricultural sector in Banyuwangi Regency.

1. Introduction

The Indonesian agricultural sector is the sector that ranks third as the sector with the most considerable contribution to the increase in the National Gross Domestic Product (GDP) with a contribution value of 12.7% [1]. Thus, this condition causes the agricultural sector to become a superior sector in several regions in Indonesia, one of which is in Banyuwangi Regency.

The agricultural sector was able to rank first in the increase in the Gross Regional Domestic Product (GRDP) of Banyuwangi Regency until 2019, with a contribution rate of 30%. However, this figure is a decrease from the previous year. The contribution of the agricultural sector decreased by 1.84% from 2018 [2].

Youth's interest in the agricultural sector is one of the problems that need special attention [3,4]. This condition can lead to non-optimal regeneration of farmers, especially young farmers, to the threat of agricultural sustainability due to the absence of rural youth who will continue the agricultural sector. According to BPS data (2020), in 2019, the number of workers in the agricultural sector fell from 33% to 29% in the last five years [1]

Unfavorable demographic structures include older farmers (over 55 years old) whose number is increasing while the younger workforce is decreasing [5]. The declining interest of youth to work in the agricultural sector is one of the critical problems being faced in developing the agricultural sector in

Indonesia, causing the regeneration of agricultural human resources not to work well [6]. The research revealed that the lack of youth participation in agricultural activities was caused by a lack of technical knowledge and experience in agriculture [7]. Whereas the sustainability of the farm requires the role of all parties and integrase between agricultural lines both on-farm and off-farm [8].

Traditional agricultural management, however, affects the lives of rural communities [9]. Traditional agricultural management is also one of the things that makes the younger generation not want to do activities in the field of agriculture; this was conveyed in a webinar held by the Ministry of National Development Planning. According to the National Development Planning Agency (Bappenas), the reduced number of farmers use traditional farming systems and have not switched to modern or digital farming systems, which are currently becoming a trend and potential for the young millennial generation.

The model of developing the interest of young farmers needs to be done to realize the regeneration of agricultural actors and maximize the role of youth in supporting the farm sector, especially in the Banyuwangi Regency. The analysis of the factors needed to develop the interest of the younger generation in the agricultural industry is carried out by identifying all the aspects required and selecting the factors that correlate with the needs of the youth interest development program in the farm sector Banyuwangi Regency. This analysis is carried out by comparing the opinions of experts (experts) using the geometric mean (geomean). After obtaining the factors that correlate with the program's needs, then analysis using Interpretative Structural Modeling (ISM) is carried out to structure these factors to find out which factors are most needed and have high power drivers, so their existence needs to be maximized.

2. Method

This research was conducted in Banyuwangi Regency for approximately 3 (three) months, starting from May 2021 to July 2021. The data used in this study were primary and secondary. The preliminary data in this study came from interviews with experts, field observations, and distributing questionnaires to experts. The secondary data in this study were sourced from the literature taken from various sources.

Respondents in the study are experts (experts) who can assess development factors. The respondents in this study were the Head of the Banyuwangi Regency Agriculture Office as the highest policyholder in the agricultural sector in Banyuwangi Regency, the Head of the Food Sector and other Heads of Division in charge of developing the interest of the younger generation towards agriculture as one of its activities, agricultural extension workers, agricultural observers, academics who have the capacity and capability in the farming sector and the development of the farm sector in Banyuwangi Regency.

Processing and analysis of research data were carried out using Interpretative Structural Modeling (ISM) analysis. The ISM method is used to formulate alternative strategies generated from the highest preferences for the younger generation. Analysis with ISM using DSS V.1 PRE-NET (Policy Research Expert Network) software (PRE-NET, 2010) to develop alternative decision models for improvement and design a structural model of the factors needed in the interest development program young people in Indonesia. Banyuwangi Regency.

Before carrying out the analysis using ISM, a multiparticipant comparison assessment needs to be carried out to determine the sub-elements that correlate with program needs. There are 18 sub-elements of program needs that have previously been determined based on interviews with experts (experts).

Assessment of the sub-elements: Interpretative Structural Modeling was carried out using a Likert scale with a range of 1 – 4, with the information: 1) Very unimportant (score = 1). 2) Not important (score = 2). Important (score = 3). Very Important (score = 4). The mean value taken from the calculation of the geo-mean is 2.75 out of a scale of 4. This rating is substantial enough to maintain the criteria below. Consideration of the median and mode values of 3 or 4 validates the importance of the selected criteria. Sub-elements with a value of 2.75 from a scale of 1-4 or above can be categorized as sub-elements considered important [10].

3. Result and Discussion

Multiparticipant Comparative Assessment was conducted to determine the participation value of the participants involved in assessing the sub-elements of each ISM element. The geomean analysis results show two invalid sub-elements, namely political and monetary stability and the participation of regional operators. Invalid sub-elements are sub-elements with geometric average values below 2.75. Invalid sub-elements have a low relevance and level of influence on the needs of such development programs. Thus, the analysis of the needs of the younger generation development program to the agricultural sector in Banyuwangi Regency does not use both sub-elements.

The 16 valid sub-elements, namely elements that have a geometric mean value above 2.75, are identified and used in the ISM analysis, namely Human Resources for Young Generation Farmers (4.00), Infrastructure (3.18), Agricultural Production Facilities (2 .93), Cultivation Technology (3.78), Technology Industry (3.78), Capital (3.37), Business Management (3.10), Financial Management (2.93), Counseling (3.78), Market Information System (3.37), Agricultural Information System (2.93), Local Government Policies (3.78), Industries Engaged in the Agricultural Sector (3.18), Assessment and Technology Transfer (3.57), Institutional Support (3.10), and Availability of Agricultural Data and Information (3.78). Table 1. shows the valid sub-elements used in ISM analysis.

Table 1. Identification of the Sub Elements of the Needs of the Program

Sub Element Code	Sub Element
E1	Human Resources of Young Generation Farmers
E2	Infrastructure
E3	Agricultural Production Facilities
E4	Cultivation Technology
E5	Industrial Technology
E6	Capital
E7	Business management
E8	Financial management
E9	Agricultural Extension
E10	Market Information System
E11	Agricultural Information System
E12	Local Government Policy
E13	Industries Engaged in the Agricultural Sector
E14	Assessment and Technology Transfer
E15	Institutional support
E16	Availability of Agricultural Data and Information

Based on identifying sub-elements, formulate contextual relationships between the elements set using the Structural Self Interaction Matrix (SSIM). The next stage is to convert the SSIM into a Reachability Matrix (RM). RM is a binary matrix. The purpose of converting SSIM into a binary matrix is to know which sub-elements have the highest and lowest driving power. In addition to knowing the sub-elements that have the highest and lowest dependence levels. Table 2. shows the Structural Self Interaction Matrix (SSIM).

The sub-element with the highest driving power (driver power) is the Local Government Policy (E12), with a driver power value of 16. The sub-element with the lowest driving power is Financial

Management, with a power driver value of 1. The sub-elements that have the highest level of dependence are Agricultural Production Facilities (E3), Business Management (E7), and Financial Management (E8). The sub-elements that have the lowest level of dependence are Local Government Policy (E12). Further formulating the Matrix of Impact Cross Multiplication Applied to Classification (MICMAC) to analyze a variable's driving power and dependency. Figure 1 describes the MICMAC Analysis.

Table 2. Matrix of SSIM

	E1	E2	E3	E4	E5	E6	E7	E8	E9	E10	E11	E12	E13	E14	E15	E16
E1		V	V	V	V	O	V	V	V	X	X	A	X	X	X	X
E2			V	V	V	A	V	O	A	A	A	A	A	A	A	A
E3				X	X	A	V	V	A	O	A	A	A	A	A	O
E4					X	A	V	V	A	O	A	A	A	A	O	O
E5						A	V	V	A	A	A	A	A	A	A	O
E6							V	V	O	O	O	A	A	A	O	O
E7								V	O	A	A	A	A	A	O	O
E8									O	A	A	A	A	A	O	O
E9										X	X	A	X	X	O	O
E10											X	A	X	A	O	O
E11												A	V	X	X	X
E12													V	V	V	V
E13														A	O	O
E14															X	X
E15																O
E16																

The results of MICMAC analysis showed that there are no sub-elements in quadrant I or quadrant of autonomous factors. This condition indicates that no element has weak driving power and dependence. Thus all parts have a relationship and can affect the development of the younger generation towards the agricultural sector in Banyuwangi Regency.

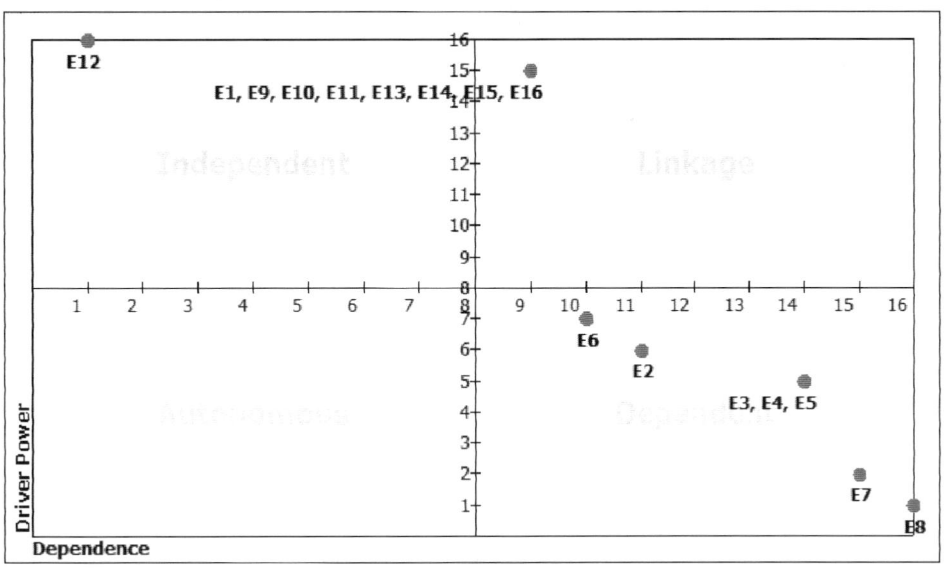

Figure 1. Analysis of MICMAC

There is a sub-element that is in an independent quadrant. This quadrant is a factor with high driving power and low dependence. Local Government Policy (E12) is in the position of a separate quadrant. This condition means that government policy is an element that has high driving power and weak dependency. Thus, government policy is a critical factor in developing the younger generation's interest in the agricultural sector in Banyuwangi Regency

There are eight sub-elements in the linkage quadrant. These quadrants are factors that are interconnected and influence each other. The eight sub-elements are the human resources of young generation farmers, agricultural extension, market information systems, agricultural information systems, industries engaged in the farming sector, assessment and transfer of technology, institutional support, and availability of farm data and information.

Seven sub-elements are in the dependent quadrant. Dependent quadrants are factors with low driving power and high dependence. Other factors very easily influence the elements in this quadrant. The factors in this quadrant are Infrastructure, Agricultural Production Facilities, Cultivation Technology, Industrial Technology, Financier, Business Management, Financial Management.

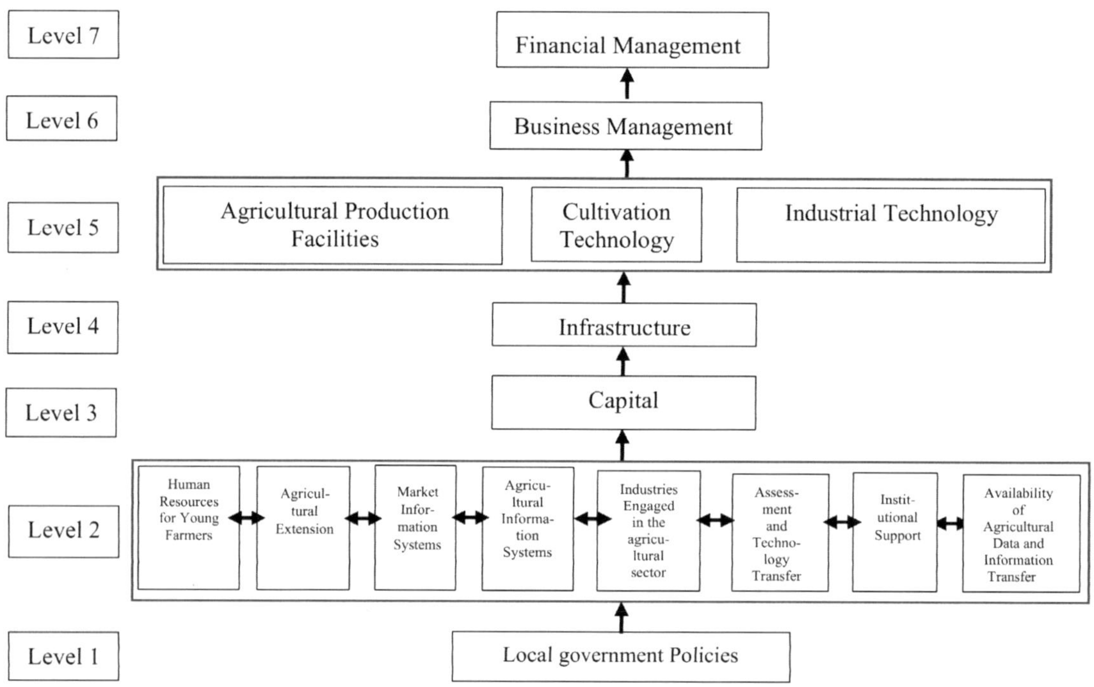

Figure 2. Digraph Program Development Objectives

Based on the analysis results, it can be seen that there are seven levels in the digraph Interpretative Structural Modeling (ISM). This level shows the strategic priority structure so that through this structure, it can be seen what elements need to be considered first to be able to influence other aspects.

At level 1 of the graph, there is one need from the program, namely business management. These needs include planning, organizing, leading, and controlling the financial cycles in an agricultural enterprise. Business management is important and influences business [11]. Conditions at this level can affect other requirements above them but can also be influenced by the needs in the level below. Thus, the fulfillment of the markets at level 1 can be met when the needs at level 7, level 6, level 5, level 4, level 3, and level 2 have been met.

At level 2 digraph, there is one need from the program: business management; this need includes planning, organizing, leading, and controlling a business in the agricultural sector. Needs at this level can affect other conditions above it but can also be influenced by requirements at the level below it. Thus, the fulfillment of the markets at level 2 can be met when the needs at level 7, level 6, level 5, level 4, and level 3 have been met.

At level 3 digraph, there are three needs of the program interconnected and mutually influence one another, namely agricultural production facilities, cultivation technology, and industrial technology. Conditions at this level can affect other needs that are above it but can also be influenced by those at the level below it. The fulfillment of the requirements at level 3 can be met when the needs at level 7, level 6, and level 5 and 4 have been met.

At level 4 digraph, there is one need from the program, namely infrastructure. This need includes the development of adequate infrastructure in the agricultural sector, such as easy road access to agricultural land, adequate lighting, good irrigation networks, etc. Needs at this level can affect other conditions that are above it but can also be influenced by requirements at the level below. Thus, the fulfillment of the needs at level 5 can be met when the needs at level 7, level 6, and level 5 have been met.

At level 5 digraph, there is one need from the program, namely capital. This need includes sources of capital for businesses in the agricultural sector. Conditions at this level can affect other elements of conditions above it but can also be influenced by aspects of needs at the level below it. Thus, the fulfillment of the requirements at level 5 can be met when the needs at levels 7 and 6 have been met.

At level 6 digraph, there are eight needs from related programs mutually influencing one another: human resources for young farmers, agricultural extension, market information systems, agricultural information systems, industries engaged in the farming sector, assessment and development technology transfer, institutional support, availability of farm data and information. A clear program is needed to influence the perception of young groups [12]. These needs can affect the conditions at the level above. However, this need can be influenced by local government policies. Thus, the fulfillment of the requirements at level 6 can be met when the needs at level 7 have been met.

At level 7 digraph, there is one need for an independent program, namely local government policies. This local government policy includes Banyuwangi Regency government policies in responding to the lack of interest of the younger generation in agriculture in Banyuwangi Regency. In this case, proposed by the Department of Agriculture and Food Crops, local government policies related to the agricultural sector must be able to take sides with farmers so that farmers and new agricultural business actors are motivated to continue doing business in agriculture.

4. Conclusion

Government policy is a sub-element of the program's needs that has the highest driving factor (driver power) in the development of young people's interest in the agricultural sector in the Banyuwangi Regency. Efforts to attract the younger generation's interest require a draft of local government policies related to the development of the farm sector. Government policies related to the agricultural industry can include agricultural and water land, agricultural and other integration especially tourism, fertilizer, seed, human resource policies, technology-related policies, post-harvest-related policies, and other policies that support regional farmers. The draft policies based on the appropriate needs sub-element will impact changing the perspective of the younger generation towards the agricultural sector to be more positive, especially in Banyuwangi Regency.

5. Acknowledgments

Acknowledgments for the completion of this research, which is annual research. Special thanks to the Ministry of Research and Technology, Politeknik Negeri Jember, and the Banyuwangi Regency Agriculture and Food Service, as institutions that have supported completing the learning process contained in this research. We also thank the younger generation, farmers, agricultural business actors, and all parties involved in Banyuwangi Regency who have helped smooth research activities. Hopefully, this research can contribute to the development of agribusiness in the Banyuwangi Regency.

References

[1] BPS 2020 Badan Pusat Statistik. PDB Indonesia Triwulanan 2016-2020 98

[2] Banyuwangi B K 2020 *Kabupaten banyuwangi dalam angka 2020* (Banyuwangi: BPS Kabupaten Banyuwangi)

[3] Muksin, Hari R, Tanti K and Titik I 2018 The main problems of na-oogst tobacco agribusiness in Jember *IOP Conf. Ser. Earth Environ. Sci.* **207** 012012

[4] Ng'atigwa A A, Hepelwa A, Yami M and Manyong V 2020 Assessment of Factors Influencing Youth Involvement in Horticulture Agribusiness in Tanzania: A Case Study of Njombe Region *Agriculture* **10** 287

[5] Insani F R, Setiawan I and Rasiska S 2018 Determinan Partisipasi Dan Peran Petani Muda Dalam Pengembangan Pertanian Ramah Lingkungan Di Desa Cisondari, Kecamatan Ciwidey, Kabupaten Bandung, Jawa Barat *Mimb. AGRIBISNIS J. Pemikir. Masy. Ilm. Berwawasan Agribisnis* **4** 153

[6] Detia Tri Yunandar, Sunarru Samsi Hariadi and Alia Bihrajihat Raya 2020 Nilai Keyakinan Diri dan Sikap Pemuda Terdidik terhadap Wirausaha Pertanian *Pros. Semin. Nas. Pembang. dan Pendidik. Vokasi Pertan.* **1** 282–9

[7] Effendy L, Maryani A and Azie A Y 2020 Faktor-Faktor yang Mempengaruhi Minat Pemuda Perdesaan pada Pertanian di Kecamatan Sindangkasih Ciamis *J. Penyul.* **16** 277–88

[8] Muksin, Rizal and Iskandar R 2021 Analysis of the Sustainable Status of Post Disaster Crop Production in Sigi Regency, Central Sulawesi Province *IOP Conf. Ser. Earth Environ. Sci.* **672** 012031

[9] Bjørkhaug H and Knickel K 2018 Rethinking the links between farm modernisation, rural development and resilience *J. Rural Stud.* **59** 194–6

[10] Mohapatra R K, Yavar A, Marg J, East B, Mohanty R P and Dhalla R S 2010 Reengineering of Logistics Value Chain of a Petroleum Products Marketing Company – Formulation of a Performance Measurement System *Int. Conf. Ind. Eng. Oper. Manag.* 190–6

[11] Cahill H and Dadvand B 2018 Re-conceptualising youth participation: A framework to inform action *Child. Youth Serv. Rev.* **95** 243–53

[12] Simões F and Brito do Rio N 2020 How to increase rural NEETs professional involvement in agriculture? The roles of youth representations and vocational training packages improvement *J. Rural Stud.* **75** 9–19

IOP Conf. Series: Earth and Environmental Science **980** (2022) 012056 doi:10.1088/1755-1315/980/1/012056

Critical factors in local food development policies, farming, and coping mechanisms mothers with stunting children in Jember

Muksin [*1], **D A Perwiraningrum**[2], **D I Amareta**[2], **D Purwoko**[3]

[1] Agribusiness Departement, Politeknik Negeri Jember, Indonesia
[2] Health Departement, Politeknik Negeri Jember, Indonesia
[3] The Indonesian Institute of Science (LIPI) Jakarta, Indonesia

[*]Corresponding email: muksin@polije.ac.id

Abstract. Jember Regency has the potential of local food diversity that has not been optimally utilized in stunting recovery. Local institutions have also not fully contributed optimal roles in rural mothers with stunting children. The research aims to determine key policy factors for local food development, farming, and coping mechanisms in mothers with stunting children. Adequate policies, especially in utilizing cheap and easily obtain nutritious local food sources, are expected to facilitate effective stunting recovery efforts. This research method uses quantitative and qualitative approaches. Description of local food sources and their processing, demographics, and maternal behavior, is carried out through focus group discussion activities involving key stakeholders. Furthermore, modeling is carried out based on key factors or elements with interpretative structural modeling methods. The study's findings suggest that increasing maternal knowledge of local food and nutritional content and health aspects is a crucial policy factor. Systematically these key factors need to be supported by strengthening maternal health in decision-making in the household, handling maternal health detection, maternal skills to process food, and maternal skills to choose food according to family needs. Policy implementation requires formal and informal institutional support that exists in the countryside.

1. Introduction

Jember Regency in 2021 became the district with the highest stunting cases in East Java at 37.94% more above the national average [1]. This condition requires handling so that the problem of stunting can be minimized or eliminated. Efforts to overcome these conditions require good cooperation between the mother's family and stunting children, rural institutions, and the government.

Family efforts in the recovery of stunting in Jember are increasingly under pressure during the Covid19 pandemic. This condition is also coupled with the background of poverty from children with stunting families. Poverty causes an influence on the ability to provide food, including getting food choices that are worth taking from the aspects of nutrition and health. Minimal food causes families to share food, including children [2].

Knowledge of personal hygiene, education, and the role of local institutions contribute to the mother's internal empowerment [3]. The quality of health policies and services in village funds that are supposed to prevent stunting overlaps its utilization with Covid19 countermeasures. Posyandu is often constrained by "Kader" (social facilitator/worker in villages) capacity and access to homes [2]. It was assisting efforts that have been implemented through government programs to tend Supplemental Feeding

Recovery from the Health Center. Efforts to utilize local food have not been optimal. At the same time, integrated agricultural management can provide adequate, sustainable, and local food used by residents [4].

Stunting children need recovery efforts by increasing the intake of macro and micronutrients and overall nutrients that are important in their body development. Food diversity is not optimal. Rice processed into rice is still seen as the primary food even though the price is relatively high [5]. Interrelationships between rural and institutional components related to dealing with stunting require an effort to determine the objectives of developing how policies regarding food access and processing, agricultural patterns, and coping mechanisms from mothers to stunting children. The proper goal provides direction so that the activity can be maximally and effectively towards the prepared achievements. Based on these considerations, the study aims to determine key factors in local food development policies, agriculture, and coping mechanisms in mothers with stunting children related to the goals set in program development. Structural goal setting will facilitate the implementation of strategies in activities.

2. Method

Maternal empowerment becomes vital by considering the overall dimensions that affect the role of the mother in stunting recovery. The design and determination of maternal empowerment goals need to be done by identifying and structuring the needs and factors of their empowerment. The research uses quantitative and qualitative approaches. The implementation of a study conducted in the highest stunting sub-districts included: Silo, Jelbuk, Wuluhan, and panti, Jember district. In unstructured interviews, descriptive quantitative methods and survey techniques are used to obtain maternal characteristics and behavior in stunting recovery. Maternal respondents were determined by purposive based on stunting children at each selected location. Based on literature studies and interviews, set elements and sub-elements of goals focused on empowering mothers with stunting children. Evaluation and measurement of comparisons between sub-elements of empowerment using are carried out with experts and key stakeholders using intellectual judgment in this interpretive Structural Modelling (ISM) [6] [7], covering related agencies, health experts, local institutions, village facilitators, and academics. Based on the assessment of experts set the critical factors of maternal empowerment goals with stunting children.

3. Result and discussion

3.1. *Factors of empowerment goals*

The development of policy cannot be separated from analyzing a condition and properly identifying the problems or constraints around the development. Policy development should consider the aspects of the main actors who will be affected and what is faced by the parties or stakeholders to develop certain conditions [8].

In efforts to develop alternative goals for the empowerment of mothers with stunting children, the purpose of the plans is evident in this activity. What main activities can be designed to arrange the actions that need and are highly considered?. Serious efforts to understand key goals and activities in local food development, including stunting, are prioritized.

Based on literature studies, discussions with the public through Focus Group Discussion and interviews with experts and researchers can be outlined several sub-elements that affect the development of maternal and community empowerment for stunting recovery. In the objective element of program development, thirteen (13) sub-elements have a relationship. Furthermore, based on the justification of experts, by obtaining geometric averages obtained sub-elements that have relatedness and that have relevance to the development of mother and community empowerment in stunting recovery. The sub-elements associated with the development of maternal and community empowerment are: Increase maternal knowledge of nutrition and health, Improve the mother's skills to choose food according to family needs, Improve the mother's skills to process food, Increase the active role of mothers in health empowerment activities, Increase maternal empowerment in decision-making in child care, Increase marketing of creations or efforts of mothers, Develop businesses or entrepreneurship that improve the welfare of mothers and families, Improving maternal and family income and well-being, driving rural communities' economies, developing diversification of food products, increasing local food added value, improving detection management in mothers with stunting children, and strengthening institutional health performance in facilitating mother and child.

3.2. Driving Power and Level of Independence between Factors

Based on these interrelated factors are arranged to perform an analysis of selected sub-elements or factors. Table 1. Displays sub-elements that are interconnected and influence each other on the goals of empowerment development programs.

Table 1. Program Objective Selected Elements and Sub-Elements

Element Code	Sub-Elements
E1	Increase the mother's knowledge of nutrition and health.
E2	Improving the mother's skills to choose food according to the needs of the family.
E3	Improve the mother's skills to process food.
E4	Increase the active role of the mother in health empowerment activities.
E5	Increases the mother's help in decision-making in childcare.
E6	Increase the marketing of the creations or efforts of mothers.
E7	Develop a business or entrepreneurship that increases the well-being of mothers and families.
E8	Increase the income and well-being of mothers and families.
E9	Driving the economy of rural communities.
E10	Develop diversification of food products.
E11	Increase national food added value.
E12	Improve detection handling in mothers with stunting children.
E13	Strengthen institutional health performance in facilitating mother and child.

Based on the determination and assessment of experts to prepare contextual relationships by modeling objectives using structural self Interaction matrix (SSIM). Next, the process of converting SSIM to Reachability Matrix (RM). Based on the results of RM data analysis, it can be known that the sub-element code (E1) increases maternal knowledge of nutrition and health has the highest Driver Power value of 8. This shows that increasing maternal knowledge of nutrition and health is the most vital driving factor in stunting recovery efforts by developing coping mechanisms and local food resources during the pandemic in Jember Regency. Increasing maternal knowledge of nutrition and health needs to be prioritized because this element will significantly impact stunting recovery by developing coping mechanisms and local food resources during the pandemic in the Jember regency.

Sub-elements Of Driving the economy of rural communities (E9) and Developing food product diversification (E10) Increase national food added value (E11) has the lowest driver power value of 1. This shows that driving the economy of rural communities, developing diversification of food products, increasing national food added value is the weakest driving factor in stunting recovery efforts by developing coping mechanisms and local food resources during the pandemic in Jember Regency. So that if this element is considered in this case, it does not have a considerable impact in stunting recovery efforts through the development of coping mechanisms and local food resources during the pandemic in Jember Regency.

In addition, when viewed from the dependency factor (dependence), it can be known that driving the economy of rural communities (E9), developing diversification of food products (E10) has the highest value (dependence) of 11, this shows that driving the economy of rural communities and developing diversification of food products is a factor that has a high level of dependency, so that in efforts to restore stunting through the development of mechanism coping and local food resources during the pandemic in Jember Regency. This element is highly dependent on other factors and is easily influenced by other factors.

While increasing the mother's help in decision-making in childcare (E5) has a low dependence of 1. This indicates that increasing maternal empowerment in childcare decision-making is a factor with the lowest dependency value. So that in the efforts of stunting recovery through the development of coping mechanisms and local food resources during the pandemic in Jember Regency, this element is not dependent on other parts and is not easily influenced by other factors.

Furthermore, based on the MICMAC analysis, the Program Objective Selected Elements can be seen in figure 1.

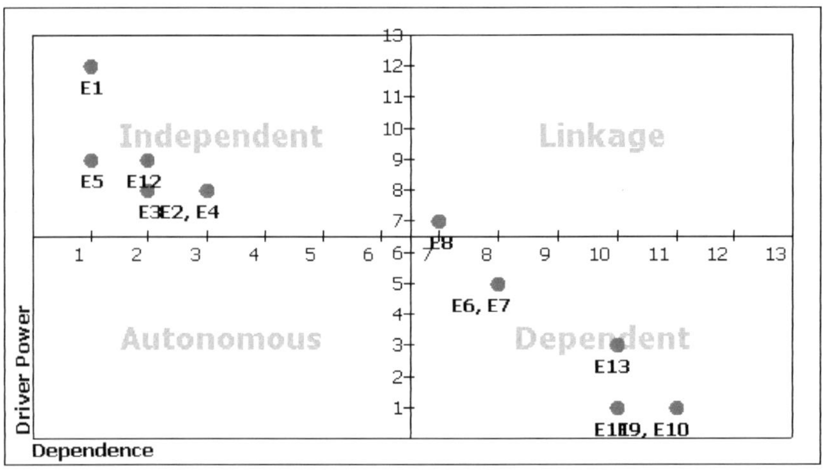

Figure 1. Sub-Element Sector Graph (MICMAC Analysis)

Based on MICMAC (Matrix of Impact Cross Multiplication Applied to Classification) analysis, it is known that none of the sub-elements are in quadrant I (autonomous quadrant). This condition indicates that none of the elements have weak driving power and dependence. All aspects of the analysis have a connection with each other.

In sector II (dependence) 5 elements include: E6, E7, E9, E10, and E11. This indicates that these elements are elements that have weak driving power and strong dependency. One element is among sector III linkage, namely E8 Increases the income and welfare of mothers and families. This variable has strong driving power and dependence, so it should be studied more carefully. In sector IV (Independence), elements include E1, E2, E3, E4, and E5. This indicates that the element has a strong driving power and weak dependence on other elements. So that these elements can become (key players) key elements in stunting recovery efforts through the development of coping mechanisms and local food resources during the pandemic in Jember Regency. Based on MICMAC analysis can be compiled structural model of program objectives as shown in figure 2.

Based on Digraph's analysis in figure 2, there are seven levels of a hierarchy consisting of Sub-Elements Selected Program Goals in stunting recovery efforts. The levels obtained come from the ranking of each element. This level indicates the priority structure of the strategy. The network formed informs the interrelationships between sub-element and their level of influence.

At the first level, one element increases the mother's knowledge of nutrition and health (E1). It is a crucial element that affects aspects of the above class. This indicates that the element must be prioritized first because the element is a priority element of improvement at the level afterward. In this case, increasing maternal knowledge of nutrition and health is an element that must be prioritized in stunting recovery efforts by developing coping mechanisms and local food resources during the pandemic in Jember Regency.

A second level, there are two elements: Increasing maternal empowerment in decision-making in childcare (E5), Improving detection handling in mothers with stunting children (E12). This shows that if the element at level 1 increases maternal knowledge of nutrition and health has been going well, then the aspects at this level will run well in stunting recovery efforts through the development of coping mechanisms and local food resources during the pandemic in Jember Regency.

At the Third level, there are three elements, among others: Improving the Mother's Skills to choose food according to family needs (E2), Improving the mother's skills to process food (E3), and Increasing the active role of mothers in health empowerment activities (E4). The mother's understanding of food sources also requires facilitation to be changed, not just rice. Some corn commodities are essential [9] but often not utilized.

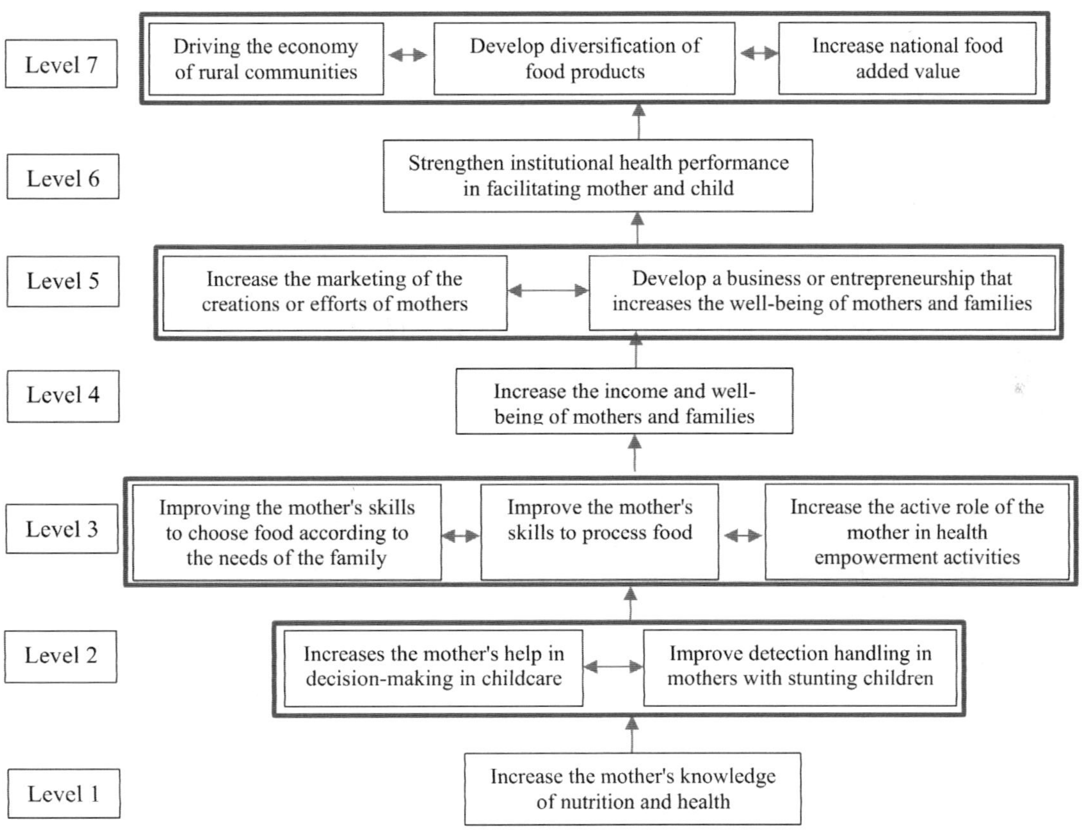

Figure 2. Digraph Program Development Objectives

Further, At the fourth level, there is one element of the program's goal of increasing the income and well-being of mothers and families (E8). In the fifth level, there are two elements: expanding the marketing of the creation or efforts of mothers (E6) and developing businesses or entrepreneurship that increase the welfare of mothers and families (E7). Structurally Sixth level, there is one element that strengthens the institutional performance of health in facilitating mother and child (E13). At the seventh level, there are three elements, namely Driving the economy of rural communities (E9), developing diversification of food products (E10), and increasing national food added value (E11).

Thus, increasing the mother's knowledge of nutrition and health is a crucial factor influencing other factors or elements in developing maternal empowerment programs with stunting children. Knowledge-related reinforcement requires counseling based on the proper scope of facts [10] and is based on the issue of maternal understanding. The following key factor is increasing maternal empowerment in decision-making in childcare and improving detection handling in mothers with stunting children. The three elements are crucial in the development of programs that boil down to the mother's empowerment in terms of individual behavior, parenting and family support skills, and facilitation of space for mothers to synergize to participate in the decision-making process in terms of childcare and family food development and management.

4. Conclusion

This study can conclude three things; namely, increasing maternal knowledge of nutrition and health is a crucial factor that should be the primary goal in maternal empowerment. Furthermore, increased knowledge must be supported by increasing maternal empowerment in decision-making in childcare and improving the handling of detection in mothers with stunting children. Motherhood represents the ability of individuals with knowledge, attitudes, and skills towards the management of nutrition and health, as well as their ability to make decisions in domestic and public spaces.

5. Acknowledgments

Politeknik Negeri Jember funds this research, Ministry of Education and Culture of Indonesia. The author thanked Jember State Polytechnic for this support. Our thanks go to the Director of Jember State Polytechnic, Health Office, Community and Village Empowerment Office, Jember District Agriculture Office, Group of Mothers with Stunting Children, Health Center officers, village organizers, facilitators, and other stakeholders who have contributed to research activities.

References

[1] Albab M U 2021 Angka Stunting, Kematian Ibu dan Anak di Jember Tertinggi di Jatim *IDN Times*

[2] Perwiraningrum D A, Elisanti A D, Amareta D I and Werdiharini A E 2021 Need Assessment of Stunted Children During Pandemic Covid-19 to Develop Nutrition Intervention Program in Jember District, East Java, Indonesia *Proceedings of the First International Conference on Social Science, Humanity, and Public Health (ICOSHIP 2020)* (Paris, France: Atlantis Press)

[3] Yappika Action Aid 2017 *Annual Report of YAPPIKA-ActionAid 2017, Making Change a Reality* (Jakarta)

[4] Muksin, Rizal and Iskandar R 2021 Analysis of the Sustainable Status of Post Disaster Crop Production in Sigi Regency, Central Sulawesi Province *IOP Conf. Ser. Earth Environ. Sci.* **672** 012031

[5] Saliem H P, Susilowati S H, Ariningsih E, Agustian A and Muksin 2021 Supporting organic rice exports: the success story of West Java organic rice exports *IOP Conf. Ser. Earth Environ. Sci.* **672** 012095

[6] A Jayant, Mohd Azhar and Priya Singh 2014 Interpretive Structural Modeling (ISM) Approach: A State of the Art Literature Review *Int. J. Res. Mech. Eng. Technol.* **5** 15–21

[7] Firmansyah I 2018 Decision Making System, For The Priority of Strategy Planning, Constraints and Policies

[8] Muksin, Hari R, Tanti K and Titik I 2018 The main problems of na-oogst tobacco agribusiness in Jember *IOP Conf. Ser. Earth Environ. Sci.* **207** 012012

[9] Susilowati S H, Ariningsih E, Saliem H P, Roosganda E, Adawiyah C R and Muksin 2021 Opportunities and challenges to increase corn export from Gorontalo province of Indonesia *IOP Conf. Ser. Earth Environ. Sci.* **672** 012027

[10] Maman U, Murodi M, Nindyantoro N, Mujahidin E, Zaenuddin D, Muhyani M and Muksin M 2021 Formulating Extension Model to Encourage Staple Food Diversification: A Solution to Food Crisis Potential *Univers. J. Agric. Res.* **9** 138–48

IOP Conf. Series: Earth and Environmental Science **980** (2022) 012057 doi:10.1088/1755-1315/980/1/012057

Modeling the Potential of Cereal Crops with a Smart Village-Based GIS Approach to Support Food Security

Denny Trias Utomo[1*] , Bety Etikasari[1] , Jumiatun[2]

[1]Information Technology Department, Politeknik Negeri Jember, Indonesia
[2] Crop Production Technology Department, Politeknik Negeri Jember, Indonesia

*denny_trias@polije.ac.id

Abstract. Since 2015, government projects related to national food security have been implemented so far. This article will write a study that uses the GIS method to model the potential of food crops. The GIS web application development method uses the prototyping method with input in the form of spatial and non-spatial data which includes data on potential rice, corn, and soybeans in the Jember district, where the data was obtained from a survey at the Jember Regency Agriculture Service in the period January - August 2021 and data statistics from the book Jember in Figures 2020 and the official website of the Jember Regency BPS.

Keywords: *Geographic Information System, smart village, Food security*

1. Introduction

The development of main cereal crops needs namely rice, corn, and soybeans (pajale) for high-level domestic products has been the government's main goal since 2015. This is an effort to achieve national food security. This achievement is highly dependent on the results of agricultural production in Indonesia. One of the factors that affect agricultural production is climate change. Research shows that one of the impacts of climate change is an increasing temperature which causes a decrease in rice production from 68.1% to 92% [1]. The impact of climate change on agriculture is important, so strategies and technologies need to be applied to maintain food availability in Indonesia.

Agricultural system solutions in Indonesia must be based on data accuracy for developing information systems in this 4.0 technology era. Research shows that new technology system uses a spatial database and use artificial intelligence (AI) to generate food insecurity information [2]. From developing AI perspective based on the Territorial Information System (GIS) in obtaining data, UAV (unmanned aerial vehicle) technology can be used for collecting data spatial in the wider area [3].

Applied research internet of things has also been carried out in smart agriculture to get real-time data on agricultural land conditions [4]. The purpose of this research is the implementation of AI and internet of things technology for agriculture which is made in the smart village concept. The Smart village program was built by integrating GIS spatial data, a database of cereal crops potential by utilizing AI on a Web-based platform. The smart village concept refers to that villages can solve their problems intelligently.

2. Literature Review

2.1. Modeling the Potential of Cereal Crops

The development of superior cereal crops commodities (rice, corn, and soybeans) based on production areas can be carried out using a GIS system. Mapping of cereal crop production areas is carried out using the Hot Spot Getis-Ord Gi* analysis method to recommend the potential of superior cereal crops in a specific area [5]. The modeling system for cereal crop production potential is influenced by several factors. So proper data and analysis are required during implementation. After doing an analysis of potential soil fertility and climate in a specific area, so that area can develop a cereal cropping system itself. Plant growth and production factors become the focus of studies to develop modeling systems. The results showed that the significant factor effect was harvested area and rainfall [6].

2.2. Smart Village-Based GIS Approach to Support Food Security

The smart village concept in this study is the application of modern agricultural technology in the specific area so the conventional agricultural systems and have less measurable results can be optimized using several technological approaches such as Geographic Information System (GIS), Decision Support System (DSS)[7]–[9].

GIS is a computer-based system that is used to store and manipulate geographic information. Geographic information system aims to collect, store and analyze objects or phenomena where geographic location is an important characteristic for analysis. GIS illustrated into a map form. It becomes an effective presentation tool and geographical data storage. The stored information will be processed and presented by changing the presentation form, a map always provides images or symbols of geographic elements with a fixed or static shape even though it is for different needed [5], [7], [8], [10].

This research is the first step for the studies series development that integrates GIS with DSS. The integration of data input into the GIS system will produce useful information that will be used for the next research to develop DSS where one system that applied to the Smart Village concept is collecting data in real-time so that the information provided to farmers is accurate.

3. Methods

These research stages use the prototype method for the software development stage. The stages include system requirements analysis, system design, system design evaluation, system coding, and alpha version system testing by the team. The prototype method is shown in Figure 1

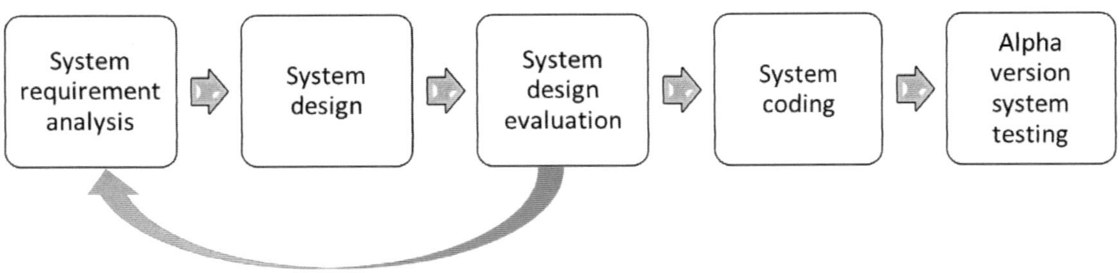

Figure 1. Prototype Method for System Development

The reason for using the prototype method is because the designing system of GIS, DSS, and aerial images for the smart village concept is complex. This smart village concept needs an evaluation stage during the coding stage for the system to become easier [11].

3.1. System requirement analysis

The system requirement analysis stage is done by using observation, interview, and documentation techniques. Observations and interviews were conducted with several farmers in Jember Regency. Observation and interview technique purpose to comparative data from the Badan Pusat Statistik (BPS) documentation to complete the data needed. The secondary data documentation technique from BPS was taken from the Jember Regency Book in 2020.

3.2. System design

The system design stage is done by using an image processing application like diagram.net and adobe illustration. Prototyping is made according to the business process of the GIS concept the smart village that will be developed. The flowchart of the GIS application is shown in Figure 2.

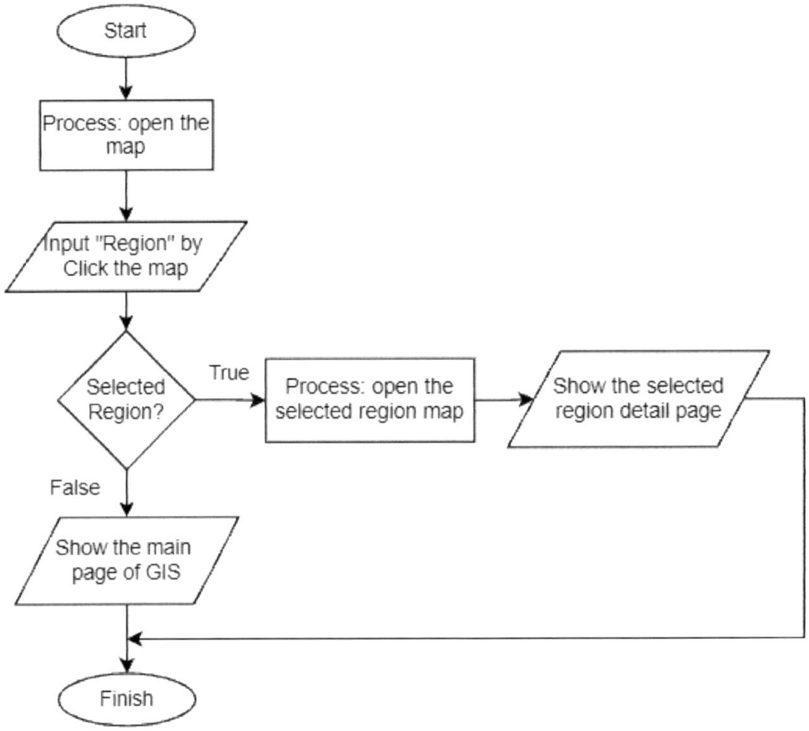

Figure 2. Flowchart of Web GIS

3.3. System design evaluation

The system design evaluation stage purpose to test the results of the prototype that has been developed. This is to determine whether there is still a need for a redesign before the system coding stage is done.

3.4. System coding

The system coding stage of the GIS system is done using PHP programming with Laravel framework, MySQL database, composer, and Github. The coding of then is system is done according to the prototype that has been developed in the system design stage.

3.5. Alpha-version system testing

The alpha version system testing stage is done using the User Acceptance Test (UAT) method. Testing is done on users who have contributed to agriculture. The form of the instrument used is a questionnaire with 5 answer choices on the Likert scale.

4. Result and Discussion

This section discusses applications development according to the research stage described in the Methods section above and also discusses the verification testing of an application program. The web-based GIS application prototype that has been developed can be accessed on the webpage https://www.smart-agrotech.com/ page.

IOP Conf. Series: Earth and Environmental Science **980** (2022) 012057 doi:10.1088/1755-1315/980/1/012057

4.1. Application development

The application development stage had two stages. The first stages are making a logical database design and physical database then the second stage is making a complete application. The logical database is shown in Figure 3 and the physical database is shown in Figures 4 to 13.

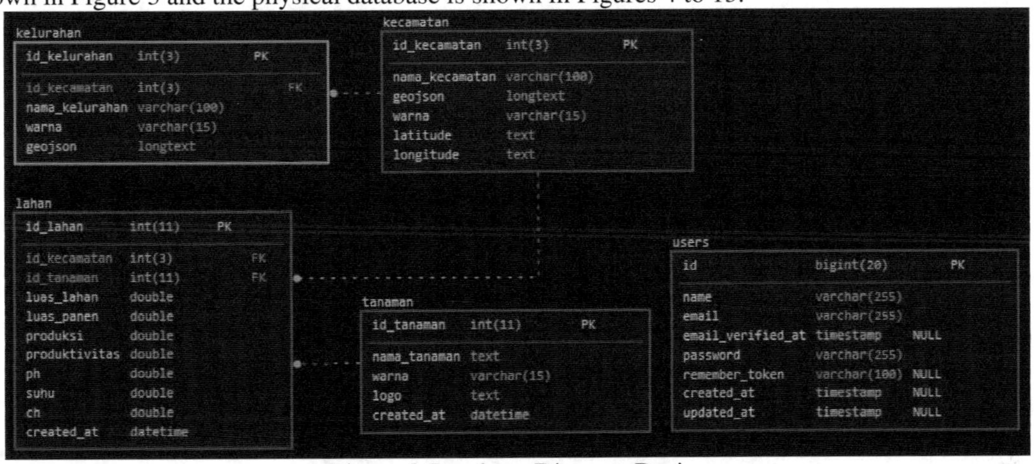

Figure 3 Database Diagram Design

Figure 3 shows the logical database is consists of 4 main tables that are interconnected with each other, namely the "lahan" table, "tanaman" table, "kecamatan" table, and "kelurahan" table. The "tanaman" table and the "kecamatan" table have a relation to "lahan" table which linked by primary key and foreign key, namely id_tanaman and id_kecamatan. The "kecamatan" table has a relation to "kecamatan" table which linked by the primary key and foreign key id_kecamatan. In addition, there is one table that is used to store user data, namely the "users" table. This logical database was made from the database design stage and fulfill the normalization principle.

	#	Name	Type	Collation	Attributes	Null	Default	Comments	Extra	Action
☐	1	**id_kecamatan**	int(2)			No	*None*		AUTO_INCREMENT	Change Drop More
☐	2	**nama_kecamatan**	varchar(100)	utf8mb4_general_ci		No	*None*			Change Drop More
☐	3	**geojson**	longtext	utf8mb4_bin		No				Change Drop More
☐	4	**warna**	varchar(15)	utf8mb4_general_ci		No	*None*			Change Drop More
☐	5	**latitude**	text	utf8mb4_general_ci		No				Change Drop More
☐	6	**longitude**	text	utf8mb4_general_ci		No				Change Drop More

Figure 4. Table structured in "kecamatan" table

	id_kecamatan	nama_kecamatan	geojson	warna	latitude	longitude
Edit Copy Delete	1	Kecamatan Ajung	{"type":"Polygon","coordinates":[[[113.68484462797...	#ffffff	-8.222823118407701	113.64428354002413
Edit Copy Delete	2	Kecamatan Ambulu	{"type":"MultiPolygon","coordinates":[[[[113.58284...	#ffffff	-8.345758650256132	113.60599757220882
Edit Copy Delete	3	Kecamatan Arjasa	{"type":"Polygon","coordinates":[[[113.77298165897...	#ffffff	-8.111865195115351	113.73682662631333
Edit Copy Delete	4	Kecamatan Balung	{"type":"Polygon","coordinates":[[[113.56882607386...	#ffffff	-8.27489299490978	113.53416051601032
Edit Copy Delete	5	Kecamatan Bangsalsari	{"type":"Polygon","coordinates":[[[113.60432420090...	#f4a100	-8.17754309803353	113.53837712042558
Edit Copy Delete	6	Kecamatan Gumukmas	{"type":"Polygon","coordinates":[[[113.45697000036...	#ffffff	-8.297230818583614	113.41630748651455
Edit Copy Delete	7	Kecamatan Jelbuk	{"type":"Polygon","coordinates":[[[113.78410960377...	#ffffff	-8.088837149671866	113.74944411466343
Edit Copy Delete	8	Kecamatan Jenggawah	{"type":"Polygon","coordinates":[[[113.69814999962...	#ffffff	-8.25841485294082	113.67433161705121
Edit Copy Delete	9	Kecamatan Jombang	{"type":"Polygon","coordinates":[[[113.39992999957...	#ffffff	-8.2445692842986 1	113.36541327037854
Edit Copy Delete	10	Kecamatan Kalisat	{"type":"Polygon","coordinates":[[[113.83298999988...	#ffffff	-8.12707487834376	113.80927658419
Edit Copy Delete	11	Kecamatan Kaliwates	{"type":"Polygon","coordinates":[[[113.70537982132...	#00ac69	-8.179242253752724	113.66609441363838
Edit Copy Delete	12	Kecamatan Kencong	{"type":"Polygon","coordinates":[[[113.38194999959...	#ffffff	-8.262491983089246	113.38058448631998
Edit Copy Delete	13	Kecamatan Ledokombo	{"type":"Polygon","coordinates":[[[113.85873834369...	#ffffff	-8.140839569527067	113.87605003732726
Edit Copy Delete	14	Kecamatan Mayang	{"type":"Polygon","coordinates":[[[113.85012142476...	#ffffff	-8.188842347461605	113.80026357111404
Edit Copy Delete	15	Kecamatan Mumbulsari	{"type":"Polygon","coordinates":[[[113.78578999960...	#ffffff	-8.285255184891327	113.75321814660737
Edit Copy Delete	16	Kecamatan Pakusari	{"type":"Polygon","coordinates":[[[113.77142679501...	#ffffff	-8.179836956543436	113.76249208254312

Figure 5. Data in "kecamatan" table

#	Name	Type	Collation	Attributes	Null	Default	Comments	Extra	Action
1	**id_kelurahan**	int(3)			No	*None*		AUTO_INCREMENT	Change Drop More
2	**id_kecamatan**	int(3)			No	*None*			Change Drop More
3	**nama_kelurahan**	varchar(100)	utf8mb4_general_ci		No	*None*			Change Drop More
4	**warna**	varchar(15)	utf8mb4_general_ci		No	*None*			Change Drop More
5	**geojson**	longtext	utf8mb4_general_ci		No				Change Drop More

Figure 6. Table structured in "kelurahan" table

	id_kelurahan	id_kecamatan	nama_kelurahan	warna	geojson
Edit Copy Delete	2	21	SIDOMULYO	#86B4D7	{ "type": "Feature", "properties": { "KECAMATAN": ...
Edit Copy Delete	4	2	KASIYAN TIMUR	#88073f	{"type":"MultiPolygon","coordinates":[[[[113.30101...
Edit Copy Delete	5	9	PADOMASAN	#594370	{ "type": "Feature", "properties": { "KECAMATAN": ...
Edit Copy Delete	6	9	SARIMULYO	#39e221	{ "type": "Feature", "properties": { "KECAMATAN": ...
Edit Copy Delete	8	12	PASEBAN	#9b6693	{ "type": "Feature", "properties": { "KECAMATAN": ...
Edit Copy Delete	9	12	KENCONG	#baef82	{ "type": "Feature", "properties": { "KECAMATAN": ...
Edit Copy Delete	10	12	WONOREJO	#745514	{ "type": "Feature", "properties": { "KECAMATAN": ...
Edit Copy Delete	11	25	JAMINTORO	#da2c43	{ "type": "Feature", "properties": { "KECAMATAN": ...
Edit Copy Delete	12	25	JATIROTO	#52dadc	{ "type": "Feature", "properties": { "KECAMATAN": ...
Edit Copy Delete	13	25	YOSORATI	#c05180	{ "type": "Feature", "properties": { "KECAMATAN": ...
Edit Copy Delete	14	23	UMBULSARI	#561fe7	{ "type": "Feature", "properties": { "KECAMATAN": ...
Edit Copy Delete	15	19	WONOSARI	#0a0366	{ "type": "Feature", "properties": { "KECAMATAN": ...
Edit Copy Delete	16	25	PRINGGOWIRAWAN	#2ba1e7	{ "type": "Feature", "properties": { "KECAMATAN": ...
Edit Copy Delete	17	6	TEMBOKREJO	#e7f9df	{ "type": "Feature", "properties": { "KECAMATAN": ...
Edit Copy Delete	18	6	PURWOASRI	#da0d13	{ "type": "Feature", "properties": { "KECAMATAN": ...
Edit Copy Delete	19	6	BAGOREJO	#0f43b0	{ "type": "Feature", "properties": { "KECAMATAN": ...
Console Copy Delete	20	6	KARANGREJO	#fc3ffa	{ "type": "Feature", "properties": { "KECAMATAN": ...

Figure 7. Data in "kelurahan" table

#	Name	Type	Collation	Attributes	Null	Default	Comments	Extra	Action
1	**id_lahan**	int(11)			No	*None*		AUTO_INCREMENT	Change Drop More
2	**id_kecamatan**	int(11)			No	*None*			Change Drop More
3	**id_tanaman**	int(11)			No	*None*			Change Drop More
4	**luas_lahan**	double			No	*None*			Change Drop More
5	**luas_panen**	double			No	*None*			Change Drop More
6	**produksi**	double			No	*None*			Change Drop More
7	**produktivitas**	double			No	*None*			Change Drop More
8	**ph**	double			No	*None*			Change Drop More
9	**suhu**	double			No	*None*			Change Drop More
10	**ch**	double			No	*None*			Change Drop More
11	**created_at**	datetime			No	*None*			Change Drop More

Figure 8. Data structure in "lahan" table

←T→	id_lahan	id_kecamatan	id_tanaman	luas_lahan	luas_panen	produksi	produktivitas	ph	suhu	ch	created_at
Edit Copy Delete	5	5	4	0	8080	51693	63.97	7	0	0	2021-09-08 20:51:37
Edit Copy Delete	8	12	4	0	6689	42958	64.23	0	0	0	2021-09-15 17:14:29
Edit Copy Delete	9	6	4	0	6885	45567	66.19	0	0	0	2021-09-15 17:27:56
Edit Copy Delete	10	19	4	0	5912	39853	67.41	0	0	0	2021-09-15 17:28:28
Edit Copy Delete	11	31	4	0	4731	35798	75.67	0	0	0	2021-09-15 17:28:55
Edit Copy Delete	12	2	4	0	3633	26004	71.58	0	0	0	2021-09-15 17:30:24
Edit Copy Delete	13	29	4	0	3441	21426	62.26	0	0	0	2021-09-15 17:31:01
Edit Copy Delete	14	22	4	0	4411	24319	55.13	0	0	0	2021-09-15 17:36:04
Edit Copy Delete	15	14	4	0	4976	30016	60.32	0	0	0	2021-09-15 17:36:33
Edit Copy Delete	16	15	4	0	6128	37714	61.54	0	0	0	2021-09-15 17:36:58
Edit Copy Delete	17	8	4	0	6934	47096	67.92	0	0	0	2021-09-15 17:37:35
Edit Copy Delete	18	1	4	0	7681	48446	63.07	0	0	0	2021-09-15 18:08:31
Edit Copy Delete	19	26	4	0	6654	40191	60.4	0	0	0	2021-09-15 18:10:15
Edit Copy Delete	20	4	4	0	5666	39466	69.65	0	0	0	2021-09-15 18:11:10
Edit Copy Delete	21	23	4	0	3078	19082	62	0	0	0	2021-09-15 18:11:41
Edit Copy Delete	22	21	4	0	4781	28625	59.87	0	0	0	2021-09-15 18:12:32
Console Copy Delete	23	9	4	0	6775	35719	52.72	0	0	0	2021-09-15 18:13:05

Figure 9. Data in "lahan" table

#	Name	Type	Collation	Attributes	Null	Default	Comments	Extra	Action
1	**id_tanaman**	int(11)			No	*None*		AUTO_INCREMENT	Change Drop More
2	**nama_tanaman**	text	utf8mb4_general_ci		No				Change Drop More
3	**warna**	varchar(15)	utf8mb4_general_ci		No	*None*			Change Drop More
4	**logo**	text	utf8mb4_general_ci		No				Change Drop More
5	**created_at**	datetime			No	*None*			Change Drop More

Figure 10. Data structured in "tanaman" table

←T→	id_tanaman	nama_tanaman	warna	logo	created_at
Edit Copy Delete	4	Padi	#f4a100	assets/tanaman/1gRJvhe1zH.png	2021-06-18 10:35:44
Edit Copy Delete	5	Jagung	#f76400	assets/tanaman/XKpS7ppDSU.png	2021-06-19 11:35:31
Edit Copy Delete	6	Kedelai	#00ac69	assets/tanaman/2eFLoE53Of.png	2021-06-19 11:38:24

Figure 11. Data in "tanaman" table

#	Name	Type	Collation	Attributes	Null	Default	Comments	Extra	Action
1	id	bigint(20)		UNSIGNED	No	None		AUTO_INCREMENT	Change Drop More
2	name	varchar(255)	utf8mb4_unicode_ci		No	None			Change Drop More
3	email	varchar(255)	utf8mb4_unicode_ci		No	None			Change Drop More
4	email_verified_at	timestamp			Yes	NULL			Change Drop More
5	password	varchar(255)	utf8mb4_unicode_ci		No	None			Change Drop More
6	remember_token	varchar(100)	utf8mb4_unicode_ci		Yes	NULL			Change Drop More
7	created_at	timestamp			Yes	NULL			Change Drop More
8	updated_at	timestamp			Yes	NULL			Change Drop More

Figure 12. The data structure in the "users" table

Options

	id	name	email	email_verified_at	password	remember_token	created_at	updated_
Edit Copy Delete	1	Yudha	yudha@gmail.com	NULL	$2y$10$90X4SqLeXPYJuuNOi3hlVOcagkFYqU/XT0lgCbnC2TA...	NULL	2021-06-12 02:55:13	2021-06-

Figure 13 The data in the "users" table

In addition to building the application databases, a GIS website is also built. The GIS website that has been developed consists of two main pages, namely the main page which contains maps of all areas of Jember Regency, and the detailed land view page which contains detailed data for each region. On the main page, a map of Jember Regency is displayed by labeling the first layer, when the label is clicked, the user can see detailed data about the conditions in the specific district. On the map, there are also different colored markings in 3 areas which are the focus of research conducted this year, namely Bangsalsari District, Panti District, and Kaliwates District. The color of the area according to the "legend" on the map that yellow indicates rice commodities, orange indicates corn commodities, and green indicates soybeans commodities.

The data displayed in this GIS get from Badan Pusat Statistik of Jember Regency and observation with farmers. The purpose of observation to farmers is to ensure that the data are real by actual field conditions. Therefore, the focus of detailed data displayed in this web GIS is still from one location point in the 3 sub-districts. In addition to the map, this home page also displays data on total harvested area, data on total rice harvested area, total corn harvested area, and total soybean harvested area. The initial view of the GIS web page is shown in Figure 14.

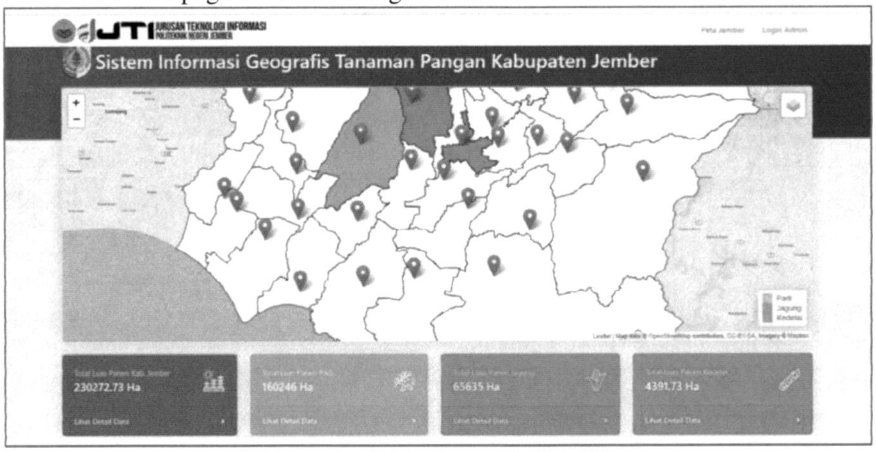

Figure 14. Main Page of WebGIS

The data detail page displays detailed sub-district data including harvested area, production, productivity, average temperature, soil pH, drainage, and soil texture. In addition, it also displays data on harvested area, production, and productivity for each cereal crop that can be produced. The data detail page display is shown in Figure 15.

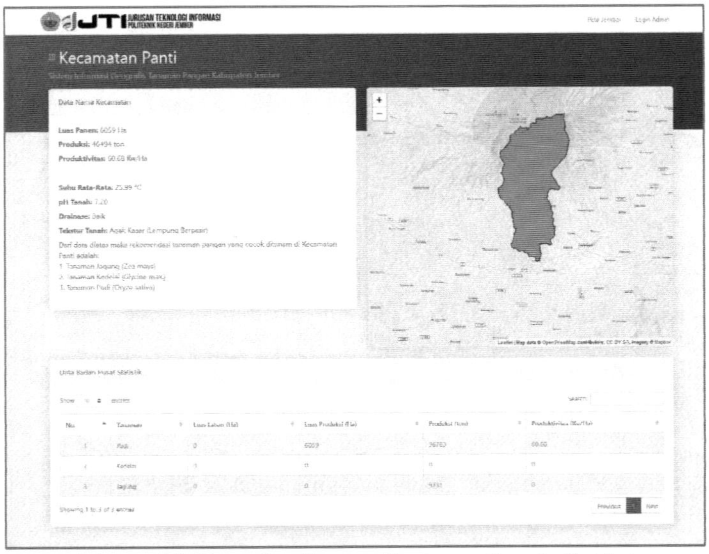

Figure 15. Data detail page displays of each district

4.2. Verification result and application testing

In the final stage of this research, testing and verification of this Web GIS application were done using the User Acceptance Testing method [12] [13] [14] [15] [16]. The test instrument was done using the google form as shown in Figure 16.

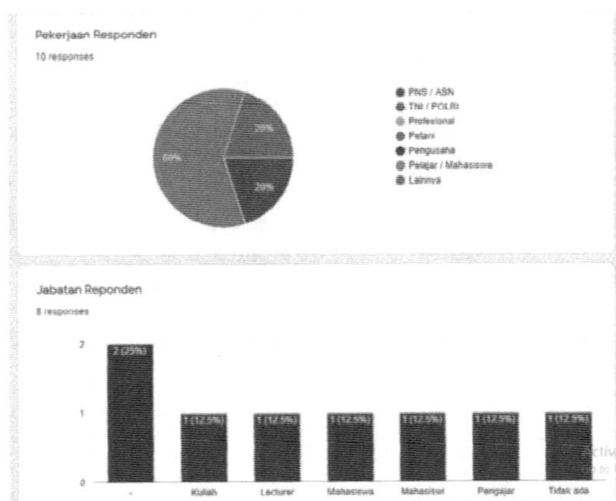

Figure 16. Verification using User Acceptance Testing in WebGIS

This test uses the Linkert scale because based on several papers it is shown that to verify web applications, you can use a the Likert scale [17]–[25] with scoring rules as shown in Table 1.

Table 1. Linkert scale in WebGIS Smart Village Evaluation

No	Linkert scale		Score
1	Very Complete	Very Simple	5
2	Complete	Simple	4
3	Quite Complete	Quite Easy	3
4	Less Complete	Less Easy	2
5	Incomplete	Not Easy	1

Based on the questionnaii\re instrument, the testing result is shown in Table 2.

Table 2. Verification Result of User Acceptance Testing

No	Question	Very complete	Complete	Quite Complete	Less complete	Incom plete	Score %
		Very Simple	Simple	Quite Easy	Less Easy	Not Easy	
		5	4	3	2	1	
1	Is the appearance of each sub-district location in Jember Regency on this geographic information system website complete according to the map?	24,4	31,7	36,6	7,3	0	100
2	Is the map display easy to identify each sub-district name with a color label?	34,1	34,1	24,4	7,3	0	100
3	Do you find it easy when you "want to see detailed data on total production land in Jember Regency on the geographic information system web"?	32,5	42,5	12,5	7,5	5	100
4	Do you find it easy when you "want to see detailed data on total rice land in Jember Regency on the geographic information system web"?	36,6	29,3	17,1	12,2	10	100
5	Do you find it easy when you "want to see detailed data on total maize land in Jember Regency on the geographic information system web"?	39	26,8	22	7	4,9	100
6	Do you find it easy when you "want to see detailed data on total soybean land in Jember Regency on the geographic information system web"?	37,65	32,5	20	5	5	100
7	Do you find it easy when you "want to find recommendations for plants that can be grown in your area of planting"?	29,3	39	14,6	12,2	2	100
8	Do you find it easy to "find cropping characteristics such as soil pH, average temperature, drainage, and soil texture in your area in the application"?	29,3	22	34	9,8	4,9	0

No	Question	Very complete / Very Simple	Complete / Simple	Quite Complete / Quite Easy	Less complete / Less Easy	Incomplete / Not Easy	Score %
		5	4	3	2	1	
9	Do you find it easy to "find production data for rice, soybeans, and maize in certain areas in this application"?	34,1	31,7	22	7,3	4,9	0
10	Does this application have an attractive appearance?	26,8	17	22	7,3	2,4	0

Based on the results of an online survey by Google Form application with 41 respondents, the following results show that: "Very Simple" and "Simple" or "very Complete" and "Complete" 80%, this indicates that the application or web that built in this study under user expectations (Verified).

5. Conclusion

This research produces another solution in agriculture to increase the production of cereal crops. Modeling the potential of cereal crops based on Web GIS in smart villages which in detail can be concluded that the Web GIS application for modeling the potential of web-based food crops by integrating the database in the display design can be developed and the results of the trial of the web GIS application using the User Acceptance method. The testing online survey results using the Google Form show that "Very Simple" and "Simple" or "Very Complete" > 80%, this indicates that the application or web that built in this study is under user expectations (Verified).

6. Suggestion

This research still requires development and improvement, including the following: (1) applicable suggestions aimed at the industry: considering that the developed application is still a prototype, it is better then this application be further developed into a professional application that will benefit the agricultural industry in Indonesia, and (2) academic suggestions for further research is that further research is needed to test the possibility of this GIS system prototype in other districts outside Jember Regency and further research is needed that can provide added value to the information generated by this GIS, such as DSS for cereal crop modeling that is most suitable in certain areas that collecting data from IoT integrated big data.

Acknowledgment

The authors would like thank to The Head of P3M State Polytechnic of Jember, Head of the Information Technology Department and its Staff, Head of the Engineering and Information Systems Laboratory, Head of the Computer Control System Laboratory, Lecturer Staff and Technicians of the Information Technology Department, and the Respondents who have helped complete the questionnaire in this study. The authors also thank the Jember State Polytechnic for providing funding for the PNBP program for the implementation of this research.

References

[1] Jumiatun, A. Junaedi, I. Lubis, C. M.A, and M. A, "Morphological, Physiological and Yield Responses of Some Rice Varieties (Oryza sativa L.) as Exposed Under High Temperature in Indonesia," *Am. J. Plant Physiol.*, vol. 11, pp. 33–41, 2016.

[2] D. T. Utomo, "Studi Pendahuluan Sistem Informasi Pangan Untuk Mengatasi Kerawanan Pangan Menggunakan Artificial Intellegence Berbasis Data Spatial," *J. Teknol. Inf. dan Terap.*,

vol. 1, pp. 17–22, 2014.
[3] B. Etikasari, H. Husin, S. Kautsar, H. Y. Riskiawan, and D. P. S. Setyohadi, "Wireless sensor network development in an unmanned aerial vehicle (UAV) for water quality monitoring system," *IOP Conf. Ser. Earth Environ. Sci.*, vol. 411, no. 1, pp. 0–7, 2020, DOI: 10.1088/1755-1315/411/1/012061.
[4] H. M. Kaidi *et al.*, "Internet of Things: A Monitoring and Control System for Rockmelon Farming," *Int. J. Integr. Eng.*, vol. 12, no. 6, pp. 55–61, 2020, DOI: 10.30880/ijie.2020.12.06.007.
[5] A. Jauhari, "Pemanfaatan SIG untuk Pemetaan Kawasan Produksi Komoditas Unggulan Tanaman Pangan di Kabupaten Pacitan," *J. Reg. Rural Dev. Plan.*, vol. 4, no. 3, pp. 154–171, 2020, DOI: 10.29244/jp2wd.2020.4.3.154-171.
[6] I. Maulana, T. A. Rumiati, and O. E. Permatasari, "Analisis Faktor-Faktor yang Mempengaruhi Produksi Padi di Provinsi Jawa Timur Menggunakan Regresi Semiparametrik Spline," *J. Sains dan Seni ITS*, vol. 5, no. 2, 2016, DOI: 10.12962/j23373520.v5i2.16992.
[7] M. Rao, G. Fan, J. Thomas, G. Cherian, V. Chudiwale, and M. Awawdeh, "A web-based GIS Decision Support System for managing and planning USDA's Conservation Reserve Program (CRP)," *Environ. Model. Softw.*, vol. 22, no. 9, pp. 1270–1280, Sep. 2007, DOI: 10.1016/j.envsoft.2006.08.003.
[8] H. Xianjin, L. Huan, H. Jinliao, and Z. Yueguang, "2.20 - Application of GIS-Based Models for Land-Use Planning in China," in *Comprehensive Geographic Information Systems*, B. Huang, Ed. Oxford: Elsevier, 2018, pp. 424–445.
[9] M. D. M. Kadiyala, S. Nedumaran, P. Singh, C. S., M. A. Irshad, and M. C. S. Bantilan, "An integrated crop model and GIS decision support system for assisting agronomic decision making under climate change," *Sci. Total Environ.*, vol. 521–522, pp. 123–134, Jul. 2015, DOI: 10.1016/j.scitotenv.2015.03.097.
[10] B. Röpke, M. Bach, and H.-G. Frede, "DRIPS—a DSS for estimating the input quantity of pesticides for German river basins," *Environ. Model. Softw.*, vol. 19, no. 11, pp. 1021–1028, Nov. 2004, DOI: 10.1016/j.envsoft.2003.11.005.
[11] M. Bolung and H. R. K. Tampangela, "Analisa Penggunaan Metodologi Pengembangan Perangkat Lunak," *J. ELTIKOM*, vol. 1, no. 1, pp. 1–10, 2017, DOI: 10.31961/eltikom.v1i1.1.
[12] F. Davis, "A technology acceptance model for empirically testing new end-user information systems: Theory and results," 1985, Accessed: Oct. 11, 2021. [Online]. Available: https://dspace.mit.edu/bitstream/handle/1721.1/15192/14927137-MIT.pdf.
[13] K. Ganesh, S. Mohapatra, S. P. Anbuudayasankar, and P. Sivakumar, "User Acceptance Test," pp. 123–127, 2014, DOI: 10.1007/978-3-319-05927-3_9.
[14] K. Ganesh, S. Mohapatra, … S. A.-E. R., undefined 2014, "User acceptance test," *Springer*, Accessed: Oct. 11, 2021. [Online]. Available: https://link.springer.com/chapter/10.1007/978-3-319-05927-3_9.
[15] H. Ho, S. Fong, Z. Y.-2008 I. I. C. one, and undefined 2008, "User acceptance testing of mobile payment in various scenarios," *ieeexplore.ieee.org*, 2008, DOI: 10.1109/ICEBE.2008.70.
[16] I. Otaduy, O. D.-J. of S. and Software, and undefined 2017, "User acceptance testing for Agile-developed web-based applications: Empowering customers through wikis and mind maps," *Elsevier*, Accessed: Oct. 11, 2021. [Online]. Available: https://www.sciencedirect.com/science/article/pii/S016412121730002X.
[17] A. Voutilainen, T. Pitkäaho, … T. K.-J. of advanced, and undefined 2016, "How to ask about patient satisfaction? The visual analog scale is less vulnerable to confounding factors and ceiling effect than asymmetric Likert scale," *Wiley Online Libr.*, Accessed: Oct. 11, 2021. [Online]. Available: https://onlinelibrary.wiley.com/DOI/abs/10.1111/jan.12875.
[18] H. T.-H. Taherdoost and undefined 2019, "What is the best response scale for a survey and questionnaire design; review of different lengths of rating scale/attitude scale/Likert scale,"

IOP Conf. Series: Earth and Environmental Science **980** (2022) 012057 doi:10.1088/1755-1315/980/1/012057

papers.ssrn.com, vol. 8, no. 1, pp. 2296–1747, 2019, Accessed: Oct. 11, 2021. [Online]. Available: https://papers.ssrn.com/sol3/papers.cfm?abstract_id=3588604.

[19] T. Nemoto, D. B.-J. 2013 conference proceedings, and undefined 2014, "Likert-scale questionnaires," *jalt-publications.org*, Accessed: Oct. 11, 2021. [Online]. Available: https://jalt-publications.org/sites/default/files/pdf-article/jalt2013_001.pdf.

[20] J. Croasmun, L. O.-J. of A. Education, and undefined 2011, "Using Likert-type scales in the social sciences.," *ERIC*, vol. 40, no. 1, 2011, Accessed: Oct. 11, 2021. [Online]. Available: https://eric.ed.gov/?id=EJ961998.

[21] H. Van Laerhoven, … H. van der Z.-A., and undefined 2004, "A comparison of Likert scale and visual analog scales as response options in children's questionnaires," *Wiley Online Libr.*, 2004, DOI: 10.1080/08035250410026572.

[22] J. B.-S. J. T. & E. S. Newsletter and undefined 2000, "What issues affect Likert-scale questionnaire formats," *jalt.org*, Accessed: Oct. 11, 2021. [Online]. Available: http://www.jalt.org/test/bro_7.htm.

[23] O. Ivanov, V. Ivanova, A. S.-I. J. of, and undefined 2018, "Likert-scale questionnaires as an educational tool in teaching discrete mathematics," *Taylor Fr.*, vol. 49, no. 7, pp. 1110–1118, Oct. 2018, DOI: 10.1080/0020739X.2017.1423121.

[24] Y. Gu, "How Often Is Often? Reference Ambiguities of the Likert-scale in Language Learning Strategy Research.," *J. Adult Educ.*, vol. 40, no. 1, 1995, Accessed: Oct. 11, 2021. [Online]. Available: https://eric.ed.gov/?id=ED391358.

[25] A. B.-J. of A. S. Research and undefined 2013, "Methods for decision-making in survey questionnaires based on Likert scale," *academia.edu*, vol. 3, no. 1, pp. 35–38, 2013, Accessed: Oct. 11, 2021. [Online]. Available: https://www.academia.edu/download/33013876/35-38.pdf.

Automatic Security System Architecture for Smart Greenhouse Using Face Recognition Approach

S Anwar [1], H Y Riskiawan [2], B Hariono [3], D P S Setyohadi [2], Yogiswara [2], A A Kurniasari [2*], and M R Hakim [2]

[1] Department of Engineering, Politeknik Negeri Jember, Indonesia
[2] Department of Information Technology, Politeknik Negeri Jember, Indonesia
[3] Department of Agricultural Technology, Politeknik Negeri Jember, Indonesia

[*] arvita@polije.ac.id

Abstract. Automation has taken over nearly all domains of operation due to the rapid progress of technology. Over the years, the evolution of human-computer interaction has sped up. With regards to technology improvements, Smart Greenhouse has come a long way. As such, it is critical to have a dependable security system in place to safeguard Greenhouse assets and employee privacy. To access Greenhouse using the conventional security system, an individual must utilize a key. However, the current security mechanism contains numerous flaws regardless of how it is cast or taken. By using face recognition, this study seeks to assist users in enhancing the door security of the Smart Greenhouse. The suggested system comprises several subsystems, including picture capture, face recognition, and automatic door access management. Face Recognition is supported by OpenCV because it uses Eigenfaces and scales down face photos without sacrificing essential features. The Doorlock Selenoid will function as the key. One relay will serve as both an active and passive solenoid. Thus, the images captured from the webcam will be compared to the stored features, and if a match is found, access will be provided to the authorized person. Otherwise, the door is not accessible.

1. Introduction

Crime rates are rising in the modern day. Numerous events, such as theft, burglary, and unwelcome invasions, occur unexpectedly. It necessitates the installation of a security system capable of preventing illegal entrance to the Greenhouse. Entryways are a critical component of the Greenhouse's security system. To open the door, we must have the appropriate key. As a result, employees often encounter problems when misplacing the key or using a programmed entrance lock framework. Employees occasionally forget the key in the house, necessitating the need to break the lock.

The stages are taken to avoid such issues by providing automatic identification that leads to biometrics. Biometric data is collected using hardware devices such as fingerprint [1], RFID [2], and OTP [3]. The disadvantage of this method is that the object must contact the hardware to collect biometric data and perform Recognition. The proposed system recognizes faces automatically through facial recognition algorithms, and the facial recognition process occurs without the user coming into direct contact with any hardware.

The initial phase in the face recognition system is face detection. The dependability of face detection affects the performance of the overall face recognition system. It can identify only the facial portion of a picture independent of the image's backdrop by utilizing face detection. The approach in this paper was mostly built on OpenCV and utilized Haar classifiers for face recognition. The Raspberry pi serves as the primary processing element. The image is captured via the webcam and compared to the stored features, and access is granted to the authorized user if a match is found. Without this, the door is inaccessible.

2. Proposed architecture for the system

The main objective of the proposed system is to design a security system that can provide great flexibility and lower cost-effectiveness by connecting all modules of the smart greenhouse automation system.

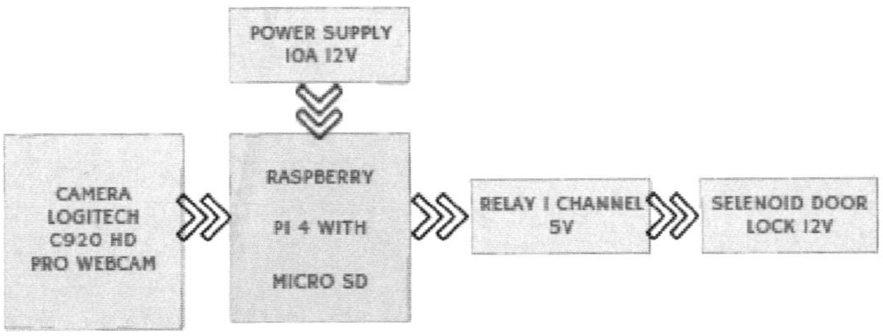

Figure 1. Block diagram of the architecture of an automated security system.

The block diagram in Figure 1 shows the Raspberry Pi 4, the Logitech camera, the relay, and the solenoid door lock. The Raspberry Pi microcontroller board acts as the master and the project's body, while the remaining hardware components function as supporting components. The system operates following the written software and automatically conducts all the above security measures and other automated activities. All physical components are required for the system to function correctly, and all of these tools operate collaboratively under a single controller.

The webcam's face recognition is handled using the haar cascade approach. The initial step, this process involves training the image to determine whether it is positive or negative. If the captured face is clear, the image is considered positive; if there is no face, the image is considered negative [4]. The Haar cascade classifier algorithm features by default are shown in Figure 2. This applies to all scales in the boosted classifier and may be quickly computed using an integral representation of the picture to be identified in [5]. The Raspberry Pi then compares the captured face to a build-in database. A signal is provided to the door lock relay when a matching face is identified, activating the solenoid.

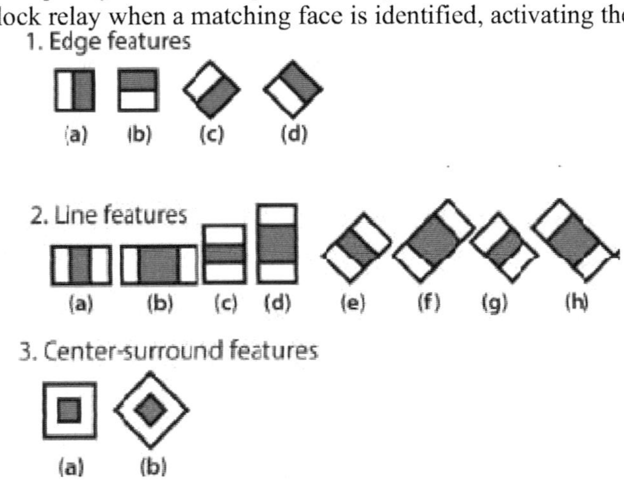

Figure 2. Haar Cascade classifier algorithm features from the OpenCV source.

The suggested system's fundamental components are as follows:

2.1. Raspberry pi 4
Raspberry Pi is a credit-card-sized single-board computer that weighs only 50g [6]. The Raspberry Pi 4 Model B is embedded with a quad-core 1.5 GHz 64-bit ARM Cortex-A72 processor, 2-8 GB RAM, 802.11ac Wi-Fi, full gigabit Ethernet (unlimited throughput), Bluetooth 5, two USB 2.0 ports, two USB 3.0 ports, and dual-monitor support via a pair of micro HDMI (HDMI Type D) ports supporting up to 4K resolution. The MicroSD card slot on the Rasberry Pi 4 can be used to expand storage space, and the HDMI port can be used to connect to the display. The Raspberry Pi 4 was created to promote the use of computer science, particularly in education. Due to the original model's primary market, robotics is substantially more popular than the projected variant. Because raspberry pi's programming language is based on Linux, it is vital to use the same environment to achieve the most performance from the raspberry pi. RASPBIAN is the optimal operating system for the raspberry pi, and it currently includes over 35,000 pre-compiled application packages that can be installed quickly and easily.

2.2. Webcam Logitech
A webcam is a video camera that streams video or images in real-time via a USB cable or a computer network via a LAN cable/ Wi-Fi connection. The resulting video resolution is pretty impressive, despite the inexpensive cost and high flexibility. Webcams are typically used to establish video connections, which enable computers to serve as video or video conferencing nodes [7]. The webcam we are using here is the Logitech C920 HD Webcam, which delivers prominent and detailed Full HD video (1080p at 30fps) with Full HD glass lens, 78° field of view, HD auto light correction, and dual microphones for stereo sound clarity.

2.3. Relay Single Channel
In our approach, we use a single channel relay to control devices from the raspberry pi. Relays are electrically regulated switches that regulate appropriate pull and work only with a holding current. It is designed to operate between 5 and 12 volts. When logic 1 is transmitted on the port pin, a relay is activated, sending a logic 0 on the port pin to turn it off. The advantages of adopting single-channel relays include their low cost, extendibility, and noiseless operation.

2.4. Door Lock
We propose to deploy a solenoid door lock with 12v power in the system. This solenoid lock is a rotary or linear actuator that controls the actuator's angle, speed, and acceleration. It utilizes a closed-loop control system attached to a relay, which automatically moves the door after the facial recognition process has been matched.

3. The proposed method for face detection
The suggested system is constructed around a Raspberry Pi with a high-performance processor, which runs a Debian-based Linux distribution commonly referred to as RASPBIAN. The proposed method's flow chart is visualized in Figure 3. When a human face is detected, the webcam captures the image. By implementing OpenCV on Raspbian, we can do face recognition using the Haar cascade Algorithm. If the face is identified using the image saved in the database system, it indicates that an authorized individual is trying to access the smart greenhouse door, at which point the door lock opens. The door does not open if the face is not recognized.

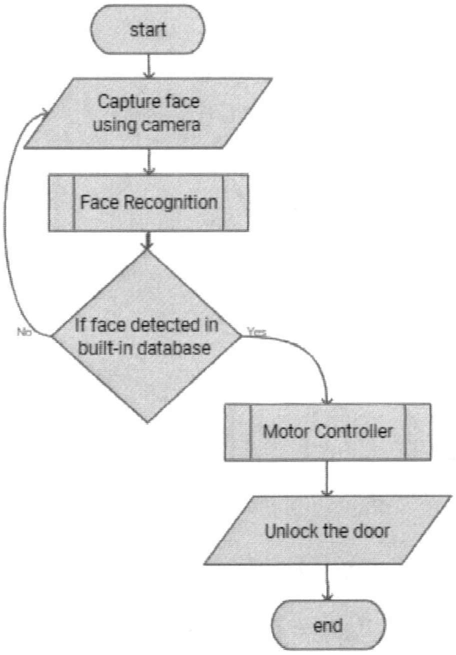

Figure 3. Flowchart for the automatic door system.

4. Experiment and Result

Our work is carried out in real-time, and our security system has demonstrated competitive outcomes compared to other door locking security systems. The GPIO pins enable the Raspberry Pi to control and monitor the outside world by connecting the Raspberry Pi to electronic circuits. The Raspberry Pi and other hardware tools have a complementary functional connection shown in figure 4. When no face is detected in front of the webcam, no signal is sent to the Raspberry Pi. Since the GPIO has not received any instructions from the Raspberry Pi, the door remains closed. The Haar cascade classifier detects the face within 2-3 seconds. The name of the recognized person is shown in the left corner of the detecting box on the Screen. Training from face recognition, we have used 50 images per face shown in figure 5. Once the face is recognized, the door is opened automatically and locked again within 10 seconds, as shown in figure 6.

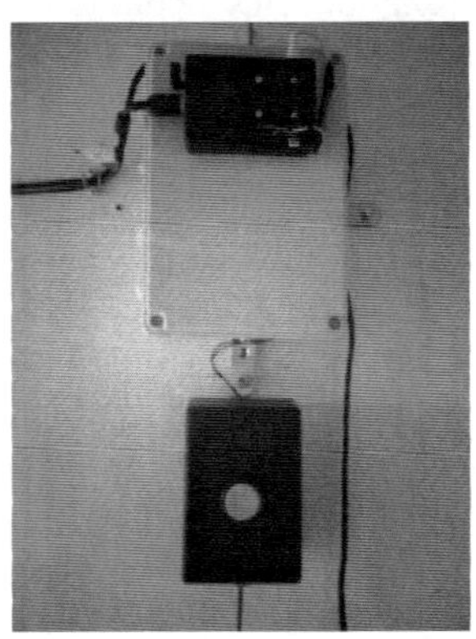

Figure 4. Hardware implementation.

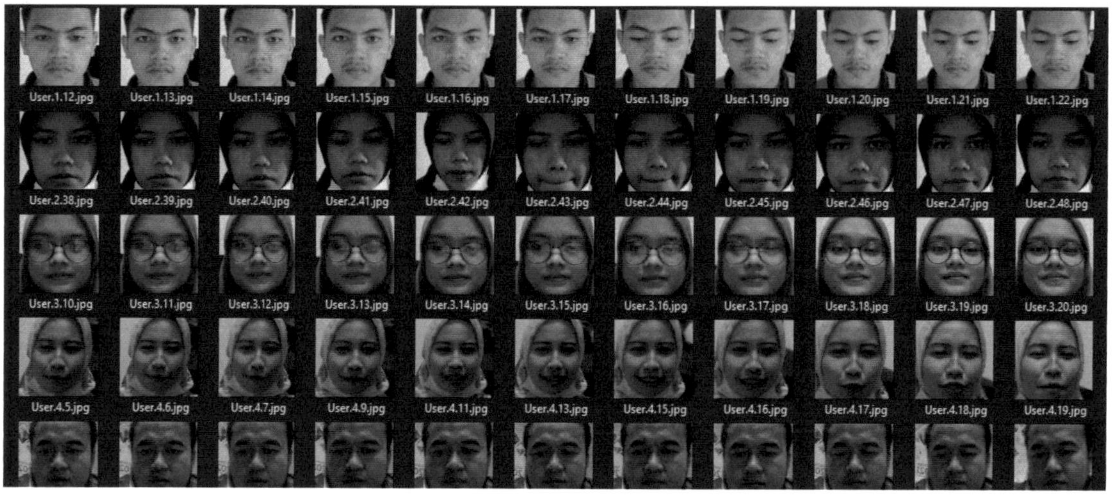

Figure 5. Face Recognition Dataset.

Figure 6. Face is detected.

The test results in Table 1 indicate that each face has a confidence rating of more than 50, meaning that the face is still recognized by the system but requires additional security. Face recognition is constrained by head orientation and lighting. The more light recorded and the proper head position used for face recognition, the higher the confidence value

Table 1. Face recognition results.

User	Face	*Confident Value (%)*
Hakim		66.67

Arvita		55.67
Nisa		64.30
Yunita		63.19
Afredo		53.60

5. Conclusion

This paper presents the implementation of a face recognition system for the Smart Greenhouse system's door lock security. The door lock solenoid is used to control the greenhouse door's opening and closing system in response to commands from the raspberry pi's GPIO pin. The door is immediately unlocked after confirming the individual's face using the Haar cascade classifier method for face recognition in an image. This approach is still limited in terms of head orientation due to the fact that the existing technology only records images of faces from the front. The system's accuracy decreases as the number of distinct faces increases. It can be avoided by training the system with other faces. For optimal performance, the system above requires adequate illumination. For the following research, the raspberry pi will send notifications to security via email or mobile phone informing them of intruders.

6. Acknowledgment

The authors like to express their appreciation to the Director and the Center for Research and Community Service at the Jember State Polytechnic for their support of international seminars, sponsorship of research, and overseeing of international food and agriculture conferences.

References

[1] Paul P, Al Achib M A, Hossain H S and Hossain M K 2019 Smart Door Lock Using Fingerprint Sensor *BRAC Univ.*
[2] Hasan Y, Abdurrahman, Wijanarko Y, Muslimin S and Maulidda R 2020 The Automatic Door Lock to Enhance Security in RFID System *J. Phys. Conf. Ser.* **1500**
[3] Kook J 2019 Design and Implementation of a OTP-based IoT Digital Door-lock System and Applications *Int. J. Eng. Res. Technol.* **12** 1841–6
[4] Roy S, Nasir Uddin M, Zahirul Haque M and Jahidul Kabir M 2018 Design and Implementation of the Smart Door Lock System with Face Recognition Method using the Linux Platform Raspberry Pi *IJCSN-International J. Comput. Sci. Netw.* **7** 382–8
[5] Li C, Qi Z, Jia N and Wu J 2017 Human face detection algorithm via Haar cascade classifier combined with three additional classifiers *ICEMI 2017 - Proc. IEEE 13th Int. Conf. Electron. Meas. Instruments* **2018-Janua** 483–7
[6] Gupta M S D, Patchava V and Menezes V 2016 Healthcare based on IoT using Raspberry Pi *Proc. 2015 Int. Conf. Green Comput. Internet Things, ICGCIoT 2015* 796–9
[7] Krishna S R, Kumar S and Kumar J 2019 Portable Text To Speech Device for Visually Impaired Using Raspberry Pi **6** 145–51

Oyster mushroom house conditions classification using K-Nearest Neighbor

I G Wiryawan[1*], E Mulyadi[2], S Choirunnisa[3], P Destarianto[4], K Agustianto[5]

[12345] Department of Information Technology, Politeknik Negeri Jember, Indonesia

*Email: wiryawan@polije.ac.id

Abstract. Internet of Things devices that were implemented to support condition monitoring and control systems in mushroom houses get a lot of temperature and humidity data from mushroom houses. The large number of temperature and humidity data can be used as the variables or indicators that affect mushroom production. Temperature and humidity data can be classified based on the production of mushrooms produced in one house or mushroom houses. This study aims to determine the classification of temperature and humidity data in mushroom houses based on mushroom production. The method used is a data mining approach based on the K-Nearest Neighbor. This research begins with determining the variables from training data or training data, taking testing data, then the testing data is reprocessed based on the K-Nearest Neighbor method with training data. Finally, evaluation of the method used was carried out by calculating the accuracy value. As a result, the accuracy of the K- Nearest Neighbor method was about 89%. These results are expected to be used to forecast the yield of mushroom production for future research. The forecast can be seen from the pattern of temperature and humidity data that is formed based on a certain period of time.

1. Introduction

The number of Internet of Things devices developed lately to support the industrial era 4.0 so that a large amount of data is produced. It happens because these IoT devices use various and many sensor devices to get data. The data that has been obtained is sometimes no longer used or only stored as temporary data. For example, there is the development of tools that measure water quality, both fresh and marine, which can then be conditioned to suit the habitats of living things [1][2][3]. In agriculture, IoT devices are used to obtain condition data from soil and plants. The utilization of these IoT devices resulted in a system called smart agriculture [4], starting from intelligent agriculture to obtaining the term precision agriculture [5].

Internet of Things devices that were implemented to support condition monitoring and control systems in mushroom houses get a lot of temperature and humidity data from mushroom houses. Oyster mushroom cultivation is one of the activities in agriculture that utilizes IoT devices. As is well known, mushrooms are plants that can be used as food and contain many minerals humans need. In addition, the planting process does not require additional chemicals. However, not all types of mushrooms can be consumed because some mushrooms contain toxins. Based on food consumption statistics in 2018, there was an increase in the average mushroom consumption per capita from 2014 of 0.087 kg/capita/year until 2017 to 0.177 kg/capita/year [6]. The average increase was also accompanied by an increase in vegetable crop production from 7,909,278 kg in 2015 to 8,605,789 in 2019. Meanwhile, for the national mushroom production in 2018, 2017, 2016, and 2015 amounted to: 31,052, 3,702 , 40,915 and 33,485 tonnes [7]. The mushroom cultivation process in Jember is more focused on cultivating white oyster mushrooms. Based on direct observations in one of the mushrooms houses or kumbung, the yield at harvest is 100 kg/month.

Many studies have discussed the development of IoT devices for monitoring purposes and also to condition the environment around the mushrooms houses (kumbung) for oyster mushroom cultivation [8][9]. The large number of temperature and humidity data can be used as the variables or indicators that affect mushroom production. This study is a continuation of the study on developing an automatic temperature and humidity control system in mushrooms houses or kumbung [10]. In this study, an IoT device has been completed that can maintain the growth of oyster mushrooms at a temperature of 16-30 Celsius degrees and humidity of 80-95%. These temperature and humidity conditions are optimal conditions for mushroom cultivation [10]. So in the previous study, temperature and humidity data were classified based on the optimal temperature and humidity conditions. The study used a Fuzzy Proportional-Integral-Derivative (PID) Controller.

The study related to this has initially been a study that discussed temperature and humidity monitoring. The first is a study that designs a real-time temperature and humidity monitoring system using ZigBee wireless network communication technology and ESP8266 WiFi technology [11]. This study succeeded in realizing simple, low-cost, and minimum-electricity data acquisition in network terminals. In addition, the embodied product also has the characteristics of easy expansion and can transmit data to the Internet. So that remote monitoring can be achieved [11]. Next, a study monitors the temperature and humidity of the server room using Lattepanda and ThingSpeak [12]. Monitoring the temperature and humidity in the server room can use Lattepanda, shown from the results of this study. Then the channel provided by ThingSpeak can be used for real-time monitoring or, more precisely, every 30 seconds [12].

The proposed research is inspired by the results of research that discuss the analysis of temperature and humidity data for predictive values [13][14][15]. In this study, we investigated the data mining technique, namely K-Nearest Neighbor, to produce a prediction for a series of numerical data. The experimental data series comes from climate data which is usually difficult to predict because of its uncertainty. This study aims to find automatic patterns to obtain satisfactory prediction results with 100 percent accuracy in prediction [15].

Based on the description of the previous background, this study proposes the processing and analysis of temperature and humidity data with a data mining approach which is then classified based on crop yields. The goals of this study are to confirm the value of the optimum temperature and humidity data that have been used in previous studies. The novelty of this proposed study lies in the approach and data associated with temperature and humidity data in mushroom houses or kumbung.

2. Related Works

The first study related to this proposed study is reference [13][14], which discusses temperature and humidity data analysis. This study produces a prediction on a series of numerical data. Then, in the end, this study continued to be developed into reference [15] which tested numerical data series obtained from climate data. This study's temperature and humidity dataset was obtained from large and numerous data, ranging from daily to yearly. The temperature and humidity data used are climatic data from observations for three years and two months, to be precise, starting from 01/01/2009 to 29/02/2012. Variation of parameter data is obtained in a file with spreadsheet format, and then the required parameter data is extracted from the file. [15]

The following related study is a study that compares several methods used in forecasting the weather [16]. The methods include Naïve Bayes, C.45, and K Nearest Neighbor classification. The research applies data mining with the three algorithms. The data used is daily data from BMKG, from January 2015 to November 2018. The data obtained consists of eight attributes, including temperature and humidity data. The results of this study indicate that the K-Nearest Neighbor classification algorithm has the highest level of accuracy, namely 71.58 percent. In comparison, the C.45 and Naïve Bayes methods have less than 70 percent accuracy rates. [16]

Reference [17] focuses on the design and application of practical matters, from forecasting crop yields to reconstructing data from missing sensors, then utilizing and comparing several machine learning techniques as a basis for providing suggestions and input on what efforts should be made. The research succeeded in showing a sufficient margin for innovation while supporting the demands and needs of companies that want to use the agricultural industry business sustainably and optimally. Investing in this means investing in technology and the knowledge and skilled workforce needed to make the best decisions. [17]

The difference between this study and previous related research lies in the correlation of the data used as one of the indicator variables or attributes of the classification. The method used in this study is still the data mining method or approach, the K-Nearest Neighbor classifier algorithm. The reason for using this algorithm is apparent because the accuracy produced is better when compared to other algorithms with case studies using numerical data series. The ultimate goal of this research is to determine the renewal of temperature and humidity levels that produce maximum oyster mushroom yields.

3. Research Method

This study uses temperature and humidity data obtained from oyster mushroom farmers in the highlands of the Jember district. In addition to temperature and humidity data, this study also collects data on oyster mushroom yields from the same farmer within a harvest time of about one and a half months. Oyster mushroom harvest data were collected from several kumbung or mushroom houses. In the harvest period, only some temperature and humidity data will be used at one time. This temperature and humidity data is taken through a microcontroller device connected to the DHT11 sensor in a random time range or interval to calculate the average by the harvest yield. The total raw data obtained in one year is thousands of temperature and humidity data, but after preprocessing, it obtains as many as 124 data. This preprocessing stage includes handling or eliminating data with a value of nil and data normalization. 105 of the 124 data were used for data training, while the remaining 19 were used for data testing.

After data collecting and data preprocessing, this study is continued to apply the K-Nearest Neighbor classifier algorithm method. The block diagram of the IoT device and proceed to classification based on the harvest data from oyster mushrooms are shown in figure 1.

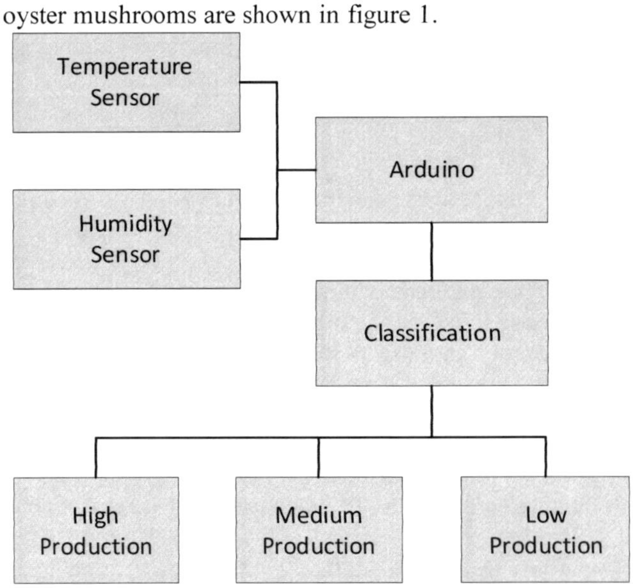

Figure 1. Block Diagram of Study

The K-Nearest Neighbor classifier algorithm determines the label or class of the data set by looking at the shortest distance between the data and the k value specified at the beginning. In this study, it has been determined that the value of k used is 1, 5, 9. The number of classes that are the results of the K-Nearest Neighbor classifier algorithm is three: the temperature and humidity data classes with high yields of oyster mushrooms, medium, and low.

4. Result and Discussion

This study begins with collecting the required data, and as a result, data on oyster mushroom harvests were obtained in six months from five mushroom houses or kumbung. In addition, the most crucial thing in this study also obtained thousands of temperature and humidity data from each kumbung. These temperature and humidity data are obtained at certain times of the day, so temperature and humidity data are taken at different times for each of these days due to the possibility that changes in temperature and humidity that occur are not too significant. Temperature and humidity data is obtained using the help of storage media. In addition, environmental conditioning in the mushroom house also causes insignificant changes in temperature and humidity. Table 1 below is a sample of data that occurred at almost close times, namely only with seconds.

Table 1. Sample temperature and humidity data with close collection times.

Timestamp	Temperature (°C)	Humidity (percent)
4/12/2021 14:29	74,71	30,61
4/12/2021 14:29	74,71	30,63
4/12/2021 14:29	74,57	30,64
4/12/2021 14:29	74,71	30,63
4/12/2021 14:29	74,57	30,64
4/12/2021 14:30	74,71	30,63
4/12/2021 14:30	74,71	30,64
4/12/2021 14:30	74,71	30,63
4/12/2021 14:30	74,57	30,64
4/12/2021 14:30	74,86	30,66
4/12/2021 14:30	74,86	30,61
4/12/2021 14:31	74,86	30,66
4/12/2021 14:31	74,86	30,66
4/12/2021 14:31	74,71	30,67
4/12/2021 14:31	74,86	30,6

The next stage in the study is the preprocessing data stage, wherein in this stage, the handling and deletion of empty data obtained from the two sensors and data normalization is carried out. After the preprocessing data stage was completed and data that could be processed was obtained, this study continued to combine 105 temperature and humidity data as a data testing with oyster mushroom harvest data from each existing mushroom house. The temperature and humidity data obtained will then be calculated on average for each oyster mushroom harvest period. The temperature data in the mushroom house is undoubtedly strongly influenced by the temperature around the mushroom house or kumbung. Weather conditions also influence humidity data at the inside of the mushroom house.

Figure 2 below the temperature and humidity data stability from one of the houses or mushroom kumbung. Stability is a challenge for this study in the selection of temperature and humidity data to be used. Conditioning of temperature and humidity causes the stability of this data. This study selects temperature and humidity data at a random time every day to solve this problem. From the graph in figure 2, there was a moment where there was an increase in the percentage of humidity. It can occur due to the delay in response to home conditioning or kumbung mushrooms.

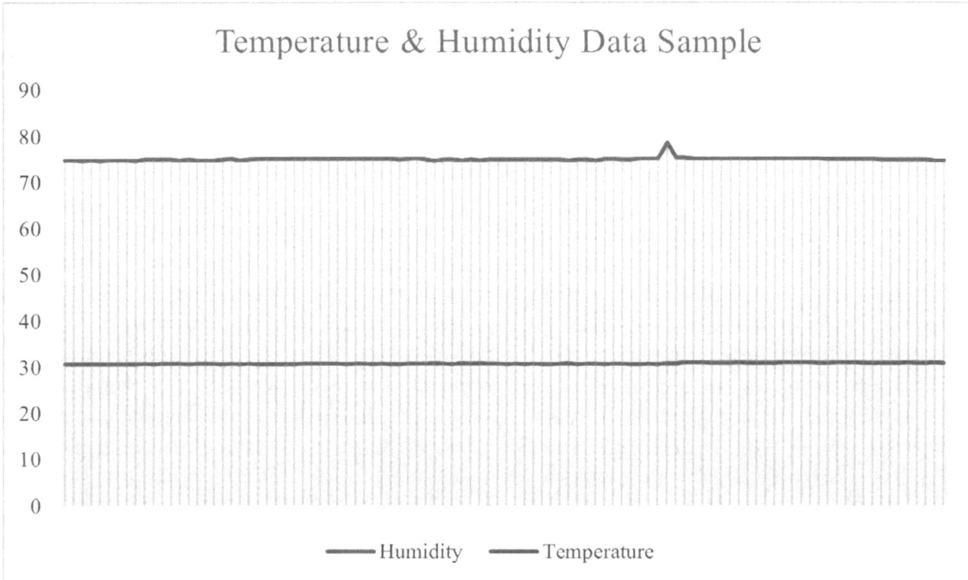

Figure 2. Temperature and Humidity Data Graphically

The application of the K-Nearest Neighbor classifier algorithm method is carried out after the testing data is formed. The algorithm method calculates the value of the Euclidean distance from all testing data. Calculation of the value of the Euclidean distance using equation (1) below.

$$d_{euclidean}(p,q) = \sqrt{\sum_{i=1}^{n}(q_i - p_i)^2} \quad (1)$$

In equation (1), the euclidean distance between p and q data is represented by $d_{euclidean}(p,q)$, then p and q are temperatures and humidity data. At the same time, i is a sequence of temperature and humidity data.

Table 2. Result of Euclidean distance calculation.

Temperature (°C)	Humidity (percent)	Production (kg)	Euclidean Distance
74,71	30,61	72	80,44
79,87	28,58	63	84,42
80	28,64	72	84,55
80	28,61	61	84,54
80,12	28,66	64	84,67
78,5	35,84	80	85,74
76,5	28,87	63	81,46
83	39,56	34	91,15
77,87	28,54	55	82,58
78,5	28,51	60	83,15
83,12	39,35	35	91,17
77,12	28,45	71	81,88
...	...	...	...
...	...	...	...
80	27,99	64	84,35
82,25	27,98	97	86,40
80,75	28,01	79	85,04

This study obtains production data directly from the owner of the oyster mushroom cultivation business. These data are data on the production of oyster mushrooms more than five times during the harvest period and from five mushroom houses owned by the same oyster mushroom cultivation entrepreneur. Then from the oyster mushroom production data, the next step is to normalize the temperature and humidity data obtained previously. Normalization of temperature and humidity data aims to facilitate the calculation of the K-Nearest Neighbor classifier by calculating the Euclidean distance later. Then the temperature and humidity data are modelled according to the K-Nearest Neighbor algorithm method [18].

Table 2 shows the results of calculating the Euclidean distance using equation (1). The calculation is carried out on 105 training data. After getting the closest distance from the training data, the next step is to rank the data with the most relative distance or have closeness in its classification category. After ranking, this study takes the highest value of many k values used in this study. If there are more classes in the highest number of values, then that class's temperature and humidity data are included.

This study initially used three k values, namely 1, 5, and 9, in the experiment. It aims to add variations in the way or method in determining the class for the test data. After assessing the value of k, the smallest Euclidean distance value will be grouped according to the training data class. The results of the calculation of the K-Nearest Neighbor classifier on 19 test data show that the value of k = 9 has the best accuracy with a value of 89%. In total, there are 17 of 19 test data that get the correct and appropriate value. The following table shows the results of the tests carried out on the three k values used.

Table 3. Evaluation Result of KNN Classifier.

k value	True value	Accuracy (%)
1	14	73,68
5	16	84,21
9	17	89,47

The evaluation results in table 3 above show that the value of k = 9 gets the best level of accuracy. Tests carried out on 19 test data managed to get the correct and appropriate value of as many as 17 data. Meanwhile, other k values only get an accuracy rate below 85%. So that in the future, it can be considered to use the value of k = 9 in developing a classification system for temperature and humidity data on the production of oyster mushrooms.

5. Conclusion

The experiment of classifying temperature and humidity data in the oyster mushroom house on the production or harvest has been carried out through several stages, from collecting temperature and humidity data for several harvest periods to preprocessing the data and continuing processing and testing data. Processing data was using the K-Nearest Neighbor classifier method by calculating the value of Euclidean distance. This study obtains production data directly from the owner of the oyster mushroom cultivation business. The test results show the accuracy of the K-Nearest Neighbor method with 9 as k value was about 89%. These results are expected to be used to forecast the yield of mushroom production for future research. The forecast can be seen from the temperature and humidity data pattern formed based on a specific period.

Acknowledgement

The paper entitled "Classification of Oyster Mushroom House Conditions Using K-Nearest Neighbor" was submitted to fulfill the output requirements, presented at an international conference organized by the Jember State Polytechnic. We express our deepest gratitude to the Ministry of Education, Culture, Research, and Technology for the financial support (funds supported by PNBP 2021 Number 371/PL17.4/PG/2021), Jember State Polytechnic so that this thesis can be carried out correctly. This research is expected to positively contribute to the development of theory or those who wish to conduct further research.

IOP Conf. Series: Earth and Environmental Science **980** (2022) 012059 doi:10.1088/1755-1315/980/1/012059

References

[1] Wibowo N S, Destarianto P, Riskiawan H Y, Agustianto K and Kautsar S 2018 Development of Low-Cost Autonomous Surface Vehicles (ASV) for Watershed Quality Monitoring *2018 6th Int. Conf. Inf. Commun. Technol.* 489–94

[2] Agustianto K, Kustiari T, Destarianto P and Wiryawan I G 2021 Development of realtime surface modeling vehicle for shrimp ponds *IOP Conference Series: Earth and Environmental Science* (IOP Publishing)

[3] Agustianto K, Destarianto P and Dewanto W K 2020 Development of real-time motion autonomous surface vehicle controlling for coral reef conservation and fisheries *IOP Conference Series: Earth and Environmental Science* vol 411

[4] Dagar R, Som S and Khatri S K 2018 Smart Farming – IoT in Agriculture *2018 Int. Conf. Inven. Res. Comput. Appl.* 1052–6

[5] Ahmed N, De D, Member S and Hussain I 2018 Internet of Things (IoT) for Smart Precision Agriculture and Farming in Rural Areas *IEEE INTERNET THINGS J.* **5** 4890–9

[6] Pusdatin 2018 *Statistik Konsumsi Pangan Tahun 2018* (Jakarta)

[7] BPS 2019 Tabel Dinamis: Produksi Tanaman Sayuran *Badan Pus. Stat.*

[8] Mat I, Kassim M R M, Harun A N and Yusoff I M 2018 Environment control for smart mushroom house *2017 IEEE Conference on Open Systems, ICOS 2017* vol 2018-Janua pp 38–42

[9] Wahab H A, Zariq M, Abdul I, Ismail A E, Pauline O, Ismon M, Zainulabidin M H and Noor F M 2019 Investigation of Temperature and Humidity Control System for Mushroom House *Int. J. Integr. Eng.* **11** 27–37

[10] Agustianto K, Wardana R, Destarianto P, Mulyadi E and Wiryawan I G 2021 Development of automatic temperature and humidity control system in kumbung (oyster mushroom) using fuzzy logic controller *IOP Conference Series: Earth and Environmental Science* (IOP Publishing)

[11] Xiao J and Li J T 2020 Design and implementation of intelligent temperature and humidity monitoring system based on ZigBee and WiFi *Procedia Comput. Sci.* **166** 419–22

[12] Nasution T H, Muchtar M A, Seniman S and Siregar I 2019 Monitoring temperature and humidity of server room using Lattepanda and ThingSpeak *J. Phys. Conf. Ser.* **1235**

[13] Badhiye S S, Chatur P N and Wakode B V 2012 Temperature and Humidity Data Analysis for Future Value Prediction using Clustering Technique : An Approach *Int. J. Emerg. Technol. Adv. Eng.* **2** 88–91

[14] Badhiye S. S and Wakode B. V. 2012 Analysis of Temperature and Humidity Data for Future value prediction *Int. J. Comput. Sci. Inf. Technol.* **3** 3012–4

[15] S.Badhiye S, U. Sambhe N and N. Chatur P 2013 KNN Technique for Analysis and Prediction of Temperature and Humidity Data *Int. J. Comput. Appl.* **61** 7–13

[16] Findawati Y, Astutik I I, Fitroni A, Indrawati I and Yuniasih N 2019 Comparative analysis of Naïve Bayes , K Nearest Neighbor and C . 45 method in weather forecast *Journal of Physics: Conference Series* (IOP Publishing)

[17] Balducci F, Impedovo D and Pirlo G 2018 Machine learning applications on agricultural datasets for smart farm enhancement *Machines* **6**

[18] Destarianto P, Etikasari B and Agustianto K 2018 Developing Automatic Student Motivation Modeling System *J. Phys. Conf. Ser.* **953**

IOP Conf. Series: Earth and Environmental Science **980** (2022) 012060 doi:10.1088/1755-1315/980/1/012060

Surface Modeling Vehicle for Shrimp Ponds using Artificial Neural Network

P Destarianto [1*], K Agustianto [1], E Rosdiana [2], I G Wiryawan [1], E Mulyadi [1]

[1] Department of Information Technology, State Polytechnic of Jember, Indonesia
[2] Department of Agricultural Production, State Polytechnic of Jember, Indonesia

* email: prawidya@polije.ac.id

Abstract. The worldwide frozen shrimp trade market in 2018 was recorded at the US $ 17.2 billion or around Rp.232.2 trillion. Indonesia is one of the principal exporters of frozen shrimp in the worldwide market. In light of this information, shrimp cultivating is a promising area, yet shrimp cultivating is particularly controlled by water quality. Water quality in shrimp cultivating consistently changes. Many variables impact changes in water quality (shrimp biomass, PH, and temperature). These components in the water should be inside the standard limit rank. Thusly, to accomplish the creation effectiveness of the shrimp business, it is important to robotize water quality control. This study aims to develop Surface Modeling Vehicles (SMV) for Shrimp Ponds utilizing an Artificial Neural Network. The test outcomes show an exactness pace of 94%, the expectation is that the instruments created by the exploration will want to acknowledge exact Vename Shrimp cultivating, so creation proficiency and expanded shrimp creation can be accomplished.

1. Introduction

The worldwide frozen shrimp trade market in 2018 was recorded at the US $ 17.2 billion or around Rp.232.2 trillion (swapping scale Rp.13,500/US$). Indonesia is one of the principal exporters of frozen shrimp in the worldwide market, to be careful, in the fourth spot after India, Ecuador, and Vietnam. The commodity worth of Indonesian frozen shrimp last year, as indicated by Worldtopexports.com, arrived at US$ 1.3 billion or Rp.17.55 trillion. Indonesia's frozen shrimp piece of the pie comes to 7.8%. Indonesia's frozen shrimp trade market incorporates, among others, the United States, Japan, and European Union nations [1]. In light of this information, shrimp cultivating is a promising area, yet shrimp cultivating is particularly controlled by water quality [2].

Water quality in shrimp cultivating consistently changes [2][3]. Many elements impact changes in water quality. Among them are the measure of broken down oxygen, the accessibility of phytoplankton as regular shrimp feed, the condition of shrimp biomass, PH, and temperature [2][3]. These elements in the water should be inside the standard limit rank. Hence, shrimp upkeep is exceptionally worried about controlling the nature of water each hour, consistently. Control of pool water quality decides the fittingness of the pool the board measures. Appropriate support of shrimp incredibly decides the usefulness of reaped shrimp.

PT. Tanjung Sumber Rejeki has been raising shrimp beginning around 2017 covering a space of 26,200 m2. Super-concentrated support of 16 shrimp lakes. A professional and helped by 20 field laborers do manual controls. Up until this point, the organization utilized a few manual devices. The low skill of the labor force in utilizing the devices causes mistaken water quality control exercises.

IOP Conf. Series: Earth and Environmental Science **980** (2022) 012060 doi:10.1088/1755-1315/980/1/012060

Regularly it neglects to get precise information with the goal that it is twofold checked, requiring a somewhat longer time and expanding costs. In this manner, we wanted a cutting-edge versatile water quality control instrument for shrimp cultivating that is proper, present-day, proficient, and precise. Determined to help the quality shrimp creation industry, arriving at the size of shrimp as per global norms, expanding efficiency, augmenting the amount of traded shrimp, it is important to foster current designing advances that can illuminate shrimp biomass. The shrimp biomass data framework is required by the industry to decide effective and precise techniques for taking care of shrimp culture.

Understanding the creation proficiency of the shrimp business, it is important to mechanize versatile data on shrimp biomass. The robotization is done by embracing a water quality control gadget as down streaming the consequences of science and innovation and data innovation of the Jember State Polytechnic as ASV [4]. ASV is a device/vehicle equipped for working on the outer layer of the water without a team. The ASV device has demonstrated its capacity to deliver precise information on PH, temperature, DO (Dioxide Oxygen), carbon dioxide and turbidity, this is conceivable because ASV in its application is furnished with sensors comprising of a PH sensor, a temperature sensor, a turbidity sensor, a DO sensor, and a carbon dioxide sensor. The information acquired from the sensor is handled in the framework utilizing the Artificial Neural Network calculation. The information acquired from the sensor readings contained in the ASV are then handled with a class model framework. Class displaying is isolated into three, specifically: Good, Medium and Dangerous for every sensor marker. Class data is communicated on a Web-based data framework

Research of the utilization of IoT has been grown, particularly compared to its capacity to help people in the entirety of their exercises, one of which is with regards to the climate. An examination led by [5][6][7][8][9]. Notwithstanding, the execution of WQM explicitly for the primary hydroponics of Vename Shrimp is as yet inadequate with regards to, this is shown by the connected examination found in IEEE [10][11].

This research is based on [2][3][10] and aim to foster Surface Modeling Vehicles (SMV) Using Artificial Neural Network Algorithm. The instrument created by this review (SMV) will utilize a similar sensor as ASV [4] however with an alternate class grouping, research is done alluding to coral reef guidelines yet in this review the reference is the ideal states of water to help development and advancement. Vename Shrimp. The expectation is that the instruments created by the examination will want to acknowledge exact Vename Shrimp cultivating, so creation productivity and expanded shrimp creation will be accomplished.

2. Related Work

Water quality in shrimp culture lakes consistently changes occasionally [2][3]. The variables that impact changes in water quality are different, including the measure of disintegrated oxygen, pH, and temperature [2][3]. These elements should be as per shrimp culture guidelines, to create great shrimp. So that shrimp upkeep is extremely worried about controlling water quality consistently, consistently (continuous). Control of pool water quality decides the suitability of pool the executive's measures. Legitimate support of shrimp enormously decides the usefulness of reaped shrimp. Hence we wanted an advanced method for controlling the Internet of Thing (IoT) water quality for shrimp cultivating that is fitting, current, effective, and exact. This is to help the quality shrimp creation industry, arriving at the size of shrimp as per global principles, boosting efficiency, expanding the amount of traded shrimp.

Research of the utilization of IoT has been grown, particularly compared to its capacity to help people in the entirety of their exercises, one of which is with regards to the climate. An examination led by [5][6][7][8][9]. Notwithstanding, the execution of WQM explicitly for the primary hydroponics of Vename Shrimp is as yet inadequate with regards to, this is shown by the connected examination found in IEEE [10][11].

Figure 1. Previous ASV Research [4][12]

This research is based on [2][3][10] and aim to foster Surface Modeling Vehicles (SMV) Using Artificial Neural Network Algorithm. The instrument created by this review (SMV) will utilize a similar sensor as ASV [4] shown by Figure 1, however with an alternate class grouping, research is done alluding to coral reef guidelines yet in this review the reference is the ideal states of water to help development and advancement. Vename Shrimp. The expectation is that the instruments created by the examination will want to acknowledge exact Vename Shrimp cultivating, so creation productivity and expanded shrimp creation will be accomplished.

3. Research Methods

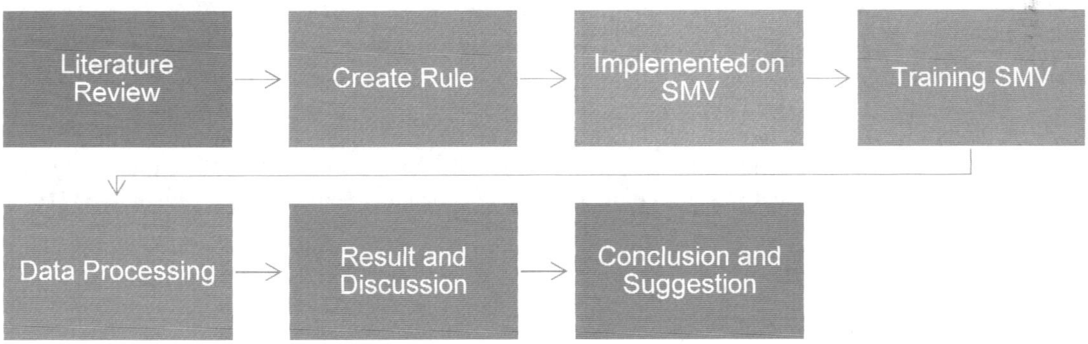

Figure 2. Research Methods

This research method begins with a literature study, shown in Figure 2, this stage becomes very important in the context of ensuring the State of the Art (research position) [13], the research position found will be compared with the problems that arise in the field, the next step is identification and resolution, problems that are part of the purpose of the study. The next research stage is making rules, this rule will be used in the Artificial Neural Network [14] process on multi-sensors [4].

The utilization of the Artificial Neural Network technique expects to deliver a more precise perusing. The following stage is trying, which is to demonstrate that the sensor readings are proper [15][16], testing in this review was completed upwards of 100 tests, at this stage, the exactness of the apparatus will likewise be acquired [17]. The test outcomes information is then acquired, this handling plans to assess the utilization of the Artificial Neural Network calculation which is carried out by the aftereffects of the perusing of the SMV water quality [12]. The eventual outcome of this exploration stage is the precision of the water quality class perusing of the SMV instrument. The aftereffects of information handling as apparatus precision are then contrasted and manual computations [18], to test the exactness by ground truth. The outcomes got at this stage will examine the variables that influence the exactness esteem got.

4. Discussion

4.1. SMV Development

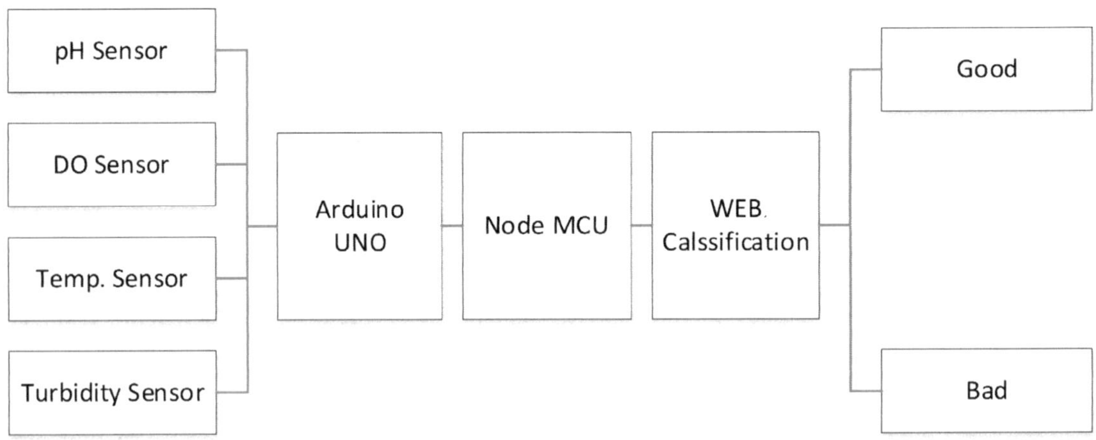

Figure 3. Block Diagram

The design of SMV based on [19][4][20] in this review is displayed in Figure 3 concerning the system block chart, Figure 3 shows a piece of the gadget containing a couple of sensors including a pH sensor what abilities to evaluate degrees Celsius of water, a temperature sensor what abilities to measure degrees Celsius of water, a turbidity sensor, what abilities to measure water turbidity even out and DO sensor what abilities to evaluate oxygen levels from water. SMV is a web-based noticing mechanical assembly that shows multi-sensor readings (pH, Dissolved Oxygen (DO), temperature, and turbidity) in shrimp lakes. The pH sensor limits the extent of the acridity of the water. Likewise, this sensor also abilities to change the non-electric sum for the present circumstance the degree of causticity (pH) into an electrical sum, specifically voltage. DO Sensors limits as an extent of the oxygen level of water, the DO regard is assessed as concentration, showing the proportion of oxygen (O2) available in water, the more essential the DO regard in water, exhibiting that the water has extraordinary quality, then again, if DO regard is low, it could be said that the water isn't satisfactory.

Then, the temperature sensor limits as an extent of the temperature of the water, the temperature showed will be changed over into Celsius which is the standard extent of temperature assessment in Indonesia [21]. Turbidity sensor limits the extent of the turbidity level of water, this sensor works by deciphering the infrared light communicated by the LED then the infrared light will go through the water and begotten by the phototransistor, the power got by the phototransistor is directly comparative with the turbidity level of the water. SMV execution using Thingspeak.

The Thingspeak application serves to show the delayed consequences of checking or data sent by sensors. The results got are taken care of again to make a graphical feature. In an instrument that scrutinizes water quality, this investigation uses Arduino Uno, which is in like manner the frontal cortex that examines and sends sensor readings similarly as giving requests to the motor driver which will move the propeller (orientation of development). This investigation uses the MCU center point. MCU center points function as connectors and for web access, as an augmentation for data transmission. Using the Thingspeak application the data that has been taken by the sensor sent from the MCU center will be displayed as a reasonable diagram containing the four sensors used similarly as the Artificial Neural Network reasoning structure what abilities to choose water quality.

4.2. Artificial Neural Network Calculation

The information utilized in this exploration is quadrant information/sensor readings of water quality. The sensors utilized in this review incorporate temperature, pH, turbidity, and Dissolved Oxygen (DO) sensors. The yield on the sensor will then, at that point, be handled as a contribution to Artificial Neural Network, by considering the cutoff points acquired from the consequences of perceptions and meetings with ranchers, the worth norm and reaches are displayed in Table 1.

Table 1. Vaname Shrimp Pond Water Quality Standards [22]

Parameter	Quality Standards
Temp.	28 °C - 32 °C
pH	7,5 - 8,5
DO	> 3,0 mg/l
Turbidity	25- 400 NTU

Class	Temp.	Range	pH	Range	Turbidity	Range	DO	Range
1	Low	≤ 26	Acid	≤ 7.5	Low	≤ 250	Low	≤ 4
2	Normal	24-32	Netral	7-9	Normal	200-450	Normal	3-7
3	High	≥ 30	Basa	≥ 8.5	High	≥ 400	High	≥ 6

Each sensor is divided into three classes, but the labels for each class differ from one another, according to the type read/according to the type of sensor. From Table 1, the research then implements the tools developed, this implementation uses the Artificial Neural Network approach, which is reflected in Figures 4.

The input layer is the layer where the artificial neural network first receives data to be processed to produce output, shown by Figure 4. Input data may consist of 2 or more data depending on the number of variables used in a study. In this study, the input data [23][24] to be processed are data from several pond water parameters consisting of 4 parameters as follows: a) X1 is a pH parameter, b) X2 is the Temperature parameter, c) X3 is a parameter of Dissolved Oxygen (DO), and d) X4 is a parameter of Turbidity (Turbidity)

The hidden layer is a layer that receives data from the input layer, but the results cannot be observed directly. Then the data will be sent to the Output Layer to observe the results of the Artificial Neural Network processing. This study uses 1 Hidden Layer with 3 Nodes (nodes). Giving the number of Nodes is based on the level of data accuracy. The output layer is a layer that contains the results of the artificial neural network process. Consists of 1 Node (node) which is the result of predicting the value of water quality from all the parameters that are input. The following is an artificial neural network architecture design with 1 Input Layer, 1 Hidden Layer, and 1 Output Layer with the following learning parameters: a) Maximum Epoch = 1000, b) Learning Rate = 0.5, and c) Bias = 1.

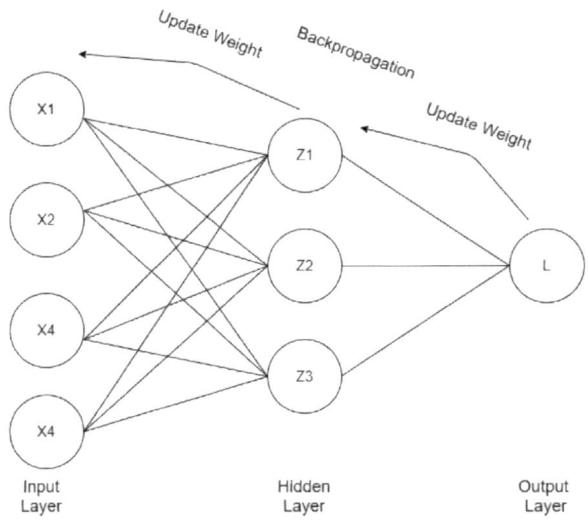

Figure 4. Backpropagation Neural Network Architecture

The training data used in this study is 70% of the total data. The maximum and minimum data values are shown in Table 2, while the target calculation is shown by Equation 1, and the result shown by Table 3.

Table 2. Maximum and Minimum of Data Training

	pH	Temp.	DO	Turbidity	Target
Maximum	8.52	38	11.98	192	232.14
Minimum	4.83	23.6	7.77	169	212

pH normalization *Normalize Target Data Training* (1)

$$y = 0.8\,(\frac{\text{x}-\text{min}}{\text{max}-\text{min}}) + 0.1$$

$$\text{y} = 0.8\,(\frac{8.21-4.83}{8.52-4.83}) + 0.1$$

$$\text{y} = 0.8\,(\frac{3.38}{3.69}) + 0.1$$

$$\text{y} = 0.8\,() + 0.1$$

$$\text{y} = 0.832791328$$

$$Target = x1 + x2 + x3 + x4$$

$$= 8.21 + 30.2 + 8.84 + 174$$

$$= 221.25$$

$$Target = 1 - (y)$$

$$Target = 1 - (0.8\,(\frac{\text{x}-\text{min}}{\text{max}-\text{min}}) + 0.1$$

$$\text{T} = 1 - (0.8\,(\frac{221.25-217.38}{232.14-217.38}) + 0.1$$

$$\text{T} = 1 - (0.8\,(\frac{221.25-217.38}{232.14-217.38}) + 0.1$$

$$\text{T} = 0.890243902$$

Table 3. Normalization Result of Data Training

pH	Temp.	DO	Turbidity	Target
0.832791328	0.706896552	0.441605839	0.273913043	0.890243902
0.856639566	0.725287356	0.540875912	0.204347826	0.94498645
0.828455285	0.725287356	0.511678832	0.204347826	0.962872629
0.828455285	0.725287356	0.548175182	0.204347826	0.949322493
0.791598916	0.743678161	0.548175182	0.204347826	0.947696477

The calculation of the new weight Testing data that is stored for test data is the weight on the 1000th Epoch obtained from the data shown in Table 4 and Table 5.

Table 4. Table of final weight values from Input Layer to Hidden Layer

	K1	K2	K3
X1	0.706896552	0.441605839	0.273913043
X2	0.725287356	0.540875912	0.204347826
X3	0.725287356	0.511678832	0.204347826
X4	0.725287356	0.548175182	0.204347826

Table 5. Final weight value from Hidden Layer to Output Layer

	L
K1	5.33
K2	-3.18
K3	-5.05

In this study, with the number of epochs and learning rates used referring to previous research, the MSE produced is classified as good: 0.000913127. So based on the results of the MSE, the research continues by processing the testing data (125 data) using the final weights that have been obtained from the Training data, calculation is shown by Equation 2.

Node Calculation *Output Calculation* (2)

$$Net\ K1 = V_{0j} + \sum_{t=1}^{n} XiVij$$
$$= Bias + X1.V_{x1k1} + X2.V_{x2k1} + X3.V_{x3k1} + X4.V_{x4k1}$$
= 1 + 0.832791328*-0.66+ 0.706896552*-0.6+0.441605839*-0.77 + 0.273913043*0.12
= −0.9809471385

$$K1 = f(Net\ K1)$$
$$= \frac{1}{1+e^{-Net\ K1}}$$
$$= \frac{1}{1+e^{-(-0.9809471385)}}$$
= 0.2727038911

$$Net\ L = V_{0j} + \sum_{t=1}^{n} XiWij$$
$$= Bias + K1.W_{k1L} + K1.W_{k2L} + K1.W_{k3L}$$
= 1 + 0.2727038911*5.33+ 0.008295373834*-3.18+ 0.4269997047* -5.05
= 0.270783942

$$L = f(Net\ L)$$
$$= \frac{1}{1+e^{-Net\ K2}}$$
$$= \frac{1}{1+e^{-0.270783942}}$$
= 0.5672853515

$$Net\ K2 = V_{0j} + \sum_{t=1}^{n} XiVij$$
$$= Bias + X1.V_{x1k2} + X2.V_{x2k2} + X3.V_{x3k2} + X4.V_{x4k2}$$
$$= 1 + 0.832791328 * -4.58 + 0.706896552 * -2.13 + 0.441605839 * -1.1 + 0.273913043 * 0.08$$
$$= -4.783727317$$

$$K2 = f(Net\ K2)$$
$$= \frac{1}{1+e^{-Net\ K2}}$$
$$= \frac{1}{1+e^{-(-5.957715025)}}$$
$$= 0.008295373834$$

$$Net\ K3 = V_{0j} + \sum_{t=1}^{n} XiVij$$
$$= Bias + X1.V_{x1k3} + X2.V_{x2k3} + X3.V_{x3k3} + X4.V_{x4k3}$$
$$= 1 + 0.832791328 * -2.96 + 0.706896552 * 0.81 + 0.441605839 * -1.43 + 0.273913043 * 4.49$$
$$= -0.2941029105$$

$$K3 = f(Net\ K3)$$
$$= \frac{1}{1+e^{-Net\ K2}}$$
$$= \frac{1}{1+e^{-(--0.8406457734)}}$$
$$= 0.4269997047$$

The output value is then processed into fuzzy to form a class (good and bad), the fuzzy equation used in this study is shown by Equation 3. The class used in this study is intentionally made in two based on the threshold value from the Ministry of Maritime Affairs and Fisheries as shown in Table 1 .

$$\mu[Bad] = \begin{cases} 0; & x \geq 0{,}4\ atau\ x \leq 0{,}6 \\ (b-x)/(b-a); & 0{,}3 < x < 0{,}4 \\ (d-x)/(d-c); & 0{,}7 < x < 0{,}8 \\ 1; & x \leq 0{,}3\ dan\ x \geq 0{,}8 \end{cases} \quad (3)$$

$$\mu[Good] = \begin{cases} 0; & x \geq 0{,}8\ dan\ x \leq 0{,}3 \\ (x-a)/(b-a); & 0{,}3 < x < 0{,}4 \\ (x-c)/(d-c); & 0{,}7 < x < 0{,}8 \\ 1; & 0{,}4 \leq x \leq 0{,}6 \end{cases}$$

4.3. Testing and Result SMV

The development of this tool was successful, this is indicated by the Black Box testing, shown in Figure 8 and Table 3, where all the results obtained by the study were all successful (100%). Tests carried out in this study use the Ground Truth approach, where the output of the system is compared with expert judgment, shown in Table 4. If the results are the same, it will be calculated as accuracy, otherwise if there is a difference it will be counted as an error. In tests carried out on 100 data generated by sensors (tools), the study produced an accuracy value of 92%.

Table 6. *Black Box Testing*

Number	Item	Func.	Result
1	Node MCU	Data Transmitting Arduino - Thingspeak	Success
2	Temperature Sensor	Temperature Value	Success
3	Turbidity Sensor	Turbidity Value	Success
4	pH Sensor	pH Value	Success
5	DO Sensor	DO Value	Success

Table 7. *Artificial Neural Network Testing Result*

Number	Temp.	pH	Turbidity	DO	Class	Expert
1	7.64	29.8	11.41	172	Good	False
2	7.64	29.8	11.42	171	Good	True
3	7.63	29.8	11.43	171	Good	True
4	7.63	29.9	11.44	170	Good	True
5	7.62	29.9	11.45	172	Good	True
6	7.61	29.9	11.46	172	Good	True
7	7.61	29.9	11.47	169	Good	True
8	7.6	29.9	11.48	175	Good	True
9	8.1	30	11.49	175	Good	True
...	..	..	..	..	..	
122	7.4	27	9	180	Bad	True
123	8.6	33	11	160	Bad	True
124	9	37	12	152	Bad	True
125	9	37	12	152	Bad	False

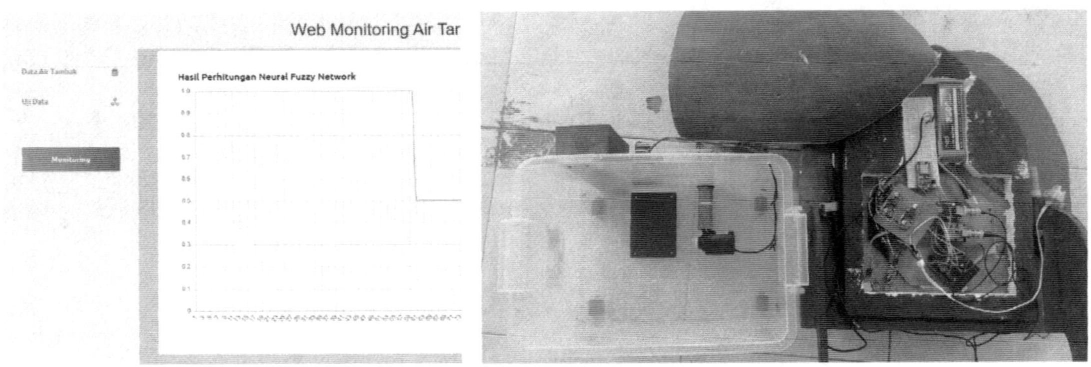

Figure 5. Testing

5. Conclusion

The conclusion obtained, the research has succeeded in developing a Real-time Surface Modeling Vehicle for Shrimp Ponds (SMV). This is indicated by an indication of the Black Box test results that result in system compatibility, with a value of 100%, or in other words all features are running well. Whereas in the test carried out with 125 data (7 false), an accuracy value of 94% was obtained. So that these results are used as a reference that the tools produced in this study can be implemented in shrimp ponds and are expected to increase the accuracy of shrimp farming.

6. Acknowledgment

The authors would like to acknowledge the financial support of this work by grants from PNBP, Politeknik Negeri Jember. The author also thanked the P3M and Jurusan Teknologi Informasi, Politeknik Negeri Jember, which has provided support and assistance in completing this research.

References

[1] Data K 2019 Indonesia Eksportir Udang Beku Terbesar Keempat di

[2] Syafaat M N, Mansyur A and Tonnek S 2012 Dinamika kualitas air pada budidaya udang vanname (Litopenaeus vannamei) semi-intensif dengan teknik pergiliran pakan *Pros. Indoaqua-Forum Inov. Teknol. Akuakultur* 487–94

[3] Fuady M F, Supardjo M N and Haeruddin 2013 Pengaruh Pengelolaan Kualitas Air terhadap Tingkat Kelulushidupan dan Laju Pertumbuhan Udang Vaname (Litopenaeus vannamei) di PT. Indokor Bangun Desa, Yogyakarta *Diponegoro J. Maquares Manag. Aquat. Resour.* **2** 155–62

[4] Wibowo N S, Destarianto P, Riskiawan H Y, Agustianto K and Kautsar S 2018 Development of low-cost autonomous surface vehicles (ASV) for watershed quality monitoring *2018 6th International Conference on Information and Communication Technology, ICoICT 2018* vol 0 (IEEE) pp 489–94

[5] Cao F, Jiang F, Liu Z, Chen B and Yang Z 2014 Application of ISFET Microsensors with Mobile Network to Build IoT for Water Environment Monitoring *Proceedings - 2014 International Conference on Intelligent Environments, IE 2014* pp 207–10

[6] Perumal T, Sulaiman N and Leong C Y 2015 Internet of Things (IoT) Enabled Water Monitoring System *2015 IEEE 4th Global Conference on Consumer Electronics (GCCE) Internet* pp 86–7

[7] Pranata A A, Lee J M and Kim D S 2017 Towards an IoT-based Water Quality Monitoring System with Brokerless Pub / Sub Architecture

[8] Myint C Z, Gopal L and Aung Y L 2017 WSN-based Reconfigurable Water Quality Monitoring System in IoT Environment *2017 14th International Conference on Electrical Engineering/Electronics, Computer, Telecommunications and Information Technology (ECTI-CON)* pp 741–4

[9] Myint C Z, Gopal L and Aung Y L 2017 Reconfigurable Smart Water Quality Monitoring System in IoT Environment *Proceedings - 16th IEEE/ACIS International Conference on Computer and Information Science, ICIS 2017* pp 435–40

[10] Muchtar E, Sanjaya E and Hariadi F I 2017 Human machine interface on e-Shrimp as smart control system for whiteleg shrimp pond *2017 Int. Symp. Electron. Smart Devices, ISESD 2017* **2018-Janua** 24–9

[11] Sacasqui M, Sanchez I and Vasquez E 2017 Adaptive predictive control of dissolved oxygen concentration in a dynamic model of whiteleg shrimp culture *2017 Chil. Conf. Electr. Electron. Eng. Inf. Commun. Technol. CHILECON 2017 - Proc.* **2017-Janua** 1–6

[12] Agustianto K, Destarianto P and Dewanto W K 2020 Development of real-time motion autonomous surface vehicle controlling for coral reef conservation and fisheries *IOP Conf. Ser. Earth Environ. Sci.* **411**

[13] Agustianto K, Permanasari A E, Kusumawardani S S and Hidayah I 2016 Design adaptive learning system using metacognitive strategy path for learning in classroom and intelligent tutoring systems *AIP Conference Proceedings* vol 1755

[14] Kurnia D W, Kautsar S, Etikasari B and Khafidurrohman A 2018 A control scheme for typist robot using Artificial Neural Network *Proc. - 2017 Int. Conf. Sustain. Inf. Eng. Technol. SIET 2017* **2018-Janua** 374–8

[15] Destarianto P, Etikasari B and Agustianto K 2018 Developing Automatic Student Motivation Modeling System *J. Phys. Conf. Ser.* **953**
[16] Dewanto W K, Agustianto K and Sari B E 2018 Developing thinking skill system for modelling creative thinking and critical thinking of vocational high school student *J. Phys. Conf. Ser.* **953**
[17] Destarianto P, Riskiawan H Y, Agustianto K and Kautsar S 2018 Developing food sensory test system with preference test (Hedonic and Hedonic quality) wheat bread case study *Proceedings - 2017 International Conference on Sustainable Information Engineering and Technology, SIET 2017* vol 2018-Janua
[18] Agustianto K and Destarianto P 2019 Imbalance Data Handling using Neighborhood Cleaning Rule (NCL) Sampling Method for Precision Student Modeling *Proc. - 2019 Int. Conf. Comput. Sci. Inf. Technol. Electr. Eng. ICOMITEE 2019* **1** 86–9
[19] Agustianto K, Destarianto P and Dewanto W K 2020 Development of real-time motion autonomous surface vehicle controlling for coral reef conservation and fisheries *IOP Conference Series: Earth and Environmental Science* vol 411
[20] Farkan M and Setiyanto D D 2021 Development of realtime surface modeling vehicle for shrimp ponds (ReSMeV-SP) *IOP Conf. Ser. Earth Environ. Sci.*
[21] Science E 2021 Development of automatic temperature and humidity control system in kumbung (oyster mushroom) using fuzzy logic controller *IOP Conf. Ser. Earth Environ. Sci.*
[22] Kkp 2016 *Peraturan Menteri Kelautan Dan Perikanan Republik Indonesia Nomor 75/Permen-Kp/2016 Tentang Pedoman Umum Pembesaran Udang Windu (Penaeus Monodon) Dan Udang Vaname (Litopenaeus Vannamei*
[23] Choirunnisa S and Lianto J 2017 Hybrid Method of Undersampling and Oversampling for Handling Imbalanced Data *2018 Int. Semin. Res. Inf. Technol. Intell. Syst.* 276–80
[24] Choirunnisa S, Meidyani B and Rochimah S 2019 Software Defect Prediction using Oversampling Algorithm: A-SUWO *2018 Electr. Power, Electron. Commun. Control. Informatics Semin. EECCIS 2018* 337–41

IOP Conf. Series: Earth and Environmental Science **980** (2022) 012061 doi:10.1088/1755-1315/980/1/012061

Agricultural Commodity Sales Recommendation System For Farmers Based on Geographic Information Systems and Price Forecasting Using Probabilistic Neural Network Algorithm

A H Utomo [1*], M A Gumilang [1], and A Ahmad [2]

[1] Information Technology Department, Politeknik Negeri Jember, Indonesia
[2] Agribusiness Management Department, Politeknik Negeri Jember, Indonesia

* email: adiheruutomo@polije.ac.id

Abstract. Since the implementation of social distancing and physical distancing due to the outbreak/pandemic of the Coronavirus (Covid-19), direct sales in the market have experienced a shortage of buyers. Farmers also share this in Indonesia, where the price game offered by collectors does not match the market price. The second problem is the mismatch of prices in each market, forcing farmers to check locations to sell their agricultural products. This problem is also experienced by the O'reng Rembangan Community Information Group (KIM), one of the community groups engaged in production to cultivate vegetable and fruit gardens in Kemuning Lor Village, Arjasa District, Jember Regency. The purpose of this research is the creation of an information system that can help farmers, especially KIM O'reng Rembangan, to obtain current market price information, receive market recommendations for agricultural products, get the nearest market from the location of farmers, and can be used by sellers to make purchases, optimize stock merchandise. This research also focuses on the prediction of agricultural commodity prices. The method used is the Probabilistic Neural Network (PNN) method to estimate the price of agricultural commodities. The resulting system in this study consists of 2 parts. The first part is the input device, which officers can use to enter the price of each agricultural commodity directly from each market. The second part is a Geographic information system used to display the forecasting results of agricultural commodity prices in each market. The forecast of agricultural commodity prices in this study has an accuracy of 98.3%.

1. Introduction

Agriculture is one of the important sectors in the national economy. National development in the 21st century will still be broadly based on agriculture, with services and business activities based on agribusiness being leading in national development. For the reconstruction of the post-covid-19 Indonesian economy towards SDG'S 2030 in the perspective of supporting business and business development, all sectors must be developed optimally. However, there are several problems faced by farmers that affect the welfare of farmers in Indonesia. The first problem faced by farmers in Indonesia is that the price game offered by collectors does not match the market price. The second problem is the incompatibility of prices in each market, making farmers have to check the location to sell their agricultural products. This problem was also experienced by the O'reng Rembangan Community Information Group (KIM). KIM O'reng Rembangan is one of the community groups engaged in production to process vegetable and fruit gardens in Kemuning Lor Village, Arjasa District, Jember Regency [1].

2. Review of Related Literature

2.1. Probabilistic Neural Network

Probabilistic Neural Network (PNN) or Probabilistic Neural Network was first developed by Donald F. Specht in 1988. PNN is an artificial neural network method that uses supervised training. PNN is included in the Feedforward structure. PNN is derived from a Bayesian network and statistical algorithm named Kernel Fisher Discriminant Analysis [2]. Bayes's rule can be used to classify a number of categories. Decision-making is based on calculating the distance between the probability density function of the feature vector. PNN has four layers, namely input layer, pattern layer, summation layer, and output layer [3], as shown in Figure 1 as follows.

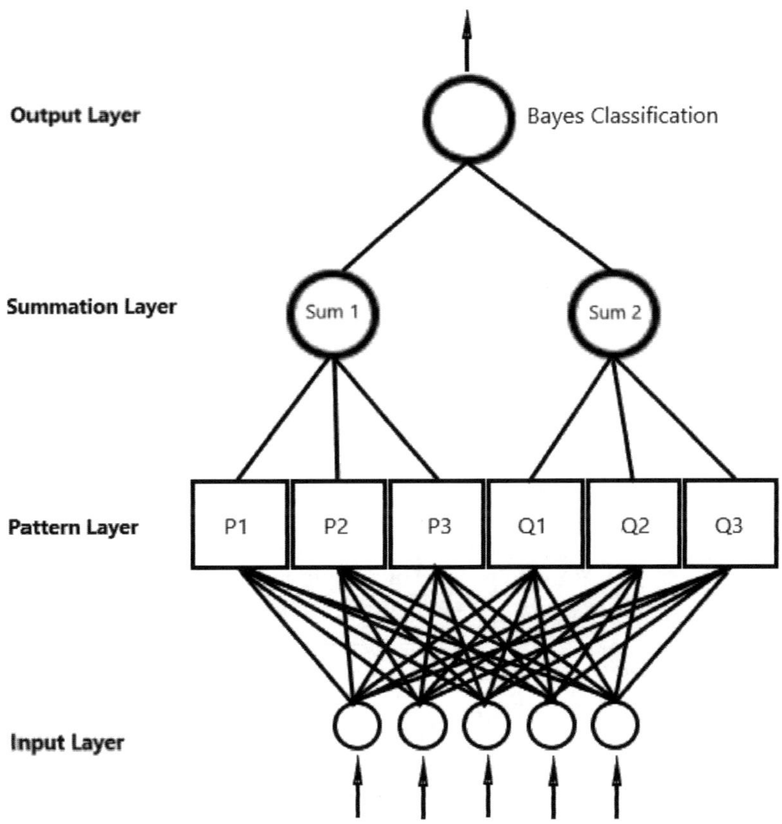

Figure 1. PNN Architecture

2.2. State of the Art

The first study using the K-NN and Particle Swarm methods [4], with an accuracy result of 98% but cannot be applied to other commodities. Meanwhile, the second study using the Arima ArchGarch method and the Single Moving Average [5] has not been implemented for other regional datasets. So it is necessary to refine the algorithm using machine learning that distinguishes the training and real datasets with the Probabilistic Neural Network (PNN) method, which is the State of The Art of this study [6][7].

IOP Conf. Series: Earth and Environmental Science **980** (2022) 012061 doi:10.1088/1755-1315/980/1/012061

3. Research Method

The methods applied in this research are shown in Figure 2 as follows

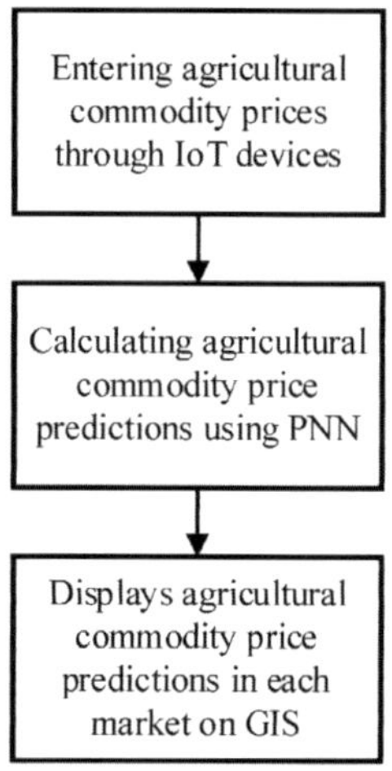

Figure 2. Research Method

3.1. Entering agricultural commodity prices through IoT devices

At this stage, an IoT is created that officers can use to enter the price of each agricultural commodity directly from each market.

3.2. Calculating agricultural commodity price predictions using PNN

The data obtained in the first dance stage is used as training data to predict prices using PNN. The data used in this predicting process is data on prices of agricultural commodities in each market since 2012. The data training is obtained from the average price of agricultural commodities every year. Every newly entered data will become training data for the prediction process. in the next period.

3.3. Displays agricultural commodity price predictions in each market on GIS

The prediction results of agricultural commodity prices in each market will be presented in a Geographic Information System.

4. Result and Discussion

4.1. IoT Devices

The device used to make this IoT device is a Raspberry Pi 3, equipped with GPS and an LCD screen. In this device, an application is built using Python, which records the price of each agricultural commodity in each market and sends the data to a web server. This device is as shown in Figure 2.

Figure 3. Raspberry Pi 3

4.2. Calculating agricultural commodity price predictions
Calculating the price predictions using PNN is carried out in the following steps [8]:

4.2.1. Input Layer
Layer input contains training data which is an input-target vector pair [9]. Table 1 shows 16 pairs of input-target data vectors to be trained.

Table 1. Input-Target Data Vector Pair to be Trained

No.	Climate/Season	Holiday	Origin of the Products	Decision
	(X_1)	(X_2)	(X3)	Class
1	Rainy season	Eid Al-Fitr	Local products	Price prediction
2	Dry season	Chinese New Year	Products from out of town	
3		New Year		
4		Normal		

Since the target consists of 3 classes, the input vector will be worth: (1, 0, 0) for class 1; (0, 1, 0) for class 2; and (0, 0, 1) for class 3.

4.2.2. Pattern Layer
The probabilistic network built has 16 neurons in the hidden layer. The network is depicted in Table 2 below.

Table 2. Probabilistic Network

Climate/Season	Holiday	Origin of the Products	Decision
Rainy season	Eid Al-Fitr	Local products	Price goes up
Rainy season	Chinese New Year	Local products	Price goes up
Rainy season	New Year	Local products	Price goes up
Rainy season	Normal	Local products	Price stays
Rainy season	Eid Al-Fitr	Products from out of town	Price goes up
Rainy season	Chinese New Year	Products from out of town	Price goes up
Rainy season	New Year	Products from out of town	Price stays
Rainy season	Normal	Products from out of town	Price goes down
Dry season	Eid Al-Fitr	Local products	Price stays
Dry season	Chinese New Year	Local products	Price stays
Dry season	New Year	Local products	Price stays
Dry season	Normal	Local products	Price stays
Dry season	Eid Al-Fitr	Products from out of town	Price stays
Dry season	Chinese New Year	Products from out of town	Price goes down
Dry season	New Year	Products from out of town	Price goes down
Dry season	Normal	Products from out of town	Decreased price

4.2.3. Summation Layer

The Summation Layer receives input from the pattern layer nodes associated with the existing class [10]. The equation used in this layer is as shown below.

$$\sum_{i=1}^{N} \exp\left[\frac{(x-x_{ij})^T\,(x-x_0)}{2\sigma^2}\right] \qquad (1)$$

This study uses training data on agricultural commodity prices starting from 2012 to 2021. In this data, the average value is sought from each year. The calculation results can be seen in Table 3 as follows.

Table 3. Training Data

Year	Amount per Year	Amount of Data	Result
2012	5.448.000	366	14.885
2013	10.847.500	365	29.719
2014	10.929.000	365	29.942
2015	10.683.000	365	29.268
2016	10.132.123	366	27.683
2017	16.692.533	365	45.732
2018	10.683.205	365	29.269
2019	11.257.551	365	30.842
2020	9.444.593	366	25.804
2021	10.654.611	176	60.537
Sum of Results			323.681

The value of h_{jk} in the Normal Pattern can be calculated as follows

$$h_{jk} = \frac{323.681}{10} = 32.368$$

4.2.4. Output Layer

The output layer produces a binary output (0,1) and only has a single weight variable Ck. Ck is calculated using the equation as follows.

$$C_k = \frac{h_{jk} l_{jk}}{h_{ik} l_{ik}} \cdot \frac{m_{ik}}{m_{jk}} \qquad (2)$$

notes:
m_{ik} = Number of pattern training from Θ_{ik} class
m_{jk} = Number of pattern training from Θ_{jk} class

Suppose there is a condition where at that time it is the rainy season, Eid al-Fitr, and the product comes from local production, then:

The average value of the last year's price minus the first year's price is

$$Average = \frac{(60.537 - 14.885)}{10} = 4.565$$

According to the pattern layer for this condition, these conditions resulting in the price increase. The h_{jk} value is taken from the enormous value from the previous ten years and then added to the average value.

$$60.537 + 4.565 = 65.102$$

Then the prediction of the commodity price is Rp. 65.102.

4.3. Prediction Result Accuracy

The evaluation carried out on this system uses three agricultural commodities in the last ten years, as described in Table 3. These data are used as a training dataset to predict the test data. Furthermore, the test data will be used to predict agricultural commodity prices for the next 60 days.

Table 4. Testing Data

	PNN Prediction								
Day	**System price prediction for commodity 1**	**Actual price of commodity 1**	**The truth of the results**	**System price prediction for commodity 2**	**Actual price of commodity 2**	**The truth of the results**	**System price prediction for commodity 3**	**Actual price of commodity 3**	**The truth of the results**
1	Up	Up	True	Up	Up	True	Up	Up	True
2	Up	Up	True	Up	Up	True	Up	Up	True
3	Up	Up	True	Up	Up	True	Stays	Stays	True
4	Up	Up	True	Stays	Stays	True	Stays	Stays	True
5	Stays	Stays	True	Stays	Stays	True	Stays	Stays	True
6	Stays	Stays	True	Stays	Stays	True	Stays	Stays	True
7	Stays	Stays	True	Stays	Stays	True	Stays	Stays	True
8	Stays	Stays	True	Stays	Stays	True	Down	Down	True
9	Stays	Stays	True	Down	Down	True	Down	Down	True
10	Down	Down	True	Down	Down	True	Down	Down	True
11	Down	Down	True	Down	Down	True	Stays	Stays	True
12	Down	Down	True	Stays	Stays	True	Stays	Stays	True
13	Stays	Up	**False**	Stays	Stays	True	Down	Down	True
14	Stays	Stays	True	Down	Down	True	Down	Down	True
15	Down	Down	True	Down	Down	True	Down	Down	True
16	Down	Down	True	Down	Down	True	Up	Up	True
17	Down	Down	True	Up	Up	True	Up	Up	True
18	Up	Up	True	Up	Up	True	Up	Up	True
19	Up	Up	True	Up	Up	True	Up	Up	True
20	Up	Up	True	Up	Up	True	Up	Up	True
21	Up	Up	True	Up	Up	True	Up	Up	True
22	Up	Up	True	Up	Up	True	Up	Up	True
23	Up	Up	True	Up	Up	True	Up	Up	True
24	Up	Up	True	Up	Up	True	Stays	Stays	True
25	Up	Up	True	Stays	Stays	True	Stays	Stays	True
26	Stays	Stays	True	Stays	Stays	True	Stays	Stays	True
					...				
58	Stays	Stays	True	Down	Down	True	Down	Down	True
59	Down	Down	True	**Down**	Up	**False**	Down	Up	**False**
60	Down	Down	True	Down	Down	True	Stays	Stays	True

Testing the system in predicting prices for 60 days showed three errors occurred on the 13th and 55th days, as seen in Table 4. So the accuracy can be calculated as follows.

$$Accuracy = \frac{Number\ of\ correct\ answers}{Amount\ of\ data}\ x\ 100\%$$

$$= \frac{\mathbf{177}}{\mathbf{180}}\boldsymbol{x}\ \mathbf{100}\%$$

$$= \mathbf{98,3}\%$$

4.4. Geographic Information System

The prediction of agricultural commodity prices produced by this system is presented in a Geographic Information System, which displays as shown in Figures 4 to 7. On the Figure 4 shows landing page result for introducing the system that already build.

Figure 4. GIS Homepage View

Figure 5 shows information some market location and commodities that already calculated for prediction analytics on this research. Recently market location were in Jember regency, there are three market's data that have been collected as training data.

Figure 5. Market Selection Menu

Figure 6 show recommendation result from PNN algorithm in three spices commodities, on the recent research prediction information show location of the market and prediction prices (up/down/fixed). After we know which one market that have best price, the user can see location of market through Google Maps API in Figure 7.

Figure 6. Product Recommendations from The Selected Market

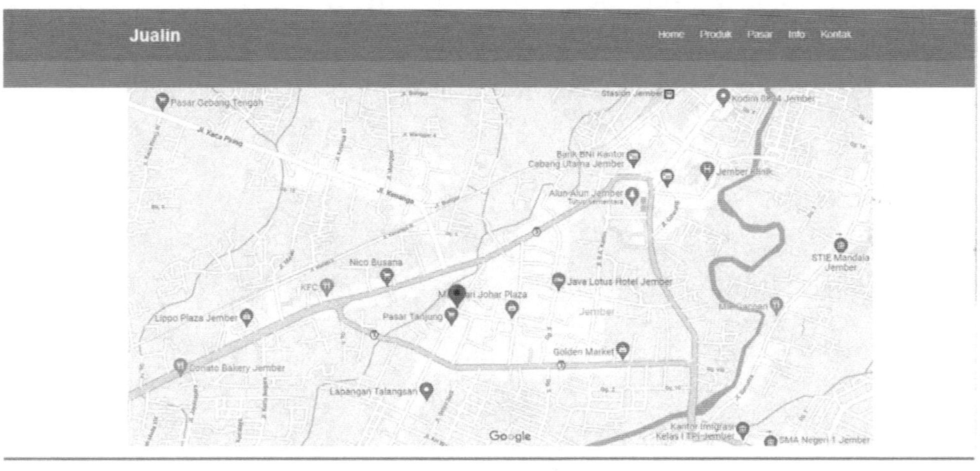

Figure 7. Display Market Location via Maps

5. Conclusion

From this research, it can be concluded that a system to obtain price information, predict prices of agricultural commodities, and get the nearest market from the location of farmers has been successfully built. The prediction of agricultural commodity prices in this study has an accuracy of 98.3%.

On the following research, prediction model can also be applied on several agricultural commodities. Big Data implementation are needed to build Datawarehouse that can store agricultural price information in real time. Furthermore, other Artificial Intelligence (AI) prediction also can be testing on various agricultural commodities because of various price's characteristic based on some factors.

IOP Conf. Series: Earth and Environmental Science **980** (2022) 012061 doi:10.1088/1755-1315/980/1/012061

References

[1] https://suaraindonesia-news.com/mie-nyaman-berbahan-organik-asal-jember, accessed on Mei 3th, 2020
[2] Friedman, N., Geiger, D., & Goldszmidt, M. (1997). Bayesian network classifiers. Machine learning, 29(2), 131-163.
[3] Wang, S. C. (2003). Artificial neural network. In Interdisciplinary computing in java programming (pp. 81-100). Springer, Boston, MA.
[4] Lasulika, M. E. (2017). Prediksi Harga Komoditi Jagung Menggunakan K-NN Dan Particle Swarm Optimazation Sebagai Fitur Seleksi. ILKOM Jurnal Ilmiah, 9(3), 233-238.
[5] Puspatika, K., & Kusumawati, Y. (2018). Peramalan Harga Cabai Dengan Metode Arima Arch-Garch dan Single Moving Average di Kota Semarang. JOINS (Journal of Information System), 3(2), 192-201.
[6] Gregorius S. Budhi, et all (2013). Penentuan Harga Jual Properti secara Otomatis menggunakan Metode Probabilistic Neural Network, Fakultas Teknologi Industri, Program Studi Teknik Informatika, Universitas Kristen Petra.
[7] A.H. Utomo and Husin (2021). Spatial Analysis and Mapping of Landslide Prone Areas in Kemuning Lor Village, Sub-District Arjasa, District of Jember. *IOP Conf. Ser.: Earth Environ. Sci.* 672 012082.
[8] Specht, D. F. (1990). Probabilistic neural networks. Neural networks, 3(1), 109-118.
[9] Coates, A., Ng, A., & Lee, H. (2011, June). An analysis of single-layer networks in unsupervised feature learning. In Proceedings of the fourteenth international conference on artificial intelligence and statistics (pp. 215-223). JMLR Workshop and Conference Proceedings.
[10] Si-ming, H. E. (2003). Modified layer-summation based on elastoplastic theory. Rock and Soil Mechanics, 24(1), 88-92.

Low-cost on-line monitoring system for agriculture based on raspberry pi zero

D P S Setyohadi [1*], S Kautsar [2], E Rosdiana [3], H Y Riskiawan [1], R Firgiyanto [3], D E Putra [4]

[1]Department of Information Technology, Politeknik Negeri Jember, Indonesia
[2]Department of Engineering, Politeknik Negeri Jember, Indonesia
[3]Department of Agriculture Production, Politeknik Negeri Jember, Indonesia
[4]Department of Agribusiness Management, Politeknik Negeri Jember, Indonesia

[*]syamsiar_kautsar@polije.ac.id

Abstract. Based on data from the National Defense Agency, the trend of conversion of agricultural land in the 1990s reached around 30,000 ha/year. However, in 2011 there was an increase of 110,000 ha/year and reached 150,000 ha/year in 2019. In fact, with the growing human population, the need for food is increasing from year to year. On the other hand, the slow regeneration of farmers and the lack of dissemination & technological innovation to increase agricultural productivity are challenges in realizing food security. In line with the ongoing industrial revolution in the industrial era 4.0, it has brought very significant changes, not only the type of technology that has shifted, but more importantly a change in mindset in this industrial era. This has a real impact on the direction of national development and has an impact on the agricultural system. The agricultural system is no longer just a cultivation activity, but agriculture is also part of the industrial system which is characterized by a transformation, both in terms of raw materials, economy, social and environment. The development of modern agriculture is marked by mindset changing of using biological and non-biological resources by the needs. The precise use of resources in agricultural production systems is then called precision agriculture. Precision agriculture is an agricultural concept with a systems approach towards low-input, high-efficiency, and sustainable agriculture. In a previous study, a CNC-based precision farming system (Farming-BOT) was developed. The device could perform vegetable crop cultivation automatically. To properly observe the development of cultivation, an online monitoring device based on the minicomputer was developed. The Raspberry Pi Zero, equipped with a 5MP camera and a temperature & humidity sensor, is used. The device is placed on the top of the Farming-BOT device. Image data, temperature, and humidity are sent to the cloud database periodically every day. Users can access the data through the website and Android application. Based on the testing result, with a stable internet connection, data on environmental conditions can be monitored perfectly.

1. Introduction

The driving wheel of the country's economy in the agricultural sector is still one of the important aspects and can lift Indonesia's image in the eyes of the world, especially as an agricultural country that is quite productive. Indonesia has very supportive climatic conditions and natural and human resources, but many farmers are still using conventional methods. It can be said that currently conventional agriculture is mostly practiced in developing countries such as Indonesia, while modern agriculture is mostly practiced in developed countries. These differences in agricultural systems affect the productivity of the results[1]. Productivity for conventional agriculture and modern agriculture is different because the inputs used for agriculture are also different, for example in terms of the tools used in conventional agriculture are still very simple such as hoes, sickles, or plowing using animals and human power as well as planting, maintaining and harvesting. In conventional agriculture, the resilience of production only lasts in the short term. This causes a country to be very dependent on imports because the availability and sustainability of agriculture are ignored. Seeing the enormous development of agriculture and this is a potential that must be maximized by all parties. However, there are challenges in increasing the potential of agriculture, several parties involved, both the government and agricultural sector actors must be more prepared to face challenges in the era of advances in information technology[2].

In line with the ongoing industrial revolution in the industrial era 4.0[3][4] and now towards the industrial era 5.0, it has brought very significant changes, not only the type of technology that has shifted, but more importantly a change in mindset in this industrial era. This has a very real impact on the direction of national development to have an impact on the agricultural system. The agricultural system is no longer just a cultivation activity, but agriculture is also part of the industrial system which is characterized by a transformation both in terms of raw materials into agricultural products that are ready to be utilized and have added value, both in terms of economic, social and environmental aspects. The development of modern agriculture is marked by a change in the mindset of using biological and non-biological resources following the needs. The precise use of resources in agricultural production systems is then called precision agriculture. Precision agriculture is an agricultural concept with a systems approach towards low-input, high-efficiency, and sustainable agriculture[5]. Or it can be said that precision agriculture is an agricultural system that optimizes the use of resources to get maximum results and also reduces the impact on the environment. The concepts that are considered include a systems approach that pays attention to inputs, processes, outputs, and outcomes.

The challenge for farmers in the era of precision agriculture is to have the ability in terms of management, be it the management of land, crops, agricultural tools, and machinery used from pre-harvest to post-harvest or processing[6], as well as human resource management. This will greatly affect the level of success in the implementation of precision agriculture. The acceleration of the development of precision agriculture in Indonesia is also inseparable from the use of modern technology today. The technology applied must be able to detect what is on agricultural land, then decide what to do, and provide treatment following the decisions that have been made so that it can provide benefits in the stages of the agricultural production system. Currently, various types of technology that can support the implementation of precision agriculture and several types of technology in precision agriculture include; (i) Geographical Position System (GPS), (ii) Geographic Information System (GIS), (iii) Variable Rate Application (VRA), (iv) Remote Sensing System, (v) Yield Mapping, (vi) Database Management System (DBMS), Spatial Variability[7].

2. Related Work

Some of the technologies that led to the development of the concept of precision agriculture, one of which was the creation of a GPS by the US Department of Defense in the late 1970s. GPS can pinpoint exact locations 24 hours a day, up to several meters. This information can provide field processing with reasonable accuracy. The application of precision agriculture is also made possible by developing sensor technology that can be combined with appropriate agricultural management measures such as cultivation, seeding, fertilization, herbicide application, and harvesting. Advances in precision farming systems have progressed along with the development and improvement of Global accuracy

Navigation Satellite System (GNSS) has also been developed since 1999. GNSS technology[8] is widely used in many livestock businesses to perform tasks related to automated control systems and geographic reference information. GNSS helps to improve the quality of auxiliary work machines, automatic controls and controlled Traffic Management Systems. In agriculture, another essential component of precision farming is Variable Rate Technology (VRT)[8] which enables precise seeding, optimization of plantings, density and better application rates, the efficiency of herbicides, pesticides and nutrients. As a result, cost efficiency and reduction of environmental pollution can be achieved. VRT can recognize various wavelengths of multi-spectral and hyperspectral cameras onboard drones and provide information on vegetation indices, such as monitoring chlorophyll content, stress levels, and their variations in space and time data. Particular attention is paid to the use of low-cost crewless aerial vehicles (UAVs)[9], or drones as they are often called. Existing and new technologies such as the Internet of Things & Big Data [10] can make processes more efficient and lead to the creation of new products and services.

An essential part of precision farming technology is the farmer and his perception. This process was started in the early 1990s by enthusiastic farmers with the best orientation, followed by despair due to lack of support and relatively low profitability. The current adoption of this approach is based almost entirely on the private sector, which offers farmers tools, products and services. Unfortunately, national advisory services in agriculture are minimal [11].

3. System Design

3.1 Hardware Design

The monitoring system is based on Raspberry Pi Zero. The Raspberry Pi Zero is a small ARM microprocessor-based mini-computer for $25. The Raspberry Pi Zero type W has a Wifi device so it can connect to the internet. In this study, the sensors used for the monitoring system are Pi Camera and DHT22. The Pi Camera is used to capture images of the environment, while the DHT22 is used to measure the temperature and humidity of the environment. Figure 1 is a system block diagram of the monitoring device used.

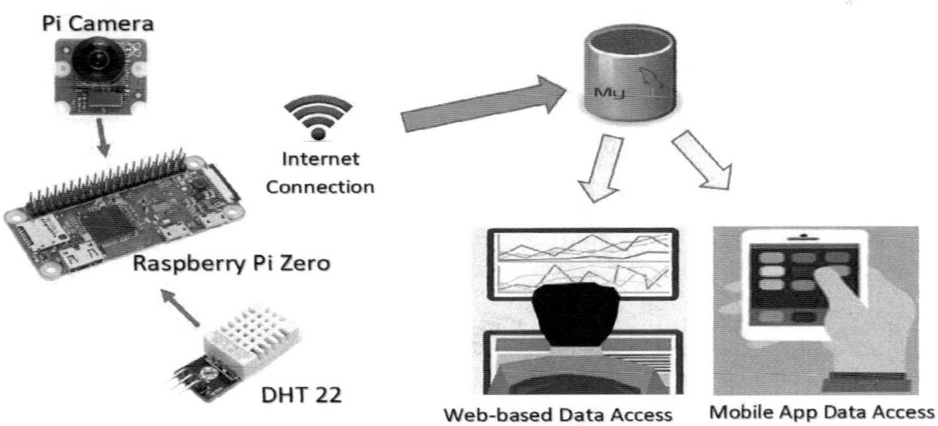

Figure 1. Block Diagram System

The Pi Camera is connected to the camera slot on the raspberry pi Zero. Whereas DHT 22 is connected to pin I2C of the Raspberry Pi Zero. DHT22 data access using one-wire serial communication. The Raspberry Pi zero connects to a wifi network that is used to remotely ssh, as well as send data to a cloud database. The whole device is powered by a 5volt DC adapter. The device is mounted on a case that is printed using 3D printing.

IOP Conf. Series: Earth and Environmental Science **980** (2022) 012062 doi:10.1088/1755-1315/980/1/012062

3.2 Software Design

For the monitoring system, there are 3 kinds of programming used: Python programming for raspberries, HTML programming for web-based monitoring, and Android programming for mobile-device-based programming. First, the raspberry pi will check the internet connection by pinging google. If an internet connection is detected, the raspberry will access the NTP server time data (GMT+7). There are 2 kinds of data sent to the database, direct data, and daily data. Live data will be sent every 10 minutes. While the daily data is sent to the database every 16.00. Direct data is used as data that is displayed on Android devices in the form of numeric values and images. While the daily data is used for web-based display in the form of graphs and images. Daily data can be used to observe the process of plant development from the beginning of planting to harvest. Figure 2 is a flowchart of raspberry pi programming. The device is also reset every 12 AM to avoid errors due to memory being full.

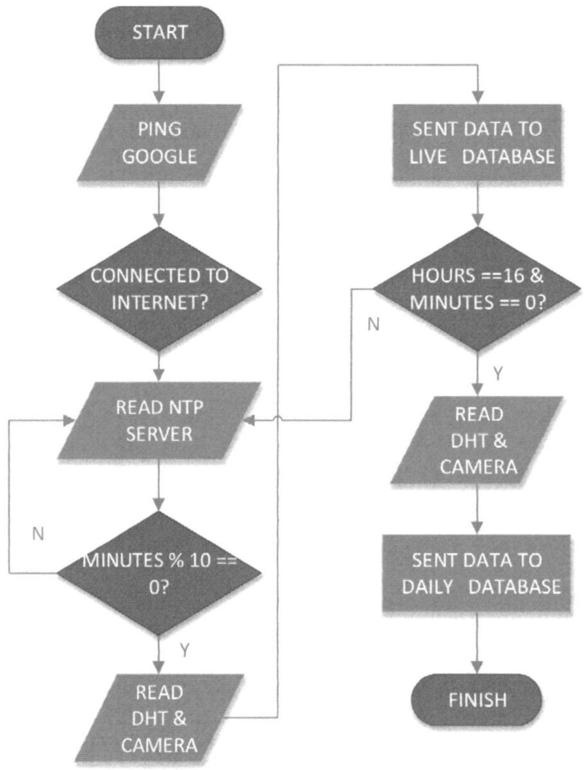

Figure 2 Flowcart system

4. Result and Discussion

4.1. Hardware Realization

The overall placement of the devices is shown in figure 3. In this study, the device is used to monitor the environmental conditions of FarmBOT [12]. The monitoring device is placed 2 meters above the FarmBOT so that the camera can capture the entire FarmBOT land. The device is also connected to the UPS as backup power in the event of a power outage.

Figure 3. Device mounting

4.2. Testing result

Once the device is installed in the FarmBOT environment, it runs for 24 hours non-stop. For direct data, the system will update the data on the MySQL database. As for daily data, the system will insert data into the MySQL database. For image data, the captured results from the pi camera are processed to DB64 format first. Figure 4 is a daily data display on an Android mobile application. The data on the mobile application will be updated every 10 minutes according to the rules on the Raspberry Pi.

Figure 4 Android based on-line monitoring system

Figure 5 is a daily data display on the web. Daily temperature and humidity data is displayed in graphical form. While the image data is displayed in RGB format per 2 days. Users can view the data on the previous day by pressing the date button displayed on the web browser. The data can be accessed at the address http://gh1.rumahkuhidroponik.com via a standard web browser..

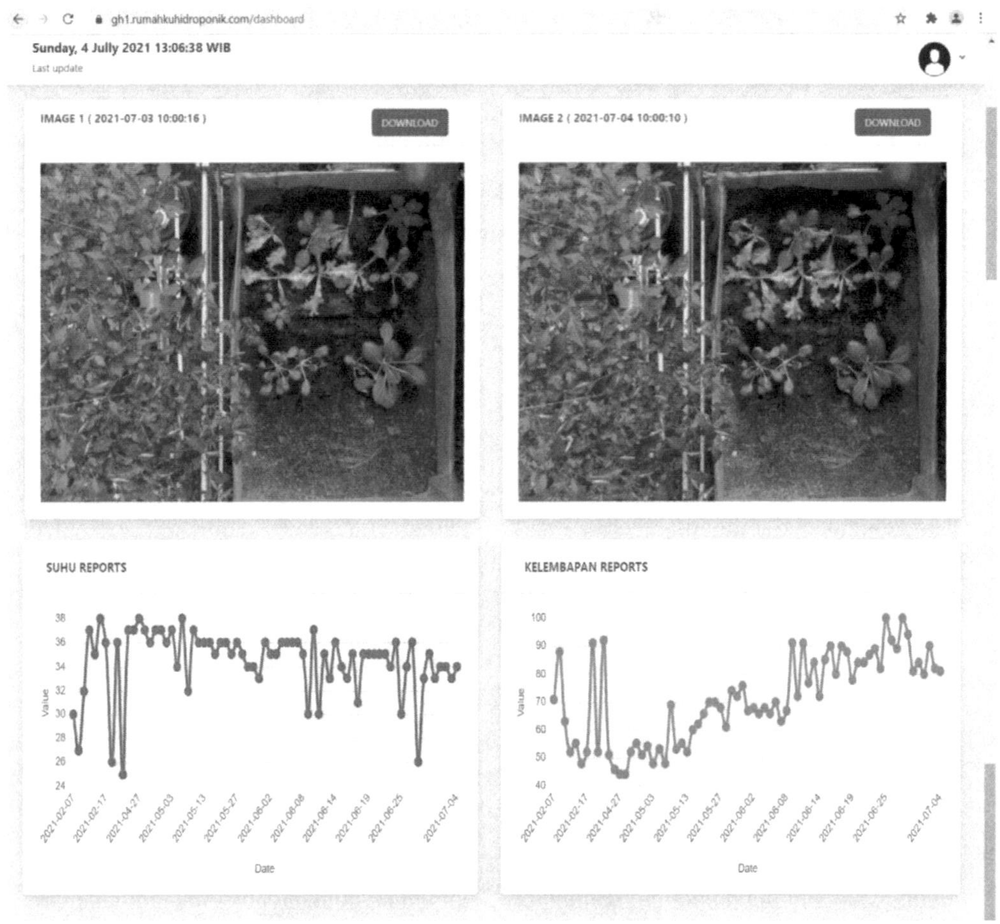

Figure 5. Web based monitoring system

The system has been running since February. However, sometimes it is still problematic. As seen in the graph there is a data storage lag from February - April. This is caused by the condition of the Raspberry which hangs every 3-4 days. On effective planting in May, the program to reset the raspberries every 12 AM had a good impact on the system. During the months of May to July, the system can store data periodically every day without experiencing a hang. Without an automatic reset system, the processing memory on the Raspberry will be full, so it can hang after 3-4 days of being turned on for 24 hours non-stop. To precision sensor measurement results, calibration is also carried out with the manufacturer's sensor. Table 1 is a comparison of measurement data between online monitoring devices and manufacturer's measuring instruments.

Table 1. Sensor data

Day	Temperature			Humidity		
	Sensor	Factory	Error	Sensor	Factory	Error
1	33	33.5	1.52	80	81	1.23
2	30	31	3.33	92	92	0.00
3	32	31.5	-1.56	82	83	1.20
4	29	29	0.00	95	96	1.04
5	29	30	3.45	95	97	2.06
6	34	34	0.00	78	78	0.00
7	31	30.5	-1.61	80	80	0.00
		Ave	0.73		Ave	0.79

5. Conclusion
Based on the research that has been done, the raspberry pi zero-based monitoring system can be an option for the low-cost version of the monitoring system. With a total price of around $50 (for pi zero, pi camera, DHT22 & 5V DC adapter) the device can fulfill the need to take sensor readings and transmit data to sensors. The system has been able to work 24 hours non-stop for more than 3 months. Power and memory management needs to be done to avoid hangs during processing. A fan can also be added to help maintain the performance of the raspberry pi, especially during the day. A stable internet connection is also required so that the process of storing data in the online database does not occur disturbances.

Acknowledgment
The authors would like to acknowledge the financial support of this work by grants from PNBP, State Polytechnic of Jember. The author also thanks to the P3M and Information Technology Department, State Polytechnic of Jember, which has provided support and assistance in completing this research.

References
[1] Chetan Dwarkani M, Ganesh Ram R, Jagannathan S, & Priyatharshini R 2015 smart farming system using sensors for agricultural task automation *IEEE Technological Innovation in ICT for Agriculture and Rural Development (TIAR)* doi:10.1109/tiar.2015.7358530
[2] Patil P, Narkhede A, Chalke A, Kalaskar H, & Rajput M, 2014 real time automation of agricultural environment *Int. Conf. for Convergence for Technology-2014* doi:10.1109/i2ct.2014.7092040
[3] Siddique T, Barua D, Ferdous Z, & Chakrabarty A, 2017 *Automated farming predictionIntelligent Systems Conference (IntelliSys)* doi:10.1109/intellisys.2017.8324214
[4] Sreeram K, Kumar R S, Bhagavath S V, Muthumeenakshi K, & Radha S, 2017 Smart farming—A prototype for field monitoring and automation in agriculture *Int. Conf. on Wireless Communications, Signal Processing and Networking (WiSPNET)* doi:10.1109/wispnet.2017.8300148
[5] Jauhari R, Jati A N, & Azmi F, 2017 Mechanical design of CNC for general farming automation *5th Int. Conf. on Instrumentation, Control, and Automation (ICA)* doi:10.1109/ica.2017.8068411
[6] B Widiawan B, C Triwidiarto, S Kautsar, R Firgiyanto, 2020 Wireless Greenhouse Monitoring System Using Tricycle Mobile-Robot Based On Rasberry PI *IOP Conference Series: Earth and Environmental Science* **411** 012058 doi:10.1088/1755-1315/411/1/012058
[7] Suakanto S, Engel V J L, Hutagalung M, & Angela D, 2016 Sensor networks data acquisition and task management for decision support of smart farming *Int. Conf. on Information Technology Systems and Innovation (ICITSI)* doi:10.1109/icitsi.2016.7858233
[8] Chen L, Yu D, Zhang H, Geng C, & Dong L, 2012 Design and implement of a modularized CNC interpreter based on the integration of tool path planning module *IEEE International Conference on Computer Science and Automation Engineering (CSAE)* doi:10.1109/csae.2012.6273027

[9] Huy Sokchea, Khieu Borin and T R Preston 2015 Carry-over effects of biochar on yield of Mustard Green vegetable (Brassica juncea) and on soil fertility *Livestock research for rural development* **27** no 6-62 Cali, Colombia

[10]D. Shinde and N. Siddiqui, 2018 IOT Based Environment change Monitoring & Controlling in Greenhouse using WSN *Int. Conf. on Information, Communication, Engineering and Technology (ICICET)* pp 1-5

[11]DPS Setyohadi, et al. 2021 application of digital image processing to predict the diameter of chrysanthemum flowers ready to *harvest food and agricultural sciences: polije proceedings series* pp 171-176

[12]S Kautsar, et al. 2020 Farming Bot: Precision Agriculture System in Limited Land Based On Computer Numerical Control (CNC) *IOP Conference Series: Earth and Environmental Science* **411** doi:10.1088/1755-1315/411/1/012059

Adaptive Temperature And Humidity Control System on Kumbung Mushroom using Fuzzy Neural Network Algorithm

D R Hartadi[1], K Agustianto[1*], P Destarianto[1], R Wardana[2], T Kustiari[3], I G Wiryawan[1], E Mulyadi[1]

[1]Department of Information Technology, State Polytechnic of Jember, Indonesia
[2]Department of Agricultural Production, State Polytechnic of Jember, Indonesia
[3]Department of Agribusiness Management, State Polytechnic of Jember, Indonesia

*Email: agustianto.khafid@gmail.com

Abstract. National mushroom production in 2018, 2017, 2016, and 2015 were: 31052, 3702, 40915 and 33485 tons. Mushroom cultivation, especially oyster mushrooms, has the potential to be developed more widely in Indonesia because it has economic value and is environmentally friendly. However, the cultivation of oyster mushrooms has challenges. The challenge that arises in the cultivation of oyster mushrooms is to grow well at a temperature of 16 – 30 °C and a relative humidity of 80 – 95%. Environmental conditioning carried out by farmers on average through spraying water in mushroom kumbung manually. The use of manual methods requires high human resources, besides that it is not effective and efficient. This study aims to develop an Adaptive Temperature and Humidity Control System on Kumbung using Backpropagation Neural Network, where to determine the temperature and humidity the system will evaluate the readings of several sensors at once. The system developed is an adaptive system with an accuracy of 97%, so it is can be used to increase the precision of mushroom cultivation, so that production efficiency and increase in mushroom production can be achieved.

1. Introduction

The Indonesian Institute of Sciences (LIPI) advances the development of food mushrooms with the goal that they can be burned through appropriately and generally by the Indonesian public. Mushrooms developed in Indonesia have high dietary benefits, even mushrooms contain 19-35 percent higher protein than rice (7.38 percent) and wheat (13.2 percent) [1]. National mushroom production in 2018, 2017, 2016 and 2015 amounted to: 31052, 3702, 40915 and 33485 tones [2]. In 2018 the per capita availability is 0.15 [3], this figure should be expanded as an elective food fixing arrangement, this depends on the surface of mushrooms that look like meat and is delightful, making this plant regularly become different sorts of handled food [1]. Development of mushrooms, particularly clam mushrooms, can be grown all the more generally in Indonesia since it has financial worth and is harmless to the ecosystem [1]. In any case, clam mushroom development has difficulties, particularly to deliver the most extreme creation in each kumbung.

The test that emerges in clam mushroom development is that it develops well at temperatures of 16-30°C and relative moistness of 80 - 95%. Natural molding is brought out by the normal rancher through the manual splashing of water in the mushroom kumbung [4]. According to Hariyanto et al, in [5] that oyster mushrooms can be harvested around the age of baglog 40 days after inoculation (planting seeds) up to a mushroom production period of about 3-4 months with a production yield of 0.6 to 1.3 kg per baglog of mushrooms. This manual strategy resolves a few issues like those found at PT. Nusantara Healthy Mushroom (JASENTRA). The utilization of the manual technique requires high HR, other than that the accuracy of arranging isn't equally appropriated. So it is important to expand the accuracy of temperature and moistness in kumbung with innovation, research identified with the utilization of innovation for mushroom development was completed by [6][7][8], where the examination utilizes the IoT way to deal with consequently screen the stickiness and temperature of the kumbung.

This study aims to develop a Temperature and Humidity Control System in Kumbung using a Fuzzy Logic Controller. This study uses a Fuzzy Proportional-Integral-Derivative (PID) Controller, sensor data (Temperature and Humidity) will be processed as a sensor array [9][10][11], which decides the temperature and humidity of the framework, will assess the readings of a few meetings immediately. This is planned to deliver a precise perusing, the consequences of this perusing will then, at that point, be utilized as a fluffy info variable/derivation motor to decide the treatment/splashing of kumbung to accomplish the objective temperature and moistness. The framework being created is a versatile framework that will adjust to the climate, so it is trusted that the arrangement produced by the device can be utilized to build the accuracy of mushroom development, so creation effectiveness and expanded mushroom creation can be accomplished.

2. Related Work

Research of the utilization of the Internet of Things (IoT) is being grown, particularly compared to its capacity to help people in the entirety of their exercises, one of which is with regards to the climate. Exploration led by [12] fostered the plan and execution of the Wireless Sensor Network (WSN) just as examination led by [13]. Examination identified with the utilization of WSN for mushroom development was done by [6][7][8], where the exploration utilized the IoT/WSN way to deal with naturally screen the dampness and temperature of the kumbung.

This study aims to develop a Temperature and Humidity Control System in Kumbung utilizing a Fuzzy Logic Controller. This review utilizes a Fuzzy Proportional-Integral-Derivative (PID) Controller, the sensor information (Temperature and Humidity) will be handled as a sensor array [9][14][15], were to decide the temperature and moistness the framework will assess the readings of a few sensors immediately. This is expected to deliver an exact perusing, the results of this perusing will then, at that point, be utilized as a fluffy information variable/deduction motor to decide the treatment/splashing of kumbung to accomplish the objective temperature and mugginess. The framework being created is a versatile framework that will adjust to the climate, so it is trusted that the design created by the apparatus can be utilized to build the accuracy of mushroom development, so creation proficiency and expanded mushroom creation can be accomplished.

3. Research Methods

Literature study on dependent on Figure 1 was completed determined to assemble data [16][17] from a few references identified with the issues to be examined [18], to frame the State of The Art. The following stage is create rule. Creating rules is a significant stage in this examination [19][20], in light of the fact that this review utilizes a few sensors (which are utilized as factors) [21]. The outcomes from the sensor perusing will then, at that point, be utilized to decide the proper water splash [22] to get the dampness and temperature as per the ideal development and advancement of Oyster Mushrooms. After the course is created, it proceeds with the execution of the ATHCS.

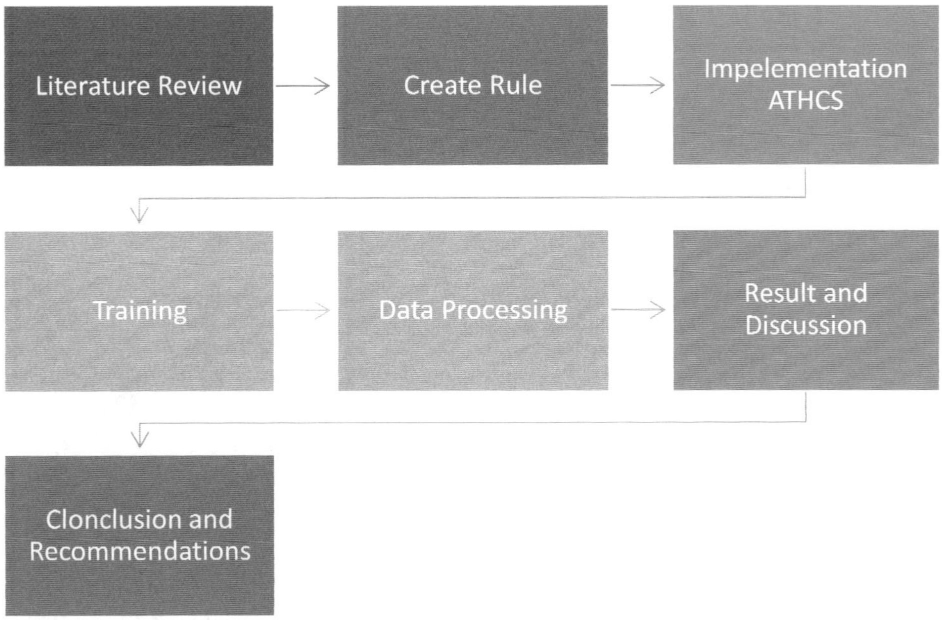

Figure 1. Research Methods

The rules that have been acquired are then executed into the ATHCS, this is finished utilizing the Fuzzy methodology. The utilization of the Fuzzy strategy plans to create a more precise perusing [22][23]. The following stage is trying, this stage is expected to demonstrate that the sensor readings are proper, at this stage the precision of the instrument will likewise be acquired. The readings will be handled to acquire the exactness of the ATHCS temperature and dampness class readings. The consequences of information handling as device exactness are then contrasted and manual computations, to test the precision by ground truth. The outcomes got at this stage will examine the elements that influence the precision esteem acquired. At this stage is the last phase of the exploration, specifically reaching inferences from the consequences of conversation investigation and giving ideas to additional research.

4. Discussion

4.1. Automatic Temperature dan Humidity Control System (ATHCS) in Kumbung using Fuzzy Logic Controller

The improvement of mushrooms is a biotechnological application that uses practical and direct development that ordinary people in provincial locales can do. Improvement of mushrooms can be a possibility for the usage of typical resources, the extension of food and sustenance, regular protection and business openings for the neighbourhood [24]. The test that arises in shellfish mushroom improvement is that it grows well at temperatures of 16 - 30 ° C and relative tenacity of 80 - 95%. A regular trim is assisted by the ordinary farmer through a manual sprinkling of water in the mushroom kumbung [4]. This manual technique settled a couple of issues like those found at PT. JASENTRA. The usage of the manual methodology requires high HR [25][26]. Other than that, the precision of organizing isn't similarly appropriated, so that some BagLog's are hurt. The Manager side can't screen the conditions of the kumbung at whatever point.

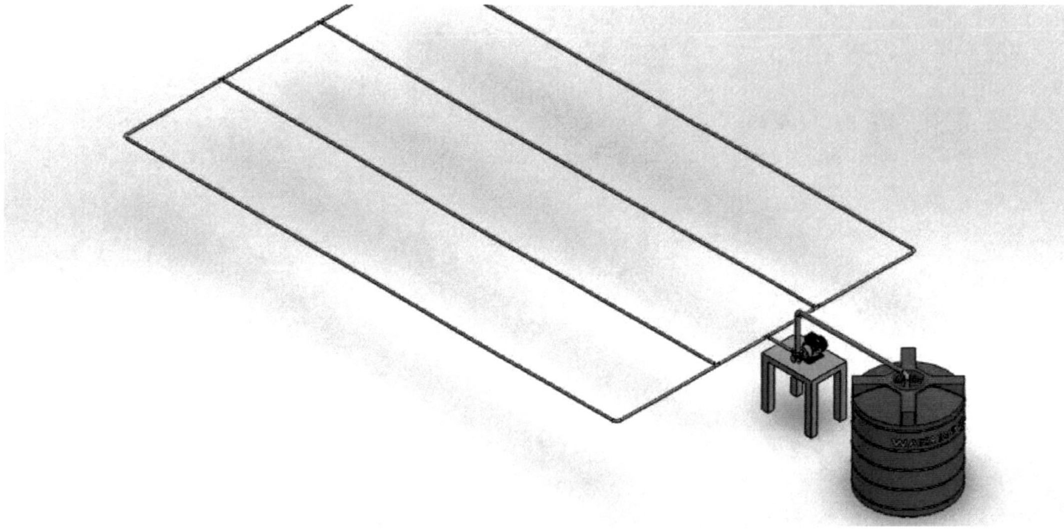

Figure 2. Design Tools

These conditions achieved the production of mushrooms at PT. JASENTRA isn't great (the resulting creation is half of the breaking point for sure might measure up to 50 kg for each kumbung). To grow the precision of temperature and tenacity in kumbung with advancement, research related to the use of development for mushroom improvement was finished by [6][7][8], where the investigation used the Internet of Things (IoT) way of managing normally screen clamminess and temperature kumbung.

This review means to support a Temperature and Humidity Control System in Kumbung utilizing a Fuzzy Logic Controller, which is displayed in Figure 2. This investigation utilizes a Fuzzy Proportional-Integral-Derivative (PID) Controller, sensor information (Temperature and Humidity) will be managed as sensors display [9], was to pick the temperature and wetness the framework will assess the readings of several sensors immediately. This method is planned to pass on accurate readings. The result of these readings will then, at that point, be utilized as a fluffy information variable/inferring motor to pick the treatment/showering of Kumbung to accomplish the actual temperature and constancy, tolerating the ideal temperature or clamminess is not developed, the specific image of the instrument is displayed in Figure 2, will begin the spout. The design being made is an adaptable framework that will adjust to the climate, so it is accepted that the approach conveyed by the instrument can be utilized to manufacture the accuracy of mushroom progression, so creation ability and broadened mushroom creation can be refined.

4.2. Testing Results and Discussion

This study develops ATHCS with a Neural Network [25] setup shown in Figure 3, in the form of 2 input layers (X) (temperature and humidity), one hidden layer (Z) with three neutrons, and finally, an output (Y) consisting of 1 neuron containing the results from the previous calculation. The next step is research to normalize and denormalize.

In the backpropagation algorithm, the conditions that must be met in the activation function used are: continuous, easily differentiated, and a function that does not descend. In this study, the activation function used is a binary sigmoid activation function where the value range is between (0 to 1). So that the data to be used needs to be normalized before being processed using the backpropagation algorithm, shown by Equation 1.

$$y = \frac{x - min}{\max - min}(0{,}8) + 0{,}1 \tag{1}$$

The next process carried out by research is to test epochs (100, 200, 300, 400, 500, 600, 700, 800, 900, 1000). This data test aims to obtain the optimal epoch value with a low error rate value. The training data used were 698 training data and the test data used were 300 test data. In this test calculation, the initial learning rate used is 0.2. Each epoch is tested ten times by initializing the initial weight using the Nguyen-Widrow algorithm so that the initial weight value is random and different. Error rate calculation using Mean Average Percentage Error (MAPE).

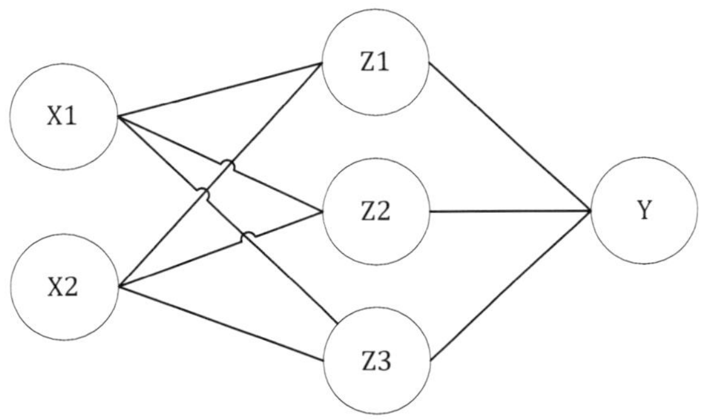

Figure 3. Neural Network Setup

Table 1. Normalization Result (Training)

Number	Temperature	Humidity	Target
1	0.53826086956522	0.41419284940412	0.57783783783784
2	0.53826086956522	0.41462621885157	0.5781981981982
3	0.53826086956522	0.41419284940412	0.57783783783784
4	0.1	0.58147345612134	0.17207207207207
5	0.53826086956522	0.41419284940412	0.57783783783784

The test results show that the best setup for this research data is Epoch 600 epoch with an average error (MAPE) of 0.599% and a Learning rate value of 0.2 with an average error (MAPE) of 0.645%. So the research using these two setups is testing the test data which is part of the data that has not been used in the study.

Table 2. Variables Range

Class	Temperature + Humidity	Range
1	Bad	≤ 0.5559
2	Optimal	0.5630 – 0.5836
3	Bad	≥ 0.5854

This study aims to develop a Temperature and Humidity Control System in Kumbung using a Fuzzy Logic Controller, shown in Figure 5. This study uses a Fuzzy Proportional-Integral-Derivative (PID) Controller. The sensor data (Temperature and Humidity) will be processed as a sensor array [9] to determine the temperature and humidity. The system will evaluate the readings of several sensors at once. This method is intended to produce an accurate reading. The results of this reading will then be used as a fuzzy input variable/inference engine to determine the treatment/spraying of kumbung to achieve the target temperature and humidity.

The value of the multi-sensor readings is determined. This worth is then utilized as a contribution to fluffy to get the enrollment level as displayed in Figure 4. The condition utilized in the exploration that addresses Figures 4 is shown by Equations 2. Separately, the condition has a similar person, where each has two classes as per Table 2. Each class is addressed by μ so that in every situation, there are two μ.

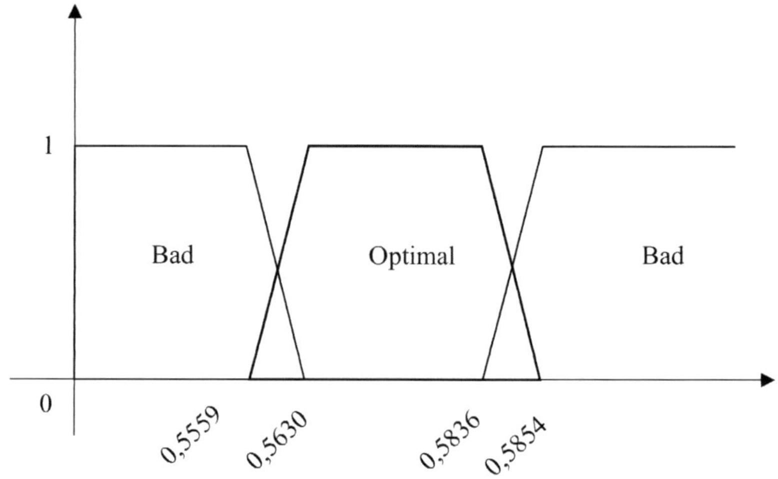

Figure 4. Temperature Membership

$$\mu[Bad] = \begin{cases} 0; \ x \geq 0.5559 \ atau \ x \leq 0.5854 \\ \dfrac{b-x}{b-a}; \ 0.5559 < x < 0.5630 \\ \dfrac{d-x}{d-c}; \ 0.5836 < x < 0.5854 \\ 1; \ x \leq 0.5559 \ atau \ x \geq 0.5854 \end{cases} \quad (2)$$

$$\mu[Optimal] = \begin{cases} 0; \ x \geq 0.5854 \ atau \ x \leq 0.5559 \\ \dfrac{x-a}{b-a}; \ 0.5559 < x < 0.5630 \\ \dfrac{x-c}{d-c}; \ 0.5836 < x < 0.5854 \\ 1; \ x \geq 0.5630 \ atau \ x \leq 0.5836 \end{cases}$$

The results from the previous process are then entered into the rules. The rules used in this study are shown in Table 2. In Table 2, it can be seen that there are 2 classes used by the study for the final evaluation, namely optimal and bad. These three classes will be used to maintain the lemur so that the temperature and humidity conditions remain optimal or in accordance with the conditions R [6] and R [9].

The outcomes from the past cycle are then gone into the guidelines. The guidelines utilized in this review are displayed in Table 2. Table 2, tends to be seen that for the last assessment, specifically ideal and terrible. These three classes will be utilized to keep up with the lemur, so the temperature and mugginess conditions stay ideal or as per the conditions R [6][9].

The rule is a rule that contains facts that are connected using the AND operator for cases that use more than one fact. In this study, the input variable used is one variable, namely the value of the calculation of the Backpropagation algorithm. Here are some predefined rules. [R1] If the Backpropagation result is less than 0.5559, then the quality is "Bad", [R2] If the Backpropagation result is more than 0.5854, then the quality is "Bad", [R3] If the Backpropagation result is less than 0.5836 and the Backpropagation result is more than 0.5630 then the quality is "Optimal". The inference stage is the stage to change the fuzzy input variable into a fuzzy output value. The following is an example of a fuzzy algorithm calculation with the input value from the calculation of the Backpropagation algorithm (Input= 0.58496436056773). The defuzzification stage is the stage to change the inference results into firm values. The final result is calculated by finding the average value if the variables used are more than one. Based on the fuzzy value from the calculation above, the fuzzy value taken is the largest value, which is 0.75797809318587, with the fuzzy output included in the Optimal category. So it can be concluded that the temperature and the pond are included in the Optimal category.

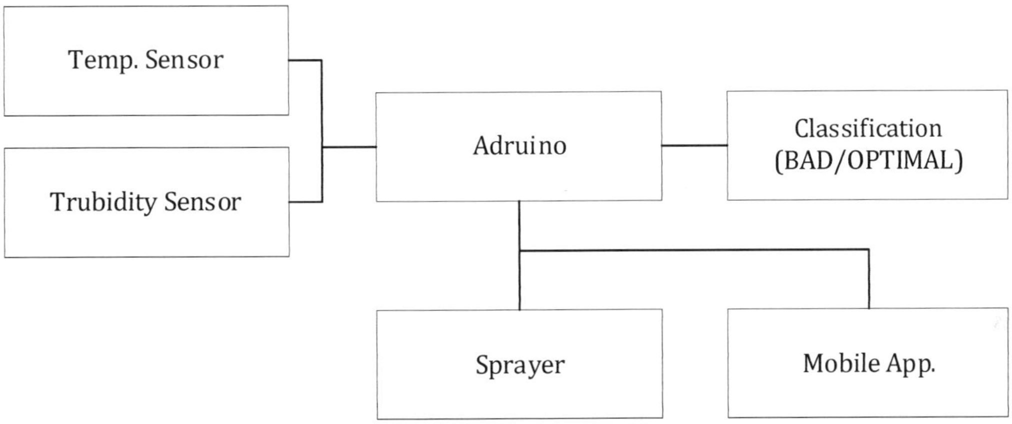

Figure 5. Diagram System

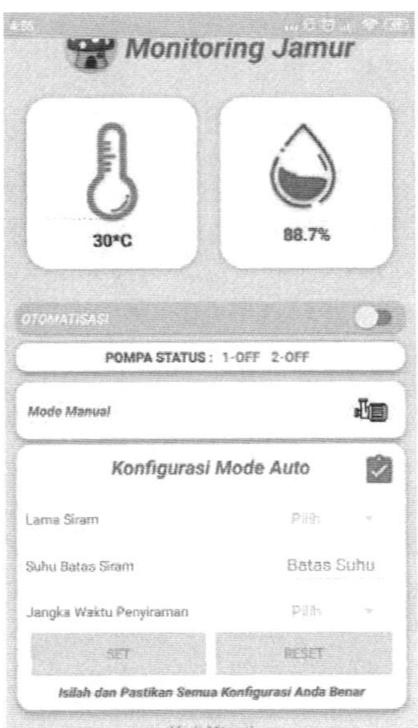

Figure 6. Android Controller

Figure 7. Kumbung

This study uses Black Box Testing to ensure that all the features are functioning properly, so the research is testing the software and hardware of the application at once, are shown in Figure 5. 6, and 7. The results of the tests carried out by the research are shown in Table 3. Not only did the test feature, this study also tested the output of the tool, where the 300 data generated by the tool were validated by experts. The method used in this research is Ground Truth, while the results obtained are an accuracy of 97%, shown in Table 4.

Table 3. *Black Box Testing*

Number	Item	Function	Result
1	Android App.	Data W/R Arduino	Success
2	Temperature Sensor	Temperature Value	Success
3	Turbidity Sensor	Turbidity Value	Success
4	Nozzle Spray	Water Spray	Success

Table 4. *Fuzzy Testing Result*

Number	Temperature + Humidity	Target	Class	Expert
1	0,715955866	107,8251782	1	True
2	0,715955866	107,8251782	1	True
3	0,708576426	107,7880965	1	True
4	0,708576426	107,7880965	1	True
...	..		..	
298	0,110262384	104,7815685	1	True
299	0,100513464	104,7325802	1	True
300	0,092142654	104,6905168	1	True

5. Conclusion

This study creates a Temperature and Humidity Control System in Kumbung utilizing Backpropagation and Fuzzy. Sensor information in this review (Temperature and Humidity) will be handled as sensor clusters to decide the temperature and dampness. The framework will assess the readings of a few sensors without a moment's delay. This strategy is expected to deliver a precise perusing. The aftereffects of this perusing will then, at that point, be utilized as a fluffy info variable/deduction motor to decide the treatment/splashing of Kumbung to accomplish the objective temperature and dampness. The test outcomes show an exactness worth of 97%. So it is expected that the design created by the device can be utilized to expand the accuracy of mushroom development, so creation productivity and expanded mushroom creation can be accomplished.

Acknowledgment

The authors would like to acknowledge the financial support of this work by grants from PNBP, Politeknik Negeri Jember. The author also thanked the P3M Politeknik Negeri Jember, which has provided support and assistance in completing this research.

References

[1] Saskiawan I 2017 LIPI Galakkan Budidaya Jamur Oangan di Indonesia
[2] BPS 2020 Produksi Tanaman Sayuran Jamur
[3] Pertanian 2020 Statistik Konsumsi Pangan Tahun 2018 2–3
[4] Waluyo S, Wahyono R E, Lanya B and Telaumbanua M 2019 Pengendalian Temperatur dan Kelembaban dalam Kumbung Jamur Tiram (Pleurotus sp) Secara Otomatis Berbasis Mikrokontroler *agriTECH* **38** 282
[5] Hariyanto B, Hendariyono J and Sukmayoga T D 2017 Efektifitas Penggunaan Peralatan Pemadat Baglog Sistem Hidrolik dan Cara Konvensional pada Produksi Jamur Tiram Putih (Pleurotus ostreatus) 136–41

[6] Kaewwiset T and Yodkhad P 2017 Automatic temperature and humidity control system by using Fuzzy Logic algorithm for mushroom nursery *2nd Jt. Int. Conf. Digit. Arts, Media Technol. 2017 Digit. Econ. Sustain. Growth, ICDAMT 2017* 396–9

[7] Najmurrokhman A, Daelami A, Kusnandar, Adiputra F and Komarudin U 2019 Design and Implementation of Temperature and Humidity Control System in Oyster Mushroom Cultivation using Fuzzy Logic Controller *2019 International Conference on Computer, Control, Informatics and its Applications (IC3INA)* (IEEE) pp 146–50

[8] Fuady G M, Turoobi A H, Majdi M N, Syaiin M, Adhitya R Y, Rachman I, Rachman F, Negara M A P, Soeprijanto A and Soelistijono R T 2018 Extreme learning machine and back propagation neural network comparison for temperature and humidity control of oyster mushroom based on microcontroller *2017 International Symposium on Electronics and Smart Devices, ISESD 2017* vol 2018-Janua pp 46–50

[9] Ghosh S, Roy A, Singh S, Saha H, Ojha V K and Dutta P 2012 Sensor Array for Manhole Gas Analysis *Proc. - ISPTS-1, 1st Int. Symp. Phys. Technol. Sensors* 9–12

[10] Ghosh S, Chatterjee S, Kundu A, Maity S and Saha H 2012 Thermal Analysis of Cantilever MEMS Based Low Power Microheater Array for The Selective Detection of Explosive and Toxic Gases *Proc. - ISPTS-1, 1st Int. Symp. Phys. Technol. Sensors* **1** 290–3

[11] Agustianto K, Destarianto P and Dewanto W K 2020 Development of real-time motion autonomous surface vehicle controlling for coral reef conservation and fisheries *IOP Conference Series: Earth and Environmental Science* vol 411

[12] Cao F, Jiang F, Liu Z, Chen B and Yang Z 2014 Application of ISFET Microsensors with Mobile Network to Build IoT for Water Environment Monitoring *Proceedings - 2014 International Conference on Intelligent Environments, IE 2014* pp 207–10

[13] Perumal T, Sulaiman N and Leong C Y 2015 Internet of Things (IoT) Enabled Water Monitoring System *2015 IEEE 4th Global Conference on Consumer Electronics (GCCE) Internet* pp 86–7

[14] Science E 2021 Development of automatic temperature and humidity control system in kumbung (oyster mushroom) using fuzzy logic controller *IOP Conf. Ser. Earth Environ. Sci.*

[15] Farkan M and Setiyanto D D 2021 Development of realtime surface modeling vehicle for shrimp ponds (ReSMeV-SP) *IOP Conf. Ser. Earth Environ. Sci.*

[16] Choirunnisa S, Meidyani B and Rochimah S 2019 Software Defect Prediction using Oversampling Algorithm: A-SUWO *2018 Electr. Power, Electron. Commun. Control. Informatics Semin. EECCIS 2018* 337–41

[17] Agustianto K and Destarianto P 2019 Imbalance Data Handling using Neighborhood Cleaning Rule (NCL) Sampling Method for Precision Student Modeling *Proceedings - 2019 International Conference on Computer Science, Information Technology, and Electrical Engineering, ICOMITEE 2019*

[18] Agustianto K, Permanasari A E, Kusumawardani S S and Hidayah I 2016 Design adaptive learning system using metacognitive strategy path for learning in classroom and intelligent tutoring systems *AIP Conference Proceedings* vol 1755

[19] Destarianto P, Etikasari B and Agustianto K 2018 Developing Automatic Student Motivation Modeling System *J. Phys. Conf. Ser.* **953**

[20] Dewanto W K, Agustianto K and Sari B E 2018 Developing thinking skill system for modelling creative thinking and critical thinking of vocational high school student *J. Phys. Conf. Ser.* **953**

[21] Destarianto P, Riskiawan H Y, Agustianto K and Kautsar S 2018 Developing food sensory test system with preference test (Hedonic and Hedonic quality) wheat bread case study *Proceedings - 2017 International Conference on Sustainable Information Engineering and Technology, SIET 2017* vol 2018-Janua

[22] Agustianto K, Destarianto P and Dewanto W K 2020 Development of real-time motion autonomous surface vehicle controlling for coral reef conservation and fisheries *IOP Conf. Ser.*

Earth Environ. Sci. **411**

[23] Agustianto K and Destarianto P 2019 Imbalance Data Handling using Neighborhood Cleaning Rule (NCL) Sampling Method for Precision Student Modeling *Proc. - 2019 Int. Conf. Comput. Sci. Inf. Technol. Electr. Eng. ICOMITEE 2019* **1** 86–9

[24] Rahayu S and Martono D S 2019 Optimizing Quality of White Oyster Mushroom Seeds Through Plant Propagation *IOP Conf. Ser. Earth Environ. Sci.* **347**

[25] Kurnia D W, Kautsar S, Etikasari B and Khafidurrohman A 2018 A control scheme for typist robot using Artificial Neural Network *Proc. - 2017 Int. Conf. Sustain. Inf. Eng. Technol. SIET 2017* **2018-Janua** 374–8

[26] Wibowo N S, Destarianto P, Riskiawan H Y, Agustianto K and Kautsar S 2018 Development of low-cost autonomous surface vehicles (ASV) for watershed quality monitoring *2018 6th International Conference on Information and Communication Technology, ICoICT 2018* vol 0 (IEEE) pp 489–94

A Prototype of Raspberry Pi Camera Morphometric Measurements on Sheep Using Digital Image Processing

Angga Rahagiyanto[1*], Bakhtiyar Hadi Prakoso[1] and M. Adhyatma[2].

[1]Medical Record Department, Politeknik Negeri Jember[3] Jalan Mastrip PO BOX 164, Jember 68121, Indonesia
[2]Livestock Department, Politeknik Negeri Jember[3] Jalan Mastrip PO BOX 164, Jember 68121, Indonesia.

rahagiyanto@polije.ac.id

Abstract.
Quantitative data are needed to identify and predict opportunities for increasing livestock productivity. The increase in livestock productivity can be assessed from the dimensions of the body of the livestock by measuring directly using morphometric. Morphometric data is an important parameter used to study livestock anatomy, productivity, growth rate, and performance quality of livestock. This research is focused on how to make a prototype development of a camera tool that can measure livestock morphometrics at a distance. The Camera use in mini camera with a raspberry pi as a microcontroller. And method to process the capture of data is Artificial Neural Network that downloaded from tensor flow. The result of this experiment show that the accuracy of the system is 50%. However, the camera has not been able to identify the object of farm animals correctly. There are several livestock objects detected as dog or cat. For the future work the library must be modified so it can only detect the sheep or goat so the morphometrics can be done.

1. Introduction

The existence of livestock, especially sheep in Indonesia, has enormous potential. In 2018, the population of sheep in Indonesia reached 17.611.000 heads. This population continues to increase in 2019 to 17.794.344 individuals [1]. The Ministry of Agriculture through the Directorate General of Animal Husbandry and Animal Health noted that the national sheep and goat population continues to increase [2]. This is in line with the vision of the Minister of Agriculture for Indonesia to become a world food barn by 2045 [3]. The role of livestock, especially sheep and goat cattle, is very important to note. Sheep and goat cattle are used as a provider of meat to meet food needs. On paper, the population of sheep and goats in Indonesia is quite large. However, the ever-increasing demand has made the supply of this small ruminant commodity in short supply. Requests come not only from hospitality (hotels, restaurants, catering) but also from religious activities such as *aqiqah* and *qurbani*. The efforts to increase the productivity of sheep and goat cattle is by collecting quantitative data (body dimensions). Quantitative data are needed to identify and predict opportunities for increasing livestock productivity. The increase in livestock productivity can be assessed from the dimensions of the body of the livestock by measuring directly using measuring instruments. This technique is called morphometrics with quantitative analysis technique (body dimensions) that includes shape and size. Morphometric data is an important parameter used to study livestock anatomy, productivity, growth rate, and performance quality of livestock. Conventional morphometric measurements are carried out by directly measuring livestock body parameters using a ruler, measuring stick, measuring tape regarding the body prominence starting from height, body length, chest circumference, hip height, and so on [4]. This technical approach causes livestock to become stressed, aggressive, and difficult to control. [5]. Based on these problems, the researchers made a prototype development of a camera tool that can measure livestock morphometrics at a distance. This method is expected to be applied for safer morphometric measurements, produce more accurate data, and livestock that can be measured at the same time and can provide more data that cannot be taken in direct measurements [6]. To support this prototype, the first thing to do is to build a system that can detect goats or sheep at a distance. This will be focus to this research. The camera prototype was made using a Raspberry Pi minicomputer. The Raspberry Pi will be connected to a mini camera that will be programmed with an algorithm to be able to recognize sheep objects. The higher the camera accuracy in recognizing sheep objects will be the first step in formulating real-time morphometric camera recognition.

2. Materials and Methods

Making a morphometric camera begins with determining the type of device to be worked on. Researchers use a mini camera with a raspberry pi as a microcontroller to create a program that is able to measure with a morphometric approach. The first step in making this morphometric camera is to assemble the camera components that will be used:

Table 1. Morphometric camera prototyping tools and materials

No.	Description
1.	Raspberry Pi Model B RAM 4 GB
2.	Camera Module for Raspberry Pi Zero – Standard
3.	Raspberry Pi case alumunium metal with dual fan
4.	Flex Cable for Raspberry Pi Camera
5.	USB cable to type-C
6.	Cable RJ45

The tools and materials used for the hardware assembly process and software creation are as shown in Table 1. Once assembled, the supporting software installation process is carried out to create programs on the hardware using MobaXterm and PuTTY.

Figure 1. Camera Raspberry pi Prototype light up

IOP Conf. Series: Earth and Environmental Science **980** (2022) 012064 doi:10.1088/1755-1315/980/1/012064

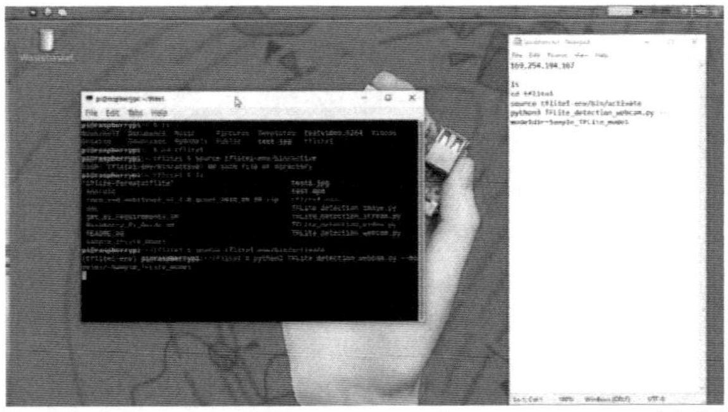

Figure 2. Build software and camera running processes with MobaXterm

The raspberry pi camera prototype assembled so that it can run properly is shown in Figure 1. The raspberry pi program is run on windows using virtual MobaXTerm and putty so that the raspberry pi can run on windows, shown in Figure 2.

2.1. Material

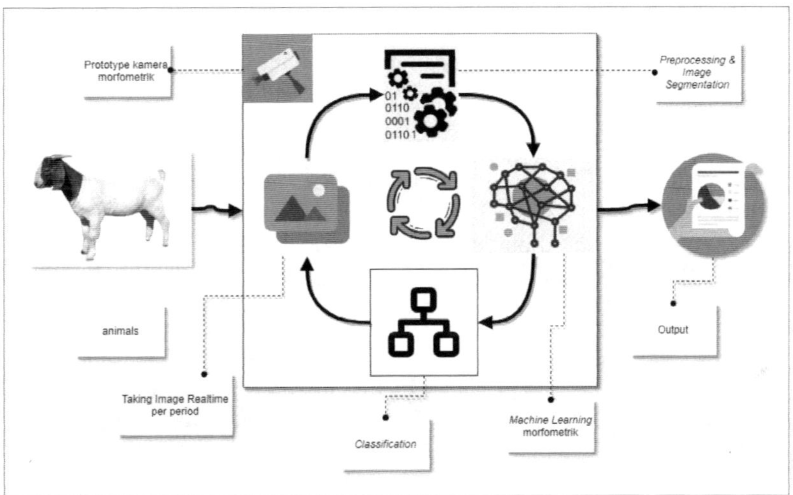

Figure 3. Block Diagram of the Proposed Work

At this stage (Figure 3) the researcher developed a camera tool that was not only able to recognize moving objects. However, it can also determine & recognize detected livestock objects for the needs of morphometric calculation techniques with a machine learning approach - Neural Network.

2.2. Methods

The fundamental idea of the Artificial Neural Network (ANN) is to adopt the thinking mechanism of a human-like system or application, both for processing various received element signals, error tolerance and parallel processing. In ANN, neurons are grouped into layers, which are called neuron layers (figure 11). Usually, each neuron of a layer is connected to all neurons in the back and front layers (except input and output). Information sent in an NN is propagated layer by layer from input to output without or through one or more hidden layers.

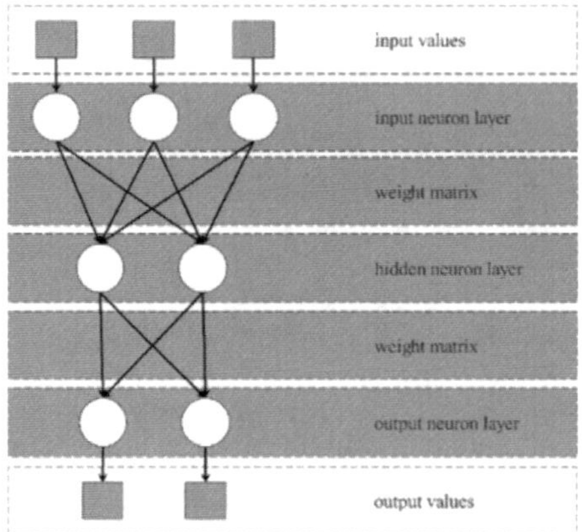

Figure 4. The Illustration of Neural Network that being used

In this research, researcher will utilize the ANN library which is derived from tensor flow. The advantage of this tensor flow is that the data set used for the data training process is quite large, resulting in the level of accuracy when detecting objects.

3. Results and Discussion

The researcher tested the camera hardware & software that had been made to the predetermined target of livestock. Camera testing was carried out on sheep and goats belonging to the Politeknik Negeri Jember. The testing process is carried out in an indoor place, as follows:

Figure 5. The process of capturing the target object with a raspberry pi camera in the environment (indoor)

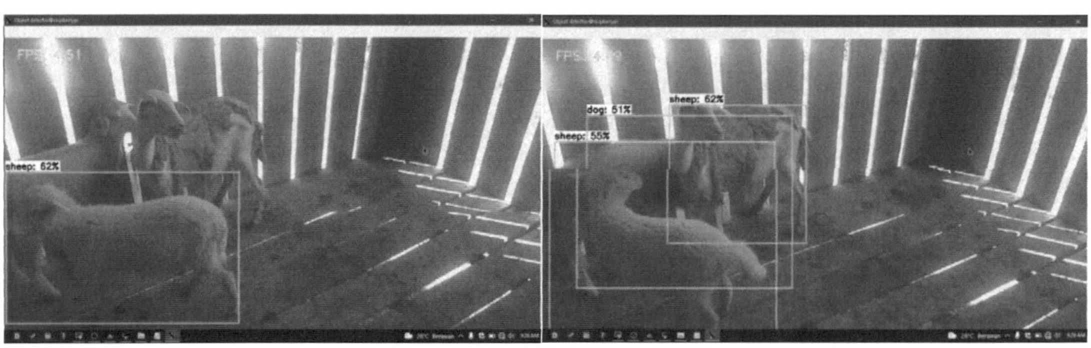

Figure 6. The accuracy of the detection results of livestock objects in the room (indoor).

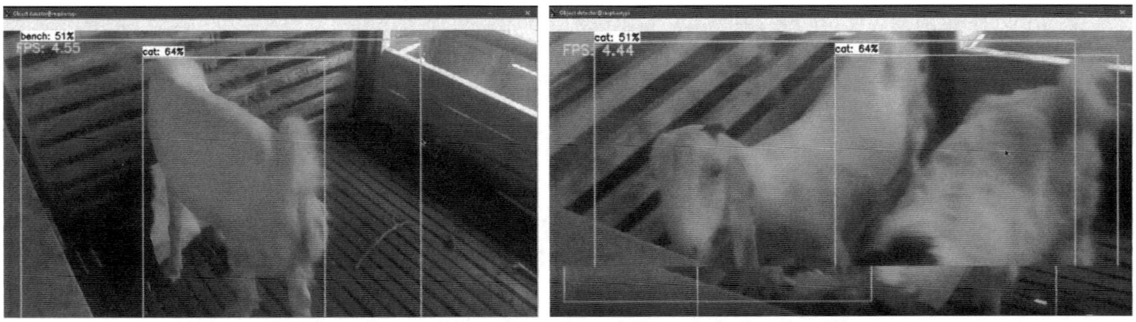

Figure 7. There is an object detection of livestock that is still wrong

The testing process is carried out in real time by researchers directly detecting livestock objects with the camera condition still alive. Thus, allowing the object of sheep and goats to move freely and it involve 10 sheep. Based on this process (Figure 6 and Figure 7), Of the 10 goats, 5 were detected correctly. The accuracy of the system is 50%. However, the camera has not been able to identify the object of farm animals correctly. There are several livestock objects detected as cats and dogs as shown in table 2.

Table 2. Analyze The Results of Classification Performance

Convolution Neural Network		
Livestock	Accuration	Detection
Sheep 1	62%	sheep
Sheep 2	61%	Sheep
Sheep 3	55%	Sheep
Sheep 4	51%	Dog
Sheep 5	58%	Sheep
Sheep 6	64%	Cat
Sheep 7	51%	Cat
Sheep 8	64%	Sheep
Sheep 9	50%	Dog
Sheep 10	53%	Cat

From that issue, it will be caused by the library of the tensor flow has many dataset of the animal both livestock an wild animal. In this case we must modified the library so it can only detect the sheep or goat.

4. Conclusions

From the tests that have been carried out on 10 goats, the results show that the accuracy of the system made is 50%. Of the 10 goats, 5 goats were detected correctly. So that morphometric measurements cannot be done. This is because the Artificial Neural Network library used to process images from raspberries can detect various kinds of objects other than goat or sheep. So that, the ANN library used needs to be modified, so the camera can detect more specific objects, in this case, goats or sheep. If that can be done, the morphometric can be calculated. On the prototype camera raspberry pi that researchers have done, there are still many things that can be developed in this morphometric measurement technique.

5. Acknowledgments

The author would like to thank the Department of Health and Department of Animal Husbandry, State Polytechnic of Jember so that this review article can be realized and beneficial for the development of health science.

References

[1] TROBOSLivestock 2020 Darurat Bibit *TROBOSLivestock Media Agribisnis Peternak.* 1–2
[2] Direktorat Jenderal Peternakan dan Kesehatan Hewan 2019 Statistik Peternakan dan Kesehatan Hewan. 2019. 2–236
[3] Irawan F 2020 Jumlah Populasi Sapi Nasional Meningkat 113% Sepanjang 2019 *Jurnalis Okezone* 1–3
[4] Rahagiyanto A, Adhyatma M and Nurkholis 2020 A Review of Morphometric Measurements Techniques on Animals Using Digital Image Processing *Proc. Third Int. Food Agric.* **3** 67–72
[5] Winiar Putra B, Maria Fuah A, Nuraini H and Priyanto R 2016 Penerapan Teknik Citra Digital Sebagai Metode Pengukuran Morfometrik Ternak pada Sapi Bali dan Peranakan Ongole (Application of Digital Image Technique for Morphometrics Measurement on Bali and Ongole Cattle) *J. Ilmu Pertan. Indones.* **21** 63–8
[6] Ozkaya S 2012 Accuracy of body measurements using digital image analysis in female Holstein calves *Anim. Prod. Sci.* **52** 917–20

IOP Conf. Series: Earth and Environmental Science **980** (2022) 012065 doi:10.1088/1755-1315/980/1/012065

The teaching factory (TEFA) pilot project of transportation system engineering as an opportunity in the department of engineering at Politeknik Negeri Jember

A Irawan[1*], A W Pratama[1] and R E Rachmanita[2]

[1]Automotive Engineering, Engineering Department, Politeknik Negeri Jember, Jember 68101, Indonesia
[2]Renewable Energy Engineering, Engineering Department, Politeknik Negeri Jember, Jember 68101, Indonesia

*e-mail: andik_irawan@polije.ac.id

Abstract. The pilot project of Teaching Factory (TEFA) at the Department of Engineering, Politeknik Negeri Jember as an effort to realize TEFA in the field of transportation system engineering, including mapping the potential and opportunities for the realization of Technical TEFA by considering several product groups plans, including the production of spare parts (Y1), services (Y2) and bio-fuel production (Y3). The problem is the need to identify with potential mapping whether TEFA pilots in the Engineering Department can be recommended to be realized. Potential mapping aims to provide an output of consideration in realizing TEFA in the Department of Engineering through several stages of the process, the first method is GADM (Grid Analysis Decision Making) as a model for providing recommendations in realizing TEFA by giving a score from the questionnaire to partners, the second method with the CIPP stage (Context, Input, Process, Product) as a consideration of the service process chain flow and products that will be produced from the three products. Potential mapping is carried out by means of questionnaires to partners spread across 4 sub-districts, Jember Regency with consideration of the radius of the mapping distance of 5 km from the Politeknik Negeri Jember. Respondent partners are used as the basis for considering feasibility in establishing TEFA in the Engineering Department. The questionnaire to productive partners in the automotive sector grouped into four sub-districts obtained a mapping of a total of 100 partners from Sumbersari, Patrang, Kaliwates, Ajung with a number of 35, 14, 17 and 34 partner respondents respectively who were considered as candidates for collaboration in pioneering TEFA in the Department of Engineering. The GADM results provide a questionnaire score for each sub-district average from product groups Y1, Y2 and Y3 with a score of 3.2, 2.9 and 3.4. The partner's choice of the highest score is bio-fuel (Y3).

1. Introduction

Politeknik Negeri Jember is one of the polytechnics that has a Teaching Factory (TEFA) such as TEFA SIP POLIJE. TEFA was developed as an implementation of business and production-oriented learning concepts to answer the challenges of future modern industry. In addition, with TEFA, students can directly improve their competence before going directly to industries outside of polytechnics. Politeknik Negeri Lampung has at least 14 TEFA [1], through the PPPTV program, Politeknik Negeri Lampung concentrates on PUT (Center for Excellence in Technology) which is available for several Study

Programs. The Lampung State Polytechnic since 2018 has established TEFA as a production and business space with a production system using a foreign language [2].

Manpower is a problem that needs to be solved to overcome the suitability of the implementation of TEFA, through the concept of Context, Input, Process, Product (CIPP) in the DKI Jakarta Vocational School which is used for evaluation in the DKI Jakarta TEFA policy in realizing the entrepreneurial spirit to students as an opportunity after graduating from school [3]. The CIPP model is also used as a TEFA evaluation to examine every aspect of Context, Input, Process, Product in evaluating the program, more than 75% have a good response in the implementation system [4]. In general, link and match is not only for polytechnics, SMK is also part of the TEFA implementation. With the right learning design and TEFA, it really supports the implementation of the Competency Test [4].

Background studies as analysis in learning models, system evaluation, competency improvement, competency evaluation and assessment and various models to support TEFA implementation in each sector using various ASPECTS. Based on the above background, it is important that this research begins with a study for mapping TEFA pilots in the Department of Engineering, Politeknik Negeri Jember to see opportunities and potential through supporting data and available equipment, technological capabilities in the laboratory and HR competencies. The TEFA pilot model will be analyzed with the Context Input Process Product (CIPP) to analyze the pilot concept and use the Grid Analysis Decision Making (GADM) decision support method before TEFA is actually made. There are three groups that will be formed including the spare-part production group (Y1), services (Y2), and fuel production (Y3) through the CIPP analysis model to realize the TEFA Pilot of Transportation System Engineering at the Engineering Department, Politeknik Negeri Jember.

2. Material and Method

In this study, there are several groups of sections as a discussion in the TEFA pilot, Table 1 contains the discussion of Context and Input. The use of plastic waste to make fuel is equivalent to premium, thermal energy from Liquid Petroleum Gas (LPG). Resin, catalyst and fiber as materials for making body parts and body variations. The questionnaire form as a data mapping includes the product of spare-part production groups (Y1), services (Y2), and bio-fuel production (Y3). The method in this study uses CIPP as a consideration of the production process and facilities, and uses GADM in the consideration of decisions on the TEFA pilot plan.

Table 1. Group Modelling Context and Input

Group	Context	Input	
		Material	Unit
Y1	Tech Roller Part Mounting	Rubber Sheet Smoke	kg
Y2	Tech Automotive	Car	pcs
	Dyno	Car-Motor	pcs
	Gas Emission Test		
Y3	Incenerator Tech	Polypropiline	kg
	Sample 1	Polypropiline	kg
	Sample 2	Polypropiline	kg
	Sample 3	Polypropiline	kg
	Sample 3		

2.1. *Context, Input, Process, Product (CIPP)*

TEFA as a learning concept that emphasizes student competence and achieves the production of goods or services [6]. TEFA is also the development of a dual system-based learning method based on competence and training-based production. Educational and research activities are part of innovation for single initiators to effectively involve industry and academia through direct practice models in realizing competencies based on production, quality and efficiency. The main purpose of the TEFA learning model is to improve the alignment of the skills, knowledge and attitude development process, through thematic alignment on normative, adaptive, and productive subjects [7]. TEFA learning adapts the concept of work-based learning. TEFA can improve cognitive competence, vocational competence, soft skills and hard skills that in the curriculum balance between knowledge, theory and analysis with manufacturing, design, business activities, and professional skills. The groups that will be formed in the TEFA Pilot of Transportation System Engineering include the production of spare parts (Y1), services (Y2) and bio-fuel (Y3) with CIPP analysis. Fuel production includes: gasoline motor fuel with the addition of an induction tube [8], EFI vehicle fuel misting system [9], pyrolyzed polypropylene plastic fuel with a premium mixture and octane booster [10], and biodiesel fuel from seed oil avocado [11]. The spare part production subsystem includes: industrial biosystem technology [12], Exhaust Pipe Fuel Preheating System (EP2FS) [13-14], and agrofiber composite rubber reinforced ramie fiber [15]. The service subsystem includes: maintenance of EFI vehicles based on diagnostic tools [16], increased efficiency and performance of the exhaust pipe fuel preheating system [17], increased fuel efficiency with engine manifold membrane engineering 2 [18-19], solar power plant installations [20]. In this case, the determination of CIPP is a stage of the process in looking at the parts of the TEFA pilot plan before implementation:

a. Context
   By looking at the facilities and infrastructure currently available in every laboratory of the Engineering Department.
b. Input
   Is the technology, human resources and raw materials available to support the pioneering TEFA Transportation System Engineering in the Department of Engineering.
c. Process
   This stage is a process that includes a description of each flow of the three groups that will be initiated.
d. Product
   Is a product that will be made and will be tested directly with consumers in the area around the Department of Engineering and Politeknik Negeri Jember.

2.2. *Grid Analysis Decision Making (GADM)*

GADM as a method to analyze decisions before the formation of a system model. Determination of the system can be based on the aspects to be studied. Various aspects are based in order to complete the stages through a questionnaire made in the form of a matrix, Table 2 shows a matrix of scoring criteria. The steps required include:

a. Assessing Problems
   Includes identification of the problem, seeking information to be carried out before the activity is carried out by generating a temporary crisis, with this condition the determination to avoid or resolve the problem can be resolved and understanding of the conflict to be faced. In the pilot of TEFA in the Department of Engineering, in this case, the context of the problem is seen which emphasizes the three main subsystems to assess the characteristics of the existing problem.
b. View Alternatives
   At this stage trying to seek input and information from those who have knowledge related to the problem and pay attention to relevant information in the mass media. The most important thing at this stage is openness and flexibility. Individuals are more interested in alternatives in the form of suggestions to solve problems. From the three subsystems, it will be obtained which context is an alternative to support the production, business, learning system which is the main basis as a support structure in the TEFA pilot system in the Department of Engineering.

c. Considering Alternatives
In considering alternative decision-making, begin to evaluate all available options based on the consequences and possibilities to do. Regarding the consequences of his actions, one sees the possible benefits and disadvantages that he must accept as well as the practicality of each alternative that he thinks is the best in his efforts to achieve the desired goal. When realizing that there is a possibility of regret in the future, this becomes the basis for careful consideration of the alternatives. This concept is to consider alternatives from the aspects studied for data to support decision making in the TEFA Pilot in the Department of Engineering.

Table 2. The Criteria of Scoring

Score	Criteria
4	Agree to realize TEFA in the Department of Engineering
3	Agree but consider the target and reach of market segments
2	Hesitate in realizing TEFA in the Engineering Department
1	Do not agree in realizing TEFA in the Department of Engineering

Scores are given by respondents, out of a total of 30 questions and divided into 10 questions in each product group. Researchers give weight to consider the decisions of respondents and produce a final score that will be decided as a sequence of decision considerations. Table 3 describes the scoring groups and the elements of the scoring model. Table 4 shows the scoring matrix and Table 5 shows the scoring weights.

Table 3. Element and Symbol

Symbol	Element	Description
R	Respondent	100 Partnership
Q_{Yn}	Questionnaire	Production Questionnaire
SQ	Score Questionnaire	1 - 4
AS_{Yn}	Average Score	Mean Production
TWR	Total Weighted Researcher	Weighted Score
FS	Final Score	Total Weighted

QYn = QY1 ; QY2 ; QY3

Assessment of scores from respondents in each production group Y1:

$$Q_{Y1} = q^{1Y1} + q^{2Y1} + q^{3Y1} \dots\dots\dots + q^{ny1} \tag{1}$$

Assessment of scores from respondents in each production group Y2:

$$Q_{Y2} = q^{1Y2} + q^{2Y2} + q^{3Y2} \dots\dots\dots + q^{ny2} \tag{2}$$

Assessment of scores from respondents for each production group Y3:

$$Q_{Y3} = q^{1Y3} + q^{2Y3} + q^{3Y3} \dots\dots\dots + q^{ny3} \tag{3}$$

Average score:

$$AS_{Yn} = \frac{QYn}{10} \tag{4}$$

Total Weighted Score from author:

$$TWR = \frac{WR_1 + WR_2 + WR_n}{\sum WR} \tag{5}$$

Final Score to consider decision:

$$FS = AS_{Yn}\ x\ TWR \tag{6}$$

Table 4. Matrix Scoring

Number	q^{1Y1}	q^{2Y1}	q^{nY1}	Q_{Yn}
R_1	q1	q2	q3	Qn
R_2	q1	q2	q3	Qn
R_n	q1	q2	q3	Qn
	AS_{Yn}			Qn/10

Table 5. Matrix Weighted

Production	Wr1	Wrn	TWR (a)	AS_{Yn} (b)	FS
Y1	w Score	w Score	w TWR	AS_{Y1}	(a + b)/2
Y2	w Score	w Score	w TWR	AS_{Y2}	(a + b)/2
Y3	w Score	w Score	w TWR	AS_{Y3}	(a + b)/2

3. Results and Discussion

Data taken from the Jember Regency Environmental Service, in Jember Regency there are 31 sub-districts, 22 urban villages and 221 villages which is one of the largest piles of waste in the Jember city area, with a population in 2020 of 2,519,000 people, waste production in Jember Regency is fairly large and has increased to now reach 1,300 m^3 which is a factor in environmental pollution which will eventually cause environmental damage. Based on the analysis of each urban village, the average daily composition of plastic waste is 6%. Whereas, the production of plastic waste fuel is very potential in utilizing plastic waste at the Politeknik Negeri Jember in particular and the environment around the campus in general. The average plastic waste disposal on campus is 2 kg per day, while outside the campus a radius of 2 km is more than 30 kg per day.

The most common type of plastic waste produced is polypropylene. The tool used to melt plastic waste is an incinerator designed in previous studies [10]. Figure 1 informs that to burn 5 kg of polypropylene plastic material to produce 0.8 liters of liquid Polypropylene fuel it takes 120 minutes at a temperature of 220˚C and to burn 10 kg of plastic waste it takes 145 minutes at a temperature of 240˚C also to produce 2.65 liters fuel require 15 kg waste it takes 220 minutes at 250° C. Meanwhile, to produce 3.7 liters of Polypropylene liquid fuel, it takes 20 kg of plastic waste and takes 310 minutes with a temperature of 280˚C.

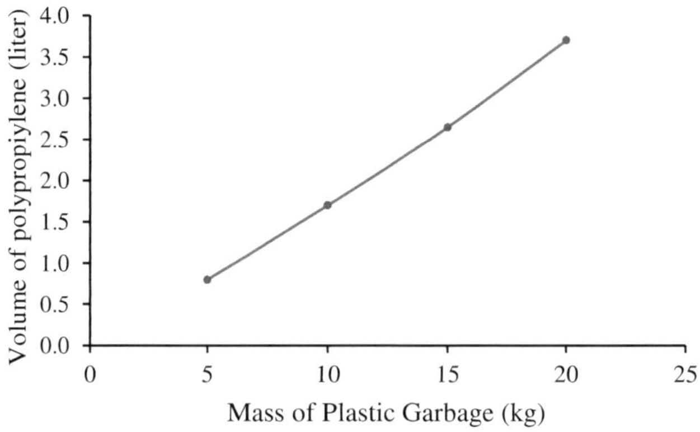

Figure 1. Volume of Polypropylene Fuel from Burning Plastic Waste

Repair of two-wheeled vehicles around the 5 km radius campus - an average of 3 to 5 vehicles per day in each workshop, a total of questionnaires spread over 100 workshops in Sumbersari, Patrang, Kaliwates and Ajung sub-districts, respectively 35, 14, 17 and 34. While four-wheeled vehicles from of 100 workshops, only 20 four-wheeled workshops were recorded. Table 6. Shows the distribution of the questionnaire in four sub-districts with an average of 2 vehicles per day for each sub-district. The service potential (Y2) in pioneering TEFA is very high. When viewed from fuel consumption, each vehicle requires 1 liter of fuel equivalent to 14 km of travel in the city. If services (Y2) require an average of 1 liter of fuel per vehicle, then 500 liters will be consumed in 4 sub-districts, this will greatly affect the fuel production plan (Y3). While in the manufacture of vehicle parts, from 4 sub-districts only 10 workshops are engaged in vehicle modification services, so the daily average spare part production rate is 1 part for complex parts and more than 3 parts for simple parts.

Table 6. Partner District

District	Quantity
Sumbersari	35
Patrang	14
Kaliwates	17
Ajung	34
Total Partnership	100

3.1. CIPP Analysis

The inventory of CIPP results in the TEFA pilot project at the Department of Engineering, Politeknik Negeri Jember is described in Tables 7 which this table is the result of an inventory to support each group, and Tables 8 is the calorific value of the results of testing the tool in group Y3 based on the calorific value. The results of the calorific value can be used as a reference for previous research in accordance with this research method. Technical data and specifications to support CIPP as a starting point for mapping the potential of supporting equipment and human resources.

Table 7. Group Modelling of CIPP

Group	Context	Input		Process	Product	
	Incinerator Tech	**Raw Material**	**Unit (Kg)**	**LPG (Kg)**	**Fuel (Litre)**	**Char (Kg)**
Y3	Sample 1	Polypropylene	5	3	0.8	0.5
	Sample 2	Polypropylene	10	6,5	1,7	0,9
	Sample 3	Polypropylene	15	10	2,65	1,45
	Sample 4	Polypropylene	20	13,5	3,7	2,4
Group	Tech Automotive	Material	Unit (pcs)	Procedure	Measure	-
Y2	Dyno	Car	2	SOP	HP; Nm	-
	Gas Emission Test	Car-Motor	2	SOP	Emission	-
Group	Tech Roller	Raw Material	Unit (Kg)	Compound (Kg)	Part (pcs)	-
		Rubber Sheet				
Y1	Part Mounting	Smoke	2	2	12	-

Y1 (Spare Part), Y2 (Service), Y3 (Bio Fuel)

Table 8. Fuel Calorie

Fuel Mixed	Cal/g
PP 85% + Pertalite 15%	10,86
PP70% + Pertalite 30%	10,83
PP 50% + Pertalite 50%	10,71
PP 100%	10,96

PP = Polypropylene

3.2. GADM Analysis

The questionnaire is used as a decision consideration from the results of the CIPP analysis to provide recommendations on whether the TEFA pilot can be considered for implementation. Table 8 is a description of the recommendations from the results of the questionnaire and the results of the research team's considerations in realizing the TEFA pilot at the Department of Engineering Politeknik Negeri Jember.

Table 9. Matrix Final Score

Production	Wr1	Wr2	Wr3	TWR (a)	AS_{Yn} (b)	FS
Y1	2.9	2.8	2.8	2.8	3	2,9
Y2	3.1	3	3.2	3.1	3.3	3,2
Y3	3.2	3.3	3.5	3.3	3.5	3,4

4. Conclusion

Based on the results of this study, it was concluded in this study with the CIPP model and GADM considerations recommendations resulted in feedback from respondents as follows Y3 > Y2 > Y1 with a score of 3.4; 3.2 and 2.9. The production of fuel (Y3) from plastic waste is recommended using 20 kg of plastic with a constant temperature of 310° C capable of producing 3.7 liters of fuel. Services (Y2) recommend performance tests and exhaust emissions, while product manufacturing (Y1) results in the manufacture of automotive components.

5. Acknowledgments

We sincerely thank to the Director of Politeknik Negeri Jember and Pusat Penelitian dan Pengabdian Masyarakat (P3M) for the funding support so this script can be carried out well. This script would hopefully give a positive contribution to the educational development or those are willing to conduct further research.

References

[1] Polinela 2020 Teaching Factory Menguatkan Keterampilan dan Kompetensi Lulusan

[2] Unila 2018 Teaching Factory Unila Pertama ada di Perguruan Tinggi Lampung

[3] Supriyantoko I Jaya A Kurnia V and Habiba P G S 2020 Evaluasi Implementasi Kebijakan TEFA dengan Model Evaluasi CIPP di SMK Negeri DKI Jakarta *JVTE: Jurnal of Vocational and Teacing Education* 02 02 1-10

[4] Pratiwi M Ridwan and Waskito 2019 Evaluasi Teaching Factory Model CIPP *Jurnal Ilmiah Pendidikan dan Pembelajaran* 3 3

[5] Fitrihana N 2018 Rancangan Pembelajaran Teaching Factory di SMK Tata Busana *HEJ (Home Economic Journal)* 2 2 56-64

[6] Manalu S R I Sogi H Rentas D J Galfri S Supriyadi Airways S P 2017 Tata Kelola Pelaksanaan Teaching Factory Jakarta Pusat: Direktorat Pembinaan Sekolah Menengah Kejuruan Direktorat Jenderal Pendidikan Dasar dan Menengah Kementerian Pendidikan dan Kebudayaan RI

[7] Kusuma 2017 Panduan Teknis Teaching Factory Bonn and Eschborn: Deutsche Gesellschaft fur Internationale Zusammenarbeit (GIZ) GmbH

[8] Irawan A and Tyagita D A 2016 Inovasi Peningkatan Efisiensi Konsumsi Bahan Bakar Motor bensin Dengan Penambahan Tabung Induksi (Studi Kasus Sepeda Motor Matic 113cc *Jurnal Ilmiah Inovasi* 16 2 94-97

[9] Irawan A Tyagita D A 2017 Rekayasa Teknologi Sistem Pengkabutan Bahan Bakar Kendaraan EFI *Seminar Nasional Hasil Penelitian 2017* 978-602-14917-5-1 24-27

[10] Budiprasojo A and Pratama AW 2016 Nilai Kalor Bahan Bakar Plastik Polypropilene (BBPP) Hasil Pyrolisis Dengan Campuran Premium Dan Octane Booster *Jurnal Ilmiah Inovasi* 16 2 122-127

[11] Rachmanita R E and Safitri A 2020 Pemanfaatan Minyak Biji Alpukat (Persea americana Mill) sebagai Bahan Baku Pembuatan Biodiesel dengan Pemurnian Water Washing *Jurnal Ilmiah Sains* 20 2 88-89

[12] Irawan A Tyagita DA and Retnowati N 2020 Aplikasi Teknologi Biosistem Industri Hilirisasi Olahan Kopi Pemeringkatan Mutu Specialti *J-Dinamika: Jurnal Pengabdian Masyarakat* 5 2 11-14

[13] Irawan A and Budiprasojo A 2017 Rancang Bangun Exhaust Pipe Fuel Preheating System (EP2FS) Serta Uji Performa dan Analisis Efisiensi Pembakaran Engine *Seminar Nasional Hasil Penelitian 2017* 978-602-14917-5-1 206-212

[14] Rachmanita R E Karimah C N and Azizah N 2021 Experimental investigations on the performance of thermoelectric generator as energy conversion system *IOP Conference Series: Earth and Environmental Science* 672 1

[15] Irawan A Tyagita D A and Nuruddin M 2017 Mechanical Properties Of Agrofiber Composite Rubber Reinforced Ramie Fiber And Banana Stem Fiber Applications In Automotive Components, *Seminar Nasional Hasil Penelitian 2017* 978-602-14917-5-1
[16] Irawan A and Tyagita D A 2017 Pelatihan Maintenance Kendaraan EFI Berbasis Diagnostic Tools, *Seminar Nasional Hasil Penelitian 2017* 978-602-14917-5-1 16-18
[17] Budiprasojo A and Irawan A 2018 Engine combustion efficiency and performance of exhaust pipe fuel preheating system *Jurnal Rekayasa Mesin* 9 1 1-7
[18] Pratama A W and Budiprasojo A 2016 Rekayasa Manifold Membrane Mesin 2 Langkah sebagai Upaya Meningkatkan Efisiensi Bahan Bakar *Jurnal Ilmiah Rotary* 1 1
[19] Rachmanita R E Hananto Y and Akbar M R 2021 Design of hydrogen reactor based on decomposition of water by aluminum as an alternative to gas fuel *IJSGCE* 10 1 8-20
[20] Febriani S D A Rachmanita R E and Nari M I 2019 Instalasi Pembangkit Listrik Tenaga Surya Berbasis Teknologi Informasi Guna Terbentuknya Pondok Mandiri Energi di PP Nurussalam Ambulu Jember *Prosiding Seminar Nasional Pengabdian Kepada Masyarakat dan Penelitian Pranata Laboratorium Pendidikan Politeknik Negeri Jember Tahun 2019* 3 1

IOP Conf. Series: Earth and Environmental Science **980** (2022) 012066 doi:10.1088/1755-1315/980/1/012066

Optimization Design for Manufacturing and Assembly Method in Multifeedstock Jember Biodiesel Processor

A Budiprasojo [1*], A T Zain [1], and C N Karimah [1]

[1] Automotive Engineering, Engineering Department, Politeknik Negeri Jember, Jember 68101, Indonesia

* email: azamataufiq@polije.ac.id

Abstract. A method used to optimize the design process in a manufacturing system is the Design For Manufacturing and Assembly Method (DFMAM). The idea of this optimization method is to minimize assembly process errors, reduce redundancy of components, save on production costs, as well as ease of operation. In this study, the method will be used on a biodiesel processor with multi feedstock type equipped with intelligent system for production perimeter control that is specifically designed to produce biodiesel from corn and coconut oil as raw materials produced by Jember, Indonesia. The design framework is made suitable for small-scale biodiesel production in MSMEs for industrial or laboratory purposes, in order to produce biodiesel that meets the biodiesel standards of Ministry of Energy and Mineral Resources of Republic of Indonesia. Processor component assembly via the DFMAM principle is simplified to reduce component complexity and redundancy. The biodiesel produced by this method has been tested and recorded a density value of 15 cel.deg. 874 - 877 kg/m3, viscosity at 40 cel.deg 4.3 – 4.7 mm2/s, Acid value 0.14 0.15 mg KOH/g, Cetane 56 – 58, Cloud Point 7 – 8 cel.deg., pour point 4 – 5, Flashpoint 171 – 178 cel.deg and distillation point 345 – 350 cel.deg.

1. Introduction

The method of optimizing a tool used in a production is an interesting research study. This method is generally carried out to obtain a simplification of the process of operating the equipment and the efficiency of the use of its operational budget. One of the optimization methods commonly used in the manufacturing process is the Design for Manufacturing and Assembly Method (DFMAM) optimization method [1].

DFMAM is a method that tries to improve the design process for a tool to make it easier for factories to carry out the process of optimizing the manufacture of a product, by integrating several kinds of model analysis, costs, energy calculations, and calculating the mechanical strength of materials into a final product in the form of a complete design of the updated tool [2].

Through this process, try to present a solution that has been optimized for an integrated manufacturing process, so that the best solution can be selected from several possible solutions that can be offered.

This will certainly reduce overall production costs and reduce complexity in the product development process, of course without compromising the standard quality of the product you want to produce.

The optimization process begins with an inventory of what components are contained in the prototype or product that we will optimize. The process is continued by developing the tool conceptually through a design-by-design process and then manufacturing assembly based on the details of the repair design [3]. Of course, the complexity of this process is greatly influenced by the orientation of the user and the difference in the geometry to be applied.

In this study, the optimization process will try to be applied to a biodiesel maker with a multi feedstock type that will be used in middle-class community business units. This biodiesel maker tool or better known as a biodiesel processor is generally made with equipment that requires many and complicated components so that the purchase price will be expensive [4]. The research target is to try to simplify the biodiesel tools on the market into a biodiesel processor that is compact, inexpensive, easy to operate, and automatic in the process.

The prototype biodiesel processor of the researcher's design has an Automation in the program controller which is integrated with data acquisition software to monitor and control its operation. The developed prototype is useful in simulating full-scale automated systems and in a new processing technology which we named the chemical low thermal biodiesel process [5]. The specific purpose of this design is to minimize manual labor and costs while ensuring worker flexibility in processing diverse raw materials. This research develops a multi-feedstock biodiesel processor with a capacity of 25 L per production.

Making a biodiesel is more environment friendly than making a pyrolysis fuel. Using a pyrolysis fuel in an internal combustion engine still need future research because its low similarity characteristic compared to fossil fuel [6]. Using a pyrolysis fuel in a internal combustion engine making a engine running roughly, increasing carbon deposit, and decrease the engine performance [7].

Biodiesel in its pure form or as a mixture with diesel is used as fuel for diesel engines and has been laboratory proven to be used to reduce air pollutants. In Indonesia a mixture of biodiesel and diesel is sold by PT. Pertamina with the name biodiesel with a mixture of 30% biodiesel and 70% diesel (B30). To meet ESDM standards, it is necessary to make a biodiesel maker (biodiesel processor) that can be used by MSMEs in the Jember area of Indonesia, using local raw materials available in the surrounding area. Departing from this background it will be used Design for manufacturing and Assembly Method (DFMAM) to make a biodiesel processor. Designing a heat exchanger-based prototype in a engineering world need a heat transfer complex analysis [8] [9].

In general, this research uses 3 main methods, namely: 1. designing an intelligent multi-feed biodiesel processor with a capacity of 25 L per production with component savings; 2. Analyze existing processors and reduce overall production costs through redesign and elimination of redundant parts; 3. Increase the versatility and flexibility of using processors for various types of vegetable oil feedstocks; 4. Testing the parameters of the fuel produced.

2. Method

This study uses the principle of design optimization in the field of manufacturing assembly which will specifically be applied to the process of developing a biodiesel maker or known as a processor but equipped with control and automation so that it will become a smart device. Optimization is carried out by carrying out the following consideration process in order to obtain an optimal design.

1. The number of components and the wishes of the user will determine how much simplicity to achieve a conformity of the user's wishes will be the initial consideration. This step will determine overall how the tool will then be assembled.
   Components or parts of the tool will be listed or inventoried in accordance with the order of assembly and will be assigned an identification number. In this initial step, the complexity factor will be calculated. The value of this complexity factor will be calculated using equation 1 [10].

$$\text{Complexity factor (Cf)} = \sqrt{\sum N_p \times \sum N_i} \tag{1}$$

   Where np is the total number of the list of all parts and ni is the total number of parts of the nth component.
2. The next step is to make a summary of the theoretical minimum required components and then identify practical requirements in a product manufacturing process. The material for each component

will be identified at the retail price in the market according to the location where the assembly process will be carried out. In this process, an analysis will be carried out whether it is possible to carry out a simplification process by redesigning or exchanging using different materials at lower prices. This step needs to be done carefully because it involves the durability and toughness of the tool. Of course, we do not want a tool that is cheap but quickly damaged. Theoretically, the minimum component that can be efficient is calculated using equation 2.

$$\frac{T_{min}}{T_n} \times 100\% \qquad (2)$$

Where T min is the minimum number of parts that are still allowed to be used and Tn is the total number of components in the old equipment that we will optimize.

3. The next step is to identify related to the properties or special characteristics of what the component will work on. In this case, because it will be used on components that can produce chemical erosion, care must be taken in removing components from iron. The main goal is safety in the handling of materials to be processed to the user.
4. The next step is the process of planning methods to combine these components. The process of merging or joining is a process that costs quite a lot. We can choose to simplify the process of combining the components with a less expensive method. Simply put, if the merging process does not require welding or bolting, why should it be forced to use that method.
5. The last step is to make a draft of the tool that has been optimized and then try to do the assembly and manufacturing process. In this step, the performance test process will also be carried out.
6. Because the product from this tool will be used in conventional diesel engines, it is necessary to test the quality of the processed product from this tool, namely how the characteristics of the biodiesel produced when compared with fuel standards used internationally and nationally. The quality standard for the type of test refers to the regulation of the ministry of trade for biodiesel fuel.

3. Results

3.1. Optimization of the minimum required components

Researchers have identified conventional biodiesel production equipment. The components needed to assemble the old biodiesel processor have been made and presented in the list. The list of components also includes a description of the function of each component to facilitate classification in determining the priority of the component being vital or not.

After identifying, compiling a list and analyzing it, a minimum list of components that must remain in an automatically operated biodiesel processor is obtained. The smart control system for a biodiesel processor must have the following components:

3.1.1. Microcontroller. Microcontroller is a component similar to a small computer that is packaged in the form of an IC (Integrated Circuit) chip and is designed to perform certain tasks or operations. Basically, a Microcontroller IC consists of one or more Processor Cores (CPU), Memory (RAM and ROM) and programmable INPUT and OUTPUT devices.

The use of this microcontroller is increasingly popular because of its ability to reduce the size and cost of a product or design when compared to designs built using microprocessors with memory and separate input and output devices.

In this equipment we recommend using a microcontroller with Arduino uno type, with consideration of price and freedom to do coding.

3.1.2. Sensor. Sensors are devices used to detect changes in physical quantities such as pressure, force, electrical quantities, light, motion, humidity, temperature, speed and other environmental phenomena. After observing the change, the detected input will be converted into output that can be understood by

IOP Conf. Series: Earth and Environmental Science **980** (2022) 012066 doi:10.1088/1755-1315/980/1/012066

humans either through the sensor device itself or transmitted electronically through the network to be displayed or processed into useful information for its users.

In this biodiesel device the minimum sensors that must be used are the following sensors:

1. Pressure Sensor (Pressure Sensor)

   Pressure Sensor or Pressure Sensor is a sensor that is used to measure the amount of pressure applied to a sensor. The pressure sensor will produce an analog output signal that is proportional to the amount of pressure applied. The pressure sensor on this tool is needed to ensure that during the process there is no overpressure in the tank.

2. Temperature Sensor (Temperature Sensor)

   Temperature Sensors are sensors that are widely available in the form of digital and analog sensors. As the temperature increases, the electrical resistance of the thermistor will increase as well. Conversely, if the temperature decreases, the resistance will also decrease.

The temperature sensor used is a thermocouple (thermocouple). Thermocouple is one of the most frequently used types of temperature sensors, this is due to the wide operating temperature range of the Thermocouple, ranging from -200°C to more than 2000°C with a relatively low price. A thermocouple is basically a Thermo-Electric temperature sensor consisting of two dissimilar metal junctions. With the difference in temperature at the two junctions, the circuit will produce a certain voltage whose value is proportional to the temperature of the heat source.

The advantage of the thermocouple is that it has a wide temperature range, is resistant to shock and vibration, and provides an immediate response to temperature changes. The thermocouple in this equipment will be useful as input data for the microcontroller for temperature regulation of biodiesel processing.

3.1.3. Heater. As many people already know, the heater has a function as a heater. This tool works by changing the normal temperature to heat. In this tool it is recommended to use an electric heater because this heater relies on electrical energy as a heating source. With an electrical energy source, it will be easier to regulate the temperature changes. Heater is an actuator to raise the temperature in the process of making biodiesel.

3.1.4. Pneumatic solenoid valve. A pneumatic solenoid valve is a valve that is driven by electrical energy through a solenoid, has a coil as its driver which functions to move a piston that can be driven by AC or DC currents, a pneumatic solenoid valve or a solenoid valve (valve) has an output hole, an input hole and an exhaust hole.

The solenoid valve is the most frequently used control element in fluid control. In this equipment the function of this component is to control so that the pressure in the tank is always at a safe limit, keep the temperature in the tank rising quickly, and drain methanol vapor from the process to be channeled to an external tube for further recycling.

3.1.5. Fluid pump. A fluid pump is a mechanical device that can flow substances from a low-pressure area to a high-pressure area. In principle, the pump converts the mechanical energy of the motor into the energy of the fluid flow. In this study, the pump is used to replace the electric motor of the mixer, fluid stirrer, belt stirrer, transmission mixer and forcing flow recirculation. Researchers chose to use a fluid pump because this tool can substitute a lot of equipment in the biodiesel manufacturing process.

3.2. Calculation of heat transfer and power for processor design efficiency

3.2.1. Demand Specification.

The design process is carried out with consideration of details, the desired production scale, as well as special technical specifications of the resulting product. The processor that we design must be able to serve the manufacture of 20 liters of a mixture of raw material oil and catalyst in one production process. The temperature of the production fluid will be maintained in the range of 60 cel. Deg. to maintain the quality of the biodiesel produced. The flow rate in the channel is kept constant at 2 LPM to keep the mixing process and the glycerol binding reaction going slowly. The solenoid valve will open when the temperature in the tank has reached 65 cel.deg and the pressure has increased to above 1 BAR.

3.2.2. Component Specification Needed.

a. Fluid pump: Flow rate = 1 -10 LPM, Head = 9-22 m, Suction Head = max 9 m, Temp Max 100 cel.deg. , Thermal protector bearing, Power 286 Watt.

b. Reactor: cylinder galvanized steel tank, well joint with electric weld, the lid of the tank must be prevented from a corrosion and can compromised with methoxide liquid fumes during the process

c. Heater: The heater is placed in the bottom of the cylinder tank and direct contact with the fluid. Its 1500 W in 220 V AC electricity. It must easily r*a*ise the temperature from 0 to 100 cel deg. The heater is controlled by Arduino microcontroller and triggered by thermocouple sensor. The heater selecting process is based by below calculation.

3.2.3. Heater Design.

Bold The fluid density of the fluid is changed and calculated by equation 3,

$$\boldsymbol{\rho} = \frac{\boldsymbol{m}}{\boldsymbol{v}} \tag{3}$$

where the volume is set to 00.027 cubic meters, and the mass is known 22.41 kg the density is expressed 830 kg/m3

The weight of the liquid to be heated is calculated by equation 4.

$$\boldsymbol{w} = \boldsymbol{m} \times \boldsymbol{g} \tag{4}$$

With a known mass and gravity using 9.81 m/s then the weight of the liquid is known to be 219.84 N

From the laboratory checking data, it was found that the maximum heat capacity for plant oil to be processed is 2340 J/Kg K. We use 3 Kw for the heater power rating, and the heater must heat the oil from 30 cels. Deg. to 60 cels. Deg. So to heat the plant oil, 1573 kJ of heat is needed which is calculated using equation 5.

$$\boldsymbol{Q} = \boldsymbol{m} \times \boldsymbol{c} \times \Delta \boldsymbol{T} \tag{5}$$

Of course, this is a theoretical calculation without heat loss, taking into account the heat loss that may arise, it is assumed that the heat conversion efficiency is 80%, so the real heat required is 1234 kJ. With a heater used of 3 kW, the processing temperature required to heat the oil to its working temperature is calculated by equation 6, which results in a time of about 500 sec.

$$P = Q/\text{time} \tag{6}$$

We must know the Heat loss coefficient by surrounding nature (alpha) is 1 J/Sm-Cel deg, A surface or the surface area of the reactor is 0.5 m3, if we already know this value, then the heat loss can be calculated by equation 7 which then gets the heat loss of 16.3 J/s.

$$\boldsymbol{Q_{loss}} = \boldsymbol{\alpha} \times \boldsymbol{A_{surface}} \times \Delta \boldsymbol{T} \tag{7}$$

This means that we will experience a heat loss of 16 watts on the heater that we use. A value that can be ignored if we use a heater of 3 kW.

3.3. Calculation of Tank Vectors.

Tank Volume Design. With equation 8, the trick is to calculate the dimensions of the reactor tank if you want the tank to be able to load 25 L of liquid, and the height of the tank must be ergonomic, so that the maximum height of the tool is a maximum of 2/3 of the average height of Indonesian humans, which is 1.7 m and the tank must be located above the surface. Bottom to tank then determined to have a height of only 0.6 m. then the radius of the rod is obtained by 0.12 m.

$$\boldsymbol{V = \pi \times r^2 \times h} \quad (8)$$

The total area of the tank that must be coated with anti-rust is calculated by equation 9, using a radius of 0.12 m, it is obtained that the area that must be coated is at least 0.55 m3

$$\boldsymbol{\sum A_{surface} = 2 \times \pi \times r^2 + 2 \times \pi \times r \times h} \quad (9)$$

3.4. Processor Fabrication

3.4.1. Assembly Method.

The assembly steps are as follows;

a. The inner cylinder is made with a diameter of 240 mm, a height of 350 mm.
b. Cylinder formation by rolling the plate to form an outer diameter of 360 mm, height 500 mm.
c. Followed by cylinder welding which produces a single cylinder with a diameter of 360 mm and an overall height of 600 mm.
d. Welding of legs 600 mm long to 2/3 x 180m – 600 mm cylinders.
e. Welding cover plate at the top of the tank complete with inlet and solenoid valve channels.
f. Installation of electric pumps and pipelines can use PVC pipes.
g. Anti-rust painting and/or coating.

3.4.2. Prototype Specifications.

According to our calculation, its will gave a good result if this prototype is can accomplish below specification

a. Size 1.53 mx 0.6 m
b. Operating temperature 20 - 90 cels. deg
c. Production capacity 20 L/production
d. Production time 3 hours/production
e. Total power 2000 watt

In this prototype, researcher try to merge the processor tank and the cleanser tank, the aim of this procedure is to simplify the number of the component then also a manufacturing cost. In first thought, we carefully concern about the quality of the biodiesel product by this simplification methods. The worry is for its draining phase. As a standard, the cleaning method for biodiesel is by mix it with an aqua distillate liquid in 30 cel.deg. If the drainage in processor not optimal, there is still residue in the bottom of the tank. So, make a bottom in a cone shape is a must for a biodiesel processor.

Using this shape design, researchers managed to prevent the accumulation of residual liquid at the bottom of the tank. Now biodiesel is ready for the next step, which is quality testing in the laboratory.

Below is the final result of our biodiesel processor.

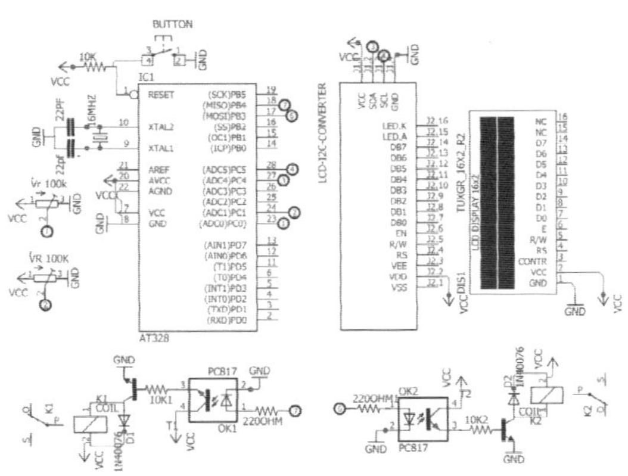

Figure 1. Microcontroller diagram

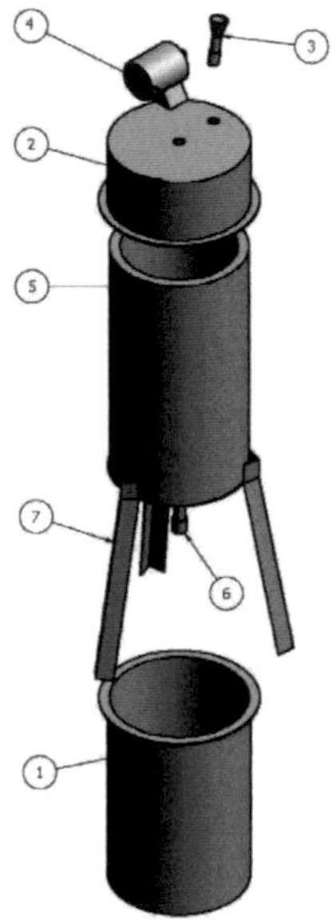

Main Reactor Part List :

1. Reaction Tank
2. Reaction Tank Top
3. Inlet Pipe
4. Solenoid Valve
5. Main Tank
6. Drainpipe
7. Stand

Figure 2. Design of Main Reactor

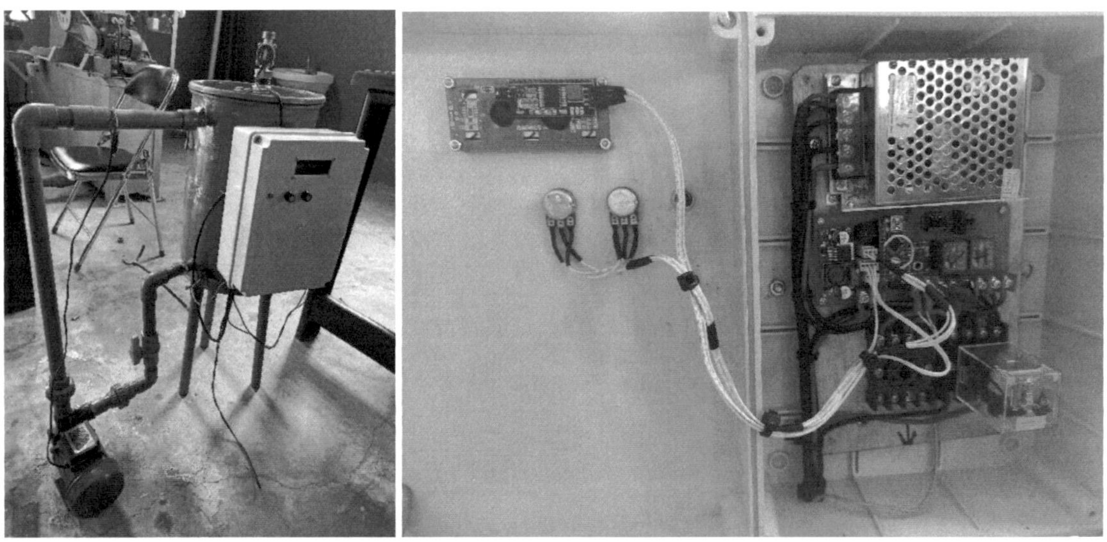

Figure 3. Redesign Prototype Biodiesel Processor By DFMAM

3.5. Laboratory Test Results

The quality of biodiesel oil processed using this prototype is then tested in the laboratory. Then the test parameters are compared with the standards of the Ministry of Energy and Mineral Resources. We tested with 2 type of common vegetable oil and also, it's a main product for researcher district. We also tested it with two types of catalyst to get a data which catalyst can produced an optimal result. The results are presented in the following table.

Table 1. Comparison of Properties of Biodiesel Production with Quality Standards of the Ministry of Energy and Mineral Resources (ESDM).

Property	Biodiesel Properties Test Value KOH	Biodiesel Properties Test Value NaOH	Biodiesel Properties Reference ESDM*
Density 15 °C (Kg/m 3)	874	877	815-880
Viscosity 40 °C (mm^2/S)	4.3	4.7	2 - 5
Acid value (mg KOH or NaOH/g)	0.14	0.15	0
Raw Cetane number	56	58	48
Cloud point (°C)	8	7	0 - 18
Pour point	4	5	0 - 18
Flashpoint (°C)	171	178	Min. 52
Distillation temperature (°C)	345	350	Max.370

* Decree of the Ministry of Energy and Mineral Resources of the Republic of Indonesia No. 0234.K/10/DJM.S/2019 [44]

The ideal condition (optimum) for biodiesel generation from oil is the truth that the ideal proportion of methanol to oil is at a blend proportion of 6:1, response temperature is 60 cel deg, response time is 60 minutes, the catalyst utilized is 1% KOH and the ideal condition for biodiesel generation is 88%; whereas the abdicate of biodiesel generation from a blend of palm oil and coconut oil beneath ideal conditions utilizing KOH and NaOH catalysts were 98.12 and 97.3%, individually. To deliver biodiesel from coconut oil employing a catalyst concentration of 1% KOH the most extreme productivity is 96%.

From table it is known that the production of biodiesel oil used in this study in general has met the quality standards of the Ministry of Energy and Mineral Resources, except for the acid value. The presence of acid values in biodiesel production is a matter of further research to make the value zero. The presence of acid values will be the cause of the acceleration of the corrosion rate in diesel engine components.

It is interesting that the biodiesel tested has a higher cetane number, namely max. 58 for production using NaOH, 10 points higher than the minimum standard reference cetane number, which is 48.

However, there are values that need to be considered to be derived through advanced refining methods, namely density and viscosity properties. If we can reduce the density and viscosity values, this biodiesel material will be safer if used in common rail type diesel engines (modern diesel engines).

4. Conclusions

This research come about within the improvement of a brilliantly small-scale biodiesel processor with a capacity of 25 L per production. The processor consolidates all the crude fabric necessities for biodiesel generation subsequently, both soluble catalyzed and corrosive catalyzed transesterification can be carried out with the processor. In expansion, processor get together operations are rearranged to decrease uncertainty and excess. In this manner, the processor is by and large cost-effective in terms of fabric prerequisites, parts generation, labor and overhead. The DFMA standards to minimize portion prerequisites without compromising quality. This diminishes the in general gathering and generation costs. Processor component assembly via the DFMAM principle is simplified to reduce component complexity and redundancy. The biodiesel produced by this method has been tested by a lab and recorded a density value of 15 cel.deg. 874 - 877 kg/m3, viscosity at 40 cel.deg 4.3 – 4.7 mm2/s, Acid value 0.14 0.15 mg KOH/g, Cetane 56 – 58, Cloud Point 7 – 8 cel.deg., pour point 4 – 5, Flashpoint 171 – 178 cel.deg and distillation point 345 – 350 cel.deg.

5. Acknowledgments
The researcher and journal authors would like to thank the Jember State Polytechnic through the Central Unit for Research and Community Service (P3M) for the financial fund and technical research support to carry out this research.

References
[1] Daniyan IA, Mpofu K. 2018. *Application of Design for Manufacturing and Assembly: Development of a Multifeedstock Biodiesel Processor*. In: Păcurar AC, editor. Applications of Design for Manufacturing and Assembly. Rijeka: InTech. pp. 5-26.
[2] Naiju CD, Jayakrishnan V, Warrier V, Rijul B. 2016. *Application of Design for Manufacturing and Assembly (DFMA) methodology in the steel furniture industry.* In: National Conference on Technological Advancements in Engineering. pp. 1-4
[3] Katja T, Miika J, Jari P. 2002. *Activity based costing and process modelling for cost-conscious product design: A case study in a manufacturing company.* International Journal of Production Economics. 79(1):75-78
[4] Olutoye MA, Wong SW, Chin LH, Asif M, Hameed BH. 2016. *Synthesis of fatty acid methyl esters via transesterification of waste cooking oil by methanol with a barium-modified Montmorillonite K10 catalyst.* Renewable Energy. 2016;86:392-398
[5] Ejaz S, Arjun MG, Naiju CD, Annamalai K. 2011. *Conceptualizatio Design for Manufacture and Assembly (DFMA) of juice mixer grinder.* In: National Conference on Advances in Mechanical Engineering. India: NCAME-2011; pp. 297-302
[6] Budiprasojo and A. W. Pratama. 2016. "*Nilai Kalor Bahan Bakar Plastik Polypropilene (Bbpp) Hasil Pyrolisis Dengan Campuran Premium Dan Octane Booster,"* J. Ilm. Inov., vol. 16, no. 2.
[7] Budiprasojo and A. G. Maliki, 2019. *"Auke-2019-012 Analisis Efektifitas Hho Carbon Cleaning Dengan Metode Pengolahan Citra Digital,"* Seminar Nasional Teknik Mesin, no. 1,pp 50-55.
[8] Budiprasojo and A. Irawan, 2018. *"Engine combustion efficiency and performance of exhaust pipe fuel preheating system,"* Rekayasa Mesin, vol. 9, no. 1, pp. 1–7.
[9] Budiprasojo and A. Irawan, 2017. *"Rancang Bangun Exhaust Pipe Fuel Preheating System (EP2FS) Serta Uji Performa dan Analisis Efisiensi Pembakaran Engine,"* Prosiding.
[10] Yusuff AS, Adeniyi OD, Olutoye MA, Akpan UG. 2018. *Development and characterization of a composite anthill-chicken eggshell catalyst for biodiesel production from waste frying oil.* International Journal of Technology. 1:110-119

The effect of changes in the fluid flow rate of Virgin Coconut Oil (VCO) on the immersion cooling process

R E Rachmanita[1], G P Utama[1], A Budiprasojo[2], S Anwar[1], B Rudiyanto[1*], T Seetawan[3]

[1]Renewable Energy Engineering, Engineering Department, Politeknik Negeri Jember, Jember 68101, Indonesia
[2]Automotive Engineering, Engineering Department, Politeknik Negeri Jember, Jember 68101, Indonesia
[3]Center of Excellence on Alternative Energy, Research and Development Institute and Program of Physics, Sakon Nakhon Rajabat University, Sakon Nakhon, 47000, Thailand

bayu_rudianto@polije.ac.id

Abstract. The cooling system in computer equipment generally uses a conventional fan, where the heat transfer occurs by radiation so that the level of heat absorption is not optimal in computer components that generate heat. In overcoming these problems, a cooling system using the immersion cooling method is used. The use of an immersion cooling system on computer equipment aims to compare the temperature drop on CPU performance using conventional cooling and immersion cooling. In testing the computer system will be operated for 24 hours with maximum CPU performance. Variations of the immersion cooling system in the form of flow rates of 2, 4 and 6 lpm with variations in the rotation of the radiator fan at 200, 400 and 600 rpm. The results of the test show that the temperature reduction that can be achieved by conventional cooling is 71.2°C, while the temperature decrease in the immersion cooling system with the best variation is at a water rate of 6 lpm and a fan rotation of 600 rpm is 42.8°C. The results of the test can be seen that the immersion cooling system can significantly reduce the working temperature of the CPU.

1. Introduction

In line with the development of information technology, the use of the internet in the world is increasing. Reporting from the digital report We Are Social in 2020 the number of internet users reached 4.5 billion million people in the world. This figure shows that internet users have reached more than 60 percent of the world's population with a total of 7.7 billion million people. This data continues to increase compared to the previous year which was only 4.3 billion. Internet users in Indonesia itself has also increased. According to the Association of Indonesian Internet Service Providers (APJII) in 2020 there were almost 197 million. This number increased by 8.9% compared to 2019 with the figure of 195 million, in addition to the increase in the number of internet usage due to the rapid development of technology, it is also influenced by the outbreak of the covid 19 virus that has hit Indonesia since March 2020 [1]. During the pandemic, all activities are required to be carried out at home. Internet access is the main choice so that activities such as education, work, and others can still be carried out.

The development of the internet which has increased significantly has an effect on the data center, because the higher the use of the internet, the greater the need for processes and internet data. A data center is a centralized repository, both physical and virtual, for the storage, management, dissemination of data and information. Data center which literally means data center, is a facility to place computer systems and related equipment, such as data communication systems and data storage. Development in the data center will require the addition of new servers in order to meet the increasing spike in access. This equipment consumes 270 TWh of energy and reaches 1.4% of the world's total energy with 60% electrical power requirements and the cooling system reaches 40% of the total operational costs of the data center [2]. Based on these facts, the data center is currently facing two major problems, how to develop new services but with the consequence that power consumption for computing and cooling continues to increase. If this trend continues, the data center's ability to add new services will be hampered [3].

The data center cooling system consumes relatively a lot of electrical energy because most data centers use cooling by air cooling method where cooling uses a conventional fan. Computer cooling system is a device used to reduce or eliminate heat in computer components or CPU, especially in the processor [4]. Excessive heat in a computer system can cause overheating where it will affect computer performance and be followed by damage to computer components [5]. In general, there are two cooling systems used in a computer, namely the water cooling and liquid cooling systems [6].

In recent years, the liquid cooling system has developed because the cooling system using air cooling method, which uses fans and heatsinks, is not able to significantly reduce CPU temperature on computers with high specifications [7]. The use of a liquid cooling system has several advantages, such as the temperature reduction is faster and more efficient because it goes directly to the intended heat source, so that it can significantly reduce the working temperature of the CPU even in a state of maximum computer workload [8]. The use of a liquid cooling system also reduces noise on computer devices [7].

There are two cooling systems in liquid cooling, indirect cooling and direct cooling. Indirect cooling is the process of removing heat without direct contact, by flowing the fluid at the heat source through a hose, while direct cooling has direct contact between the heat source and the cooling fluid by immersing it directly into the coolant [6]. The direct cooling system began to be developed on electronic components, especially in the data center [9]. The advantage of the direct cooling system is that components that are immersed directly into the coolant experience direct heat transfer or convection so that the heat generated by the components can be directly transferred to the coolant [6]. Direct cooling or immersion cooling system is a cooling method by immersing the motherboard components of a computer into a non-conducting liquid so that the heat generated from the components can be transferred directly into the cooling liquid [10]. The dielectric fluid used in this test as a cooling fluid is virgin coconut oil (VCO).

The immersion cooling system works by absorbing heat from components that are immersed in coolant in a vessel, the fluid that absorbs heat from the components will be flowed into a heat exchanger in the form of a radiator, the fluid that has dropped its temperature in the radiator is then flowed back into the vessel to absorb heat back from the components. The process takes place in a circulation with the help of a pump [11,12]. In this study, the testing was conducted on the central processing unit (CPU) using virgin coconut oil (VCO) fluid with variations in flow rates of 2, 4 and 6 lpm and variations in radiator fan rotation of 200, 400 and 800 rpm. This study aims to compare the conventional cooling system and immersion cooling on CPU performance and determine the effect of immersion cooling with variations in flow rate using Virgin Coconut Oil on CPU temperature.

IOP Conf. Series: Earth and Environmental Science **980** (2022) 012067 doi:10.1088/1755-1315/980/1/012067

2. Methodology

2.1. Design of Immersion Cooling System

The research method used is an experimental method which is carried out in several stages of research. The first stage is the design of the immersion cooling system on a central processing unit (CPU) as shown in Figure 1. The working system of the tool is to put the motherboard into a glass vessel with a volume of 11,250 cm^3 filled with VCO liquid. The radiator with the fan functions as a liquid cooling device for the VCO that carries heat away from the motherboard components. The flow meter is used as a regulator of the size of the flow rate while the thermocouple is used to measure the temperature of the VCO before and after passing through the radiator. after passing through the radiator the VCO will flow back into the glass vessel. The process takes place in circulation with the help of a submersible pump.

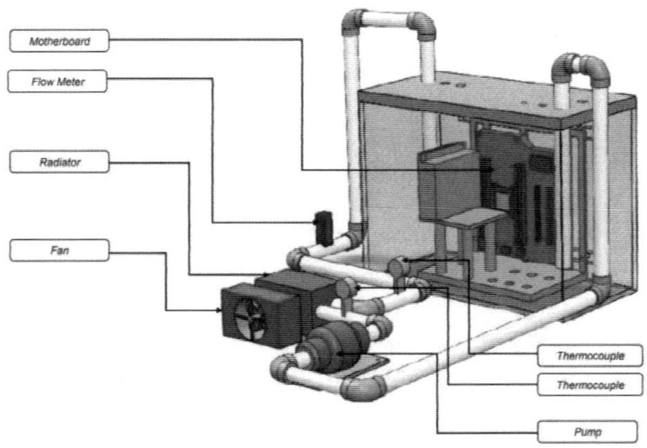

Figure 1. Design of immersion cooling system

2.2. Tools and Materials

The tools needed in this research include a glass vessel (30 x 25 x 15 cm), flow meter, thermocouple, hose 1.5 inch in diameter, radiator, radiator fan and submersible pump. While the materials needed are Virgin Coconut Oil (VCO) with specifications in Table 1 and Central Processing Unit (CPU) with specifications in Table 2.

Processing of instructions on the CPU consists of the operation of reading instructions (fetch) and instructions for controlling work functions or execute. On the CPU, the part that tends to generate heat when the computer is operated is the motherboard. On the motherboard there are several important components such as Processor, RAM, VGA, VGA slot, AGP, AGP slot and additional card slots [9].

Table 1. Specifications of Virgin Coconut Oil (VCO)

Parameter	
Kinematic Viscosity	49.5 mm^2/s
Flash Point	232°C
Pour Point	15.5°C
Kinematic Viscosity Temperature	40°C

Table 2. Specifications of Central Processing Unit (CPU)

Operating System	Microsoft windows 7 ultimateCopyright © 2009
System Manufacture	PCCHIPS
System Type	64-bit operating system
BIOS	Default System Bios
Processor	Intel (R), Core (TM), Dual CPUE28300, @2.83GHz 2.83 GHz
Memory	2,00 GB RAM
Direct Version	DirectX 9.0c (4.09.0000.0904)

2.3. Data Retrieval and Analysis

The variations used in testing the immersion cooling system are the fluid flow rate and the rotational speed of the radiator fan. Variations in flow rates are 2, 4 and 6 lpm while the fan speeds used are 200, 400 and 600 rpm which aims to determine the rate of decrease in computer temperature in each variation. Testing the computer system will be operated for 24 hours by running various software with the assumption of daily use so that the CPU works in maximum condition. CPU temperature data retrieval is done by running the HW monitor application on the computer which is carried out every 3 hours. The measurement data are CPU temperature, inlet channel temperature, and outlet channel temperature. The data obtained will be analyzed and presented in graphical form.

3. Results and Discussion

3.1. CPU Temperature with Conventional Cooling System

In this study, the computer system is operated for 24 hours with the assumption of daily use. In a period of 24 hours the CPU will work optimally so that the temperature measurement of the CPU can be carried out using a conventional cooling system and an immersion cooling system with a predetermined variable flow rate and radiator fan rotation. The aim is to find out how effective the influence of each variable in lowering the CPU temperature when working optimally. The conventional cooling temperature is the basis for comparison with the maximum temperature using immersion cooling. Figure 2 shows the initial CPU temperature reaching 86.4°C. The temperature that can be achieved by a conventional cooling system within 24 hours is the final temperature reaching 71.2°C. The room temperature tends to be normal, 31°C during the day and 28°C at night.

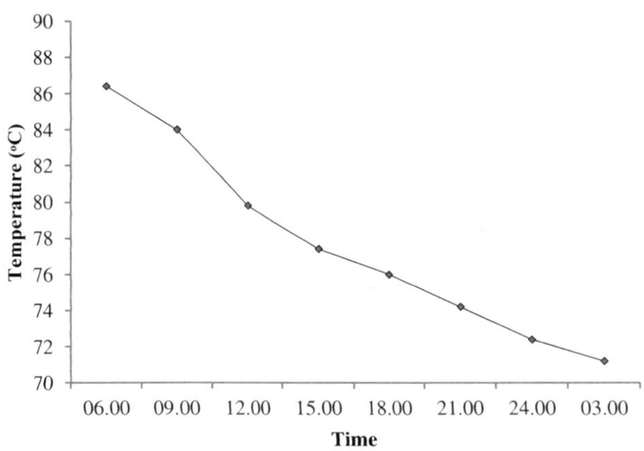

Figure 2. Conventional cooler CPU temperature graph

3.2. CPU temperature at a flow rate of 2 lpm

CPU temperature using an immersion cooling system at a flow rate of 2 lpm showed a greater temperature drop than conventional coolers. This shows that the immersion cooling system is able to reduce the CPU working temperature more significantly. The graph of the decrease in CPU temperature is shown in Figure 3. At a fan rotation speed of 200 rpm there is a decrease in temperature, but the decrease in temperature is not significantly different compared to conventional coolers, the temperature is recorded at 72.8 °C. Fan rotation with 400 rpm temperature drop is more stable and there is no spike when compared to the rotation speed of 200 rpm. The final temperature that can be achieved on this variable is 64.8 °C. At a speed of 600 rpm, it shows that the decrease in temperature is more significant than the previous variation with a final temperature of 63.2 °C. The speed of 600 rpm cools the VCO faster so that the VCO has a lower temperature after passing through the radiator.

The spike in CPU temperature at a fan rotation of 200 rpm occurred in the 3rd to 6th measurements. The spike occurred because the VCO temperature in the vessel continued to increase because it absorbed heat from the CPU components when at maximum workload. The recorded VCO temperature at the time of the CPU temperature spike was 42ºC to 46ºC. It can be concluded that the reduced flow rate will impede the flow circulation so that heat tends to be trapped in a vessel, the slow fan rotation also causes the fluid heat transfer from the system to the environment not to be maximal [9].

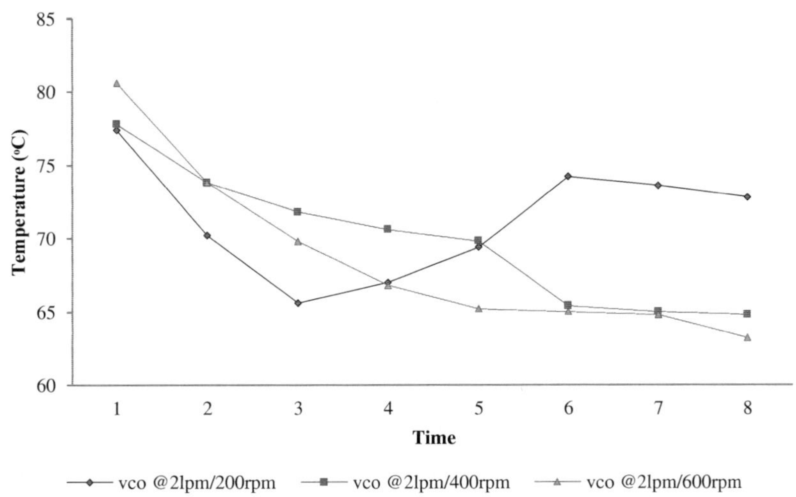

Figure 3. Graph of CPU temperature at a flow rate of 2 lpm

3.3. CPU temperature at a flow rate of 4 lpm

Measurement of CPU temperature at a flow rate of 4 lpm showed a greater decrease in temperature compared to a flow rate of 2 lpm. The maximum working temperature that can be achieved at a flow rate of 4 lpm is shown in Figure 4. The decrease in CPU temperature at a flow rate of 4 lpm with a fan rotation speed of 200 rpm indicates that the decrease that can be achieved is 64.2 °C. The decrease in temperature is smaller than the flow rate of 2 lpm with a fan rotation of 600 rpm, which is 63.2 °C because the rotation of the radiator fan at 600 rpm is more significant to restore the VCO temperature after carrying heat from the glass vessel.

In variations of fan rotation with a speed of 400 rpm the temperature decrease is more stable with an average decrease of 2 °C every 3 hours. At the initial measurement, the CPU temperature reached 85.2 °C. The initial temperature tends to be higher because the computer system does not experience a pause or off after testing at 200 rpm. The maximum temperature reduction that can be achieved on this variable is 62.8 °C. The CPU temperature at a flow rate of 4 lpm with a fan rotation speed of 600 rpm indicates that the maximum temperature that can be achieved is 61.8 °C. The decrease in temperature at 09.00-15.00 West Indonesia time tends to be slower than the previous two variables because the room temperature during the day increases by 32.2 °C compared to the normal temperature which is in the range of 30-31 °C. The decrease in CPU temperature is stable at 18.00-24.00 West Indonesia time because the temperature is back to normal, which is 28 °C so that the VCO is able to absorb heat from the motherboard optimally. At a flow rate of 4 lpm with 3 variations of fan speed, 200, 400, 600 rpm, it can be seen that the maximum temperature that can be lowered is 61.8 °C with a radiator fan speed of 600 rpm.

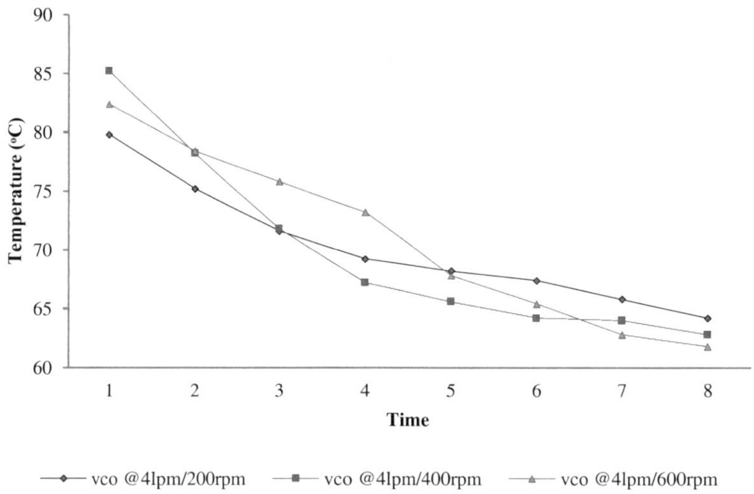

Figure 4. Graph of CPU temperature at a flow rate of 4 lpm

3.4. CPU temperature at a flow rate of 6 lpm

Figure 5, graph of CPU temperature at 6 lpm flow rate shows that the temperature drop is more significant than 2 lpm and 4 lpm flow rates. CPU temperature at a flow rate of 6 lpm decreased more rapidly than at a flow rate of 2 or 4 lpm. At a flow rate of 6 lpm with a radiator fan rotation of 200 rpm the temperature can be lowered to 56.4 °C, meanwhile, the decrease in temperature in the CPU with a radiator fan rotation speed of 400 rpm is greater than the previous variable. At 400 rpm the CPU temperature decreased with an initial temperature of 77.2 C to 49.4 °C with the same CPU workload.

At 600 rpm fan rotation, CPU temperature decreased more significantly than the previous variable. The temperature that can be achieved in this variable is 42.8 °C. The increase in flow rate causes the VCO to carry heat away from the CPU in the vessel. The glass will quickly flow to the radiator, while increasing the fan rotation on the radiator will quickly restore the VCO temperature. At a flow rate of 6 lpm the maximum working temperature of the CPU experienced a much more significant decrease than the previous variable which indicated that the flow rate and rotational speed of the enlarged radiator fan had a significant effect on the decrease in the working temperature of the CPU.

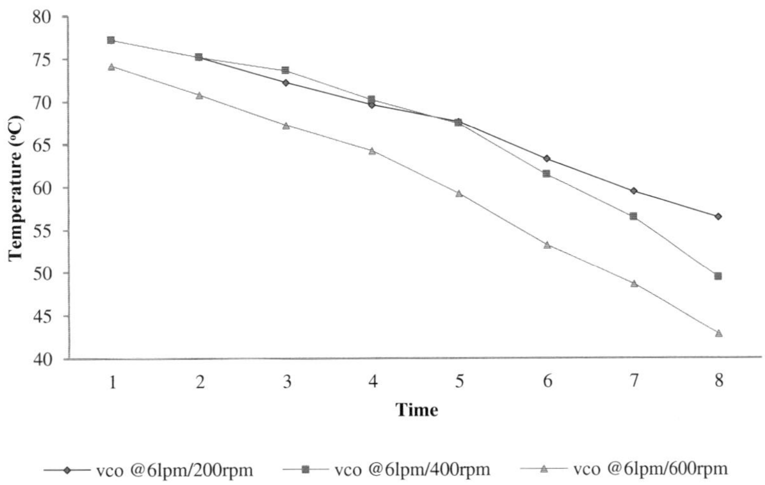

Figure 5. Graph of CPU temperature at a flow rate of 6 lpm

3.5. Evaluation of Radiator Fan Flow Rate and Rotation Speed

Immersion cooling is a cooling system by immersing the motherboard and other components that function to reduce heat in the Central Processing Unit (CPU). The heat generated by the CPU will be absorbed by convection by Virgin Coconut Oil (VCO) and then flowed to the radiator to be cooled by the fan. In this study, the data obtained show that the use of an immersion cooling system is better at lowering the temperature on the CPU compared to a conventional system. The maximum temperature that can be achieved within 24 hours when using a conventional cooling system is 71.2 °C, while the maximum temperature that can be achieved by an immersion cooling system is 42.8 °C at the highest flow rate and fan rotation.

The heat transfer system in the immersion cooling process occurs by convection. Convection heat transfer is heat transfer that occurs due to a temperature difference followed by the flow of medium particles. The faster the fluid movement, the greater the rate of convection heat transfer that occurs. In an immersion cooling process, the CPU temperature has a higher temperature than the fluid (VCO) so that the VCO will absorb heat from the CPU work, then the VCO will carry heat from the vessel through the inlet channel to the radiator, the VCO which has been cooled by the radiator is then flowed through the outlet channel. to the vessel to carry out the process of re-absorption of heat.

In the Immersion Cooling system, the flow rate and rotational speed of the fan affect the decrease in the working temperature of the CPU. At a flow rate of 2 lpm the maximum temperature that can be achieved is 63.2 °C, at a flow rate of 4 lpm the maximum temperature is 61.8 °C, while at a flow rate of 6 lpm the decrease in CPU temperature is more significant, namely 42.8 °C. The difference in temperature reduction is due to the heat generated by the CPU will be absorbed by the VCO but with a small flow rate, the VCO flowing to the radiator will be slow, resulting in the VCO temperature increasing and the circulation process not optimal. This is inversely proportional to if the flow rate is accelerated, the process occurs is the working heat generated by the CPU will be quickly flowed by the fluid to the radiator so that the heat absorption process and circulation process take place optimally.

Each flow rate with a variation of the speed of the fan rotation occurs a different temperature reduction process. At a flow rate of 4 lpm with a fan rotation of 600 rpm, the temperature drop is unstable at 09.00-15.00 Western Indonesia time, this is due to the environmental temperature increasing from 30 °C to 32 °C so that the temperature reduction process is slower. At night the temperature drop is faster because the room temperature is relatively lower 28 °C, such as at a flow rate of 6 lpm at a fan rotation of 600 rpm at 18.00-03.00 West Indonesia time, the CPU temperature can be lowered quickly from 59 °C to 42 °C. The ambient temperature affects the process of decreasing the temperature.

The flow rate factor and the rotation of the radiator cooling fan in the immersion cooling system have also been proven in the research of Krisian et al (2018)[8], where the variation of the flow rate used is 0.5 lpm, 1 lpm and 1.5 lpm with a maximum fan rotation of 800 rpm which resulted in a decrease in CPU temperature of 51 °C, 48 °C, and 47 °C. The practical implication obtained from this research is the use of immersion cooling with a flow rate variation of 1.5 lpm to achieve the highest temperature reduction with the same room conditions.

4. Conclusion

The immersion cooling system is proven to be more significant in lowering the working temperature of the CPU than cooling systems with air cooling or conventional fans. The maximum temperature drop that can be achieved with the immersion cooling system with the best variation is 42.8 °C, while the temperature drop that can be achieved with the conventional cooling system is only 71.2°C. Variation of flow rate 6 lpm with fan rotation of 600 rpm is the best combination in reducing CPU temperature significantly in this study. The increased flow rate will speed up the circulation process in the VCO to the radiator, while the accelerated fan rotation can cool the VCO more optimally.

5. Acknowledgments

We sincerely thank to the Director of Politeknik Negeri Jember and Pusat Penelitian dan Pengabdian Masyarakat (P3M) for the funding support so this script can be carried out well. This script would hopefully give a positive contribution to the educational development or those are willing to conduct further research.

References

[1] Anon Asosiasi Penyelenggara Jasa Internet Indonesia
[2] Amirullah A, Ijtihadie R M and Studiawan H Optimasi Daya Data Center Cloud Computing Pada Workload High Performance Computing (HPC) Dengan Scheduling Prediktif Secara Realtime | Amirullah | JUTI: Jurnal Ilmiah Teknologi Informasi
[3] Anon Central Processing Unit (CPU) Sebuah mesin tipe von neumann
[4] Testi D 2018 Heat transfer enhancement by an impinging ionic jet in a viscous transformer coolant *Int. Commun. Heat Mass Transf.* **91** 256–61
[5] Septiadi W N, Budiarsa I N, Tnunay I A and Desy I 2018 Sistem Pendingin Central Processing Unit (CPU) Berbasis Cascade Straight Heat Pipe 7
[6] Habibi Khalaj A and Halgamuge S K 2017 A Review on efficient thermal management of air- and liquid-cooled data centers: From chip to the cooling system *Appl. Energy* **205** 1165–88
[7] Kheirabadi A C and Groulx D 2016 Cooling of server electronics: A design review of existing technology *Appl. Therm. Eng.* **105** 622–38
[8] Krisian S N 2018 Analisis Efek Variasi Flow Rate Terhadap Performa CPU dan PSU pada Immersion Cooling dengan Mineral Oil
[9] Kuncoro I W, Pambudi N A, Biddinika M K, Widiastuti I, Hijriawan M and Wibowo K M 2019 Immersion cooling as the next technology for data center cooling: A review *J. Phys. Conf. Ser.* **1402** 044057
[10] Kuncoro I W 2019 Performa Pendinginan Immersion Cooling Menggunakan Mineral Oil Pada Server Rackmount Tipe 1u
[11] Yusuf A M 2018 Analisis Performa Pendinginan Celup (Immersion Cooling) Terhadap Data Center dengan Variasi Cairan Pendingin
[12] Wang Y, Pambudi N A, Bugis H, Kuncoro I W, Setiawan N D, Hijriawan M, Rudiyanto B and Basori B 2019 Preliminary experimental of GPU immersion-cooling *E3S Web Conf.* **93**

Thermo-environmental economic study on energy conversion system using thermoelectric generators (TEG SP1848) in cold storage condenser

R E Rachmanita[1*], D Prasetyo[1]

[1]Renewable Energy Engineering, Engineering Department, Politeknik Negeri Jember, Jember 68101, Indonesia

risse_rachmanita@polije.ac.id

Abstract. Thermoelectric generator is an alternative energy that utilizes temperature differences into electrical energy. The use of a thermoelectric generator (TEG SP1848) by utilizing a heat source from the exhaust heat of a cold storage condenser requires thermodynamic, environmental thermal, and techno-economic analysis. The method used in this study is a qualitative approach with descriptive and evaluative methods. Thermodynamic analysis of the amount of heat wasted at each position of the engine pipe obtained T_1 pipe 38613.47 Watts, T_2 pipe 3960.35 Watts, and T_3 pipe 39651.80 Watts. The SP1848-based TEG energy conversion device absorbs 11.14 watts of heat energy placed in the T_3 pipe and has an efficiency of 0.23%. Analysis of the environmental conditions of the cold storage engine room based on the Predicted Mean Vote (PMV), Predicted Percentage Discomfort (PPD) and Temperature Humidity Index (THI) values shows that the engine room environment is not comfortable for human activities and can disrupt the balance of human metabolism. Techno-economic feasibility evaluation based on Net Present Value (NPV) and Payback Period (PBP) produces a negative value, while Benefit Cost Ratio (BCR) is 0.0012, indicating that the conversion tool is not feasible to run.

1. Introduction

The increasing demand for electrical energy has created a global problem, namely the depletion of potential reserves of fossil fuels. The price of crude oil fluctuates every second, causing instability in the world economy. These problems arouse many people to create and develop alternative energy sources. Indonesia has several sources of energy that can be used to generate electricity which are divided into two types. First, power plants with macro capacities that usually utilize water, steam, gas, and nuclear. Second, power plants with micro capacities, one of which is utilizing heat energy. The heat energy can come from sunlight and objects that release heat, such as home electronic equipment, motor vehicle exhaust and industrial kitchen combustion heat [1].

Thermoelectric generator is an alternative energy that utilizes the difference temperature into electrical energy and has a high economic value, as well as does not cause pollution so it is very environmentally friendly. A thermoelectric generator is a small plate-shaped device that operates as a heating engine by converting heat directly into electricity [2,3]. Thermoelectricity is influenced by three thermodynamic principles, Seebeck, Peltier and Thomson effects. Thermoelectric generator technology has been used by several researchers, in an effort to utilize waste heat energy from several heat sources in the surrounding environment converted into electrical energy [2]. Ryanuargo et al. (2013) have conducted research on centralized air conditioning systems (AC Central) in office buildings and shopping centers. Utilizing waste heat energy from the central AC condenser pipe which is designed to be winding and equipped with fins using the thermoelectric method [4]. Experiments on the condenser pipe have an average temperature of 34°C which can produce a voltage of 3.14 Volts and a power of 0.16 Watts. Busthomy and Widyartono (2020) have investigated the use of a thermoelectric generator on the heat wasted from the combustion fire for charging mobile phone batteries, using 8 thermoelectric arranged in a mixture of series and parallel, producing a voltage of 4.2 Volts and a battery charging current of 0.34 Ampere with a difference flame temperature 48-50.1°C [5]. The use of thermoelectric for charging batteries is also carried out by [6,7].The researcher used a thermoelectric TEG SP1848 and used a heat source from an exhaust heat of a cold storage condenser. The results of this study indicate that the 4 modules produce the highest output power of 2.07 Watts in a series circuit with an iron temperature of 110°C and a resistance of 33 Ω. While the power of 0.144 Watts and efficiency of 0.24% is generated from the exhaust heat source of the condenser input pipe with a temperature difference of 24.17 K, using 5 modules connected in series.

Cold storage is a cooling machine that accommodates objects to undergo a cooling process. Cold storages are commonly used in everyday life to cool or preserve foods such as meat, vegetables and fruits as well as beverages [8]. Politeknik Negeri Jember has 5 cold storage units that operate for 24 hours, the cold storage is used to store fish and edamame. Utilization of alternative energy in waste heat energy from a hot object requires several aspects that need to be considered, such as aspects of the thermal and economic environment. The wasted heat certainly increases the temperature into the environment, with high ambient temperatures can interfere with thermal comfort for activities. According to [9], ideally aesthetic values are owned in a building, which provides a sense of security and provides thermal, visual and audio comfort.

Economic analysis is needed to compare the use of alternative energy technologies with existing conventional energy, as well as to determine the economic value invested in the development of thermoelectric technology. The Ministry of Energy and Mineral Resources and PLN 2020 determine the basic electricity tariff for the October-December 2020 period for the R-1/TR group with a power limit of 900 VA-RTM of IDR 1,352 per kWh and class I-4/TT with a power limit of 30,000 kVA and above of IDR 996.74 per kWh. Suryaningrat (2011) reveals that the principles of technical economics are used to analyze the uses of money, especially in relation to the physical assets and operations of an organization to help make decisions [10]. Based on the description of the background, this study aims to conduct an analysis in terms of thermodynamics, thermal environment, and economic aspects of the use of thermoelectric. This study is to create a comfortable working environment by developing thermoelectric technology that is economically efficient.

2. Methodology

The research method used is a qualitative approach with descriptive and evaluative methods. Heat transfer analysis data retrieval is carried out directly based on observations. The thermodynamic data retrieval on the cold storage condenser input pipe includes the size of the input pipe and the temperature of the condenser input pipe.Thermal comfort analysis data retrieval is carried out directly based on observations. Data retrieval of thermal comfort in the environment around the cold storage condenser includes space sketches, wind speed, air temperature, air humidity, clothing insulation and metabolic values based on activities carried out in that place. The techno-economic analysis data was collected based on the performance data of the thermoelectric energy conversion system. Data collection includes investment costs, replacement of thermoelectric components, depreciation, PBP (Payback Period), NPV (Net Present Value) and BCR (Benefit Cost Ratio).

This study uses several supporting tools including temperature data logger, anemometer, hygrometer, roll meter, timer, Global Positioning System (GPS), caliper, thermoelectric devices in the cold storage condenser input pipe and the environment around the cold storage condenser. Cold storage at Politeknik Negeri Jember has two operating machines and three inactive machines. The machine has a capacity of 40 tons with a room size of 3 × 14 × 4 meters and does not run for full 24 hours, but automatically turns on when the cold storage temperature is -19°C and the machine turns off when the cold storage temperature is -27°C. The operation of energy conversion system-based TEG SP1848 utilizes a heat source from the cold storage condenser input pipe. The condenser input pipe has several turns that start from the compressor then pass through the separator and end up entering the condenser. The condenser input pipe is made of copper and has two different diameter sizes, 28.60 mm (the pipe from the compressor output to before the condenser) and 34.9 mm (the condenser input and output pipe/pipe adjacent to the condenser). The size of pipe and position of sensor can be seen in Figure 1.

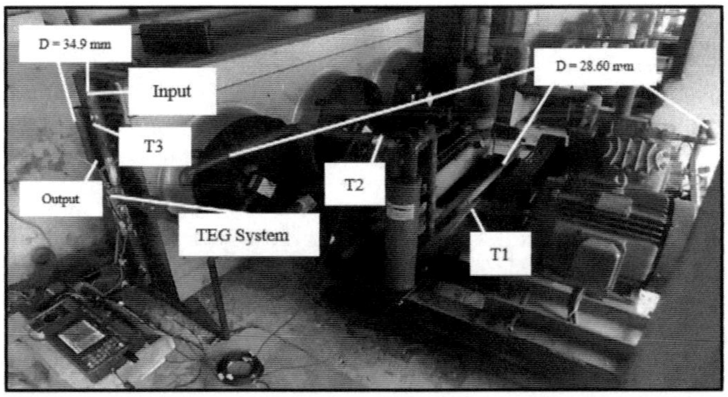

Figure 1. TEG system in cold storage condenser (sensor's position and pipe size)

2.1. Thermodynamic Analysis on Thermoelectric Generator System
Thermodynamic analysis includes [6]:

a. Heat transfer rate at the condenser input pipe

$$q = \frac{\Delta T}{R_{th}} = 2\pi k L \frac{T_i - T_o}{\ln\left(\frac{r_i}{r_o}\right)} \quad (1)$$

b. The amount of heat absorbed by the thermoelectric generator

$$Qh = N\,(\propto . T_h . I - 0{,}5 . I^2 . R + K . \Delta T) \quad (2)$$

2.2. Thermo Environmental Analysis
Thermo environmental analysis includes [11]:

a. Predicted Mean Vote (PMV) thermal comfort index

$$PMV = (0.303e^{-0.036}M + 0.028)((M - W) - H - E_c - C_{res} - E_{res}) \quad (3)$$

b. Predicted Percentage of Dissatisfied (PPD) Index

$$PD = 100 - 95e^{-(0.03353PMV^4 + 0.2179PMV^2)} \quad (4)$$

c. Temperature Humidity Index (THI)

$$THI = 0.8T_a + \frac{(RH \times T_a)}{500} \quad (5)$$

2.3. Techno-economic Analysis
Analysis of the feasibility of investing in energy conversion system includes [10]:

a. Depreciation

$$\text{Depreciation} = \frac{P - s}{n} \quad (6)$$

b. Net Present Value (NPV)

$$NPV = \sum_{t=0}^{n} \frac{(C)t}{(1+i)t} \sum_{t=0}^{n} \frac{(Co)t}{(1+i)^t} \quad (7)$$

c. Benefit Cost Ratio (BCR)

$$BCR = \frac{(PV)B}{(PV)C} \quad (8)$$

d. Payback Period (PBP)

$$PBP = \frac{C_r}{A} \quad (9)$$

Nomenclature:

q = Heat transfer rate (Watts)
k = Thermal conductivity (W/m°C)
T_i = Internal temperature (°C)
T_o = Outside temperature (°C)
L = Cylinder length (m)
Qh = Heat absorbed by TEG (Watts)
N = Number of thermoelectric thermocouples
α = Seebeck coefficient (Watt/ΔT)
Th = Thermoelectric hot side temperature (K)
K = Conductance of thermocouple (Watt/K)
M = Metabolic rate (W/m^2)
W = Effective mechanical power (W/m^2)
E_c = Evaporative heat exchange (W/m^2)
C_{res} = Respiratory convective heat exchange (W/m^2)
E_{res} = Respiratory evaporative heat exchange (W/m^2)
T_{sk} = Skin temperature (°C)
H = Heat loss through the skin (Joule)
Ta = Air temperature (°C)
RH = Humidity (%)

3. Results and Discussion

3.1. Thermodynamic Analysis on Thermoelectric Generator System

Thermodynamic analysis on the condenser input pipe was carried out at 08.00-16.00 West Indonesia Time for 7 days using a temperature data logger to record any temperature changes in the condenser input pipe every 1 minute. The data logger records the temperature of the condenser pipe at 3 points, the input pipe after the compressor (T1), the input pipe after the separator (T2), and the input pipe adjacent to the condenser (T3). The following data points for the temperature of the condenser pipe are presented in Figure 2.

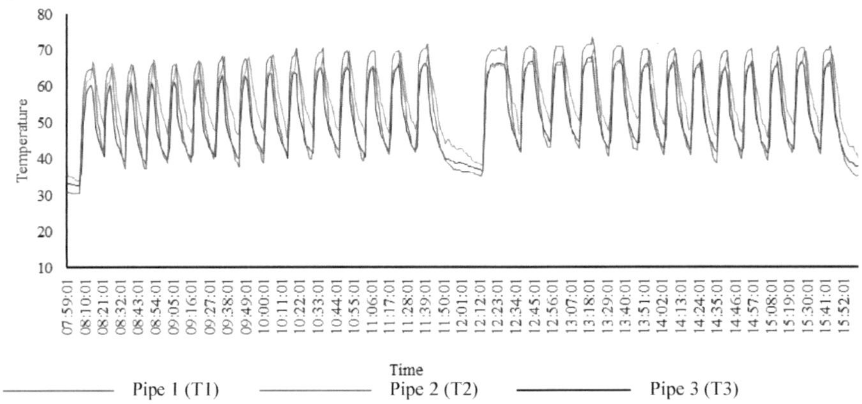

Figure 2. Temperature of condenser input pipe

The temperature of the condenser input pipe at the 3 observation points has different temperatures. The temperature of the input pipe after the compressor (T1) has an average temperature of 55.01˚C, with the highest temperature of 73.86˚C, and the lowest temperature of 29.25˚C. The temperature of the input pipe after the separator (T2) has an average temperature of 57.13˚C, with the highest temperature of 70.45˚C, and the lowest temperature of 31.62˚C. The temperature of the input pipe adjacent to the condenser (T3) has an average temperature of 54.39˚C, with the highest temperature of 71.78˚C, and the lowest temperature of 31.03˚C. The input pipe T1 has the highest temperature of 73.86˚C, when compared to the highest temperature of the T2 pipe of 70.45˚C and the highest temperature of the T3 pipe is 71.78˚C. The high temperature that occurs in the T1 pipe and T2 pipe is influenced by the wind of condenser fan when operating because the position of the two pipes is in front of the condenser fan, while the T3 pipe which is next to the condenser fan is not affected by the wind from the fan condenser. The position of the pipe that is affected by the condenser fan can result in a fairly rapid temperature drop in the T1 and T2 pipes, so that the two pipes cannot maintain the heat they receive.

The decrease temperature in pipe can be seen from the recorded data that has been observed for 7 days, when the three pipes reached their peak within ± 2 minutes, the temperature of T1 pipe decreased

±5°C, the T2 pipe decreased ±6°C, and the T3 pipe decreased ±4°C. Observational data is used to calculate the amount of heat released from each pipe position. The heat released by pipes T1, T2 and T3 are 38613.47Watts, 3960.35Watts and 39651.80Watts. Pipes that have a high temperature, are stable and can maintain temperatures are one of the considerations for the placement and design of the energy conversion system-based TEG SP1848. The energy conversion system-based TEG SP1848 has been designed to utilize the heat wasted in the T3 pipe of 39651.80 Watts, which is one of the pipes that emits a large amount of waste heat because it is not affected by wind from the condenser fan.

The conductor in the energy conversion system is designed using aluminum material with a thermal conductivity of 202W/m°C with a thickness of 1.8 cm. The conductor distributes heat to 5 thermoelectric of 55.71Watts. The exhaust heat produced by the T3 pipe is only 0.14% which can be absorbed by the thermoelectric to be used as electrical energy because the energy conversion system only utilizes a small part of the total length of the pipe. The operational results of the energy conversion system using 5 TEG SP1848 modules in series produce the highest power of 0.132 Watts, and have an efficiency of 0.23%. Utilization of exhaust heat in the cold storage condenser pipe optimally using TEG SP1848 can be reviewed based on the length of the condenser pipe that can be occupied by the energy conversion system based-TEG SP1848. The results of the identification of the measurement of the pipe length, the position of the pipe that can allow the energy conversion system to be occupied, namely the position of the T1 pipe and T3 pipe because the T1 pipe has a length of 105 cm and the length of the T3 pipe is 82 cm. TEG SP1848 module component has a size of 40 x 40 mm, so it can occupy as many as 26 modules on the T1 pipe and 20 modules on the T3 pipe. Optimizing the utilization of condenser exhaust heat by using the total length of the pipe can make it possible to reduce waste heat that is wasted in the environment and can generate considerable electrical energy. In the calculation of the heat that can be absorbed by 1 TEG SP1848 module is 11.14 Watts, then when using 26 modules can absorb heat of 289.64 Watts and 222.8 Watts when using 20 TEG SP1848 modules.

3.2. Thermo Environmental Analysis

a. Analysis of the relationship between PMV and PPD values

The results of measurement data, such as air temperature, relative humidity, air velocity, metabolism, and clothing resistance are used to calculate the PMV and PPD values. The CBE Thermal Comfort Tool For ASHRAE-55 program is a program that helps calculate PMV and PPD values. The results of the calculation show that the highest PMV value is 3.46 with the highest PPD 100% and the lowest PMV value 0.51 with the lowest PPD 10%. The average values of PMV and PPD are 2.37 and 83.96%, respectively. Figure 3 is a graph of the relationship between PMV and PPD.

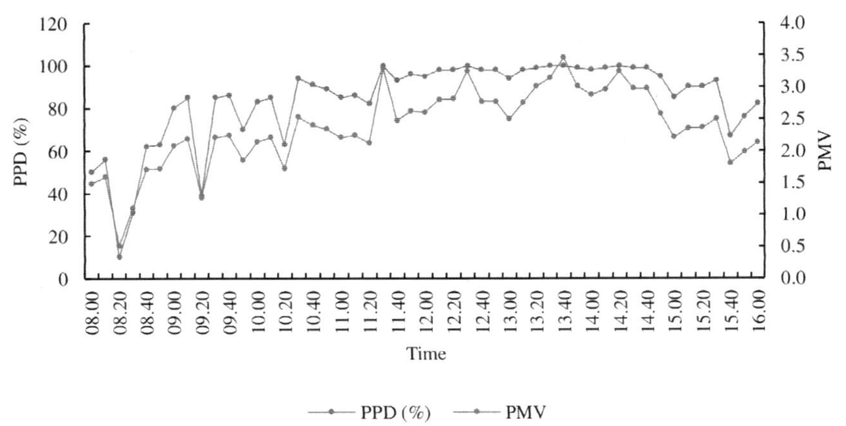

Figure 3. Relationship between PMV and PPD

The graph shows the PMV value from 0.51 to 3.46 while the PPD value is from 10% to 100%. This identifies that the thermal impression felt in the space is in the rather comfortable to hot category based on [12]. The room is dominated by the engine operating at the same time releasing exhaust heat to the environment, resulting in a fairly high indoor air temperature. The large number of ventilations and the high ceiling are not able to neutralize the room air temperature which is high enough so that it can immediately return to standard comfortable conditions, this is due to the small amount of natural air from the surrounding environment and the amount of heat that is wasted into the environment as a result of the reaction of the condenser engine.

These environmental conditions can greatly affect the balance of the human body's metabolism. The normal body temperature of a healthy human being mentioned by [13] is 37˚C. An increase in body temperature which is influenced by environmental temperature will result in a faster decrease in performance, besides that the content of lactic acid in the blood is also lower. As mentioned by [13] that during exercise at sub-maximal and maximal intensity, inhaled oxygen and pulse rate are at lower levels when exercising at ambient temperature is 22±1˚C from 10±1˚C and 35±1˚C. This statement means that at temperatures lower than 22°C or higher, the oxygen uptake process is less than optimal, and the amount of lactic acid released due to metabolic processes is higher.

Someone who performs activities in the engine room will also experience an increase in dehydration, as a side effect of high air humidity from high environmental temperatures. This is an adjustment of body temperature to the environment, by enlarging the skin pores and also sweating more. It can be seen that high environmental temperature and humidity result in excessive sweat secretion, changes in body physiology also occur (changes in the metabolic/energy system), increased lactic acid in muscles, thereby accelerating fatigue in muscles and the body [14]. Incidences of heat exhaustion, stroke, asthenia and even heart attack may occur if a person is in the engine room long enough, without any additional body fluids.

b. Analysis of the relationship between PMV and THI values

The PMV and THI index parameters are two parameters that have the same function, namely to measure thermal comfort in a room or area. Despite having the same function, the two parameters have significant differences in measuring thermal comfort. The PMV index parameter was introduced by Fanger, 1982[11]. The PMV index uses 5 parameters, namely air temperature, relative humidity, wind speed, clothing insulation and value metabolism of an activity. The PMV index has a value limit that is from -3 to +3. While the THI index was developed by [12]. in 2003 which uses a combination method of temperature and humidity to determine the effect of hot conditions on human comfort, and has a value limit of 20-26. The results of the study of thermal comfort in the cold storage engine room, correlated the value of thermal comfort between the PMV and THI indexes. The graph of the relationship between PMV and THI is shown in Figure 4.

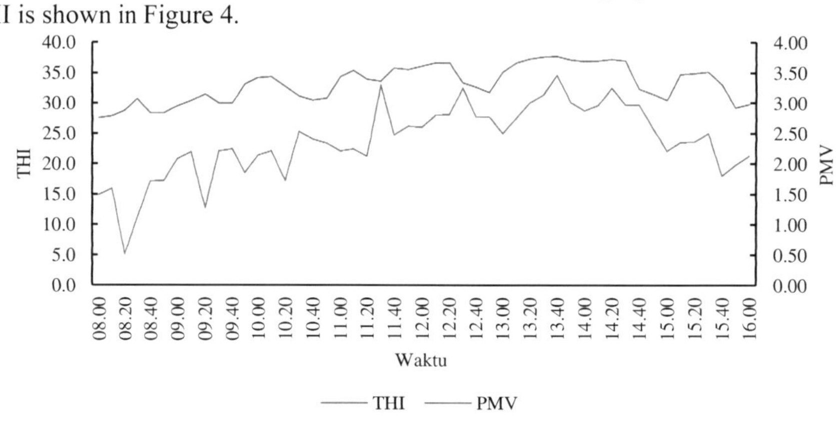

Figure 4. Relationship between PMV and THI

The PMV and THI index values presented in graphical form show that there is an opposite interaction, when the PMV value has decreased but the THI value has increased, the PMV value has increased but the THI value has decreased. Although the two thermal comfort indices have different value limits, there is still an interrelated relationship. The results of the PMV and THI index calculations also show the same category of environmental conditions, namely uncomfortable environmental conditions. The PMV index value in the cold storage engine room which is calculated at 0.51 – 3.46 indicates that the environment is in the rather comfortable to hot category. While the THI index value in the cold storage engine room which is calculated at 26.4 – 37.7 indicates that the environment exceeds the value limit for the comfortable condition category.

3.3. Techno-economic Analysis

The initial investment spent in the manufacture of energy conversion systems includes the purchase of TEG SP1848, water pumps, cooling radiators, adapters, radiator fans, power banks, cost of lathe, cost of cast aluminum, and other components. The total investment cost obtained is IDR 1,792,170.00. The investment costs incurred are calculated on the 0 year at the age of TEG SP1848.

The depreciation applied to energy conversion system is functional depreciation. Based on information on the electroniscomp.com, the TEG SP1848 has decreased function by 0.2% every year. The calculation of the cost of replacing components also needs to be taken into account because the age of the energy conversion system is based on the life time of the TEG SP1848 which is 200,000 hours or 23 years 1 month. There are 2 components that have a shorter life time than the TEG SP1818, including the water pump and radiator fan. The cost of replacing water pump components which is carried out every 3 years 5 months, and there are 6 replacements for 23 years. Meanwhile, the replacement of radiator fan components is carried out every 4 years 7 months, and there are 5 replacements for 23 years. The replacement cost of each of these components is included in the cost of expenditure or cash outflows at each of these times, the total replacement of components for 23 years is IDR 666,800.00.

NPV analysis was carried out after knowing the net cash flow from the energy conversion system-based TEG SP1848. Net cash flow is the difference between the cash inflows and cash flows out of the equipment over the life of the equipment. The cash inflow is an effort to save electricity from the amount of electrical power that can be produced by the TEG SP1848, while the cash outflow is the need for electrical power to operate the energy conversion tool, namely the use of a 12 W adapter to run the water pump and radiator fan. By using the standard reference of the PLN R-1/TR class TDL price with a power limit of 900 VA-RTM, which is IDR 1,352.00/kWh. Cash inflows and outflows are assumed to experience a price increase of 20% every 5 years because annually there is an increase in the value of inflation [15]. Cash inflow is an effort to save electrical energy from energy conversion electricity production by TEG SP1848 in 23 years amounting to 1,890 kWh with a cost that can be saved as much as IDR 2,809.34. The cash flow that comes out in year 0 is the investment cost used in the design of the energy conversion tool. The total cash outflow for 23 years is IDR 2,264,317,537.00. The NPV value that has been calculated from the energy conversion system using TEG SP1848 is the difference between cash inflows and cash outflows is IDR -2,261,508.2. The NPV value on the tool is negative, this indicates that the conversion tool is not feasible to be developed [15]. The NPV value is negative in the design of this system, because the cost of expenditure or cash outflow is greater than the cost of saving or cash inflow. The electrical power produced is very small so it is not optimal if it is used to save conventional electrical energy.

By using formula (8) and comparing the value of income with the value of costs to obtain the BCR value. According to [16], if the value of BCR>1 then the project deserves to be selected, but if the value of BCR<1 then the project is not feasible to be selected and run. The B/C value in the conversion system is 0.0012, so the energy conversion system is not feasible to run.

PBP analysis is obtained using the formula 3.7, then the PBP value from the energy conversion system using TEG SP1848 is -18.84 years or rounded up to -19 years. The negative sign in the payback period means that the investment in the energy conversion equipment has suffered a loss or cannot recover the investment cost.

4. Conclusion

The wasted heat from the 3 measured pipe positions including T1 pipe, T2 pipe, and T3 pipe is 38613.47Watts, 3960.35 Watts and 39651.80 Watts. The energy conversion system-based TEG SP1848 can absorb heat energy of 11.14 Watts which is placed in the T3 pipe and has an efficiency value of 0.23%. The environmental conditions of the cold storage engine room have a PMV value of 0.51-3.46, a PPD value of 10%-100% and a THI value of 26.4-37.7. These three parameter values indicate that the engine room environment is not comfortable for human activities, and can disrupt the balance of human metabolism which can cause heat exhaustion, stroke, asthenia and heart attack. The energy conversion system using TEG SP1848 which utilizes the cold storage condenser input pipe is declared unfeasible to run using the NPV, BCR and PBP methods.

Acknowledgments

We sincerely thank to cold storage Politeknik Negeri Jember, P3M and Energy and Mechanics Workshop for the facility so this script can be carried out well. This script would hopefully give a positive contribution to the educational development or those are willing to conduct further research.

References

[1] Khalid M, Syukri M and Gapy M 2016 Pemanfaatan Energi Panas Sebagai Pembangkit Listrik Alternatif Berskala Kecil Dengan Menggunakan Termoelektrik *J. Komput. Inf. Teknol. Dan Elektro* **1**

[2] Orr B and Akbarzadeh A 2017 Prospects of Waste Heat Recovery and Power Generation Using Thermoelectric Generators *Energy Procedia* **110** 250–5

[3] Muntini M S Rachmanita R E Anggoro D Rahayu I Afiyanti M M Pilasuta P Paengson S Namhongsa W and Ruamruk S 2017 Comparison of Electrical Power for Thermoelectric Oxide Module 5

[4] Ryanuargo Anwar S and Sari S P 2013 Generator Mini dengan Prinsip Termoelektrik dari Uap Panas Kondensor pada Sistem Pendingin *J. Rekayasa Elektr.* **10** 180–5

[5] Busthomy P L and Widyartono M 2020 Generator Termoelektrik Dengan Memanfaatkan Panas Yang Terbuang Dari Api Pembakaran Untuk Pengisian Baterai Handphone *J. Tek. ELEKTRO* **9**

[6] Rachmanita R E and Azizah N 2021 Utilization of condenser waste heat of cold storage as energy conversion system based on thermoelectric generators *J. Phys.: Conf. Ser.* **2013** 012007

[7] Rachmanita R E, Karimah C N and Azizah N 2021 Experimental investigations on the performance of thermoelectric generator as energy conversion system *IOP Conf. Ser.: Earth Environ. Sci.* **672** 012103

[8] Rahmat M R 2015 Perancangan Cold Storage Untuk Produk Reagen *J. Ilm. Tek. Mesin Unisma Quot45quot Bekasi* **3** 97583

[9] Talarosha B 2009 Menciptakan Kenyamanan Thermal Dalam Bangunan

[10] Suryaningrat I B 2009 *Ekonomi Teknik: Teori dan Aplikasi untuk Agroindustri* (UPT Penerbitan & Percetakan Universitas Jember)

[11] Fanger P O 1970 Thermal comfort. Analysis and applications in environmental engineering. *Therm. Comf. Anal. Appl. Environ. Eng.*

[12] Muhammad R R 2016 *Kajian Kenyamanan Termal Ruang Gambar Paket Keahlian Teknik Gambar Bangunan Smk Negeri 2 Pengasih* skripsi (U N Y)

[13] No M and Kwak H-B 2016 The Effects of Environmental Temperature on Physiological Responses during Submaximal and Maximal Exercises in Soccer Players *Integr. Med. Res.* **5**

[14] Mintarto E and Fattahilah M 2019 Efek Suhu Lingkungan Terhadap Fisiologi Tubuh pada saat Melakukan Latihan Olahraga *JSES J. Sport Exerc. Sci.* **2** 9–13

[15] Anon Buku Ajar Ekonomi Manajerial Book by Usep Sudrajat, SE. MM Suwaji, SE.MM - Gramedia Digital

AUTHOR INDEX

Institute of Physics
Dirac House, Temple Back
Bristol BS1 6BE UK

ISSN: 1755-1307
ISBN 978-1-7138-4961-2

CURRAN ASSOCIATES INC.
proceedings.com